Lecture Notes in Physics

Volume 843

Founding Editors

W. Beiglböck
J. Ehlers
K. Hepp
H. Weidenmüller

Editorial Board

B.-G. Englert, Singapore
U. Frisch, Nice, France
F. Guinea, Madrid, Spain
P. Hänggi, Augsburg, Germany
W. Hillebrandt, Garching, Germany
M. Hjorth-Jensen, Oslo, Norway
R. A. L. Jones, Sheffield, UK
H. v. Löhneysen, Karlsruhe, Germany
M. S. Longair, Cambridge, UK
M. L. Mangano, Geneva, Switzerland
J.-F. Pinton, Lyon, France
J.-M. Raimond, Paris, France
A. Rubio, Donostia, San Sebastian, Spain
M. Salmhofer, Heidelberg, Germany
D. Sornette, Zurich, Switzerland
S. Theisen, Potsdam, Germany
D. Vollhardt, Augsburg, Germany
W. Weise, Garching, Germany

For further volumes:
http://www.springer.com/series/5304

The Lecture Notes in Physics

The series Lecture Notes in Physics (LNP), founded in 1969, reports new developments in physics research and teaching—quickly and informally, but with a high quality and the explicit aim to summarize and communicate current knowledge in an accessible way. Books published in this series are conceived as bridging material between advanced graduate textbooks and the forefront of research and to serve three purposes:

- to be a compact and modern up-to-date source of reference on a well-defined topic
- to serve as an accessible introduction to the field to postgraduate students and nonspecialist researchers from related areas
- to be a source of advanced teaching material for specialized seminars, courses and schools

Both monographs and multi-author volumes will be considered for publication. Edited volumes should, however, consist of a very limited number of contributions only. Proceedings will not be considered for LNP.

Volumes published in LNP are disseminated both in print and in electronic formats, the electronic archive being available at springerlink.com. The series content is indexed, abstracted and referenced by many abstracting and information services, bibliographic networks, subscription agencies, library networks, and consortia.

Proposals should be sent to a member of the Editorial Board, or directly to the managing editor at Springer:

Christian Caron
Springer Heidelberg
Physics Editorial Department I
Tiergartenstrasse 17
69121 Heidelberg/Germany
christian.caron@springer.com

Daniel C. Cabra · Andreas Honecker
Pierre Pujol
Editors

Modern Theories
of Many-Particle Systems
in Condensed Matter Physics

 Springer

Prof. Daniel C. Cabra
Departamento de Física/IFLP
Universidad Nacional de La Plata
C.C. 67 La Plata, Argentina
e-mail: cabra@fisica.unlp.edu.ar

Privatdozent Dr. Andreas Honecker
Institut für Theoretische Physik
Georg-August-Universität Göttingen
Friedrich-Hund-Platz 1
37077 Göttingen
Germany
e-mail: ahoneck@uni-goettingen.de

Prof. Pierre Pujol
IRSAMC, Laboratoire de Physique
  Théorique
Université Paul Sabatier
Route de Narbonne 118
Toulouse Cedex 4, 31062
France
e-mail: pierre.pujol@irsamc.ups-tlse.fr

ISSN 0075-8450
ISBN 978-3-642-10448-0
DOI 10.1007/978-3-642-10449-7
Springer Heidelberg Dordrecht London New York

e-ISSN 1616-6361
e-ISBN 978-3-642-10449-7

Library of Congress Control Number: 2011938389

© Springer-Verlag Berlin Heidelberg 2012
This work is subject to copyright. All rights are reserved, whether the whole or part of the material is concerned, specifically the rights of translation, reprinting, reuse of illustrations, recitation, broadcasting, reproduction on microfilm or in any other way, and storage in data banks. Duplication of this publication or parts thereof is permitted only under the provisions of the German Copyright Law of September 9, 1965, in its current version, and permission for use must always be obtained from Springer. Violations are liable to prosecution under the German Copyright Law.
The use of general descriptive names, registered names, trademarks, etc. in this publication does not imply, even in the absence of a specific statement, that such names are exempt from the relevant protective laws and regulations and therefore free for general use.

Printed on acid-free paper

Springer is part of Springer Science+Business Media (www.springer.com)

Preface

The discovery in 1986 of superconductors with a high critical temperature has initiated a dramatic renewal in modern condensed-matter physics. The richness of the phase diagram in such materials illustrates the complexity in the underlying physics. One key ingredient behind such a complex behavior is the interaction between particles, as many phases observed in experiments cannot be explained within a context of conventional solid-state physics based on a single-particle picture. This simple observation is the core of the subject of strongly correlated systems.

Systems where interactions are strong and play a crucial role are inherently difficult to analyze theoretically. The situation is particularly interesting in low-dimensional systems, where on the one hand quantum fluctuations and interactions play a crucial role, and on the other hand many sophisticated techniques have been developed. More precisely, from an analytical perspective, the development of non-perturbative methods and the study of integrable field theory have facilitated the understanding of the behavior of many quasi one-dimensional strongly correlated systems. This progress on analytical techniques was accompanied by a comparable advance in the development of numerical techniques.

From an experimental point of view, the study of modern condensed-matter physics has also been enriched by the development of devices of sizes in the nanometer region. Furthermore, cold-atom and novel solid-state systems such as graphene have emerged as new testing grounds for theoretical ideas. Both analytical and numerical techniques have been adapted accordingly such that the current understanding of condensed matter systems differs considerably from textbooks.

The aim of the 2009 Les Houches school on "Modern theories of correlated electron systems" was to provide an overview about recent developments in the theory of strongly correlated electrons and related problems. On the occasion of this school, it was generally felt that it would be useful to edit the Lecture Notes as a book and you are holding the result in your hands.

Nevertheless, in order to be of lasting value, some adjustments were made concerning the content of this book as compared to the 2009 school. In particular, this book only has eight chapters as compared to the 12 series of lectures in the school.

The lectures of Subir Sachdev led to a general introduction to quantum phase transitions of antiferromagnets and the cuprate-based high-temperature super-conductors. Eduardo Fradkin's lectures and the corresponding chapter on electronic liquid crystal phases in strongly correlated systems may appear a more specialized topic, but still constitute a unique overview of this subject. Antonio Castro Neto covered selected topics in graphene physics. Even if this is just one material, this has become one of the biggest efforts in contemporary Condensed Matter Physics (which was even honored with the Nobel Prize in Physics 2010). Accordingly, it is very difficult if not impossible to provide a complete overview of the field, but this chapter covers the basic quantum chemistry and elastic properties of a sheet of graphene in a manner which is complementary to other reviews.

In the school there were two lectures by Antoine Georges and Alexander Lichtenstein on Dynamical Mean-Field Theory and applications to correlated materials. These two authors joined forces with Hartmut Hafermann, Frank Lechermann, Alexei N. Rubtsov, and Mikhail I. Katsnelson to write one chapter on this topic. Two further lectures were given by Alexander Altland and Reinhold Egger on disordered electronic systems and transport through quantum dots. There is one joint chapter in this volume by these two authors which focuses on the second topic.

Further lectures in the school and the corresponding three final chapters of this volume make contact with related fields. Maciej Lewenstein gave a quantum information perspective on many-body physics; the corresponding chapter was written with the help of Remigiusz Augusiak and Fernando Cucchietti. Roderich Moessner covered the wide field of frustrated magnetism in his lectures; Chris Laumann, Antonello Scardicchio, and Shivaji Sondhi contributed to the resulting chapter on the statistical mechanics of classical and quantum computational complexity which focuses on a specific new development in this field. The lectures of Giuseppe Mussardo covered integrable methods in statistical field theory and the corresponding chapter concludes this volume.

Last but not least, we would like to thank the contributing authors for the effort which they have put in the individual chapters as well as the INSTANS network of the European Science Foundation, the Deutsch-Französische Hochschule/Université franco-allemande, and the Centre national de la recherche scientifique (CNRS) for the financial support of the 2009 school which ultimately made this volume possible.

Les Houches, April 2011   Daniel C. Cabra
                                                              Andreas Honecker
                                                                  Pierre Pujol

Reinhold Egger

Andreas Honecker   Pierre Pujol   **Alexander Altland
Eduardo Fradkin**

Giuseppe Mussardo
Daniel Cabra

Antonio Castro Neto

Subir Sachdev

Maciej Lewenstein

Alexander Lichtenstein

Antoine Georges

Contents

Chapter 1
Quantum Phase Transitions of Antiferromagnets and the Cuprate Superconductors

Subir Sachdev

Abstract I begin with a proposed global phase diagram of the cuprate superconductors as a function of carrier concentration, magnetic field, and temperature, and highlight its connection to numerous recent experiments. The phase diagram is then used as a point of departure for a pedagogical review of various quantum phases and phase transitions of insulators, superconductors, and metals. The bond operator method is used to describe the transition of dimerized antiferromagnetic insulators between magnetically ordered states and spin-gap states. The Schwinger boson method is applied to frustrated square lattice antiferromagnets: phase diagrams containing collinear and spirally ordered magnetic states, Z_2 spin liquids, and valence bond solids are presented, and described by an effective gauge theory of spinons. Insights from these theories of insulators are then applied to a variety of symmetry breaking transitions in d-wave superconductors. The latter systems also contain fermionic quasiparticles with a massless Dirac spectrum, and their influence on the order parameter fluctuations and quantum criticality is carefully discussed. I conclude with an introduction to strong coupling problems associated with symmetry breaking transitions in two-dimensional metals, where the order parameter fluctuations couple to a gapless line of fermionic excitations along the Fermi surface.

1.1 Introduction

The cuprate superconductors have stimulated a great deal of innovative theoretical work on correlated electron systems. On the experimental side, new experimental techniques continue to be discovered and refined, leading to striking advances over 20 years after the original discovery of high temperature superconductivity [1].

S. Sachdev (✉)
Department of Physics, Harvard University,
Cambridge, MA 02138, USA
e-mail: sachdev@physics.harvard.edu

D. C. Cabra et al. (eds.), *Modern Theories of Many-Particle Systems in Condensed Matter Physics*, Lecture Notes in Physics 843, DOI: 10.1007/978-3-642-10449-7_1,
© Springer-Verlag Berlin Heidelberg 2012

1

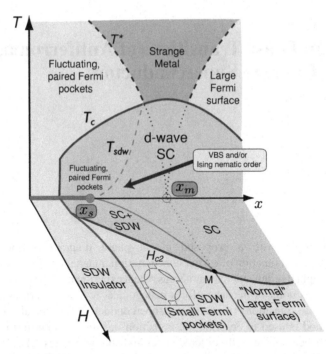

Fig. 1.1 Proposed global phase diagram for the hole- and electron-doped cuprates [3–5]. The axes are carrier concentration (x), temperature (T), and a magnetic field (H) applied perpendicular to the square lattice. The regions with the SC label have d-wave superconductivity. The strange metal and the "pseudogap" regime are separated by the temperature T^*. *Dashed lines* indicate crossovers. After accounting for the valence bond solid (VBS) or Ising nematic orders that can appear in the regime $x_s < x < x_m$, the *dashed* T^* line and the *dotted line* connecting x_m to the point M become true phase transitions. There can also be fractionalized phases in the region $x_s < x < x_m$, as discussed recently in Refs. [28, 29]

In the past few years, a number of experiments, and most especially the discovery of quantum oscillations in the underdoped regime [2], have shed remarkable new light on the origins of cuprate superconductivity. I believe these new experiments point to a synthesis of various theoretical ideas, and that a global theory of the rich cuprate phenomenology may finally be emerging. The ingredients for this synthesis were described in Refs. [3–5], and are encapsulated in the phase diagram shown in Fig. 1.1. Here I will only highlight a few important features of this phase diagram, and use those as motivations for the theoretical models described in these lectures. The reader is referred to the earlier papers [4, 5] for a full discussion of the experimental support for these ideas. Throughout the lectures, I will refer back to Fig. 1.1 and point out the relevance of various field theories to different aspects of this rich phase diagram.

It is simplest to examine the structure of Fig. 1.1 beginning from the regime of large doping. There, ample evidence has established that the ground state is a conventional Fermi liquid, with a single "large" Fermi surface enclosing the area demanded by the Luttinger theorem. Because of the underlying band structure, this large Fermi surface is hole-like (for both electron and hole-doping), and so encloses an area $1 + x$ for hole density x, and an area $1 - p$ for doped electron density p. The central quantum phase transition (QPT) in Fig. 1.1 is the onset of spin density wave (SDW) order in this large Fermi surface metal at carrier concentration $x = x_m$, shown in Fig. 1.1 near the region where T_c is largest (the subscript m refers to the fact that the transition takes place in a metal); we will describe this transition in more detail in Sect. 1.4. Because of the onset of superconductivity, the QPT at $x = x_m$ is revealed only at magnetic fields strong enough to suppress superconductivity, i.e., at $H > H_{c2}$. For $x < x_m$, we then have a Fermi liquid metal with SDW order. Close to the transition, when the SDW order is weak, the large Fermi surface is generically broken up by the SDW order into "small" electron and hole pockets, each enclosing an area of order x (see Fig. 1.17 later in the text). Note that electron pockets are present *both* for hole and electron doping: such electron pockets in the hole-doped cuprates were first discussed in Ref. [6]. There is now convincing experimental evidence for the small Fermi pockets in the hole underdoped cuprates, including accumulating evidence for electron pockets [7, 8]. The QPT between the small and large Fermi surface metals is believed to be at $x_m \approx 0.24$ in the hole-doped cuprates [9, 10], and at $p_m \approx 0.165$ in the electron-doped cuprates [11, 12]. One of the central claims of Fig. 1.1 is that it is the QPT at $x = x_m$ which controls the non-Fermi liquid "strange metal" behavior in the normal state above the superconductivity T_c. We leave open the possibility [13] that there is an extended non-Fermi liquid phase for a range of densities with $x > x_m$: this is not shown in Fig. 1.1, and will be discussed here only in passing.

The onset of superconductivity near the SDW ordering transition of a metal has been considered in numerous previous works [14, 15]. These early works begin with the large Fermi surface found for $x > x_m$, and consider pairing induced by exchange of SDW fluctuations; for the cuprate Fermi surface geometry, they find an attractive interaction in the d-wave channel, leading to d-wave superconductivity. Because the pairing strength is proportional to the SDW fluctuations, and the latter increase as $x \searrow x_m$, we expect T_c to increase as x is decreased for $x > x_m$, as is shown in Fig. 1.1. Thus for $x > x_m$, stronger SDW fluctuations imply stronger superconductivity, and the orders effectively attract each other.

It was argued in Ref. [3] that the situation becomes qualitatively different for $x < x_m$. This becomes clear from an examination of Fig. 1.1 as a function of decreasing T for $x < x_m$. It is proposed [3, 16] in the figure that the Fermi surface already breaks apart locally into the small pocket Fermi surfaces for $T < T^*$. So the onset of superconductivity at T_c involves the pairing of these small Fermi surfaces, unlike the large Fermi surface pairing considered above for $x > x_m$. For $x < x_m$, an increase in local SDW ordering is not conducive to stronger superconductivity: the SDW order 'eats up' the Fermi surface, leaving less room for the Cooper pairing instability on the Fermi surface. Thus in this regime we find a *competition* between SDW ordering and superconductivity for 'real estate' on the Fermi surface [3, 17]. As we expect the SDW ordering to increase as x is decreased for $x < x_m$, we should have a decrease in T_c with decreasing x, as is indicated in Fig. 1.1.

We are now ready to describe the second important feature of Fig. 1.1. The complement of the suppression of superconductivity by SDW ordering is the suppression of SDW ordering by superconductivity. The competition between superconductivity and SDW order moves [3] the actual SDW onset at $H = 0$ and $T = 0$ to a lower carrier concentration $x = x_s$ (or $p = p_s$ for electron doping). The QPT at $x = x_s$ controls the criticality of spin fluctuations within the superconducting phase (and hence the subscript s), while that $x = x_m$ continues to be important for $T > T_c$ (as is indicated in Fig. 1.1). There is now a line of SDW-onset transitions within the superconducting phase [18, 19], connecting the point x_s to the point M, for which there is substantial experimental evidence [20–25]. The magnitude of the shift from x_m to x_s depends a great deal upon the particular cuprate: it is largest in the materials with the strongest superconductivity and the highest T_c. In the hole-doped YBCO series we estimate $x_s \approx 0.085$ [25] and in the hole-doped LSCO series we have $x_s \approx 0.14$ [22] (recall our earlier estimate $x_m \approx 0.24$ in the hole-doped cuprates [9, 10]), while in the electron-doped cuprate $Nd_{2-x}Ce_xCuO_4$, we have $p_s = 0.145$ [26] (recall $p_m \approx 0.165$ in the electron-doped cuprates [11, 12]).

With the shift in SDW ordering from x_m to x_s, the need for the crossover line labeled T_{sdw} in Fig. 1.1 becomes evident. This is the temperature at which the electrons finally realize that they are to the 'disordered' of the actual SDW ordering transition at $x = x_s$, rather than to the 'ordered' side of the transition at $x = x_m$. Thus, for $T < T_{sdw}$, the large Fermi surface re-emerges at the lowest energy scales, and SDW order is never established. This leaves us with an interesting superconducting state at $T = 0$, where the proximity to the Mott insulator can play an important role. Other orders linked to the antiferromagnetism of the Mott insulator can appear here, such as valence bond solid (VBS) and Ising-nematic order [27], or even topologically ordered phases [28, 29]. Experimental evidence for such orders has appeared in a number of recent experiments [10, 30–32], and we will study these orders in the sections below.

The shift in the SDW ordering from x_m to x_s has recently emerged as a generic property of quasi-two-dimensional correlated electron superconductors, and is not special to the cuprates. Knebel et al. [33] have presented a phase diagram for CeRhIn₅ as a function of temperature, field, and pressure (which replaces carrier concentration) which is shown in Fig. 1.2. Notice the very similar structure to Fig. 1.1: the critical pressure for the onset of antiferromagnetism shifts from the metal to the superconductor, so that the range of antiferromagnetism is smaller in the superconducting state. In the pnictides, the striking observations by the Ames group [34–36] on $Ba[Fe_{1-x}Co_x]_2As_2$ show a 'back-bending' in the SDW onset temperature upon entering the superconducting phase: see Fig. 1.2. This is similar to the back-bending of the line T_{sdw} from T^* in Fig. 1.1, and so can also be linked to the shift in the SDW onset transition between the metal and the superconductor.

It is clear from Fig. 1.2 that the shift in the SDW order between the metal and the superconductor is relatively small in the non-cuprate materials, and may be overlooked in an initial study without serious consequences. Similar comments apply to the electron-doped cuprates. However, the shift is quite large in the hole-doped cuprates: this can initially suggest that the cuprates are a different class of materials,

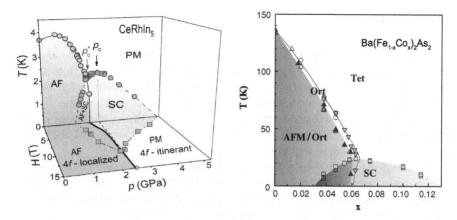

Fig. 1.2 Phase diagrams for CeRhIn₅ from Ref. [33] and for Ba[Fe₁₋ₓCoₓ]₂As₂ from Refs. [34–36]. For CeRhIn₅, the shift from p_c to p_c^* is similar to the shift from x_m to x_s in Fig. 1.1; this shift is significantly larger in the cuprates (and especially in YBCO) because the superconductivity is stronger. In the ferropnictide Ba[Fe₁₋ₓCoₓ]₂As₂, the back-bending of the SDW ordering transition in the superconducting phase is similar to that of T_{sdw} in Fig. 1.1

with SDW ordering playing a minor role in the physics of the superconductivity. One of the main claims of Fig. 1.1 is that after accounting for the larger shift in the SDW transition, all of the cuprates fall into a much wider class of correlated electron superconductors for which the SDW ordering transition in the metal is the central QPT controlling the entire phase diagram (see also the recent discussion by Scalapino [37]).

Our discussion will be divided into three sections, dealing with the nature of quantum fluctuations near SDW ordering in insulators, d-wave superconductors, and metals respectively. These cases are classified according to the increasing density of states for single-electron excitations. We will begin in Sect. 1.2 by considering a variety of Mott insulators, and describe their phase diagrams. The results will apply directly to experiments on insulators not part of the cuprate family. However, we will also gain insights, which will eventually be applied to various aspects of Fig. 1.1 for the cuprates. Then we will turn in Sect. 1.3 to d-wave superconductors, which have a Dirac spectrum of single-electron excitations as described in Sect. 1.3.1. Their influence on the SDW ordering transition at $x = x_s$ will be described using field-theoretical methods in Sect. 1.3.2. Section 1.3.3 will describe the Ising-nematic ordering at or near $x = x_s$ indicated in Fig. 1.1. Finally, in Sect. 1.4, we will turn to metals, which have a Fermi surface of low-energy single-particle excitations. We will summarize the current status of QPTs of metals: in two dimensions most QPTs lead to strong coupling problems which have not been conquered. It is clear from Fig. 1.1 that such QPTs are of vital importance to the physics of metallic states near $x = x_m$.

Significant portions of the discussion in the sections below have been adapted from other review articles by the author [38–40].

1.2 Insulators

The insulating state of the cuprates at $x = 0$ is a $S = 1/2$ square lattice antifer-romagnet, which is known to have long-range Néel order. We now wish to study various routes by which quantum fluctuations may destroy the Néel order. In this section, we will do this by working with undoped insulators in which we modify the exchange interactions. These do not precisely map to any of the transitions in the phase diagram in Fig. 1.1, but we will see in the subsequent sections that closely related theories do play an important role.

The following subsections will discuss two distinct routes to the destruction of Néel order in two-dimensional antiferromagnet. In Sect. 1.2.1 we describe coupled dimer antiferromagnets, in which the lattice has a natural dimerized structure, with $2S = 1/2$ spins per unit cell which can pair with each other. These are directly relevant to experiments on materials like TlCuCl$_3$. We will show that these anti-ferromagnets can be efficiently described by a bond-operator method. Then in Sect. 1.2.2 we will consider the far more complicated and subtle case where the lattice has full square lattice symmetry with only a single $S = 1/2$ spin per unit cell, and the Néel order is disrupted by frustrating exchange interactions. We will explore the phase diagram of such antiferromagnets using the Schwinger boson method. These results have direct application to experimental and numerical studies of a variety of two-dimensional Mott insulators on the square, triangular, and kagome lattices; such applications have been comprehensively reviewed in another recent article by the author [41], and so will not be repeated here.

1.2.1 Coupled Dimer Antiferromagnets: Bond Operators

We consider the "coupled dimer" Hamiltonian [42]

$$H_d = J \sum_{\langle ij \rangle \in \mathcal{A}} \mathbf{S}_i \cdot \mathbf{S}_j + gJ \sum_{\langle ij \rangle \in \mathcal{B}} \mathbf{S}_i \cdot \mathbf{S}_j, \tag{1.1}$$

where $\mathbf{S}_j$ are spin-1/2 operators on the sites of the coupled-ladder lattice shown in Fig. 1.3, with the $\mathcal{A}$ links forming decoupled dimers while the $\mathcal{B}$ links couple the dimers as shown. The ground state of H_d depends only on the dimensionless coupling g, and we will describe the low temperature (T) properties as a function of g. We will restrict our attention to $J > 0$ and $0 \leq g \leq 1$. A three-dimensional model with the same structure as H_d describes the insulator TlCuCl$_3$ [43–45].

Note that exactly at $g = 1$, H_d is identical to the square lattice antiferromagnet, and this is the only point at which the Hamiltonian has only one spin per unit cell. At all other values of g, H_d has a pair of $S = 1/2$ spins in each unit cell of the lattice.

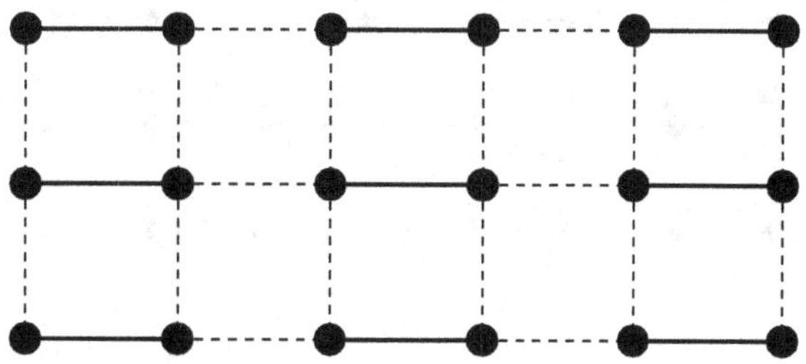

Fig. 1.3 The coupled dimer antiferromagnet. Spins ($S = 1/2$) are placed on the sites, the $\mathcal{A}$ links are shown as *full lines*, and the $\mathcal{B}$ links as *dashed lines*

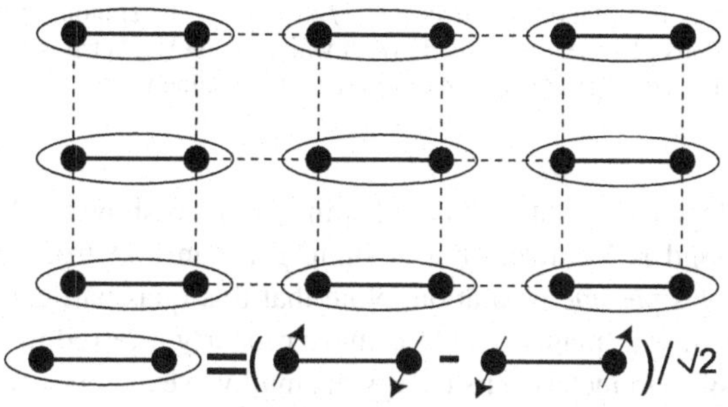

Fig. 1.4 Schematic of the quantum paramagnet ground state for small g. The *ovals* represent singlet valence bond pairs

1.2.1.1 Phases and Their Excitations

Let us first consider the case where g is close to 1. Exactly at $g = 1$, H_d is identical to the square lattice Heisenberg antiferromagnet, and this is known to have long-range, magnetic Néel order in its ground state, i.e., the spin-rotation symmetry is broken and the spins have a non-zero, staggered, expectation value in the ground state with

$$\langle S_j \rangle = \eta_j N_0 n, \tag{1.2}$$

where n is some fixed unit vector in spin space, η_j is ± 1 on the two sublattices, and N_0 is the Néel order parameter. This long-range order is expected to be preserved for a finite range of g close to 1. The low-lying excitations above the ground state consist of slow spatial deformations in the orientation n : these are the familiar spin waves, and they can carry arbitrarily low energy, i.e., the phase is 'gapless'. The spectrum of the spin waves can be obtained from a text-book analysis of small fluctuations about the ordered Néel state using the Holstein–Primakoff method [46]: such an analysis yields *two* polarizations of spin waves at each wavevector $k = (k_x, k_y)$ (measured from the antiferromagnetic ordering wavevector), and they have excitation energy $\varepsilon_k = (c_x^2 k_x^2 + c_y^2 k_y^2)^{1/2}$, with c_x, c_y the spin-wave velocities in the two spatial directions.

Let us turn now to the vicinity of $g = 0$. Exactly at $g = 0$, H_d is the Hamiltonian of a set of decoupled dimers, with the simple exact ground state wavefunction shown in Fig. 1.4: the spins in each dimer pair into valence bond singlets, leading to a paramagnetic state which preserves spin rotation invariance and all lattice symmetries. Excitations are now formed by breaking a valence bond, which leads to a *threefold*

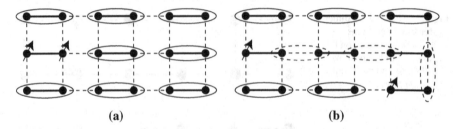

Fig. 1.5 a Cartoon picture of the bosonic $S = 1$ excitation of the paramagnet. **b** Fission of the $S = 1$ excitation into two $S = 1/2$ spinons. The spinons are connected by a "string" of valence bonds (denoted by *dashed ovals*) which lie on weaker bonds; this string costs a finite energy per unit length and leads to the confinement of spinons

degenerate state with total spin $S = 1$, as shown in Fig. 1.5a. At $g = 0$, this broken bond is localized, but at finite g it can hop from site-to-site, leading to a triplet quasiparticle excitation. Note that this quasiparticle is *not* a spin-wave (or equivalently, a 'magnon') but is more properly referred to as a spin 1 *exciton* or a *triplon*. We parameterize its energy at small wavevectors k (measured from the minimum of the spectrum in the Brillouin zone) by

$$\varepsilon_k = \Delta + \frac{c_x^2 k_x^2 + c_y^2 k_y^2}{2\Delta}, \tag{1.3}$$

where Δ is the spin gap, and c_x, c_y are velocities; we will provide an explicit derivation of (1.3) in Sect. 1.2.1.2. Figure 1.5 also presents a simple argument which shows that the $S = 1$ exciton cannot fission into two $S = 1/2$ 'spinons'.

The very distinct symmetry signatures of the ground states and excitations between $g \approx 1$ and $g \approx 0$ make it clear that the two limits cannot be continuously connected. It is known that there is an intermediate second-order phase transition at [42, 47] $g = g_c = 0.52337(3)$ between these states as shown in Fig. 1.6. Both the spin gap Δ and the Néel order parameter N_0 vanish continuously as g_c is approached from either side.

1.2.1.2 Bond Operators and Quantum Field Theory

In this section we will develop a continuum description of the low energy excitations in the vicinity of the critical point postulated above. There are a number of ways to obtain the same final theory: here we will use the method of *bond operators* [48, 49], which has the advantage of making the connection to the lattice degrees of freedom most direct. We rewrite the Hamiltonian using bosonic operators which reside on the centers of the $\mathcal{A}$ links so that it is explicitly diagonal at $g = 0$. There are 4 states on each $\mathcal{A}$ link ($|\uparrow\uparrow\rangle$, $|\uparrow\downarrow\rangle$, $|\downarrow\uparrow\rangle$, and $|\downarrow\downarrow\rangle$) and we associate these with the canonical singlet boson s and the canonical triplet bosons t_a ($a = x, y, z$) so that

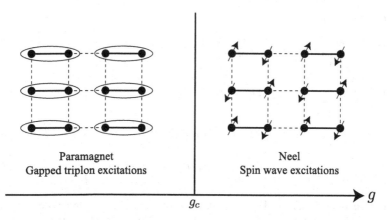

Paramagnet
Gapped triplon excitations

Neel
Spin wave excitations

Fig. 1.6 Ground states of H_d as a function of g. The quantum critical point is at [47] $g_c = 0.52337(3)$. The compound TlCuCl$_3$ undergoes a similar quantum phase transition under applied pressure [43, 45]

$$|s\rangle \equiv s^\dagger|0\rangle = \frac{1}{\sqrt{2}}\left(|\uparrow\downarrow\rangle - |\downarrow\uparrow\rangle\right) \; ; \quad |t_x\rangle \equiv t_x^\dagger|0\rangle = \frac{-1}{\sqrt{2}}\left(|\uparrow\uparrow\rangle - |\downarrow\downarrow\rangle\right) \; ;$$

$$|t_y\rangle \equiv t_y^\dagger|0\rangle = \frac{i}{\sqrt{2}}\left(|\uparrow\uparrow\rangle + |\downarrow\downarrow\rangle\right) \; ; \quad |t_z\rangle \equiv t_z^\dagger|0\rangle = \frac{1}{\sqrt{2}}\left(|\uparrow\downarrow\rangle + |\downarrow\uparrow\rangle\right).$$

$$(1.4)$$

Here $|0\rangle$ is some reference vacuum state which does not correspond to a physical state of the spin system. The physical states always have a single bond boson and so satisfy the constraint

$$s^\dagger s + t_a^\dagger t_a = 1. \tag{1.5}$$

By considering the various matrix elements $\langle s|S_1|t_a\rangle$, $\langle s|S_2|t_a\rangle$, $\ldots$, of the spin operators $S_{1,2}$ on the ends of the link, it follows that the action of S_1 and S_2 on the singlet and triplet states is equivalent to the operator identities

$$S_{1a} = \frac{1}{2}\left(s^\dagger t_a + t_a^\dagger s - i\epsilon_{abc}t_b^\dagger t_c\right),$$

$$S_{2a} = \frac{1}{2}\left(-s^\dagger t_a - t_a^\dagger s - i\epsilon_{abc}t_b^\dagger t_c\right). \tag{1.6}$$

where a, b, c take the values x, y, z, repeated indices are summed over and ϵ is the totally antisymmetric tensor. Inserting (1.6) into (1.1), and using (1.5), we find the following Hamiltonian for the bond bosons:

$$H_d = H_0 + H_1$$

$$H_0 = J \sum_{\ell \in \mathcal{A}} \left(-\frac{3}{4} s_\ell^\dagger s_\ell + \frac{1}{4} t_{\ell a}^\dagger t_{\ell a} \right)$$

$$H_1 = gJ \sum_{\ell, m \in \mathcal{A}} \left[a(\ell, m) \left(t_{\ell a}^\dagger t_{ma} s_m^\dagger s_\ell + t_{\ell a}^\dagger t_{ma}^\dagger s_m s_\ell + \text{H.c.} \right) + b(\ell, m) \right.$$

$$\times \left(i\epsilon_{abc} t_{ma}^\dagger t_{\ell b}^\dagger t_{\ell c} s_m + \text{H.c.} \right)$$

$$\left. + c(\ell, m) \left(t_{\ell a}^\dagger t_{ma}^\dagger t_{mb} t_{\ell b} - t_{\ell a}^\dagger t_{mb}^\dagger t_{ma} t_{\ell b} \right) \right], \tag{1.7}$$

where ℓ, m label links in $\mathcal{A}$, and a, b, c are numbers associated with the lattice couplings which we will not write out explicitly. Note that $H_1 = 0$ at $g = 0$, and so the spectrum of the paramagnetic state is fully and exactly determined. The main advantage of the present approach is that application of the standard methods of many body theory to (1.7), while imposing the constraint (1.5), gives a very satisfactory description of the phases with $g \neq 0$, including across the transition to the Néel state. In particular, an important feature of the bond operator approach is that the simplest mean field theory already yields ground states and excitations with the correct quantum numbers; so a strong fluctuation analysis is not needed to capture the proper physics.

A complete numerical analysis of the properties of (1.7) in a self-consistent Hartree–Fock treatment of the four boson terms in H_1 has been presented in Ref. [48]. In all phases the s boson is well condensed at zero momentum, and the important physics can be easily understood by examining the structure of the low-energy action for the t_a bosons. For the particular Hamiltonian (1.1), the spectrum of the t_a bosons has a minimum at the momentum $(0, \pi)$, and for large enough g the t_a condense at this wavevector: the representation (1.6) shows that this condensed state is the expected Néel state, with the magnetic moment oscillating as in (1.2). The condensation transition of the t_a is therefore the quantum phase transition between the paramagnetic and Néel phases of the coupled dimer antiferromagnet. In the vicinity of this critical point, we can expand the t_a bose field in gradients away from the $(0, \pi)$ wavevector: so we parameterize

$$t_{\ell,a}(\tau) = t_a(r_\ell, \tau) e^{i(0,\pi) \cdot r_\ell} \tag{1.8}$$

where τ is imaginary time, $r \equiv (x, y)$ is a continuum spatial coordinate, and expand the effective action in spatial gradients. In this manner we obtain

$$S_t = \int d^2r d\tau \left[t_a^\dagger \frac{\partial t_a}{\partial \tau} + C t_a^\dagger t_a - \frac{D}{2} (t_a t_a + \text{H.c.}) + K_{1x} |\partial_x t_a|^2 + K_{1y} |\partial_y t_a|^2 \right.$$

$$\left. + \frac{1}{2} \left(K_{2x} (\partial_x t_a)^2 + K_{2y} (\partial_y t_a)^2 + \text{H.c.} \right) + \cdots \right]. \tag{1.9}$$

Here C, D, $K_{1,2x,y}$ are constants that are determined by the solution of the self-consistent equations, and the ellipses represent terms quartic in the t_a. The action $\mathcal{S}_t$ can be easily diagonalized, and we obtain a $S = 1$ quasiparticle excitation with the spectrum

$$\varepsilon_k = \left[\left(C + K_{1x}k_x^2 + K_{1y}k_y^2 \right)^2 - \left(D + K_{2x}k_x^2 + K_{2y}k_y^2 \right)^2 \right]^{1/2}. \tag{1.10}$$

This is, of course, the triplon (or spin exciton) excitation of the paramagnetic phase postulated earlier in (1.3); the latter result is obtained by expanding (1.10) in momenta, with $\Delta = \sqrt{C^2 - D^2}$. This value of Δ shows that the ground state is paramagnetic as long as $C > D$, and the quantum critical point to the Néel state is at $C = D$.

The critical point and the Néel state are more conveniently described by an alternative formulation of $\mathcal{S}_t$ (although an analysis using bond operators directly is also possible [50]). It is useful to decompose the complex field t_a into its real and imaginary parts as follows

$$t_a = Z(\varphi_a + i\pi_a), \tag{1.11}$$

where Z is a normalization chosen below. From (1.8) and the connection to the lattice spin operators, it is not difficult to show that the vector φ_a is proportional to the Néel order parameter $\boldsymbol{n}$ in Eq. 1.2. Insertion of (1.11) into (1.9) shows that the field π_a has a quadratic term $\sim (C + D)\pi_a^2$, and so the coefficient of π_a^2 remains large even as the spin gap Δ becomes small. Consequently, we can safely integrate π_a out, and the resulting action for the Néel order parameter φ_a takes the form

$$\mathcal{S}_\varphi = \int d^2r\,d\tau \left[\frac{1}{2} \left\{ (\partial_\tau \varphi_a)^2 + c_x^2 (\partial_x \varphi_a)^2 + c_y^2 (\partial_y \varphi_a)^2 + s\varphi_a^2 \right\} + \frac{u}{24} \left(\varphi_a^2 \right)^2 \right]. \tag{1.12}$$

Here we have chosen Z to fix the coefficient of the temporal gradient term, and $s = C^2 - D^2$.

The action $\mathcal{S}_\varphi$ gives a simple picture of excitations across the quantum critical point, which can be quantitatively compared to neutron scattering experiments [45] on TlCuCl$_3$. In the paramagnetic phase ($s > 0$), a triplet of gapped excitations is observed, corresponding to the three normal modes of φ_a oscillating about $\varphi_a = 0$; as expected, this triplet gap vanishes upon approaching the quantum critical point. In a mean field analysis, the field theory in Eq. 1.12 has a triplet gap of $\sqrt{s}$ (mean field theory is applicable to TlCuCl$_3$ because this antiferromagnet is three dimensional). In the Néel phase, the neutron scattering detects two gapless spin waves, and one gapped longitudinal mode [51] (the gap to this longitudinal mode vanishes at the quantum critical point), as is expected from fluctuations in the inverted 'Mexican hat' potential of $\mathcal{S}_\varphi$ for $s < 0$. The longitudinal mode has a mean-field energy gap of $\sqrt{2|s|}$. These mean field predictions for the energy of the gapped modes on the two sides of the transition are tested in Fig. 1.7: the observations are in good agreement

Fig. 1.7 Energies of the gapped collective modes across the pressure (p) tuned quantum phase transition in TlCuCl$_3$ observed by Ruegg et al. [45]. We test the description by the action S_φ in Eq. 1.12 with $s \propto (p_c - p)$ by comparing $\sqrt{2}$ times the energy gap for $p < p_c$ with the energy of the longitudinal mode for $p > p_c$. The lines are the fits to a $\sqrt{|p - p_c|}$ dependence, testing the 1/2 exponent

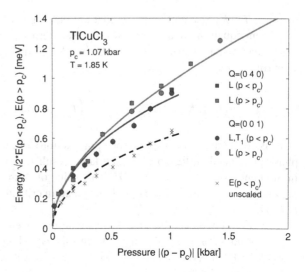

with the 1/2 exponent and the predicted [41, 52] $\sqrt{2}$ ratio, providing a non-trivial experimental test of the S_φ field theory.

We close this subsection by noting that all of the above results have a direct generalization to other lattices. One important difference that emerges in such calculations on some frustrated lattices [53] is worth noting explicitly here: the minimum of the t_a spectrum need not be at special wavevector like $(0, \pi)$, but can be at a more generic wavevector $\mathbf{K}$ such that $\mathbf{K}$ and $-\mathbf{K}$ are not separated by a reciprocal lattice vector. A simple example which we consider here is an extension of (1.1) in which there are additional exchange interactions along all diagonal bonds oriented 'north-east' (so that the lattice has the connectivity of a triangular lattice). In such cases, the structure of the low energy action is different, as is the nature of the magnetically ordered state. The parameterization (1.8) must be replaced by

$$t_{\ell a}(\tau) = t_{1a}(r_\ell, \tau)e^{i\mathbf{K}\cdot\mathbf{r}_\ell} + t_{2a}(r_\ell, \tau)e^{-i\mathbf{K}\cdot\mathbf{r}_\ell}, \tag{1.13}$$

where $t_{1, 2a}$ are independent complex fields. Proceeding as above, we find that the low-energy effective action (1.12) is replaced by

$$S_\Phi = \int d^2 r d\tau \left[|\partial_\tau \Phi_a|^2 + c_x^2 |\partial_x \Phi_a|^2 + c_y^2 |\partial_y \Phi_a|^2 + s |\Phi_a|^2 \right.$$
$$\left. + \frac{u}{2}\left(|\Phi_a|^2\right)^2 + \frac{v}{2}\left|\Phi_a^2\right|^2 \right], \tag{1.14}$$

where now Φ_a is a *complex* field such that $\langle\Phi_a\rangle \sim \langle t_{1a}\rangle \sim \langle t_{2a}^\dagger\rangle$. Notice that there is now a second quartic term with coefficient v. If $v > 0$, configurations with $\Phi_a^2 = 0$ are preferred: in such configurations $\Phi_a = n_{1a} + in_{2a}$, where $n_{1, 2a}$ are two equal-length orthogonal vectors. Then from (1.13) and (1.6) it is easy to see that the physical spins

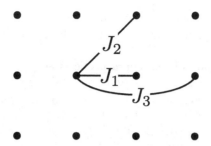

Fig. 1.8 The J_1-J_2-J_3 antiferromagnet. Spin S spins are placed on each site of the square lattice, and they are coupled to all first, second, and third neighbors as shown. The Hamiltonian has the full space group symmetry of the square lattice, and there is only one spin per unit cell

possess *spiral* order in the magnetically ordered state in which Φ_a is condensed. For the case $v<0$, the optimum configuration has $\Phi_a = n_a e^{i\theta}$ where n_a is a real vector: this leads to a magnetically ordered state with spins polarized *collinearly* in a spin density wave at the wavevector $\mathbf{K}$. The critical properties of the model in Eq. 1.14 have been described in Ref. [54].

1.2.2 Frustrated Square Lattice Antiferromagnets: Schwinger Bosons

As discussed at the beginning of Sect. 1.2, the more important and complex cases of quantum antiferromagnets are associated with those that have a single $S = 1/2$ spin per unit cell. Such models are more likely to have phases in which the exotic spinon excitations of Fig. 1.5 are deconfined, i.e., their ground states possess neutral $S = 1/2$ excitations and 'topological' order. We will meet the earliest established examples [55, 56] of such phases below.

We are interested in Hamiltonians of the form

$$\mathcal{H} = \sum_{i,j} J_{ij} \mathbf{S}_i \cdot \mathbf{S}_j \tag{1.15}$$

where we consider the general case of $\mathbf{S}_i$ being spin S quantum spin operators on the sites, i, of a two-dimensional lattice. The J_{ij} are short-ranged antiferromagnetic exchange interactions. We will mainly consider here the so-called square lattice J_1-J_2-J_3 model, which has first, second, and third neighbor interactions (see Fig. 1.8). Similar results have also been obtained on the triangular and kagome lattices [57, 58].

The main direct applications of the results here are to experiments on a variety of two-dimensional Mott insulators on the square, triangular, and kagome lattices. As noted earlier, we direct the reader to Ref. [41] for a discussion of these experiments. There have also been extensive numerical studies, also reviewed in the previous article [41], which are in good accord with the phase diagrams presented below. Applications to the cuprates, and to Fig. 1.1, will be discussed in the following sections.

A careful examination of the non-magnetic 'spin-liquid' phases requires an approach which is designed explicitly to be valid in a region well separated from Néel long range order, and preserves $SU(2)$ symmetry at all stages. It should also be designed to naturally allow for neutral $S = 1/2$ excitations. To this end, we introduce the Schwinger boson description [59, 60], in terms of elementary $S = 1/2$ bosons. For the group $SU(2)$ the complete set of $(2S + 1)$ states on site i are represented as follows

$$|S, m\rangle \equiv \frac{1}{\sqrt{(S+m)!(S-m)!}} (b_{i\uparrow}^{\dagger})^{S+m} (b_{i\downarrow}^{\dagger})^{S-m} |0\rangle, \tag{1.16}$$

where $m = -S, \ldots S$ is the z component of the spin ($2m$ is an integer). We have introduced two flavors of bosons on each site, created by the canonical operator $b_{i\alpha}^{\dagger}$, with $\alpha = \uparrow, \downarrow$, and $|0\rangle$ is the vacuum with no bosons. The total number of bosons, n_b is the same for all the states; therefore

$$b_{i\alpha}^{\dagger} b_i^{\alpha} = n_b \tag{1.17}$$

with $n_b = 2S$ (we will henceforth assume an implied summation over repeated upper and lower indices). It is not difficult to see that the above representation of the states is completely equivalent to the following operator identity between the spin and boson operators

$$S_{ia} = \frac{1}{2} b_{i\alpha}^{\dagger} \sigma^{a\alpha}{}_{\beta} b_i^{\beta}, \tag{1.18}$$

where $a = x, y, z$ and the σ^a are the usual 2×2 Pauli matrices. The spin-states on two sites i, j can combine to form a singlet in a unique manner–the wavefunction of the singlet state is particularly simple in the boson formulation:

$$\left(\varepsilon^{\alpha\beta} b_{i\alpha}^{\dagger} b_{j\beta}^{\dagger} \right)^{2S} |0\rangle. \tag{1.19}$$

Finally we note that, using the constraint (1.17), the following Fierz-type identity can be established

$$\left(\varepsilon^{\alpha\beta} b_{i\alpha}^{\dagger} b_{j\beta}^{\dagger} \right) \left(\varepsilon_{\gamma\delta} b_i^{\gamma} b_j^{\delta} \right) = -2S_i \cdot S_j + n_b^2/2 + \delta_{ij} n_b \tag{1.20}$$

where ε is the totally antisymmetric 2×2 tensor

$$\varepsilon = \begin{pmatrix} 0 & 1 \\ -1 & 0 \end{pmatrix}. \tag{1.21}$$

This implies that $\mathcal{H}$ can be rewritten in the form (apart from an additive constant)

$$\mathcal{H} = -\frac{1}{2}\sum_{\langle ij\rangle} J_{ij}\left(\varepsilon^{\alpha\beta}b_{i\alpha}^{\dagger}b_{j\beta}^{\dagger}\right)\left(\varepsilon_{\gamma\delta}b_i^{\gamma}b_j^{\delta}\right). \tag{1.22}$$

This form makes it clear that $\mathcal{H}$ counts the number of singlet bonds.

We have so far defined a one-parameter (n_b) family of models $\mathcal{H}$ for a fixed realization of the J_{ij}. Increasing n_b makes the system more classical and a large n_b expansion is therefore not suitable for studying the quantum-disordered phase. For this reason we introduce a second parameter—the flavor index α on the bosons is allowed to run from $1 \ldots 2N$ with N an arbitrary integer. This therefore allows the bosons to transform under $SU(2N)$ rotations. However, the $SU(2N)$ symmetry turns out to be too large. We want to impose the additional restriction that the spins on a pair of sites be able to combine to form a singlet state, thus generalizing the valence-bond structure of $SU(2)$— this valence-bond formation is clearly a crucial feature determining the structure of the quantum disordered phase. It is well-known that this is impossible for $SU(2N)$ for $N>1$—there is no generalization of the second-rank, antisymmetric, invariant tensor ε to general $SU(2N)$.

The proper generalization turns out to be to the group $Sp(N)$ [55]. This group is defined by the set of $2N \times 2N$ unitary matrices U such that

$$U^T \mathcal{J} U = \mathcal{J} \tag{1.23}$$

where

$$\mathcal{J}_{\alpha\beta} = \mathcal{J}^{\alpha\beta} = \begin{pmatrix} & 1 & & & & \\ -1 & & & & & \\ & & & 1 & & \\ & & -1 & & & \\ & & & & \ddots & \\ & & & & & \ddots \end{pmatrix} \tag{1.24}$$

is the generalization of the ε tensor to $N>1$; it has N copies of ε along the diagonal. It is clear that $Sp(N) \subset SU(2N)$ for $N>1$, while $Sp(1) \cong SU(2)$. The b_i^{α} bosons transform as the fundamental representation of $Sp(N)$; the "spins" on the lattice therefore belong to the symmetric product of n_b fundamentals, which is also an irreducible representation. Valence bonds

$$\mathcal{J}^{\alpha\beta}b_{i\alpha}^{\dagger}b_{j\alpha}^{\dagger} \tag{1.25}$$

can be formed between any two sites; this operator is a singlet under $Sp(N)$ because of (1.23). The form (1.22) of $\mathcal{H}$ has a natural generalization to general $Sp(N)$:

$$\mathcal{H} = -\sum_{i>j}\frac{J_{ij}}{2N}\left(\mathcal{J}^{\alpha\beta}b_{i\alpha}^{\dagger}b_{j,\beta}^{\dagger}\right)\left(\mathcal{J}_{\gamma\delta}b_i^{\gamma}b_j^{\delta}\right) \tag{1.26}$$

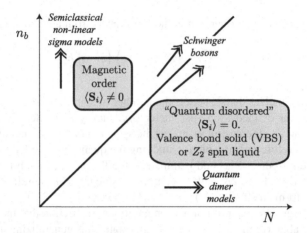

Fig. 1.9 Phase diagram of the 2D $Sp(N)$ antiferromagnet $\mathcal{H}$ as a function of the "spin" n_b; from Refs. [55, 61–64] The "quantum disordered" region preserves $Sp(N)$ spin rotation invariance, and there is no magnetic long-range order; however, the ground states here have new types of emergent order (VBS or Z_2 topological order), which are described in the text. On the square lattice, the Z_2 spin liquid phases also break a global lattice rotational symmetry, and so they have 'Ising-nematic' order; the Z_2 spin liquids on the *triangular* and *kagome lattices* do not break any lattice symmetry

where the indices α, β, γ, δ now run over $1 \ldots 2N$. We recall also that the constraint (1.17) must be imposed on every site of the lattice.

We now have a two-parameter (n_b, N) family of models $\mathcal{H}$ for a fixed realization of the J_{ij}. It is very instructive to consider the phase diagram of $\mathcal{H}$ as a function of these two parameters (Fig. 1.9).

The limit of large n_b, with N fixed leads to the semi-classical theory. For the special case of $SU(2)$ antiferromagnets with a two-sublattice collinear Néel ground state, the semiclassical fluctuations are described by the $O(3)$ non-linear sigma model. For other models [61, 65–73], the structure of the non-linear sigma models is rather more complicated and will not be considered here.

A second limit in which the problem simplifies is N large at fixed n_b [61, 74]. It can be shown that in this limit the ground state is quantum disordered. Further, the low-energy dynamics of $\mathcal{H}$ is described by an effective quantum-dimer model [61, 75], with each dimer configuration representing a particular pairing of the sites into valence-bonds. There have been extensive studies of such quantum dimer models which we will not review here. All the quantum dimer model studies in the "quantum disordered" region of Fig. 1.9 have yielded phases which were obtained earlier [55] by the methods to be described below.

The most interesting solvable limit is obtained by fixing the ratio of n_b and N

$$\kappa = \frac{n_b}{N} \tag{1.27}$$

and subsequently taking the limit of large N [59, 60]; this limit will be studied in this section in considerable detail. The implementation of $\mathcal{H}$ in terms of bosonic operators also turns out to be naturally suited for studying this limit. The parameter κ is arbitrary; tuning κ modifies the slope of the line in Fig. 1.9 along which the large N limit is taken. From the previous limits discussed above, one might expect that the ground state of $\mathcal{H}$ has magnetic long range order (LRO) for large κ and is quantum-disordered for small κ. We will indeed find below that for any set of J_{ij} there is a critical value of $\kappa = \kappa_c$ which separates the magnetically ordered and the quantum disordered phase.

The transition at $\kappa = \kappa_c$ is second-order at $N = \infty$, and is a powerful feature of the present large-N limit. In the vicinity of the phase transition, we expect the physics to be controlled by long-wavelength, low-energy spin fluctuations; the large-N method offers an unbiased guide in identifying the proper low-energy degress of freedom and determines the effective action controlling them. Having obtained a long-wavelength continuum theory near the transition, one might hope to analyze the continuum theory independently of the large-N approximation and obtain results that are more generally valid.

We will discuss the structure of the $N = \infty$ mean-field theory , with $n_b = \kappa N$ in Sect. 1.2.2.1. The long-wavelength effective actions will be derived and used to describe general properties of the phases and the phase transitions in Sect. 1.2.2.2.

1.2.2.1 Mean-Field Theory

We begin by analyzing $\mathcal{H}$ at $N = \infty$ with $n_b = \kappa N$. As noted above, this limit is most conveniently taken using the bosonic operators. We may represent the partition function of $\mathcal{H}$ by

$$
Z = \int \mathcal{D}Q \mathcal{D}b \mathcal{D}\lambda \exp\left(-\int_0^\beta \mathcal{L} d\tau \right),
\tag{1.28}
$$

where

$$
\mathcal{L} = \sum_i \left[b_{i\alpha}^\dagger \left(\frac{d}{d\tau} + i\lambda_i \right) b_i^\alpha - i\lambda_i n_b \right]
$$
$$
+ \sum_{\langle i,j \rangle} \left[N \frac{J_{ij}|Q_{i,j}|^2}{2} - \frac{J_{ij}Q_{i,j}^*}{2} \mathcal{J}_{\alpha\beta} b_i^\alpha b_j^\beta + H.c. \right].
\tag{1.29}
$$

Here the λ_i fix the boson number of n_b at each site; τ-dependence of all fields is implicit. The complex field Q was introduced by a Hubbard–Stratonovich decoupling of $\mathcal{H}$: performing the functional integral over Q reproduces the exchange coupling in Eq. 1.26. An important feature of the lagrangian $\mathcal{L}$ is its $U(1)$ gauge invariance under which

$$b_{i\alpha}^{\dagger} \to b_{i\alpha}^{\dagger}(i)\exp\left(i\rho_i(\tau)\right)$$
$$Q_{i,j} \to Q_{i,j}\exp\left(-i\rho_i(\tau) - i\rho_j(\tau)\right) \tag{1.30}$$
$$\lambda_i \to \lambda_i + \frac{\partial \rho_i}{\partial \tau}(\tau).$$

The functional integral over $\mathcal{L}$ faithfully represents the partition function apart from an overall factor associated with this gauge redundancy.

The $1/N$ expansion of the free energy can be obtained by integrating out of $\mathcal{L}$ the $2N$-component b, $\bar{b}$ fields to leave an effective action for Q, λ having coefficient N (because $n_b \propto N$). Thus the $N \to \infty$ limit is given by minimizing the effective action with respect to "mean-field" values of $Q = \bar{Q}$, $i\lambda = \bar{\lambda}$ (we are ignoring here the possibility of magnetic LRO which requires an additional condensate $x^\alpha = \langle b^\alpha \rangle$—this has been discussed elsewhere [55, 64]). This is in turn equivalent to solving the mean-field Hamiltonian

$$\mathcal{H}_{MF} = \sum_{\langle i,j \rangle}\left(N\frac{J_{ij}|\bar{Q}_{ij}|^2}{2} - \frac{J_{ij}\bar{Q}_{i,j}^*}{2}\mathcal{J}_{\alpha\beta}b_i^\alpha b_j^\beta + H.c.\right)$$
$$+ \sum_i \bar{\lambda}_i(b_{i\alpha}^{\dagger}b_i^\alpha - n_b). \tag{1.31}$$

This Hamiltonian is quadratic in the boson operators and all its eigenvalues can be determined by a Bogoliubov transformation. This leads in general to an expression of the form

$$\mathcal{H}_{MF} = E_{MF}[\bar{Q}, \bar{\lambda}] + \sum_\mu \omega_\mu[\bar{Q}, \bar{\lambda}]\gamma_{\mu\alpha}^{\dagger}\gamma_\mu^\alpha. \tag{1.32}$$

The index μ extends over $1 \ldots$ number of sites in the system, E_{MF} is the ground state energy and is a functional of $\bar{Q}$, $\bar{\lambda}$, ω_μ is the eigenspectrum of excitation energies which is a also a function of $\bar{Q}$, $\bar{\lambda}$, and the γ_μ^α represent the bosonic eigenoperators. The excitation spectrum thus consists of non-interacting spinor bosons. The ground state is determined by minimizing E_{MF} with respect to the $\bar{Q}_{ij}$ subject to the constraints

$$\frac{\partial E_{MF}}{\partial \bar{\lambda}_i} = 0. \tag{1.33}$$

The saddle-point value of the $\bar{Q}$ satisfies

$$\bar{Q}_{ij} = \langle \mathcal{J}_{\alpha\beta}b_i^\alpha b_j^\beta \rangle. \tag{1.34}$$

Note that $\bar{Q}_{ij} = -\bar{Q}_{ji}$ indicating that $\bar{Q}_{ij}$ is a directed field—an orientation has to be chosen on every link.

We now describe the ground state configurations of the $\bar{Q}$, $\bar{\lambda}$ fields and the nature of the bosonic eigenspectrum for the J_1-J_2-J_3 model. We examined the values of the energy E_{MF} for $\bar{Q}_{ij}$ configurations which had a translational symmetry with two sites per unit cell. For all parameter values configurations with a single site per unit cell were always found to be the global minima. We will therefore restrict our attention to such configurations. The $\bar{\lambda}_i$ field is therefore independent of i, while there are six independent values of $\bar{Q}_{ij}$:

$$
\begin{aligned}
\bar{Q}_{i,i+\hat{x}} &\equiv Q_{1,x} \\
\bar{Q}_{i,i+\hat{y}} &\equiv Q_{1,y} \\
\bar{Q}_{i,i+\hat{y}+\hat{x}} &\equiv Q_{2,y+x} \\
\bar{Q}_{i,i+\hat{y}-\hat{x}} &\equiv Q_{2,y-x} \\
\bar{Q}_{i,i+2\hat{x}} &\equiv Q_{3,x} \\
\bar{Q}_{i,i+2\hat{y}} &\equiv Q_{3,y}.
\end{aligned}
\tag{1.35}
$$

For this choice, the bosonic eigenstates are also eigenstates of momentum with momenta k extending over the entire first Brillouin zone. The bosonic eigenenergies are given by

$$
\begin{aligned}
\omega_k &= \left(\bar{\lambda}^2 - |A_k|^2\right)^{1/2} \\
A_k &= J_1 \left(Q_{1,x}\sin k_x + Q_{1,y}\sin k_y\right) \\
&\quad + J_2 \left(Q_{2,y+x}\sin(k_y+k_x) + Q_{2,y-x}\sin(k_y-k_x)\right) \\
&\quad + J_3 \left(Q_{3,x}\sin(2k_x) + Q_{3,y}\sin(2k_y)\right).
\end{aligned}
\tag{1.36}
$$

We have numerically examined the global minima of E_{MF} as a function of the three parameters J_2/J_1, J_3/J_1, and N/n_b [55, 64]. The values of the $\bar{Q}_{ij}$ at any point in the phase diagram can then be used to classify the distinct classes of states. The results are summarized in Figs. 1.10 and 1.11 which show two sections of the three-dimensional phase diagram. All of the phases are labeled by the wavevector at which the spin structure factor has a maximum. This maximum is a delta function for the phases with magnetic LRO, while it is simply a smooth function of k for the quantum disordered phases (denoted by SRO in Figs. 1.10 and 1.11). The location of this maximum will simply be twice the wavevector at which ω_k has a mimimum: this is because the structure factor involves the product of two bosonic correlation functions, each of which consists of a propagator with energy denominator ω_k.

Each of the phases described below has magnetic LRO for large n_b/N and is quantum disordered for small n_b/N. The mean-field result for the structure of all of the quantum disordered phases is also quite simple: they are featureless spin fluids with free spin-1/2 bosonic excitations ("spinons") with energy dispersion ω_k which is gapped over the entire Brillouin zone. Some of the quantum disordered phases break the lattice rotation symmetry (leading to 'Ising-nematic' order) even at $N = \infty$—these will be described below. The mininum energy spinons lie at a wavevector k_0 and ω_{k_0} decreases as n_b/N. The onset of magnetic LRO occurs at the value of n_b/N at which the gap first vanishes: $\omega_{k_0} = 0$. At still larger values of

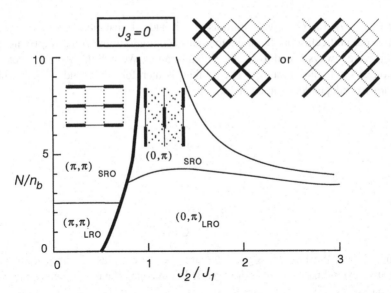

Fig. 1.10 Ground states of the J_1-J_2-J_3 model for $J_3 = 0$ as a function of J_2/J_1 and N/n_b ($n_b =$ $2S$ for $SU(2)$). *Thick (thin) lines* denote first (second) order transitions at $N = \infty$. Phases are identified by the wavevectors at which they have magnetic long-range-order (LRO) or short-range-order (SRO); the SRO phases are "quantum disordered" as in Fig. 1.9: The links with $Q_p \neq 0$ in each SRO phase are shown. The large N/n_b, large J_2/J_1 phase has the two sublattices decoupled at $N = \infty$. All LRO phases above have two-sublattice collinear Néel order. All the SRO phases above have valence bond solid (VBS) order at finite N for odd n_b; this is illustrated by the *thick, thin* and *dotted lines*

n_b/N, we get macroscopic bose condensation of the b quanta at the wavevector $\boldsymbol{k}_0$, leading to magnetic LRO at the wavevector $2\boldsymbol{k}_0$.

We now turn to a description of the various phases obtained. They can be broadly classified into two types:

Commensurate collinear phases

In these states the wavevector $\boldsymbol{k}_0$ remains pinned at a commensurate point in the Brillouin zone, which is independent of the values of J_2/J_1, J_3/J_1 and n_b/N. In the LRO phase, the spin condensates on the sites are either parallel or anti-parallel to each other, which we identify as collinear ordering. This implies that the LRO phase remains invariant under rotations about the condensate axis and the rotation symmetry is not completely broken.

Three distinct realizations of such states were found

a) (π, π)

This is the usual two-sublattice Néel state of the unfrustrated square lattice and its quantum-disordered partner. These states have

$$Q_{1,x} = Q_{1,y} \neq 0, \quad Q_{2,y+x} = Q_{2,y-x} = Q_{3,x} = Q_{3,y} = 0. \tag{1.37}$$

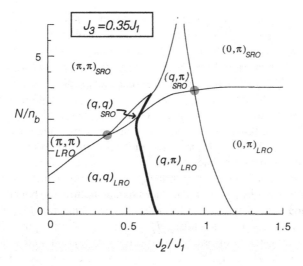

Fig. 1.11 As in Fig. 1.10, but for $J_3/J_1 = 0.35$. The $(0, \pi)_{SRO}$ and $(\pi, \pi)_{SRO}$ phases have VBS order as illustrated in Fig. 1.10. The $(q, q)_{SRO}$ and $(q, \pi)_{SRO}$ phases are Z_2 spin liquids: they have topological order, and a topological fourfold degeneracy of the ground state on the torus. The Z_2 spin liquids here also have Ising-nematic order, i.e., they break the $90°$ rotation symmetry of the square lattice, which leads to an additional twofold degeneracy. The $(q, q)_{LRO}$ and $(q, \pi)_{LRO}$ have magnetic long-range order in the form of an incommensurate spiral. The two *shaded circles* indicate regions which map onto the generalized phase diagram in Fig. 1.12

From (1.36), the minimum spinon excitation occurs at $k_0 = \pm(\pi/2, \pi/2)$. The SRO states have no broken symmetry at $N = \infty$. The boundary between the LRO and SRO phases occurs at $N/n_b < 2.5$, independent of J_2/J_1 (Fig. 1.10). This last feature is surely an artifact of the large N limit. Finite N fluctuations should be stronger as J_2/J_1 increases, causing the boundary to bend a little downwards to the right.

b) $(\pi, 0)$ or $(0, \pi)$

The $(0, \pi)$ states have

$$Q_{1,x} = 0, \ Q_{1,y} \neq 0, \ Q_{2,y+x} = Q_{2,y-x} \neq 0, \quad \text{and} \quad Q_{3,x} = Q_{3,y} = 0 \quad (1.38)$$

and minimum energy spinons at $k_0 = \pm(0, \pi/2)$. The degenerate $(\pi, 0)$ state is obtained with the mapping $x \leftrightarrow y$. The SRO state has a two-field degeneracy due to the broken $x \leftrightarrow y$ lattice symmetry: the order associated with this symmetry is referred to as 'Ising-nematic' order. We can use the Q variables here to define an Ising nematic order parameter

$$\mathcal{I} = |Q_{1x}|^2 - |Q_{1y}|^2. \quad (1.39)$$

This is a gauge-invariant quantity, and the square lattice symmetry of the Hamiltonian implies that $\langle \mathcal{I} \rangle = 0$ unless the symmetry is spontaneously broken. The sign of $\langle \mathcal{I} \rangle$ chooses between the $(\pi, 0)$ and $(0, \pi)$ states. The LRO state again has two-sublattice collinear Néel order, but the assignment of the sublattices is different from the (π, π) state. The spins are parallel along the x-axis, but anti-parallel along the y-axis.

An interesting feature of the LRO state here is the occurrence of "order-from-disorder" [76]. The classical limit $(n_b/N = \infty)$ of this model has an accidental degeneracy for $J_2/J_1 > 1/2$: the ground state has independent collinear Néel order on each of the A and B sublattices, with the energy independent of the angle between the spins on the two sublattices. Quantum fluctuations are included self-consistently in the $N = \infty$, n_b/N finite, mean-field theory, and lead to an alignment of the spins on the sublattices and LRO at $(0, \pi)$. The orientation of the ground state has thus been selected by the quantum fluctuations.

The $(0, \pi)$ states are separated from the (π, π) states by a first-order transition. In particular, the spin stiffnesses of both states remain finite at the boundary between them. This should be distinguished from the classical limit in which the stiffness of both states vanish at their boundary $J_2 = J_1/2$; the finite spin stiffnesses are thus another manifestation of order-from-disorder. At a point well away from the singular point $J_2 = J_1/2$, $n_b/N = \infty$ in Fig. 1.10, the stiffness of both states is of order $N(n_b/N)^2$ for $N = \infty$ and large n_b/N; near this singular point however, the stiffness is of order $N(n_b/N)$ and is induced purely by quantum fluctuations. These results have also been obtained by a careful resummation of the semiclassical expansion [77, 78].

c) "Decoupled"

For J_2/J_1 and N/n_b both large, we have a "decoupled" state (Fig. 1.10) with

$$Q_{2,y+x} = Q_{2,y-x} \neq 0 \quad \text{and} \quad Q_1 = Q_3 = 0. \tag{1.40}$$

In this case Q_p is non-zero only between sites on the same sublattice. The two sublattices have Néel type SRO which will be coupled by finite N fluctuations. The $N = \infty$ state does not break any lattice symmetry. This state has no LRO partner.

Incommensurate phases

In these phases the wavevector k_0 and the location of the maximum in the structure factor move continuously with the parameters. The spin-condensate rotates with a period which is not commensurate with the underlying lattice spacing. Further the spin condensate is *coplanar*: the spins rotate within a given plane in spin space and are not collinear. There is no spin rotation axis about which the LRO state remains invariant.

Further, no states in which the spin condensate was fully three dimensional ("double-spiral" or chiral states) were found; these would be associated with complex values of Q_p. All the saddle points possessed a gauge in which all the Q_p were real. Time-reversal symmetry was therefore always preserved in all the SRO phases of Figs. 1.10 and 1.11.

The incommensurate phases occur only in models with a finite J_3 (Fig. 1.11), at least at $N = \infty$. There were two realizations:

d) (π, q) or (q, π)

Here q denotes a wavevector which varies continuously between 0 and π as the parameters are changed. The (q, π) state has

$$Q_{1,x} \neq Q_{1,y} \neq 0, \ Q_{2,x+y} = Q_{2,y-x} \neq 0, \ Q_{3,x} \neq 0 \ \text{ and } \ Q_{3,y} = 0; \quad (1.41)$$

the degenerate (π, q) helix is obtained by the mapping $x \leftrightarrow y$. The SRO state has a two-fold degeneracy due to the broken $x \leftrightarrow y$ lattice symmetry, and so this state has Ising-nematic order. The order parameter in Eq. 1.39 continues to measure this broken symmetry.

e) (q, q) or $(q, -q)$

The (q, q) state has

$$Q_{1,x} = Q_{1,y} \neq 0, \ Q_{2,x+y} \neq 0, \ Q_{2,y-x} = 0, \ Q_{3,x} = Q_{3,y} \neq 0; \quad (1.42)$$

this is degenerate with the $(q, -q)$ phase. The SRO state therefore has a twofold degeneracy due to a broken lattice reflection symmetry, and so it also has Ising nematic order. However, the Ising symmetry now corresponds to reflections about the principle square axes, and the analog of Eq. 1.39 is now

$$\mathcal{I} = |Q_{2,x+y}|^2 - |Q_{2,y-x}|^2. \quad (1.43)$$

As we noted above, the broken discrete symmetries in states with SRO at $(0, \pi)$ and (q, π) are identical: both are two-fold degenerate due to a breaking of the $x \leftrightarrow y$ symmetry. The states are only distinguished by a non-zero value of Q_3 in the (q, π) phase and the accompanying incommensurate correlations in the spin-spin correlation functions. However, Q_3 is gauge-dependent and so somewhat unphysical as an order parameter. In the absence of any further fluctuation-driven lattice symmetry breaking, the transition between SRO at $(0, \pi)$ and (q, π) is an example of a *disorder line* [79]; these are lines at which incommensurate correlations first turn on. However, we will see that quantum fluctuations clearly distinguish these two phases, which have confined and deconfined spinons respectively, and the associated topological order requires a phase transitions between them.

An interesting feature of Fig. 1.11 is that the commensurate states squeeze out the incommensurate phases as N/n_b increases. We expect that this suppression of incommensurate order by quantum fluctuations is a general feature of frustrated antiferromagnets.

1.2.2.2 Fluctuations: Long Wavelength Effective Actions

We now extend the analysis of Sect. 1.2.2.1 beyond the mean-field theory and examine the consequences of corrections at finite N. The main question we hope to address are:

- The mean-field theory yielded an excitation spectrum consisting of free spin-1/2 bosonic spinons. We now want to understand the nature of the forces between these spinons and whether they can lead to confinement of half-integer spin excitations.
- Are there any collective excitations and does their dynamics modify in any way the nature of the mean field ground state?

The structure of the fluctuations will clearly be determined by the low-energy excitations about the mean-field state. We have already identified one set of such excitations: spinons at momenta near mimima in their dispersion spectrum, close to the onset of the magnetic LRO phase whence the spinon gap vanishes. An additional set of low-lying spinless excitations can arise from the fluctuations of the Q_{ij} and λ_i fields about their mean-field values. The gauge-invariance (1.30) will act as a powerful restriction on the allowed terms in the effective action for these spinless fields. We anticipate that the only such low-lying excitations are associated with the λ_i and the *phases* of the Q_{ij}. We therefore parametrize

$$Q_{i,i+\hat{e}_p} = \bar{Q}_{i,i+\hat{e}_p} \exp\left(-i\Theta_p\right), \tag{1.44}$$

where the vector $\hat{e}_p$ connects the two sites of the lattice under consideration, $\bar{Q}$ is the mean-field value, and Θ_p is a real phase. The gauge invariance (1.30) implies that the effective action for the Θ_p must be invariant under

$$\Theta_p \rightarrow \Theta_p + \rho_i + \rho_{i+\hat{e}_p}. \tag{1.45}$$

Upon performing a Fourier transform, with the link variables Θ_p placed on the center of the links, the gauge invariance takes the form

$$\Theta_p(\boldsymbol{k}) \rightarrow \Theta_p(\boldsymbol{k}) + 2\rho(\boldsymbol{k})\cos(k_p/2), \tag{1.46}$$

where $k_p = \boldsymbol{k} \cdot \hat{e}_p$. This invariance implies that the effective action for the Θ_p, after integrating out the b quanta, can only be a function of the following gauge-invariant combinations:

$$I_{pq} = 2\cos(k_q/2)\Theta_p(\boldsymbol{k}) - 2\cos(k_p/2)\Theta_q(\boldsymbol{k}). \tag{1.47}$$

We now wish to take the continuum limit at points in the Brillouin zone where the action involves only gradients of the Θ_p fields and thus has the possibility of gapless excitations. This involves expanding about points in the Brillouin zone where

$$\cos(k_p/2) = 0 \text{ for the largest numbers of } \hat{e}_p. \tag{1.48}$$

We now apply this general principle to the J_1-J_2-J_3 model.

Commensurate collinear phases

We begin by examining the (π, π) − SRO phase. As noted in (1.37), this phase has the mean field values $Q_{1,x} = Q_{1,y} \neq 0$, and all other $\bar{Q}_{ij}$ zero. Thus we need only examine the condition (1.48) with $\hat{e}_p = \hat{e}_x, \hat{e}_y$. This uniquely identifies the point $k = G = (\pi, \pi)$ in the Brillouin zone. We therefore parametrize

$$\Theta_x(r) = e^{iG \cdot r} A_x(r) \tag{1.49}$$

and similarly for Θ_y; it can be verified that both Θ and A_x are real in the above equation. We will also be examining invariances of the theory under gauge transformations near G : so we write

$$\rho(r) = e^{iG \cdot r} \zeta(r). \tag{1.50}$$

It is now straightforward to verify that the gauge transformations (1.46) are equivalent to

$$A_x \rightarrow A_x + \partial_x \zeta \tag{1.51}$$

and similarly for A_y. We will also need in the continuum limit the component of λ near the wavevector G. We therefore write

$$i\lambda_i = \bar{\lambda} + ie^{iG \cdot r} A_\tau(r_i). \tag{1.52}$$

Under gauge transformations we have

$$A_\tau \rightarrow A_\tau + \partial_\tau \zeta. \tag{1.53}$$

Thus A_x, A_y, A_τ transform as components of a continuum $U(1)$ vector gauge field.

We will also need the properties of the boson operators under the gauge transformation ζ. From (1.30) and (1.50) we see that the bosons on the two sublattices (A, B) carry opposite charges ± 1 :

$$\begin{aligned} b_A &\rightarrow b_A e^{i\zeta} \\ b_B &\rightarrow b_B e^{-i\zeta}. \end{aligned} \tag{1.54}$$

Finally, we note that the bosonic eigenspectrum has a minimum near $k = k_0 = (\pi/2, \pi/2)$; we therefore parametrize

$$\begin{aligned} b_{Ai}^\alpha &= \psi_1^\alpha(r_i) e^{ik_0 \cdot r_i} \\ b_{Bi}^\alpha &= -i\mathcal{J}^{\alpha\beta} \psi_{2\beta}(r_i) e^{ik_0 \cdot r_i}. \end{aligned} \tag{1.55}$$

We insert the continuum parameterizations (1.49), (1.52) and (1.55) into the functional integral (1.29), perform a gradient expansion, and transform the Lagrangian $\mathcal{L}$ into

$$\begin{aligned} \mathcal{L} = \int \frac{d^2r}{a^2} \Big[&\psi_{1\alpha}^* \left(\frac{d}{d\tau} + iA_\tau \right) \psi_1^\alpha + \psi_2^{\alpha*} \left(\frac{d}{d\tau} - iA_\tau \right) \psi_{2\alpha} \\ &+ \bar{\lambda} \left(|\psi_1^\alpha|^2 + |\psi_{2\alpha}|^2 \right) - 4J_1 \bar{Q}_1 \left(\psi_1^\alpha \psi_{2\alpha} + \psi_{1\alpha}^* \psi_2^{\alpha*} \right) \\ &+ J_1 \bar{Q}_1 a^2 \big[(\nabla + iA) \psi_1^\alpha (\nabla - iA) \psi_{2\alpha} \\ &+ (\nabla - iA) \psi_{1\alpha}^* (\nabla + iA) \psi_2^{\alpha*} \big] \Big]. \end{aligned} \tag{1.56}$$

We now introduce the fields

$$z^\alpha = (\psi_1^\alpha + \psi_2^{\alpha*})/\sqrt{2}$$
$$\pi^\alpha = (\psi_1^\alpha - \psi_2^{\alpha*})/\sqrt{2}.$$

Following the definitions of the underlying spin operators, it is not difficult to show that the Néel order parameter φ_a (which is proportional to $\boldsymbol{n}$ in (1.2)) is related to the z_α by

$$\varphi_a = z_\alpha^* \sigma^{a\alpha}{}_\beta z^\beta. \tag{1.57}$$

From Eq. 1.56, it is clear that the π fields turn out to have mass $\bar\lambda + 4J_1 \bar{Q}_1$, while the z fields have a mass $\bar\lambda - 4J_1 \bar{Q}_1$ which vanishes at the transition to the LRO phase. The π fields can therefore be safely integrated out, and $\mathcal{L}$ yields the following effective action, valid at distances much larger than the lattice spacing [62, 63]:

$$S_{\mathrm{eff}} = \int \frac{d^2r}{\sqrt{8}a} \int\limits_0^{c\beta} d\tilde\tau \left\{ |(\partial_\mu - iA_\mu)z^\alpha|^2 + \frac{\Delta^2}{c^2}|z^\alpha|^2 \right\}. \tag{1.58}$$

Here μ extends over x, y, τ, $c = \sqrt{8}J_1\bar{Q}_1a$ is the spin-wave velocity, $\Delta = (\bar\lambda^2 - 16J_1^2\bar{Q}_1^2)^{1/2}$ is the gap towards spinon excitations, and $A_{\tilde\tau} = A_\tau/c$. Thus, in its final form, the long-wavelength theory consists of a massive, spin-1/2, relativistic, boson z^α (spinon) coupled to a *compact* $U(1)$ gauge field. By 'compact' we mean that values A_μ and $A_\mu + 2\pi$ are identified with each other, and the gauge field lives on a circle: this is clearly required by Eq. 1.44.

At distances larger than c/Δ, we may safely integrate out the massive z quanta and obtain a a compact $U(1)$ gauge theory in $2+1$ dimensions. This theory was argued by Polyakov [80, 81] to be permanently in a confining phase, with the confinement driven by "monopole" tunnelling events. The compact $U(1)$ gauge force will therefore confine the z^α quanta in pairs. So the conclusion is that the $(\pi, \pi)_{SRO}$ does *not* possess $S = 1/2$ spinon excitations, as was the case in the mean field theory. Instead, the lowest-lying excitations with non-zero spin will be triplons, similar to those in Sect. 1.2.1. A further important effect here, not present in the $U(1)$ gauge theories considered by Polyakov, is that the monopole tunnelling events carry Berry phases. The influence of these Berry phases has been described [62, 63] and reviewed [39] elsewhere, and so will not be explained here. The result is that the condensation of monopoles with Berry phases leads to valence bond solid (VBS) order in the ground state. This order is associated with the breaking of the square lattice space group symmetry, as illustrated in Figs. 1.10 and 1.12 below. For the $(\pi, \pi)_{SRO}$ phase, this means that the singlet spin correlations have a structure similar to that in Fig. 1.4. In other words, the square lattice antiferromagnet *spontaneously* acquires a ground state with a symmetry similar to that of the paramagnetic phase of coupled-dimer antiferromagnet. Because the VBS order is spontaneous, the ground state is fourfold degenerate (associated with 90° rotations about a lattice site), unlike the

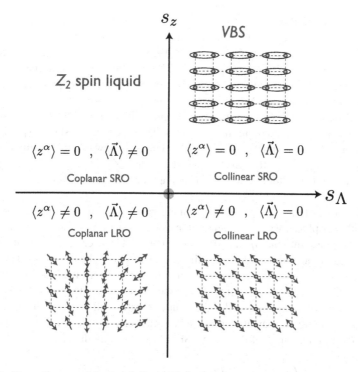

Fig. 1.12 Phase diagram of the theory S_{eff} in Eq. 1.60 for the bosonic spinons z_α and the charge -2 spinless boson Λ. Figure 1.10 contains examples of the region $s_\Lambda > 0$. Figure 1.11 contains two separate instances of four phases meeting at a point as above, with the four phases falling into the classes labeled above; these points are labeled in both figures by the *shaded circles*

non-degenerate ground state of the dimerized antiferromagnet of Sect. 1.2.1. VBS states with a plaquette ordering pattern can also appear, but are not shown in the figures.

The quantum phase transition between the $(\pi, \pi)_{SRO}$ and $(\pi, \pi)_{LRO}$ phases has been the topic of extensive study. The proposal of Refs. [82, 83] is that monopoles are suppressed precisely at the quantum critical point, and so the continuum action in Eq. 1.58 constitutes a complete description of the critical degrees of freedom. It has to be supplemented by a quartic non-linearity $\left(|z^\alpha|^2\right)^2$, because such short-range interactions are relevant perturbations at the critical point. A review of this deconfined criticality proposal is found elsewhere [41].

The properties of the $(0, \pi)$ phase are very similar to those of the (π, π) phase considered above. It can be shown quite generally that any quantum disordered state which has appreciable commensurate, collinear spin correlations will have similar properties: confined spinons, a collective mode described by a compact $U(1)$ gauge field, and VBS order for odd n_b.

Incommensurate phases

We now turn to a study of the incommensurate phases. It is not difficult to show that in this case it is not possible to satisfy the constraints (1.48) at any point in the Brillouin zone for all the non-zero Q_p. This implies that, unlike the commensurate phases, there is no gapless collective gauge mode in the gaussian fluctuations of the incommensurate SRO phases. This has the important implication that the mean-field theory is stable: the structure of the mean-field ground state, and its spinon excitations will survive fluctuation corrections. Thus we obtain a stable 'spin liquid' with bosonic $S = 1/2$ spinon excitations. We will now show that these spinons carry a Z_2 gauge charge, and so this phase is referred to as a Z_2 spin liquid. The Z_2 gauge field also accounts for 'topological order' and a fourfold ground state degeneracy on the torus.

The structure of the theory is simplest in the vicinity of a transition to a commensurate collinear phase: we now examine the effective action as one moves from the (π, π)-SRO phase into the (q, q)-SRO phase (Fig. 1.11; a very similar analysis can be performed at the boundary between the (π, π)-SRO and the (π, q)-SRO phases). This transition is characterized by a continuous turning on of non-zero values of $Q_{i,i+\hat{y}+\hat{x}}$, $Q_{i,i+2\hat{x}}$ and $Q_{i,i+2\hat{y}}$. It is easy to see from Eq. 1.30 that these fields transform as scalars of charge ± 2 under the gauge transformation associated with A_μ. Performing a gradient expansion upon the bosonic fields coupled to these scalars we find that the Lagrangian $\mathcal{L}$ of the (π, π)-SRO phase gets modified to

$$\mathcal{L} \to \mathcal{L} + \int \frac{d^2r}{a} \left(\mathbf{\Lambda}_A \cdot \left(\mathcal{J}_{\alpha\beta} \psi_1^\alpha \nabla \psi_1^\beta \right) + \mathbf{\Lambda}_B \cdot \left(\mathcal{J}^{\alpha\beta} \psi_{2\alpha} \nabla \psi_{2\beta} \right) + \text{c.c.} \right), \quad (1.59)$$

where $\mathbf{\Lambda}_{A,B}$ are two-component scalars $\equiv (J_3 Q_{3,x} + J_2 Q_{2,y+x}, J_3 Q_{3,y} + J_2 Q_{2,y+x})$ with the sites on the ends of the link variables on sublattices A, B. Finally, as before, we transform to the z, π variables, integrate out the π fluctuations and obtain [64]

$$S_{\text{eff}} = \int \frac{d^2r}{\sqrt{8}a} \int\limits_0^{c\beta} d\tilde{\tau} \left\{ |(\partial_\mu - iA_\mu)z^\alpha|^2 + s_z|z^\alpha|^2 + \mathbf{\Lambda} \cdot \left(\mathcal{J}_{\alpha\beta} z^\alpha \nabla z^\beta \right) \right.$$

$$+ \text{c.c.} + K_\Lambda |(\partial_\mu + 2iA_\mu)\mathbf{\Lambda}|^2 + s_\Lambda \mathbf{\Lambda}^2$$

$$\left. + \text{terms quartic in } z^\alpha, \, \mathbf{\Lambda} \right\}. \quad (1.60)$$

Here $s_z = \Delta^2/c^2$, $\mathbf{\Lambda} = (\mathbf{\Lambda}_A + \mathbf{\Lambda}_B^*)/(2J_1\bar{Q}_1a)$ is a complex scalar of charge -2, and K_Λ is a stiffness. We have explicitly written the quadratic terms in the effective action for the $\mathbf{\Lambda}$: these are generated by short wavelength fluctuations of the b^α quanta. We have omitted quartic and higher order terms which are needed to stabilize the theory when the 'masses' s_z or s_Λ are negative, and are also important near the quantum phase transitions. This effective action is also the simplest theory that can be written down which couples a spin-1/2, charge 1, boson z^α, a compact $U(1)$ gauge field A_μ, and a two spatial component, charge -2, spinless boson $\mathbf{\Lambda}$. It is the main result of this section and summarizes essentially all of the physics we are trying to describe.

We now describe the various phases of S_{eff}, which are summarized in Fig. 1.12.

1. *Commensurate, collinear, LRO:* $\langle z^\alpha \rangle \neq 0$, $\langle \Lambda \rangle = 0$. This is the $(\pi, \pi)_{LRO}$ state with commensurate, collinear, magnetic LRO.
2. *Commensurate, collinear, SRO:* $\langle z^\alpha \rangle = 0$, $\langle \Lambda \rangle = 0$. This is the $(\pi, \pi)_{SRO}$ "quantum-disordered" state with collinear spin correlations peaked at (π, π). Its properties were described at length above. The compact $U(1)$ gauge force confines the z^α quanta. The spinless collective mode associated with the gauge fluctuations acquires a gap from monopole condensation, and the monopole Berry phases induce VBS order for odd n_b.
3. *Incommensurate, coplanar, SRO:* $\langle z^\alpha \rangle = 0$, $\langle \Lambda \rangle \neq 0$. This is the incommensurate phase with SRO at (q, q) which we want to study. It is easy to see that condensation of Λ necessarily implies the appearance of incommensurate SRO: ignore fluctuations of Λ about $\langle \Lambda \rangle$ and diagonalize the quadratic form controlling the z^α fluctuations; the minimum of the dispersion of the z^α quanta is at a non-zero wavevector

$$k_0 = (\langle \Lambda_x \rangle, \langle \Lambda_y \rangle)/2. \tag{1.61}$$

The spin structure factor will therefore have a maximum at an incommensurate wavevector. This phase also has a broken lattice rotation symmetry due to the choice of orientation in the $x - y$ plane made by Λ condensate, i.e., it has Ising-nematic order.

The condensation of Λ also has a dramatic impact on the nature of the force between the massive z^α quanta. Detailed arguments have been presented by Fradkin and Shenker [84] that the condensation of a doubly charged Higgs scalar quenches the confining compact $U(1)$ gauge force in 2+1 dimensions between singly charged particles. We can see this here from Eq. 1.60 by noticing that the condensation of Λ expels A_μ by the Meissner effect: consequently, monopoles in A_μ are connected by a flux tube whose action grows linearly with the separation between monopoles. The monopoles are therefore confined, and are unable to induce the confinement of the z^α quanta. From Eq. 1.60 we also see that once Λ is condensed, the resulting theory for the spinons only has an effective Z_2 gauge invariance: Ref. [84] argued that there is an effective description of this free spinon phase in terms of a Z_2 gauge theory. The excitation structure is therefore very similar to that of the mean-field theory: spin-1/2, massive bosonic spinons and spinless collective modes which have a gap. The collective mode gap is present in this case even at $N = \infty$ and is associated with the condensation of Λ.

This state is also 'topologically ordered'. We can see this by noticing [55, 56] that it carries stable point-like excitations which are 2π vortices in either of Λ_x or Λ_y. Because of the screening by the A_μ gauge field, the vortices carry a finite energy (this is analogous to the screening of supercurrents by the magnetic field around an Abrikosov vortex in a superconductor). Because the Λ carry charge -2, the total A_μ flux trapped by a vortex is π. Thus the vortices are also stable to monopole tunneling events, which change the A_μ flux by integer multiples of 2π.

A z_α spinon circumnavigating such a vortex would pick up an Aharanov–Bohm phase factor of π (because the spinons have unit charge), and this is equivalent to the statement that the vortex and the spinon obey mutual 'semionic' statistics. All these characteristics identify the vortex excitation as one dubbed later [85] as the *vison*. The vison also allows us to see the degeneracy of the gapped ground state on surfaces of non-trivial topology. We can insert a vison through any of the 'holes' in surface, and obtain a new candidate eigenstate. This eigenstate has an energy essentially degenerate with the ground state because the core of the vortex is within the hole, and so costs no energy. The 'far field' of the vison is within the system, but it costs negligible energy because the currents have been fully screened by A_μ in this region. Thus we obtain a factor of two increase in the degeneracy for every 'hole' in the surface (which is in turn related to the genus of the surface). The state so obtained is now referred to as a Z_2 spin liquid, and has been labeled as such in the figures. As we noted above, the present theory only yields Z_2 spin liquids with Ising-nematic order, associated with broken symmetry of $90°$ lattice rotations. We also note an elegant exactly solvable model described by Kitaev [86], which has spinon and vison excitations with the characteristics described above, but without the Ising-nematic order.

4. *Incommensurate, coplanar, LRO:* $\langle z^\alpha \rangle \neq 0$, $\langle \Lambda \rangle \neq 0$. The condensation of the z quanta at the wavevector k_0 above leads to incommensurate LRO in the $(q, q)_{LRO}$ phase, with the spin condensate spiraling in the plane.

We also note a recent work [41, 87, 88] which has given a dual perspective on the above phases, including an efficient description of the phase transitions between them, and applied the results to experiments on $\kappa - (ET)_2 Cu_2 (CN)_3$.

1.3 d-Wave Superconductors

In our discussion of phase transitions in insulators we found that the low-energy excitations near the critical point were linked in some way to the broken symmetry of the magnetically ordered state. In the models of Sect. 1.2.1 the low-energy excitations involved long wavelength fluctuations of the order parameter. In Sect. 1.2.2 the connection to the order parameter was more subtle but nevertheless present: the field z_α in Eq. 1.60 is a 'fraction' of the order parameter as indicated in (1.57), and the gauge field A_μ represents a non-coplanarity in the local order parameter orientation.

We will now move from insulators to the corresponding transitions in d-wave superconductors. Thus we will directly address the criticality of the magnetic QPT at $x = x_s$ in Fig. 1.1. We will also consider the criticality of the 'remnant' Ising-nematic ordering at $x = x_m$ within the superconducting phase. A crucial property of d-wave superconductors is that they generically contain gapless, fermionic Bogoliubov excitations, as we will review below. These gapless excitations have a massless Dirac spectrum near isolated points in Brillouin zone. While these fermionic excitations are present in the non-critical d-wave superconductor, it is natural to ask whether they modify the theory of the QPT. Even though they may not be directly related to the order parameter, we can ask if the order parameter and fermionic excitations couple in interesting ways, and whether this coupling modifies the universality class of the transition. These questions will be answered in the following subsections.

We note that symmetry breaking transitions in graphene are also described by field theories similar to those discussed in this section [89, 90].

1.3.1 Dirac Fermions

We begin with a review of the standard BCS mean-field theory for a d-wave superconductor on the square lattice, with an eye towards identifying the fermionic Bogoliubov quasiparticle excitations. For now, we assume we are far from any QPT associated with SDW, Ising-nematic, or other broken symmetries. We consider the generalized Hamiltonian

$$H_{tJ} = \sum_k \varepsilon_k c_{k\alpha}^\dagger c_{k\alpha} + J_1 \sum_{\langle ij \rangle} S_i \cdot S_j, \qquad (1.62)$$

where $c_{j\alpha}$ is the annihilation operator for an electron on site j with spin $\alpha = \uparrow, \downarrow$, $c_{k\alpha}$ is its Fourier transform to momentum space, ε_k is the dispersion of the electrons (it is conventional to choose $\varepsilon_k = -2t_1(\cos(k_x) + \cos(k_y)) - 2t_2(\cos(k_x + k_y) + \cos(k_x - k_y)) - \mu$, with $t_{1,2}$ the first/second neighbor hopping and μ the chemical potential), and the J_1 term is the same as that in Eq. 1.15 with

$$S_{ja} = \frac{1}{2} c_{j\alpha}^\dagger \sigma_{\alpha\beta}^a c_{j\beta} \qquad (1.63)$$

and σ^a the Pauli matrices. We will consider the consequences of the further neighbor exchange interactions in (1.15) for the superconductor in Sect. 1.3.3.1 below. Applying the BCS mean-field decoupling to H_{tJ} we obtain the Bogoliubov Hamiltonian

$$H_{BCS} = \sum_k \varepsilon_k c_{k\alpha}^\dagger c_{k\alpha} - \frac{J_1}{2} \sum_{j\mu} \Delta_\mu \left(c_{j\uparrow}^\dagger c_{j+\hat\mu,\downarrow}^\dagger - c_{j\downarrow}^\dagger c_{j+\hat\mu,\uparrow}^\dagger \right) + \text{h.c.}. \qquad (1.64)$$

For a wide range of parameters, the ground state energy is optimized by a $d_{x^2-y^2}$ wavefunction for the Cooper pairs: this corresponds to the choice $\Delta_x = -\Delta_y = \Delta_{x^2-y^2}$. The value of $\Delta_{x^2-y^2}$ is determined by minimizing the energy of the BCS state

$$E_{BCS} = J_1 |\Delta_{x^2-y^2}|^2 - \int \frac{d^2k}{4\pi^2} [E_k - \varepsilon_k], \qquad (1.65)$$

where the fermionic quasiparticle dispersion is

$$E_k = \left[\varepsilon_k^2 + \left|J_1\Delta_{x^2-y^2}(\cos k_x - \cos k_y)\right|^2\right]^{1/2}. \tag{1.66}$$

The energy of the quasiparticles, E_k, vanishes at the four points $(\pm Q, \pm Q)$ at which $\varepsilon_k = 0$. We are especially interested in the low-energy quasiparticles in the vicinity of these points, and so we perform a gradient expansion of H_{BCS} near each of them. We label the points $Q_1 = (Q, Q)$, $Q_2 = (-Q, Q)$, $Q_3 = (-Q, -Q)$, $Q_4 = (Q, -Q)$ and write

$$c_{j\alpha} = f_{1\alpha}(r_j)e^{iQ_1 \cdot r_j} + f_{2\alpha}(r_j)e^{iQ_2 \cdot r_j} + f_{3\alpha}(r_j)e^{iQ_3 \cdot r_j} + f_{4\alpha}(r_j)e^{iQ_4 \cdot r_j}, \tag{1.67}$$

while assuming that the $f_{1-4,\alpha}(r)$ are slowly varying functions of r. We also introduce the bispinors $\Psi_1 = (f_{1\uparrow}, f_{3\downarrow}^\dagger, f_{1\downarrow}, -f_{3\uparrow}^\dagger)$, and $\Psi_2 = (f_{2\uparrow}, f_{4\downarrow}^\dagger, f_{2\downarrow}, -f_{4\uparrow}^\dagger)$, and then express H_{BCS} in terms of $\Psi_{1,2}$ while performing a spatial gradient expansion. This yields the following effective action for the fermionic quasiparticles:

$$S_\Psi = \int d\tau d^2r \left[\Psi_1^\dagger \left(\partial_\tau - i\frac{v_F}{\sqrt{2}}(\partial_x + \partial_y)\tau^z - i\frac{v_\Delta}{\sqrt{2}}(-\partial_x + \partial_y)\tau^x \right) \Psi_1 \right.$$
$$\left. + \Psi_2^\dagger \left(\partial_\tau - i\frac{v_F}{\sqrt{2}}(-\partial_x + \partial_y)\tau^z - i\frac{v_\Delta}{\sqrt{2}}(\partial_x + \partial_y)\tau^x \right) \Psi_2 \right], \tag{1.68}$$

where the $\tau^{x,z}$ are 4×4 matrices which are block diagonal, the blocks consisting of 2×2 Pauli matrices. The velocities $v_{F,\Delta}$ are given by the conical structure of E_k near the Q_{1-4} : we have $v_F = \left|\nabla_k \varepsilon_k|_{k=Q_a}\right|$ and $v_\Delta = |J_1\Delta_{x^2-y^2}\sqrt{2}\sin(Q)|$. In this limit, the energy of the Ψ_1 fermionic excitations is $E_k = (v_F^2(k_x + k_y)^2/2 + v_\Delta^2(k_x - k_y)^2/2)^{1/2}$ (and similarly for Ψ_2), which is the spectrum of massless Dirac fermions.

1.3.2 Magnetic Ordering

We now focus attention on the QPT involving loss of magnetic ordering within the d-wave superconductor at $x = x_s$ in Fig. 1.1. As in Sect. 1.2, we have to now consider the fluctuations of the SDW order parameter. We discussed two routes to such a magnetic ordering transition in Sect. 1.2: one involving the vector SDW order parameter in Sect. 1.2.1, and the other involving the spinor z_α in Sect. 1.2.2. In principle, both routes also have to be considered in the d-wave superconductor. The choice between the two routes involves subtle questions on the nature of fractionalized excitations at intermediate scales which we will not explore further here. These questions were thoroughly addressed in Ref. [27] in the context of simple toy models: it was found that either route could apply, and the choice depended sensitively on microscopic details. In particular, it was found that among the fates of the non-magnetic superconductor was that it acquired VBS or Ising-nematic ordering, as was found in the models explored in Sect. 1.2.2. This is part of the motivation for the expectation of such ordering in the regime $x_s < x < x_m$, as indicated in Fig. 1.1.

Table 1.1 Transformations of the fields under operations which generate the symmetry group

	T_x	T_y	R	I	$\mathcal{T}$
Φ_{xa}	$e^{iq}\Phi_{xa}$	$-\Phi_{xa}$	Φ_{ya}	Φ_{xa}^*	$-\Phi_{xa}$
Φ_{ya}	$-\Phi_{ya}$	$e^{iq}\Phi_{ya}$	Φ_{xa}^*	Φ_{ya}^*	$-\Phi_{ya}$
$\Psi_{1\alpha}$	$e^{iQ}\Psi_{1\alpha}$	$e^{iQ}\Psi_{1\alpha}$	$i\tau^z\Psi_{2\alpha}$	$\Psi_{2\alpha}$	$-\tau^y\Psi_{1\alpha}$
$\Psi_{2\alpha}$	$e^{-iQ}\Psi_{2\alpha}$	$e^{iQ}\Psi_{2\alpha}$	$-i\varepsilon_{\alpha\beta}\left[\Psi_{1\beta}^{\dagger}\tau^x\right]^T$	$\Psi_{1\alpha}$	$-\tau^y\Psi_{2\alpha}$

$T_{x,y}$ = translation by a lattice spacing in the x, y directions, R = rotation about a lattice site by 90°, I = reflection about the y axis on a lattice site, and $\mathcal{T}$ = time reversal. The theory is also invariant under spin rotations, with i a vector index and α, β spinor indices. We define $\mathcal{T}$ as an invariance of the imaginary time path integral, in which $\Phi_{1,2i}^*$ transform as the complex conjugates of $\Phi_{1,2i}$, while $\Psi_{1,2\alpha}^{\dagger}$ are viewed as independent complex Grassman fields which transform as $\Psi_{1,2\alpha}^{\dagger} \rightarrow \Psi_{1,2\alpha}^{\dagger}\tau^y$

In the interests of brevity and simplicity, we will limit our discussion of the SDW ordering transition here to the vector formulation analogous to that in Sect. 1.2.1. We have full square lattice symmetry, and so allow for incommensurate SDW ordering similar to the (q, q), $(q, -q)$ and (π, q), (q, π) states of Sect. 1.2.2. Because there are two distinct but degenerate ordering wavevectors, the complex order parameter Φ_a in Eq. 1.14 is now replaced by two complex order parameters Φ_{xa} and Φ_{ya} for orderings along (π, q) and (q, π) (the orderings along $(q, \pm q)$ can be treated similarly and we will not describe it explicitly). These order parameters are related to the spin operator by

$$S_a(r) = \Phi_{xa}e^{iK_x \cdot r} + \Phi_{ya}e^{iK_y \cdot r} + \text{c.c.} \tag{1.69}$$

where $K_x = (q, \pi)$ and $K_y = (\pi, q)$. As discussed below Eq. 1.14, depending upon the structure of the complex numbers Φ_{xa}, Φ_{ya}, the SDW ordering can be either collinear (i.e., stripe-like) or spiral. Also, as in Eqs. 1.39 and 1.43, we can use these SDW order parameters to also define a subsidiary Ising-nematic order parameter

$$\mathcal{I} = |\Phi_{xa}|^2 - |\Phi_{ya}|^2 \tag{1.70}$$

to measure the breaking of $x \leftrightarrow y$ symmetry.

Symmetry considerations will play an important role in our analysis of the $\Phi_{x,ya}$ order parameters and their coupling to the Dirac fermions. In Table 1.1 we therefore present a table of transformations under important operations of the square lattice space group: these are easily deduced from the representations in Eqs. 1.67 and 1.69.

The effective action for the SDW order parameters has a direct generalization from (1.14): it can be obtained by requiring invariance under the transformations in Table 1.1, and has many more allowed quartic nonlinearities [18, 19]:

$$\mathcal{S}_\Phi = \int d^2 r d\tau \Bigg[|\partial_\tau \Phi_{xa}|^2 + c_x^2 |\partial_x \Phi_{xa}|^2$$

$$+ c_y^2 |\partial_y \Phi_{xa}|^2 + |\partial_\tau \Phi_{ya}|^2 + c_x^2 |\partial_y \Phi_{ya}|^2$$

$$+ c_y^2 |\partial_x \Phi_{ya}|^2 + s \left(|\Phi_{xa}|^2 + |\Phi_{xa}|^2 \right)$$

$$+ \frac{u_1}{2} \left[\left(|\Phi_{xa}|^2 \right)^2 + \left(|\Phi_{ya}|^2 \right)^2 \right]$$

$$+ \frac{u_2}{2} \left[\left| (\Phi_{xa})^2 \right|^2 + \left| (\Phi_{ya})^2 \right|^2 \right]$$

$$+ w_1 |\Phi_{xa}|^2 |\Phi_{ya}|^2 + w_2 |\Phi_{xa} \Phi_{ya}|^2$$

$$+ w_3 \left| \Phi_{xa} \Phi_{ya}^* \right|^2 \Bigg]. \tag{1.71}$$

Remarkably, a fairly complete 5-loop renormalization group analysis of this model has been carried out by De Prato et al. [91], and reliable information on its critical properties is now available.

(We note parenthetically that Eq. 1.71 concerns the theory of the transition at x_s from an SDW ordered state to a d-wave superconductor with the full symmetry of the square lattice. However, as we have discussed in Sect. 1.1 and in the beginning of Sect. 1.3, there could be Ising nematic order in the regime $x_s < x < x_m$. In this case one of Φ_{xa} or Φ_{ya} orderings would be preferred, and we need only consider the critical fluctuations of this preferred component. The resulting action for this preferred component would then be identical to Eq. 1.14, with critical properties as in Ref. [54].)

Now we turn to the crucial issue of the coupling between the $\Phi_{x,ya}$ order parameter degrees of freedom in $\mathcal{S}_\Phi$ and the massless Dirac fermions $\Psi_{1,2}$ in Eq. 1.68. Again a great deal follows purely from symmetry considerations. The simplest possible terms are cubic 'Yukawa' interaction terms like $\Phi_{xa} \Psi_1^\dagger \Psi_2$ etc. However, these are generically forbidden by translational invariance, or equivalently, momentum conservation. In particular, the transformation of the $\Phi_{x,ya}$ under translation by one lattice spacing follows from (1.69), while those of the $\Psi_{1,2}$ follow from (1.67). Unless the SDW ordering wavevectors $\boldsymbol{K}_{x,y}$ and the positions of the Dirac nodes $\boldsymbol{Q}_{1,2,3,4}$ satisfy certain commensurability conditions, the Yukawa coupling will not be invariant under this translation operation. This is illustrated schematically in Fig. 1.13. The observed values of the wavevectors are not commensurate, and so we can safely neglect the Yukawa term.

The absence of the Yukawa coupling suggests that the fixed point theory describing the QPT at $x = x_s$ in the superconductor may be $\mathcal{S}_\Phi$ in Eq. 1.71 alone, i.e., the transition is in the same universality class as the insulator. However, to ensure this, we have to examine the influence of higher terms coupling the degrees of freedom of $\mathcal{S}_\phi$ and $\mathcal{S}_\Psi$. The simplest couplings not prohibited by translational invariance are associated with operators which are close to net zero momentum in both sectors. These are further constrained by the other square lattice space group

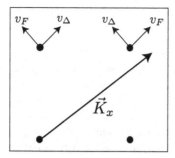

Fig. 1.13 The *filled circles* indicate the positions of the gapless Dirac fermions in the square lattice Brillouin zone: these are at wavevectors $Q_{1,2,3,4}$. An SDW fluctuation scatters a fermion at one of the nodes by wavevector K_x to a generic point in the Brillouin zone. The final state of the fermion has a high energy, and so such processes are suppressed

operations in Table 1.1; requiring invariance under them shows that the simplest allowed terms are [92]

$$\mathcal{S}_1 = \vartheta_1 \int d\tau d^2 r \left(|\Phi_{xa}|^2 + |\Phi_{ya}|^2 \right) \left(\Psi_1^\dagger \tau^z \Psi_1 + \Psi_2^\dagger \tau^z \Psi_2 \right)$$
$$\mathcal{S}_2 = \vartheta_2 \int d\tau d^2 r \left(|\Phi_{xa}|^2 - |\Phi_{ya}|^2 \right) \left(\Psi_1^\dagger \tau^x \Psi_1 + \Psi_2^\dagger \tau^x \Psi_2 \right). \tag{1.72}$$

The first term is a fairly obvious 'density–density' coupling between the energies of the two systems. The second is more interesting: it involves the Ising nematic order $\mathcal{I}$, as measured in the order parameter sector by (1.70), and in the fermion sector by the bilinear shown above.

Now we can ask if the fixed point described by the decoupled theory $\mathcal{S}_\Phi + \mathcal{S}_\Psi$ is stable under the perturbations in $\mathcal{S}_1$ and $\mathcal{S}_2$. This involves a computation of the scaling dimensions of the couplings $\vartheta_{1,2}$ at the decoupled theory fixed point. These scaling dimensions were computed to 5-loop order in Ref. [92], and it was found that $\dim[\vartheta_1] \approx -1.0$, and $\dim[\vartheta_2] \approx -0.1$. Thus both couplings are irrelevant, and we can indeed finally conclude that the SDW onset transition is described by the same theory as in the insulator. However, the scaling dimension of the ϑ_2 coupling is quite small, indicating that it will lead to appreciable effects. Thus we have demonstrated quite generally that it is the Ising-nematic order $\mathcal{I}$ which is most efficient in coupling the SDW order parameter fluctuations to the Dirac fermions. Note that $\mathcal{I}$ was not chosen by hand, but was selected by the theory among all other possible composite orders of the SDW field $\Phi_{x,ya}$. The near zero scaling dimension of ϑ_2 implies that it will induce a linewidth $\sim T$ in the Dirac fermion spectrum. Moreover, because this broadening is mediated by $\mathcal{I}$, the broadening will be strongly anisotropic in space [93].

1.3.3 Ising Transitions

Now we turn our attention to the vicinity of the point x_m in Fig. 1.1. Although x_m was defined in terms of the SDW transition in the metal at high magnetic fields, we have also argued in Sect. 1.1 and in the beginning of Sect. 1.3 that there can also be transitions associated with VBS or Ising nematic order near x_m but within the superconducting phase at zero field. Strong evidence for a nematic order transition near x_m has emerged in recent experiments [10, 31, 32].

This section will therefore consider the theory of Ising-nematic ordering within a d-wave superconductor. Unlike the situation in Sect. 1.3.2, we will find here that the order parameter and the Dirac fermions are strongly coupled, and the universality class of the transition is completely changed by the presence of the Dirac fermions. In Sect. 1.3.2 we found that although the fermions were moderately strongly coupled to the critical theory, they were ultimately reduced to spectators to the asymptotic critical behavior.

Before considering the Ising nematic transition, we will take a short detour in Sect. 1.3.3.1 and describe another Ising transition associated with the breaking of time-reversal symmetry in a d-wave superconductor. This leads to a model which has a somewhat simpler structure, and for which conventional renormalization group techniques work easily. We will return to Ising-nematic ordering in Sect. 1.3.3.2.

1.3.3.1 Time-Reversal Symmetry Breaking

We will consider a simple model in which the pairing symmetry of the superconductor changes from $d_{x^2-y^2}$ to $d_{x^2-y^2} \pm i d_{xy}$. The choice of the phase between the two pairing components leads to a breaking of time-reversal symmetry. Studies of this transition were originally motivated by the cuprate phenomenology, but we will not explore this experimental connection here because the evidence has remained sparse.

The mean field theory of this transition can be explored entirely within the context of BCS theory, as we will review below. However, fluctuations about the BCS theory are strong, and lead to non-trivial critical behavior involving both the collective order parameter and the Bogoliubov fermions: this is probably the earliest known example [94–96] of the failure of BCS theory in two (or higher) dimensions in a superconducting ground state. At $T > 0$, this failure broadens into the "quantum critical" region.

We extend H_{tJ} in Eq. 1.62 so that BCS mean-field theory permits a region with d_{xy} superconductivity. It turns out that the frustrating interactions as in Eq. 1.15 are precisely those needed. With a J_2 interaction, Eq. 1.62 is modified to:

$$\tilde{H}_{tJ} = \sum_k \varepsilon_k c_{k\sigma}^\dagger c_{k\sigma} + J_1 \sum_{\langle ij \rangle} S_i \cdot S_j + J_2 \sum_{\text{nnn } ij} S_i \cdot S_j. \tag{1.73}$$

We will follow the evolution of the ground state of $\tilde{H}_{tJ}$ as a function of J_2/J_1.

The mean-field Hamiltonian is now modified from Eq. 1.64 to

Fig. 1.14 Values of the pairing amplitudes, $-\langle c_{i\uparrow}c_{j\downarrow} - c_{i\downarrow}c_{j\uparrow}\rangle$ with i the *central site*, and j is one of its eight nearest neighbors

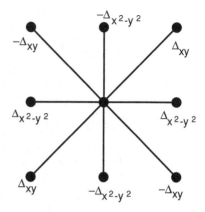

$$\widetilde{H}_{BCS} = \sum_k \varepsilon_k c_{k\sigma}^\dagger c_{k\sigma} - \frac{J_1}{2}\sum_{j,\mu} \Delta_\mu (c_{j\uparrow}^\dagger c_{j+\hat{\mu},\downarrow}^\dagger - c_{j\downarrow}^\dagger c_{j+\hat{\mu},\uparrow}^\dagger) + \text{h.c.}$$

$$- \frac{J_2}{2}\sum_{j,\nu}' \Delta_\nu (c_{j\uparrow}^\dagger c_{j+\hat{\nu},\downarrow}^\dagger - c_{j\downarrow}^\dagger c_{j+\hat{\nu},\uparrow}^\dagger) + \text{h.c.}, \tag{1.74}$$

where the second summation over ν is along the diagonal neighbors $\hat{x}+\hat{y}$ and $-\hat{x}+\hat{y}$. To obtain d_{xy} pairing along the diagonals, we choose $\Delta_{x+y} = -\Delta_{-x+y} = \Delta_{xy}$. We summarize our choices for the spatial structure of the pairing amplitudes (which determine the Cooper pair wavefunction) in Fig. 1.14. The values of $\Delta_{x^2-y^2}$ and Δ_{xy} are to be determined by minimizing the ground state energy (generalizing Eq. 1.65)

$$E_{BCS} = J_1 |\Delta_{x^2-y^2}|^2 + J_2 |\Delta_{xy}|^2 - \int \frac{d^2k}{4\pi^2} [E_k - \varepsilon_k], \tag{1.75}$$

where the quasiparticle dispersion is now (generalizing Eq. 1.66)

$$E_k = \left[\varepsilon_k^2 + \left|J_1 \Delta_{x^2-y^2}(\cos k_x - \cos k_y) + 2J_2\Delta_{xy}\sin k_x \sin k_y\right|^2\right]^{1/2}. \tag{1.76}$$

Notice that the energy depends upon the relative phase of $\Delta_{x^2-y^2}$ and Δ_{xy} : this phase is therefore an observable property of the ground state.

It is a simple matter to numerically carry out the minimization of Eq. 1.76, and the results for a typical choice of parameters are shown in Fig. 1.15 as a function J_2/J_1. One of the two amplitudes $\Delta_{x^2-y^2}$ or Δ_{xy} is always non-zero and so the ground state is always superconducting. The transition from pure $d_{x^2-y^2}$ superconductivity to pure d_{xy} superconductivity occurs via an intermediate phase in which *both* order parameters are non-zero. Furthermore, in this regime, their relative phase is found to be pinned to $\pm\pi/2$, i.e.,

$$\arg(\Delta_{xy}) = \arg(\Delta_{x^2-y^2}) \pm \pi/2. \tag{1.77}$$

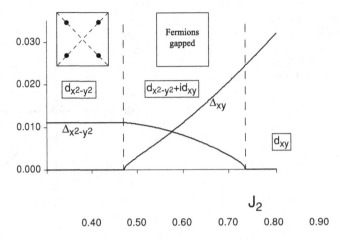

Fig. 1.15 BCS solution of the phenomenological Hamiltonian $\widetilde{H}_{tJ}$ in Eq. 1.73. Shown are the optimum values of the pairing amplitudes $|\Delta_{x^2-y^2}|$ and $|\Delta_{xy}|$ as a function of J_2 for $t_1 = 1$, $t_2 = -0.25$, $\mu = -1.25$, and J_1 fixed at $J_1 = 0.4$. The relative phase of the pairing amplitudes was always found to obey Eq. 1.77. The *dashed lines* denote locations of phase transitions between $d_{x^2-y^2}$, $d_{x^2-y^2} + id_{xy}$, and d_{xy} superconductors. The pairing amplitudes vanishes linearly at the first transition corresponding to the exponent $\beta_{BCS} = 1$ in Eq. 1.80. The Brillouin zone location of the gapless Dirac points in the $d_{x^2-y^2}$ superconductor is indicated by *filled circles*. For the dispersion ε_k appropriate to the cuprates, the d_{xy} superconductor is fully gapped, and so the second transition is ordinary Ising

The reason for this pinning can be intuitively seen from Eq. 1.76: only for these values of the relative phase does the equation $E_k = 0$ never have a solution. In other words, the gapless nodal quasiparticles of the $d_{x^2-y^2}$ superconductor acquire a finite energy gap when a secondary pairing with relative phase $\pm\pi/2$ develops. By a level repulsion picture, we can expect that gapping out the low-energy excitations should help lower the energy of the ground state. The intermediate phase obeying Eq. 1.77 is called a $d_{x^2-y^2} + id_{xy}$ superconductor.

The choice of the sign in Eq. 1.77 leads to an overall twofold degeneracy in the choice of the wavefunction for the $d_{x^2-y^2} + id_{xy}$ superconductor. This choice is related to the breaking of time-reversal symmetry, and implies that the $d_{x^2-y^2} + id_{xy}$ phase is characterized by the non-zero expectation value of a Z_2 Ising order parameter; the expectation value of this order vanishes in the two phases (the $d_{x^2-y^2}$ and d_{xy} superconductors) on either side of the $d_{x^2-y^2} + id_{xy}$ superconductor. As is conventional, we will represent the Ising order by a real scalar field ϕ. Fluctuations of ϕ become critical near both of the phase boundaries in Fig. 1.15. As we will explain below, the critical theory of the $d_{x^2-y^2}$ to $d_{x^2-y^2} + id_{xy}$ transition is *not* the usual ϕ^4 field theory which describes the ordinary Ising transition in three spacetime dimensions. (For the dispersion ε_k appropriate to the cuprates, the d_{xy} superconductor is fully gapped, and so the $d_{x^2-y^2} + id_{xy}$ to d_{xy} transition in Fig. 1.15 will be ordinary Ising.)

Near the phase boundary from $d_{x^2-y^2}$ to $d_{x^2-y^2} + id_{xy}$ superconductivity it is clear that we can identify

$$\phi = i\Delta_{xy}, \tag{1.78}$$

(in the gauge where $\Delta_{x^2-y^2}$ is real). We can now expand E_{BCS} in Eq. 1.75 for small ϕ (with $\Delta_{x^2-y^2}$ finite) and find a series with the structure [97, 98]

$$E_{BCS} = E_0 + s\phi^2 + v|\phi|^3 + \ldots, \tag{1.79}$$

where s, v are coefficients and the ellipses represent regular higher order terms in even powers of ϕ; s can have either sign, whereas v is always positive. Notice the non-analytic $|\phi|^3$ term that appears in the BCS theory— this arises from an infrared singularity in the integral in Eq. 1.75 over E_k at the four nodal points of the $d_{x^2-y^2}$ superconductor, and is a preliminary indication that the transition differs from that in the ordinary Ising model, and that the Dirac fermions play a central role. We can optimize ϕ by minimizing E_{BCS} in Eq. 1.79— this shows that $\langle\phi\rangle = 0$ for $s>0$, and $\langle\phi\rangle \neq 0$ for $s<0$. So $s \sim (J_2/J_1)_c - J_2/J_1$ where $(J_2/J_1)_c$ is the first critical value in Fig. 1.15. Near this critical point, we find

$$\langle\phi\rangle \sim (s_c - s)^\beta, \tag{1.80}$$

where we have allowed for the fact that fluctuation corrections will shift the critical point from $s = 0$ to $s = s_c$. The present BCS theory yields the exponent $\beta_{BCS} = 1$; this differs from the usual mean-field exponent $\beta_{MF} = 1/2$, and this is of course due to the non-analytic $|\phi|^3$ term in Eq. 1.79.

We have laid much of the ground work for the required field theory of the onset of d_{xy} order in Sect. 1.3.2. In addition to the order parameter ϕ, the field theory should also involve the low-energy nodal fermions of the $d_{x^2-y^2}$ superconductor, as described by S_Ψ in Eq. 1.68. For the ϕ fluctuations, we write down the usual terms permitted near a phase transition with Ising symmetry, and similar to those in Eq. 1.12:

$$S_\phi = \int d^2r d\tau \left[\frac{1}{2}\left((\partial_\tau\phi)^2 + c^2(\partial_x\phi)^2 + c^2(\partial_y\phi)^2 + s\phi^2 \right) + \frac{u}{24}\phi^4 \right]. \tag{1.81}$$

Note that, unlike Eq. 1.79, we do not have any non-analytic $|\phi|^3$ terms in the action: this is because we have not integrated out the low-energy Dirac fermions, and the terms in Eq. 1.81 are viewed as arising from high-energy fermions away from the nodal points. Finally, we need to couple the ϕ and $\Psi_{1,2}$ excitations. Their coupling is already contained in the last term in Eq. 1.74: expressing this in terms of the $\Psi_{1,2}$ fermions using Eq. 1.67 we obtain

$$S_{\Psi\phi} = \vartheta_{xy} \int d^2r d\tau \left[\phi \left(\Psi_1^\dagger \tau^y \Psi_1 - \Psi_2^\dagger \tau^y \Psi_2 \right) \right], \tag{1.82}$$

where ϑ_{xy} is a coupling constant. This coupling also has been obtained by symmetry considerations, by examining invariants under the transformations of Table 1.1. The partition function of the full theory is now

$$Z = \int \mathcal{D}\phi \mathcal{D}\Psi_1 \mathcal{D}\Psi_2 \exp\left(-\mathcal{S}_\Psi - \mathcal{S}_\phi - \mathcal{S}_{\Psi\phi}\right), \tag{1.83}$$

where $\mathcal{S}_\Psi$ was in Eq. 1.68. It can now be checked that if we integrate out the $\Psi_{1,2}$ fermions for a spacetime independent ϕ, we do indeed obtain a $|\phi|^3$ term in the effective potential for ϕ.

We begin our analysis of Z in Eq. 1.83 by following the procedure of Sect. 1.3.2. Assume that the transition is described by a fixed point with $\vartheta_{xy} = 0$: then as in Sect. 1.3.2, the theory for the transition would be the ordinary ϕ^4 field theory $\mathcal{S}_\phi$, and the nodal fermions would again be innocent spectators. The scaling dimension of ϕ at such a fixed point is $(1 + \eta_I)/2$ (where η_I is the anomalous order parameter exponent at the critical point of the ordinary three dimensional Ising model), while that of $\Psi_{1,2}$ is 1. Consequently, the scaling dimension of ϑ_{xy} is $(1 - \eta_I)/2 > 0$. This positive scaling dimension implies that ϑ_{xy} is relevant and the $\vartheta_{xy} = 0$ fixed point is unstable: the Dirac fermions are fully involved in the critical theory.

Determining the correct critical behavior now requires a full renormalization group analysis of Z. This has been described in some detail in Ref. [96], and we will not reproduce the details here. The main result we need for our purposes is that couplings ϑ_{xy}, u, v_F/c and v_Δ/c all reach *non-zero* fixed point values which define a critical point in a new universality class. These fixed point values, and the corresponding critical exponents, can be determined in expansions in either $(3 - d)$ [94–96] (where d is the spatial dimensionality) or $1/N$ [99] (where N is the number of fermion species). An important simplifying feature here is that the fixed point is actually relativistically invariant. Indeed the fixed point has the structure of the so-called Higgs–Yukawa model which has been studied extensively in the particle physics literature [100] in a different physical context: quantum Monte Carlo simulation of this model also exist [101], and provide probably the most accurate estimate of the exponents.

The non-trivial fixed point has strong implications for the correlations of the Bogoliubov fermions. The fermion correlation function $G_1 = \langle \Psi_1 \Psi_1^\dagger \rangle$ obeys

$$G_1(k, \omega) = \frac{\omega + v_F k_x \tau^z + v_\Delta \tau^x}{(v_F^2 k_x^2 + v_\Delta^2 k_y^2 - \omega^2)^{(1-\eta_f)/2}} \tag{1.84}$$

at low frequencies for $s \geq s_c$. Away from the critical point in the $d_{x^2-y^2}$ superconductor with $s > s_c$, Eq (1.84) holds with $\eta_f = 0$, and this is the BCS result, with sharp quasi-particle poles in the Green's function. At the critical point $s = s_c$ Eq. 1.84 holds with the fixed point values for the velocities (which satisfy $v_F = v_\Delta = c$) and with the anomalous dimension $\eta_f \neq 0$–the $(3-d)$ expansion [94, 95] estimate is $\eta_f \approx (3 - d)/14$, and the $1/N$ expansion estimate [99] is $\eta_f \approx 1/(3\pi^2 N)$, with $N = 2$. This is clearly non-BCS behavior, and the fermionic quasiparticle pole in the spectral function has been replaced by a branch-cut representing the continuum of critical excitations. The corrections to BCS extend also to correlations of the Ising order ϕ : its expectation value vanishes as Eq. 1.80 with the Monte Carlo estimate $\beta \approx 0.877$ [101]. The critical point correlators of ϕ have the anomalous dimension $\eta \approx 0.754$ [101], which is clearly different from the very small value of the exponent η_I at the unstable $\vartheta_{xy} = 0$ fixed point. The value of β is related to η by the usual scaling law $\beta = (1 + \eta)\nu/2$, with $\nu \approx 1.00$ the correlation length exponent (which also differs from the exponent ν_I of the Ising model).

1.3.3.2 Nematic Ordering

We now turn, as promised, to the case of Ising-nematic ordering within the d-wave superconductor at $x = x_m$.

The ingredients of such an ordering are actually already present in our simple review of BCS theory in Sect. 1.3.1. In Eq. 1.64, we introduce two variational pairing amplitudes Δ_x and Δ_y. Subsequently, we assumed that the minimization of the energy led to a solution with $d_{x^2-y^2}$ pairing symmetry with $\Delta_x = -\Delta_y = \Delta_{x^2-y^2}$. However, it is possible that upon including the full details of the microscopic interactions we are led to a minimum where the optimal solution also has a small amount of s-wave pairing. Then $|\Delta_x| \neq |\Delta_y|$, and we would expect all physical properties to have distinct dependencies on the x and y coordinates. So, as in Eqs. 1.39, 1.43 and 1.70, we can define the Ising-nematic order parameter by

$$\mathcal{I} = |\Delta_x|^2 - |\Delta_y|^2. \tag{1.85}$$

The derivation of the field theory for this transition follows closely our presentation in Sect. 1.3.3.1. We allow for small Ising-nematic ordering by introducing a scalar field ϕ and writing

$$\Delta_x = \Delta_{x^2-y^2} + \phi \; ; \; \Delta_y = -\Delta_{x^2-y^2} + \phi; \tag{1.86}$$

note that $\mathcal{I} \propto \phi$. The evolution of the Dirac fermion spectrum under such a change is indicated in Fig. 1.16. We now develop an effective action for ϕ and the Dirac fermions $\Psi_{1,2}$. The result is essentially identical to that in Sect. 1.3.3.1, apart from a change in the structure of the Yukawa coupling. Thus we obtain a theory $\mathcal{S}_\Psi + \mathcal{S}_\phi + \overline{\mathcal{S}}_{\Psi\phi}$, defined by Eqs. 1.68 and 1.81, and where Eq. 1.82 is now replaced by

$$\overline{\mathcal{S}}_{\Psi\phi} = \vartheta_I \int d^2r d\tau \left[\phi \left(\Psi_1^\dagger \tau^x \Psi_1 + \Psi_2^\dagger \tau^x \Psi_2 \right) \right]. \tag{1.87}$$

Not surprisingly, the fermion bilinear coupling to the nematic order parameter ϕ is identical to that in Eq. 1.72, as is expected from the transformations of Table 1.1.

The seemingly innocuous change between Eqs. 1.82 and 1.87 however, has strong consequences. This is partly linked to the fact with $\overline{\mathcal{S}}_{\Psi\phi}$ cannot be relativistically

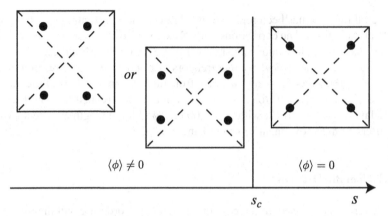

Fig. 1.16 Phase diagram of Ising nematic ordering in a d-wave superconductor as a function of the coupling s in $\mathcal{S}_\phi$. The *filled circles* indicate the location of the gapless fermionic excitations in the Brillouin zone. The two choices for $s < s_c$ are selected by the sign of $\langle \phi \rangle$

invariant even after all velocities are adjusted to be equal. A weak-coupling renormalization group analysis in powers of the coupling ϑ_I was performed in $(3 - d)$ dimensions in Refs. [94–96], and led to flows to strong coupling with no accessible fixed point: thus no firm conclusions on the nature of the critical theory were drawn.

This problem remained unsolved until the recent works of Refs. [93, 102]. It is essential that the coupling ϑ_I not be used as a perturbative expansion parameter. This is because it leads to strongly non-analytic changes in the structure of the ϕ propagator, which have to be included at all stages. In a model with N fermion flavors, the $1/N$ expansion does avoid any expansion in ϑ_I. The renormalization group analysis has to be carried out within the context of the $1/N$ expansion, and this involves some rather technical analysis which is explained in Ref. [102]. In the end, an asymptotically exact description of the vicinity of the critical point was obtained. It was found that the velocity ratio v_F/v_Δ diverged logarithmically with energy scale, leading to strongly anisotropic 'arc-like' spectra for the Dirac fermions. Associated singularities in the thermal conductivity have also been computed [103].

1.4 Metals

We finally turn to the transition in the metal at x_m, which anchored our discussion of the cuprate phase diagram in Sect. 1.1. This controls the high field transition line in Fig. 1.1 between the large Fermi surface and small Fermi pocket states. We also argued that this transition was a key ingredient in a theory of the strange metal.

In addition to the order parameter ingredients we met in Sect. 1.2, we now have to also account for fermion excitations as in Sect. 1.3. In Sect. 1.3 the fermionic excitations had vanishing energy only at isolated nodal points in the Brillouin zone:

see Fig. 1.13. In the present section we are dealing with metals, which have fermionic excitations with vanishing energy along an entire line in the Brillouin zone. Thus we can expect them to have an even stronger effect on the critical theory. This will indeed be the case, and we will be led to problems with a far more complex structure. Unlike the situation in insulators and d-wave superconductors, many basic issues associated with ordering transitions in two dimensional metals have not been fully resolved. The problem remains one of active research and is being addressed by many different approaches.

As discussed in Sects. 1.2 and 1.3.2, we can describe magnetic ordering by using either vector or spinor variables for the order parameter, and these lead to very different phases and critical points. For metals, the relationship between these two approaches, and their distinct physical properties have been described recently in Ref. [13]. The spinor route is more 'exotic' and leads to intermediate non-Fermi liquid critical phases between the small and large Fermi surface Fermi liquid phases. These intermediate critical phases could well be important for the experiments and for Fig. 1.1, but we will not describe them here. We will limit our present discussion to the more conventional vector mode description of the SDW ordering transition.

In recent papers [104, 105] Metlitski and the author have argued that the problem of symmetry breaking transitions in two-dimensional metals is strongly coupled, and proposed field theories and scaling structures for the vicinity of the critical point. Here we will be satisfied with a simple description of the effective action and its mean field theory [106]: the reader is referred to the recent papers [104, 105] for further analyses.

As in Sect. 1.3, let us begin by a description of the non-critical fermionic sector, before its coupling to the order parameter fluctuations. We use the band structure describing the cuprates in the over-doped region, well away from the Mott insulator. Here the electrons $c_{k\alpha}$ are described by the kinetic energy in Eq. 1.62, which we write in the following action

$$S_c = \int d\tau \sum_k c_{k\alpha}^\dagger \left(\frac{\partial}{\partial \tau} + \varepsilon_k \right) c_{k\alpha}. \tag{1.88}$$

This band structure leads to the Fermi surface shown in the right-most panel of Fig. 1.17, and also later in Fig. 1.18.

1.4.1 SDW Ordering

As discussed in Sect. 1.3, we must now couple the fermions of Eq. 1.88 to the bosonic modes associated with the SDW ordering transition. As noted above, we will use the more conventional vector mode description of the SDW ordering transition using the order parameters in Eq. 1.69. The analog of the coupling in Eq. 1.82 now leads to the interaction term

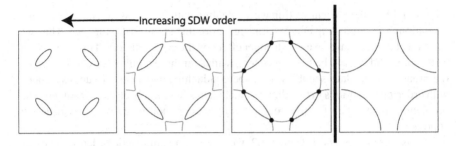

Fig. 1.17 Evolution of the Fermi surface of the hole doped cuprates in a conventional SDW theory [6] as a function of the magnitude of the SDW order $|\varphi_a|$, obtained from Eq. 1.90. The *right panel* is the large Fermi surface state with no SDW order, with states contiguous to $k = 0$ occupied by electrons. The "hot spots" are indicated by the *filled circles* in the *second panel* from the *right*. The onset of SDW order induces the formation of electron and hole pockets (the hole pockets are the ones intersecting the diagonals of the Brillouin zone). With further increase of $|\varphi_a|$, the electron pockets disappear and only hole pockets remain (the converse happens in the last step for the electron-doped cuprates)

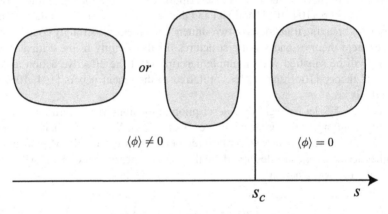

Fig. 1.18 Phase diagram of Ising nematic ordering in a metal as a function of the coupling s in $\mathcal{S}_\phi$. The Fermi surface for $s>0$ is the same as that in the *right-most panel* of Fig. 1.17, but with the $k = 0$ point shifted from the center to the edge of the Brillouin zone. The interior regions are the occupied hole (or empty electron) states. The choice between the two quadrapolar distortions of the Fermi surface is determined by the sign of $\langle \phi \rangle$

$$\mathcal{S}_{c\Phi} = \int d\tau \sum_{k,q} \Phi_{xa}(q) c_{k+K_x+q,\alpha}^\dagger \sigma_{\alpha\beta}^a c_{k\beta} + \text{c.c.} + \quad x \to y, \qquad (1.89)$$

where q is a small momentum associated with a long-wavelength SDW fluctuation, while the sum over the momentum k extends over the entire Brillouin zone. The complete theory for the SDW transition is now contained, in principle, in $\mathcal{S}_c + \mathcal{S}_{c\Phi} + \mathcal{S}_\Phi$, with $\mathcal{S}_\Phi$ contained in Eq. 1.71.

Let us consider the mean-field predictions of this theory in the SDW ordered state. For simplicity, we consider ordering at $K = (\pi, \pi)$, in which case Φ_{xa} and Φ_{ya} both reduce to the Néel order φ_a in Eq. 1.12. In the state with SDW order, we can take $\varphi_a = (0, 0, \varphi)$ a constant. Then $\mathcal{S}_c + \mathcal{S}_{c\Phi}$ is a bilinear in the fermions and can be diagonalized to yield a fermion band structure (the analog of Eq. 1.66)

$$E_k = \frac{\varepsilon_k + \varepsilon_{k+K}}{2} \pm \left(\left(\frac{\varepsilon_k + \varepsilon_{k+K}}{2} \right)^2 + \varphi^2 \right)^{1/2}. \tag{1.90}$$

Filling the lowest energy bands of this dispersion leads to the Fermi surface structure [6] shown in Fig. 1.17. The second panel from the right shows the Fermi surface obtained by translating the original Fermi surface by K, and the remaining panels show the consequences of mixing between the states at momentum k and $k + K$. Note that the Fermi surface has split apart into "small" electron and hole pockets, as discussed in Sect. 1.1.

Let us now attempt to move beyond this simple mean field theory. As written, the action $\mathcal{S}_c + \mathcal{S}_{c\Phi} + \mathcal{S}_\Phi$ is not conducive to a field-theoretic analysis: this is mainly because the sum over k in Eq. 1.89 extends over the entire Brillouin zone, and there are low-energy fermionic excitations along an entire line of k close to the Fermi surface. However, one simplifying feature here is that most of these low-energy fermions do not couple efficiently to the SDW order parameter, and their situation is similar to the fate of the Dirac fermions illustrated in Fig. 1.13—upon scattering with the wavevector K, they end up at generic points in the Brillouin zone at which there are only high-energy fermionic states. There are now eight special "hot spots" on the Fermi surface which do connect via the wavevector K to other spots directly on the Fermi surface: these are illustrated in Fig. 1.17. These "hot spots" are thus similar to the Dirac hot spots we met in Sect. 1.3.3 upon considering Ising transitions with a zero-momentum order parameter in a d-wave superconductor. However, the present situation is more complex because we also have "cold lines" of zero energy fermionic excitations coming into the hot spots.

A successful theory of the fermionic hot spots was reviewed in Sect. 1.3.3. A natural idea is to apply the same approach to the present situation with fermionic hot spots and cold lines. This leads to a problem of considerably complexity, which remains strongly coupled even within the context of the $1/N$ expansion: see Refs. [105, 107] for further details.

1.4.2 Nematic Ordering

For completeness, we also consider the case of the Ising-nematic ordering in the presence of the large Fermi surface metal. Then we will have an Ising order parameter represented by the real scalar field ϕ, which is described as before by Eq. 1.81. Its coupling to the electrons can be deduced by symmetry considerations, and the most natural coupling (the analog of Eqs. 1.87 and 1.89) is

$$\mathcal{S}_{c\phi} = \int d\tau \sum_{k,q} (\cos k_x - \cos k_y)\phi(q)c^{\dagger}_{k+q/2,\alpha}c_{k-q/2,\alpha}. \qquad (1.91)$$

The momentum dependent form factor is the simplest choice with changes sign under $x \leftrightarrow y$, as is required by the symmetry properties of ϕ. Again, the sum over q is over small momenta, while that over k extends over the entire Brillouin zone. The theory for the nematic ordering transition is now described by $\mathcal{S}_c + \mathcal{S}_\phi + \mathcal{S}_{c\phi}$. The evolution of the Fermi surface as a function of the Ising coupling in $\mathcal{S}_\phi$ is shown in Fig. 1.18.

Note that Eq. 1.91 does not have any large momentum transfer associated with K. Consequently, at any generic point on the Fermi surface, there can be scattering to other nearby low-energy fermionic excitations by long wavelength modes of ϕ. In other words, the entire Fermi surface is "hot". Thus we are faced with a third case of a "hot line" of fermions coupled to the critical order parameter mode of the transition. This case has been analyzed in Ref. [104], where it is proposed that the critical point is actually described by an infinite number of 2+1 dimensional field theories, labeled by points on the Fermi surface. The reader is referred to Ref. [104] for further results on this complex problem— a review of the main results appears in Ref. [108].

Acknowledgments I thank Eun Gook Moon for valuable comments on the manuscript and for a collaboration [3] which led to Fig. 1.1, R. Fernandes, J. Flouquet, G. Knebel, and J. Schmalian, for providing the plots shown in Fig. 1.2, C. Ruegg for the plot shown in Fig. 1.7, and the participants of the schools for their interest, and for stimulating discussions. This research was supported by the National Science Foundation under grant DMR-0757145, by the FQXi foundation, and by a MURI grant from AFOSR.

References

1. Bednorz, J.G., Müller, K.A.: Possible high T_c superconductivity in the Ba-La-Cu-O system. Z. Phys. B **64**, 188 (1986)
2. Doiron-Leyraud, N., Proust, C., LeBoeuf, D., Levallois, J., Bonnemaison, J.-B., Liang, R., Bonn, D.A., Hardy, W.N., Taillefer, L.: Quantum oscillations and the Fermi surface in an underdoped high-T_c superconductor. Nature **447**, 565 (2007)
3. Moon, E.G., Sachdev, S.: Competition between spin density wave order and superconductivity in the underdoped cuprates. Phys. Rev. B **80**, 035117 (2009)
4. Sachdev, S.: Where is the quantum critical point in the cuprate superconductors? Phys. status solidi B **247**, 537 (2010)
5. Sachdev, S.: *Quantum criticality and the phase diagram of the cuprates*, 9th International Conference on Materials and Mechanisms of Superconductivity, Tokyo, Sep 7–12, 2009, Physica C **470**, S4 (2010)
6. Sachdev, S., Chubukov, A.V., Sokol, A.: Crossover and scaling in a nearly antiferromagnetic Fermi liquid in two dimensions. Phys. Rev. B **51**, 14874 (1995)
7. LeBoeuf, D., Doiron-Leyraud, N., Levallois, J., Daou, R., Bonnemaison, J.-B., Hussey, N.E., Balicas, L., Ramshaw, B.J., Liang, R., Bonn, D.A., Hardy, W.N., Adachi, S., Proust, C., Taillefer, L.: Electron pockets in the Fermi surface of hole-doped high-T_c superconductors. Nature **450**, 533 (2007)

8. Sebastian, S.E., Harrison, N., Goddard, P.A., Altarawneh, M.M., Mielke, C.H., Liang, R., Bonn, D.A., Hardy, W.N., Andersen, O.K., Lonzarich, G.G.: Compensated electron and hole pockets in an underdoped high-T_c superconductor. Phys. Rev. B **81**, 214524 (2010)
9. Daou, R., Doiron-Leyraud, N., LeBoeuf, D., Li, S.Y., Laliberté, F., Cyr-Choinière, O., Jo, Y.J., Balicas, L., Yan, J.-Q., Zhou, J.-S., Goodenough, J.B., Taillefer, L.: Linear temperature dependence of resistivity and change in the Fermi surface at the pseudogap critical point of a high-T_c superconductor. Nat. Phys. **5**, 31 (2009)
10. Daou, R., Chang, J., LeBoeuf, D., Cyr-Choiniere, O., Laliberte, F., Doiron-Leyraud, N., Ramshaw, B.J., Liang, R., Bonn, D.A., Hardy, W.N., Taillefer, L.: Broken rotational symmetry in the pseudogap phase of a high-T_c superconductor. Nature **463**, 519 (2010)
11. Helm, T., Kartsovnik, M.V., Bartkowiak, M., Bittner, N., Lambacher, M., Erb, A., Wosnitza, J., Gross, R.: Evolution of the Fermi surface of the electron-doped high-temperature superconductor $Nd_{2-x}Ce_xCuO_4$ revealed by Shubnikov-de Haas oscillations. Phys. Rev. Lett. **103**, 157002 (2009)
12. Helm, T., Kartsovnik, M.V., Sheikin, I., Bartkowiak, M., Wolff-Fabris, F., Bittner, N., Biberacher, W., Lambacher, M., Erb, A., Wosnitza, J., Gross, R.: Magnetic breakdown in the electron-doped cuprate superconductor $Nd_{2-x}Ce_xCuO_4$: the reconstructed Fermi surface survives in the strongly overdoped regime. Phys. Rev. Lett. **105**, 247002 (2010)
13. Sachdev, S., Metlitski, M.A., Qi, Y., Xu, C.: Fluctuating spin density waves in metals. Phys. Rev. B **80**, 155129 (2009)
14. Scalapino, D.J.: The case for $d_{x^2-y^2}$ pairing in the cuprate superconductors. Phys. Rep. **250**, 329 (1995)
15. Abanov, Ar., Chubukov, A.V., Schmalian, J.: Quantum-critical theory of the spin-fermion model and its application to cuprates: normal state analysis. Adv. Phys. **52**, 119 (2003)
16. Galitski, V., Sachdev, S.: Paired electron pockets in the hole-doped cuprates. Phys. Rev. B **79**, 134512 (2009)
17. Kato, M., Machida, K.: Superconductivity and spin-density waves: application to heavy-fermion materials. Phys. Rev. B **37**, 1510 (1988)
18. Demler, E., Sachdev, S., Zhang, Y.: Spin-ordering quantum transitions of superconductors in a magnetic field. Phys. Rev. Lett. **87**, 0067202 (2001)
19. Zhang, Y., Demler, E., Sachdev, S.: Competing orders in a magnetic field: Spin and charge order in the cuprate superconductors. Phys. Rev. B **66**, 094501 (2002)
20. Lake, B., Aeppli, G., Clausen, K.N., McMorrow, D.F., Lefmann, K., Hussey, N.E., Mangkorntong, N., Nohara, M., Takagi, H., Mason, T.E., Schröder, A.: Spins in the vortices of a high-temperature superconductor. Science **291**, 1759 (2001)
21. Lake, B., Rønnow, H.M., Christensen, N.B., Aeppli, G., Lefmann, K., McMorrow, D.F., Vorderwisch, P., Smeibidl, P., Mangkorntong, N., Sasagawa, T., Nohara, M., Takagi, H., Mason, T.E.: Antiferromagnetic order induced by an applied magnetic field in a high-temperature superconductor. Nature **415**, 299 (2002)
22. Khaykovich, B., Wakimoto, S., Birgeneau, R.J., Kastner, M.A., Lee, Y.S., Smeibidl, P., Vorderwisch, P., Yamada, K.: Field-induced transition between magnetically disordered and ordered phases in underdoped $La_{2-x}Sr_xCuO_4$. Phys. Rev. B **71**, 220508 (2005)
23. Chang, J., Niedermayer, Ch., Gilardi, R., Christensen, N.B., Rønnow, H.M., McMorrow, D.F., Ay, M., Stahn, J., Sobolev, O., Hiess, A., Pailhes, S., Baines, C., Momono, N., Oda, M., Ido, M., Mesot, J.: Tuning competing orders in $La_{2-x}Sr_xCuO_4$ cuprate superconductors by the application of an external magnetic field. Phys. Rev. B **78**, 104525 (2008)
24. Chang, J., Christensen, N.B., Niedermayer, Ch., Lefmann, K., Rønnow, H.M., McMorrow, D.F., Schneidewind, A., Link, P., Hiess, A., Boehm, M., Mottl, R., Pailhes, S., Momono, N., Oda, M., Ido, M., Mesot, J.: Magnetic-field-induced soft-mode quantum phase transition in the high-temperature superconductor $La_{1.855}Sr_{0.145}CuO_4$: an inelastic neutron-scattering study. Phys. Rev. Lett. **102**, 177006 (2009)
25. Haug, D., Hinkov, V., Suchaneck, A., Inosov, D.S., Christensen, N.B., Niedermayer, Ch., Bourges, P., Sidis, Y., Park, J.T., Ivanov, A., Lin, C.T., Mesot, J., Keimer, B.:

Magnetic-field-enhanced incommensurate magnetic order in the underdoped high-temperature superconductor $YBa_2Cu_3O_{6.45}$. Phys. Rev. Lett. **103**, 017001 (2009)

26. Motoyama, E.M., Yu, G., Vishik, I.M., Vajk, O.P., Mang, P.K., Greven, M.: Spin correlations in the electron-doped high-transition-temperature superconductor $Nd_{2-x}Ce_xCuO_{4\pm\delta}$. Nature **445**, 186 (2007)

27. Kaul, R.K., Metlitski, M.A., Sachdev, S., Xu, C.: Destruction of Néel order in the cuprates by electron doping. Phys. Rev. B **78**, 045110 (2008)

28. Qi, Y., Sachdev, S.: Effective theory of Fermi pockets in fluctuating antiferromagnets. Phys. Rev. B **81**, 115129 (2010)

29. Moon, E.G., Sachdev, S.: Underdoped cuprates as fractionalized Fermi liquids: Transition to superconductivity. Phys. Rev. B **83**, 224508 (2011)

30. Kohsaka, Y., Taylor, C., Fujita, K., Schmidt, A., Lupien, C., Hanaguri, T., Azuma, M., Takano, M., Eisaki, H., Takagi, H., Uchida, S., Davis, J.C.: An intrinsic bond-centered electronic glass with unidirectional domains in underdoped cuprates. Science **315**, 1380 (2007)

31. Ando, Y., Segawa, K., Komiya, S., Lavrov, A.N.: Electrical resistivity anisotropy from self-organized one dimensionality in high-temperature superconductors. Phys. Rev. Lett. **88**, 137005 (2002)

32. Hinkov, V., Haug, D., Fauqué, B., Bourges, P., Sidis, Y., Ivanov, A., Bernhard, C., Lin, C.T., Keimer, B.: Electronic liquid crystal state in the high-temperature superconductor $YBa_2Cu_3O_{6.45}$. Science **319**, 597 (2008)

33. Knebel, G., Aoki, D., Flouquet, J.: Magnetism and superconductivity in $CeRhIn_5$ arXiv:0911.5223

34. Ni, N., Tillman, M.E., Yan, J.-Q., Kracher, A., Hannahs, S.T., Bud'ko, S.L., Canfield, P.C.: Effects of Co substitution on thermodynamic and transport properties and anisotropic H_{c2} in $Ba(Fe_{1-x}C_x)_2As_2$ single crystals. Phys. Rev. B **78**, 214515 (2008)

35. Nandi, S., Kim, M.G., Kreyssig, A., Fernandes, R.M., Pratt, D.K., Thaler, A., Ni, N., Bud'ko, S.L., Canfield, P.C., Schmalian, J., McQueeney, R.J., Goldman, A.I.: Anomalous suppression of the orthorhombic lattice distortion in superconducting $Ba(Fe_{1-x}Co_x)_2As_2$ single crystals. Phys. Rev. Lett. **104**, 057006 (2010)

36. Fernandes, R.M., Pratt, D.K., Tian, W., Zarestky, J., Kreyssig, A., Nandi, S., Kim, M.G., Thaler, A., Ni, N., Bud'ko, S.L., Canfield, P.C., McQueeney, R.J., Schmalian, J., Goldman, A.I.: Unconventional pairing in the iron arsenide superconductors. Phys. Rev. B **81**, 140501(R) (2010)

37. Scalapino, D.J.: A common thread. Phys. C **470**, S1 (2010)

38. Sachdev, S.: Quantum antiferromagnets in two dimensions. In: Lu, Y., Lundqvist, S., Morandi, G. (eds) Low Dimensional Quantum Field Theories for Condensed Matter Physicists, World Scientific, Singapore (1995) cond-mat/9303014

39. Sachdev, S.: Quantum phases and phase transitions of Mott insulators in quantum magnetism. In: Schollwöck, U., Richter, J., Farnell, D.J.J., Bishop, R.F. (eds) Lecture Notes in Physics, Springer, Berlin (2004) cond-mat/0401041

40. Sachdev, S.: Quantum phase transitions of correlated electrons in two dimensions, Lectures at the international summer school on fundamental problems in statistical physics X, Phys. A, vol. 313, pp. 252. Altenberg, Germany (2002), Aug–Sept 2001, cond-mat/0109419

41. Sachdev, S.: Exotic phases and quantum phase transitions: model systems and experiments, 24th Solvay Conference on Physics, Quantum Theory of Condensed Matter, Brussels, 11–13 Oct 2008, arXiv:0901.4103

42. Gelfand, M.P., Singh, R.R.P., Huse, D.A.: Zero-temperature ordering in two-dimensional frustrated quantum Heisenberg antiferromagnets. Phys. Rev. B **40**, 10801 (1989)

43. Oosawa, A., Fujisawa, M., Osakabe, T., Kakurai, K., Tanaka, H.: Neutron diffraction study of the pressure-induced magnetic ordering in the spin gap system $TlCuCl_3$. J. Phys. Soc. Jpn. **72**, 1026 (2003)

44. Rüegg, Ch., Cavadini, N., Furrer, A., Güdel, H.-U., Krämer, K., Mutka, H., Wildes, A., Habicht, K., Vorderwisch, P.: Bose-Einstein condensation of the triplet states in the magnetic insulator $TlCuCl_3$. Nature **423**, 62 (2003)

45. Rüegg, Ch., Normand, B., Matsumoto, M., Furrer, A., McMorrow, D.F., Krämer, K.W., Güdel, H.-U., Gvasaliya, S.N., Mutka, H., Boehm, M.: Quantum magnets under pressure: controlling elementary excitations in $TlCuCl_3$. Phys. Rev. Lett. **100**, 205701 (2008)

46. Callaway, J.: Quantum Theory of the Solid State. Academic Press, New York (1974)

47. Matsumoto, M., Yasuda, C., Todo, S., Takayama, H.: Ground-state phase diagram of quantum Heisenberg antiferromagnets on the anisotropic dimerized square lattice. Phys. Rev. B **65**, 014407 (2002)

48. Sachdev, S., Bhatt, R.N.: Bond-operator representation of quantum spins: mean-field theory of frustrated quantum Heisenberg antiferromagnets. Phys. Rev. B **41**, 9323 (1990)

49. Chubukov, A.V., Jolicoeur, Th.: Dimer stability region in a frustrated quantum Heisenberg antiferromagnet. Phys. Rev. B **44**, 12050 (1991)

50. Sommer, T., Vojta, M., Becker, K.W.: Magnetic properties and spin waves of bilayer magnets in a uniform field. Eur. Phys. J. B **23**, 329 (2001)

51. Normand, B., Rice, T.M.: Dynamical properties of an antiferromagnet near the quantum critical point: application to $LaCuO_{2.5}$. Phys. Rev. B **56**, 8760 (1997)

52. Sachdev, S.: Theory of finite-temperature crossovers near quantum critical points close to,or above, their upper-critical dimension. Phys. Rev. B **55**, 142 (1997)

53. Carpentier, D., Balents, L.: Field theory for generalized shastry-sutherland models. Phys. Rev. B **65**, 024427 (2002)

54. Calabrese, P., Parruccini, P., Pelissetto, A., Vicari, E.: Critical behavior of $O(2) \otimes O(N)$ symmetric models. Phys. Rev. B **70**, 174439 (2004)

55. Read, N., Sachdev, S.: Large-N expansion for frustrated quantum antiferromagnets. Phys. Rev. Lett. **66**, 1773 (1991)

56. Wen, X.G.: Mean-field theory of spin-liquid states with finite energy gap and topological orders. Phys. Rev. B **44**, 2664 (1991)

57. Sachdev, S.: Kagomé- and triangular-lattice Heisenberg antiferromagnets: ordering from quantum fluctuations and quantum-disordered ground states with unconfined bosonic spinons. Phys. Rev. B **45**, 12377 (1992)

58. Wang, F., Vishwanath, A.: Spin-liquid states on the triangular and Kagomé lattices: a projective-symmetry-group analysis of Schwinger boson states. Phys. Rev. B **74**, 174423 (2006)

59. Arovas, D.P., Auerbach, A.: Functional integral theories of low-dimensional quantum Heisenberg models. Phys. Rev. B **38**, 316 (1988)

60. Arovas, D.P., Auerbach, A.: Spin dynamics in the square-lattice antiferromagnet. Phys. Rev. Lett. **61**, 617 (1988)

61. Read, N., Sachdev, S.: Some features of the phase diagram of the square lattice $SU(N)$ antiferromagnet. Nucl. Phys. B **316**, 609 (1989)

62. Read, N., Sachdev, S.: Valence-bond and spin-peierls ground states of low-dimensional quantum antiferromagnets. Phys. Rev. Lett. **62**, 1694 (1989)

63. Read, N., Sachdev, S.: Spin-peierls, valence-bond solid, and Néel ground states of low-dimensional quantum antiferromagnets. Phys. Rev. B **42**, 4568 (1990)

64. Sachdev, S., Read, N.: Large N expansion for frustrated and doped quantum antiferromagnets. Int. J. Mod. Phys. B **5**, 219 (1991) cond-mat/0402109

65. Affleck, I.: The quantum Hall effects, σ-models at $\Theta = \pi$ and quantum spin chains. Nucl. Phys. B **257**, 397 (1985)

66. Affleck, I.: Exact critical exponents for quantum spin chains, non-linear σ-models at $\theta = \pi$ and the quantum Hall effect. Nucl. Phys. B **265**, 409 (1985)

67. Einarsson, T., Johannesson, H.: Effective-action approach to the frustrated Heisenberg antiferromagnet in two dimensions. Phys. Rev. B **43**, 5867 (1991)

68. Einarsson, T., Frojdh, P., Johannesson, H.: Weakly frustrated spin-1/2 Heisenberg antiferromagnet in two dimensions: thermodynamic parameters and the stability of the Néel state. Phys. Rev. **45**, 13121 (1992)
69. Chandra, P., Coleman, P., Larkin, A.I.: Ising transition in frustrated Heisenberg models. Phys. Rev. Lett. **64**, 88 (1990)
70. Chandra, P., Coleman, P.: Twisted magnets and twisted superfluids. Int. J. Mod. Phys. B **3**, 1729 (1989)
71. Halperin, B.I., Saslow, W.M.: Hydrodynamic theory of spin waves in spin glasses and other systems with noncollinear spin orientations. Phys. Rev. B **16**, 2154 (1977)
72. Dombre, T., Read, N.: Nonlinear σ models for triangular quantum antiferromagnets. Phys. Rev. B **39**, 6797 (1989)
73. Azaria, P., Delamotte, B., Jolicoeur, T.: Nonuniversality in helical and canted-spin systems. Phys. Rev. Lett. **64**, 3175 (1990)
74. Affleck, I., Marston, J.B.: Large-n limit of the Heisenberg–Hubbard model: implications for high-T_c superconductors. Phys. Rev. B **37**, 3774 (1988)
75. Rokhsar, D., Kivelson, S.: Superconductivity and the quantum hard-core dimer gas. Phys. Rev. Lett. **61**, 2376 (1988)
76. Henley, C.: Ordering due to disorder in a frustrated vector antiferromagnet. Phys. Rev. Lett. **62**, 2056 (1989)
77. Chubukov, A.: First-order transition in frustrated quantum antiferromagnets. Phys. Rev. B **44**, 392 (1991)
78. Mila, F., Poilblanc, D., Bruder, C.: Spin dynamics in a frustrated magnet with short-range order. Phys. Rev. B **43**, 7891 (1991)
79. Stephenson, J.: Range of order in antiferromagnets with next-nearest neighbor coupling. Can. J. Phys. **48**, 2118–1724 (1970)
80. Polyakov, A.M.: Gauge Fields and Strings. Harwood, New York (1987)
81. Polyakov, A.M.: Quark confinement and topology of gauge theories. Nucl. Phys. B **120**, 429 (1977)
82. Senthil, T., Vishwanath, A., Balents, L., Sachdev, S., Fisher, M.P.A.: Deconfined quantum critical points. Science **303**, 1490 (2004)
83. Senthil, T., Balents, L., Sachdev, S., Vishwanath, A., Fisher, M.P.A.: Quantum criticality beyond the Landau–Ginzburg–Wilson paradigm. Phys. Rev. B **70**, 144407 (2004)
84. Fradkin, E., Shenker, S.H.: Phase diagrams of lattice gauge theories with Higgs fields. Phys. Rev. D **19**, 3682 (1979)
85. Senthil, T., Fisher, M.P.A.: Z_2 gauge theory of electron fractionalization in strongly correlated systems. Phys. Rev. B **62**, 7850 (2000)
86. Kitaev, A.Y.: Fault-tolerant quantum computation by anyons. Ann. Phys. **303**, 2 (2003)
87. Xu, C., Sachdev, S.: Global phase diagrams of frustrated quantum antiferromagnets in two dimensions: doubled Chern–Simons theory. Phys. Rev. B **79**, 064405 (2009)
88. Qi, Y., Xu, C., Sachdev, S.: Dynamics and transport of the Z_2 spin liquid: application to $\kappa - (ET)_2Cu_2(CN)_3$. Phys. Rev. Lett. **102**, 176401 (2009)
89. Herbut, I.F., Juričič, V., Roy, B.: Theory of interacting electrons on the honeycomb lattice. Phys. Rev. B **79**, 085116 (2009)
90. Juričič, V., Herbut, I.F., Semenoff, G.W.: Coulomb interaction at the metal-insulator critical point in graphene. Phys. Rev. B **80**, 081405(R) (2009)
91. De Prato, M., Pelissetto, A., Vicari, E.: Spin-density-wave order in cuprates. Phys. Rev. B **74**, 144507 (2006)
92. Pelissetto, A., Sachdev, S., Vicari, E.: Nodal quasiparticles and the onset of spin-density-wave order in cuprate superconductors. Phys. Rev. Lett. **101**, 027005 (2008)
93. Kim, E.-A., Lawler, M.J., Oreto, P., Sachdev, S., Fradkin, E., Kivelson, S.A.: Theory of the nodal nematic quantum phase transition in superconductors. Phys. Rev. B **77**, 184514 (2008)
94. Vojta, M., Zhang, Y., Sachdev, S.: Quantum phase transitions in d-wave superconductors. Phys. Rev. Lett. **85**, 4940 (2000)

95. Vojta, M., Zhang, Y., Sachdev, S.: Quantum phase transitions in d-wave superconductors. Phys. Rev. Lett. **100**, 089904(E) (2008)
96. Vojta, M., Zhang, Y., Sachdev, S.: Renormalization group analysis of quantum critical points in d-wave superconductors. Int. J. Mod. Phys. B **14**, 3719 (2000)
97. Laughlin, R.B.: Magnetic induction of $d_{x^2-y^2} + i d_{xy}$ order in high-T_c superconductors. Phys. Rev. Lett. **80**, 5188 (1998)
98. Li, M.-R., Hirschfeld, P.J., Woelfle, P.: Vortex state of a d-wave superconductor at low temperatures. Phys. Rev. B **63**, 054504 (2001)
99. Khveshchenko, D.V., Paaske, J.: Incipient nodal pairing in planar d-wave superconductors. Phys. Rev. Lett. **86**, 4672 (2001)
100. Rosenstein, B., Warr, B.J., Park, S.H.: Dynamical symmetry breaking in four-fermion interaction models. Phys. Rep. **205**, 59 (1991)
101. Kärkkäinen, L., Lacaze, R., Lacock, P., Petersson, B.: Critical behaviour of the three-dimensional Gross–Neveu and Higgs–Yukawa models. Nucl. Phys. B **415**, 781 (1994)
102. Huh, Y., Sachdev, S.: Renormalization group theory of nematic ordering in d-wave superconductors. Phys. Rev. B **78**, 064512 (2008)
103. Fritz, L., Sachdev, S.: Signatures of the nematic ordering transitions in the thermal conductivity of d-wave superconductors. Phys. Rev. B **80**, 144503 (2009)
104. Metlitksi, M.A., Sachdev, S.: Quantum phase transitions of metals in two spatial dimensions: I. Ising-nematic order. Phys. Rev. B **82**, 075127 (2010)
105. Metlitski, M.A., Sachdev, S.: Quantum phase transitions of metals in two spatial dimensions: II. Spin density wave order. Phys. Rev. B **82**, 075128 (2010)
106. Löhneysen, H.v., Rosch, A., Vojta, M., Wölfle, P.: Fermi-liquid instabilities at magnetic quantum phase transitions. Rev. Mod. Phys. **79**, 1015 (2007)
107. Metlitski, M.A., Sachdev, S.: Instabilities near the onset of spin density wave order in metals. New J. Phys. **12**, 105007 (2010)
108. Sachdev, S.: Condensed matter and AdS/CFT. Lect. Notes Phys. **828**, 273 (2011)

Chapter 2
Electronic Liquid Crystal Phases in Strongly Correlated Systems

Eduardo Fradkin

Abstract I discuss the electronic liquid crystal (ELC) phases in correlated electronic systems, what these phases are and in what context they arise. I will go over the strongest experimental evidence for these phases in a variety of systems: the two-dimensional electron gas (2DEG) in magnetic fields, the bilayer material $Sr_3Ru_2O_7$ (also in magnetic fields), and a set of phenomena in the cuprate superconductors (and more recently in the pnictide materials) that can be most simply understood in terms of ELC phases. Finally we will go over the theory of these phases, focusing on effective field theory descriptions and some of the known mechanisms that may give rise to these phases in specific models.

2.1 Electronic Liquid Crystal Phases

Electronic liquid crystal phases [1] are states of correlated quantum electronic systems that break spontaneously either rotational invariance or translation invariance. Since most correlated electronic systems arise in a solid state environment the underlying crystal symmetry plays a role as it is the unbroken symmetry of the system. Thus in practice these phases break the point group symmetry of the underlying lattice, in addition of the possible breaking of the lattice translation symmetry.

This point of view is commonplace in the classification of phases of *classical* liquid crystals [2]. Classical liquid crystal systems are assemblies of a macroscopically large number of molecules with various shapes. The shapes of the individual molecules (the "nematogens") affect their mutual interactions, as well as enhancing entropically-driven interactions ("steric forces") which, when combined, give rise to the dazzling phase diagrams of liquid crystals and the fascinating properties of their phases [2].

E. Fradkin (✉)
Department of Physics, University of Illinois at Urbana-Champaign,
1110 West Green Street, Urbana, IL 61801-3080 USA
e-mail: efradkin@illinois.edu

D. C. Cabra et al. (eds.), *Modern Theories of Many-Particle Systems in Condensed Matter Physics*, Lecture Notes in Physics 843, DOI: 10.1007/978-3-642-10449-7_2,
© Springer-Verlag Berlin Heidelberg 2012

The physics of liquid crystals is normally regarded as part of "soft" condensed matter physics, while the physics of correlated electrons is usually classified as part of "hard" condensed matter physics. The necessity to use both points of view clearly brings to the fore the underlying unity of Physics as a science. Thus, one may think of this area as "soft quantum matter" or "quantum soft matter" depending on to which tribe you belong to.

These lectures are organized as follows. In Sect. 2.1 ELC phases and their symmetries are described. In Sect. 2.2 I cover the main experimental evidence for these phases in 2DEGs, in $Sr_3Ru_2O_7$ and in high temperature superconductors. In Sect. 2.3 I present the theories of stripe phases, in Sect. 2.4 the relation between electronic inhomogeneity and high temperature superconductivity is discussed, and Sect. 2.5 is devoted to the theory of the pair density wave (the striped superconductor). Section 2.6 is devoted to the theories of nematic phases and a theory of nematic electronic order in the strong coupling regime is discussed in Sect. 2.7 The stripe-nematic quantum phase transition is discussed in Sect. 2.8.

2.1.1 Symmetries of Electronic Liquid Crystal Phases

We will follow Ref.[1] and classify the ELC phases of strongly correlated electrons[1] following the symmetry-based scheme used in classical liquid crystals [2, 3]:

1. *Crystalline phases*: phases that break all continuous translation symmetries and rotational invariance.
2. *Smectic ("stripe") phases*: phases break one translation symmetry and rotational invariance.
3. *Nematic and hexatic phases*: uniform (liquid) phases that break rotational invariance.
4. *Isotropic*: uniform and isotropic phases.

A cartoon of the real space structure of these ordered phases is shown in Fig. 2.1.

Unlike classical liquid crystals, electronic systems carry charge and spin, and have strong quantum mechanical effects (particularly in the strong correlation regime). This leads to a host of interesting possibilities of ordered states in which the liquid crystalline character of the spatial structure of these states becomes intertwined with the "internal" degrees of freedom of electronic systems. These novel ordered phases will be the focus of these lectures. One of the aspects that we will explore is the structure of their phase diagrams. Thus in addition of considering the thermal melting of these phases, we will also be interested in the *quantum* melting of these states and the associated quantum phase transitions (see a sketch in Fig. 2.2).

---

[1]   You may call the ELC phases the anisotropic states of point particles!

Fig. 2.1 Cartoon of liquid crystal phases

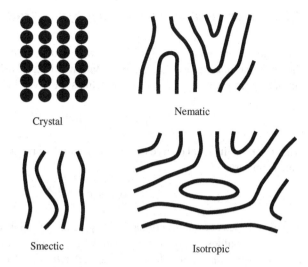

Crystal

Nematic

Smectic

Isotropic

Fig. 2.2 Schematic phase diagram of electronic liquid crystal phases. Here T is temperature and r denotes a tuning parameter the controls the strength of the quantum fluctuations. In practice it can represent doping, magnetic field, pressure or even material. The full dots are quantum critical points

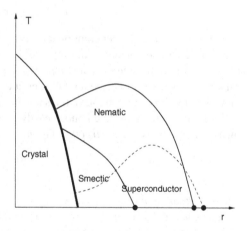

In this context, the *crystalline phases* are either insulating or "almost insulating", e.g. multiple charge density waves (CDW) ordered states either commensurate or sliding (incommensurate). However, these phases may also be superconducting either by coexistence or, more interestingly, by modulating the superconducting states themselves. Similarly, electron nematics are anisotropic metallic or superconducting states, while the isotropic phases are also either metallic or superconducting. As we will see these phases display a set of rather striking and unusual behaviors, some of which have been observed in recent experiments.

2.1.2 Order Parameters and Their Symmetries

The order parameters of ELC phases are well known [1, 4]. In the crystalline phases, the order parameters are ρ_K, the expectation values of the density operators at the set of ordering wave vectors $\{K\}$ that defines the crystal [3].

$$\rho_K = \int d\boldsymbol{r}\, \rho(\boldsymbol{r})\, e^{i K \cdot \boldsymbol{r}} \tag{2.1}$$

where $\rho(\boldsymbol{r})$ is the local charge density. Thus, under an uniform translation by $\boldsymbol{R}$, ρ_K transforms as

$$\rho_K \to \rho_K\, e^{i K \cdot \boldsymbol{R}} \tag{2.2}$$

Smectic phases are unidirectional density waves and their order parameters are also expectation values ρ_K but for *only one* wave vector K. For charged systems, $\rho(\boldsymbol{r})$ is the *charge density*, and the order parameter ρ_K is the *charge density wave* order parameter. Since $\rho(\boldsymbol{r})$ is *real*, $\rho_K = \rho^*_{-K}$, and the density can be expanded as

$$\rho(\boldsymbol{r}) = \rho_0(\boldsymbol{r}) + \rho_K(\boldsymbol{r}) e^{i K \cdot \boldsymbol{r}} + \text{c.c.} \tag{2.3}$$

where $\rho_0(\boldsymbol{r})$ are the Fourier components *close* to zero wave vector, $k = 0$, and $\rho_K(\boldsymbol{r})$ are the Fourier components with wave vectors *close* to $k = K$. Hence, a density wave (a *smectic*) is represented by a *complex* order parameter field, in this case $\rho_K(\boldsymbol{r})$. This is how we will describe a CDW and a charge stripe (which from the point of view of symmetry breaking have the same description).[2]

Smectic order is detected most easily in scattering experiments through the measurement of the *static structure factor*, usually denoted by $S(k)$,

$$S(\boldsymbol{k}) = \int \frac{d\omega}{2\pi} S(\boldsymbol{k}, \omega) \tag{2.5}$$

where $S(\boldsymbol{k}, \omega)$ is the *dynamical structure factor*, i.e. the Fourier transform of the (in this case) density-density correlation function. The signature of smectic order is the existence of a delta-function component of $S(\boldsymbol{k})$ at the ordering wave vector, $k = K$, with a prefactor that is equal to $|\langle \rho_K \rangle|^2$ [3].

In the case of a *spin density wave* (a "spin stripe") the picture is the same except that the order parameter field is multi-component, $\boldsymbol{S}_K(\boldsymbol{r})$, corresponding to different spin polarizations. Thus, the local spin density $\boldsymbol{S}(\boldsymbol{r})$ has the expansion

$$\boldsymbol{S}(\boldsymbol{r}) = \boldsymbol{S}_0(\boldsymbol{r}) + \boldsymbol{S}_K(\boldsymbol{r})\, e^{i K \cdot \boldsymbol{r}} + \text{c.c.} \tag{2.6}$$

---

[2] On the other hand, in the case of a crystal phase, the expansion is

$$\rho(\boldsymbol{r}) = \rho_0(\boldsymbol{r}) + \sum_{K \in \Gamma} \rho_K(\boldsymbol{r}) e^{i K \cdot \boldsymbol{r}} + \text{c.c.} \tag{2.4}$$

where Γ denotes the set of primitive lattice vectors of the crystal phase [3].

where $S_0(r)$ denotes the local (real) *ferromagnetic* order parameter and $S_K(r)$ is the (complex) SDW (or spin stripe) order parameter field, a complex vector in spin space.

One of the questions we will want to address is the connection between these orders and superconductivity. The superconducting order parameter, a pair conden- sate, is the *complex* field $\Delta(r)$. It is natural (and as we will see it is borne out by current experiments) to consider the case in which the superconducting order is also modulated, and admits an expansion of the form

$$\Delta(r) = \Delta_0(r) + \Delta_K(r)\, e^{iK \cdot r} + \Delta_{-K}(r)\, e^{-iK \cdot r} \tag{2.7}$$

where the uniform component Δ_0 is the familiar BCS order parameter, and $\Delta_K(r)$ is the *pair-density-wave* (PDW) order parameter [5, 6], closely related to the Fulde-Ferrell-Larkin-Ovchinnikov (FFLO) order parameter [7, 8] (but without an external magnetic field). Since $\Delta(r)$ is complex, $\Delta_K(r) \neq \Delta_{-K}(r)^*$, the PDW state has two complex order parameters.[3]

In contrast, nematic phases are translationally invariant but break rotational invari- ance. Their order parameters transform irreducibly under the rotation group for a continuous system, or under the point (or space) group of the lattice. Hence, the order parameters of a nematic phase (hexatic and their generalizations) are symmetric traceless tensors, that we will denote by Q_{ij} [2]. In 2D, as most of the problems we will be interested in are 2D systems (or quasi2D systems), the order parameter takes the form (with $i, j = x, y$)

$$Q_{ij} = \begin{pmatrix} Q_{xx} & Q_{xy} \\ Q_{xy} & -Q_{xx} \end{pmatrix} \tag{2.8}$$

which, alternatively, can be written in terms of a *director N*,

$$N = Q_{xx} + i Q_{xy} = |N|\, e^{i\varphi}. \tag{2.9}$$

Under a rotation by a fixed angle θ, N transforms as[4]

$$N \to N\, e^{i2\theta}. \tag{2.10}$$

Hence, it changes sign under a rotation by $\pi/2$ and it is *invariant* under a rotation by π (hence the name *director*, a headless vector). On the other hand, it is invariant under uniform translations by R.

In practice we will have great latitude when choosing a nematic order parameter since any symmetric traceless tensor in space coordinates will transform properly under rotations. In the case of a charged metallic system, a natural choice to describe

---

[3]   I will not discuss the case of spiral order here.

[4]   For a lattice system, rotational symmetries are those of the point (or space) group symmetry of the lattice. Thus, nematic order parameters typically become Ising-like (on a square lattice) or three-state Potts on a triangular lattice (and so forth).

a metallic nematic state is the traceless symmetric component of the *resistivity* (or conductivity) tensor [9–12]. In 2D we will use the traceless symmetric tensor

$$Q_{ij} = \begin{pmatrix} \rho_{xx} - \rho_{yy} & \rho_{xy} \\ \rho_{xy} & \rho_{yy} - \rho_{xx} \end{pmatrix} \tag{2.11}$$

where ρ_{xx} and ρ_{yy} are the *longitudinal* resistivities and $\rho_{xy} = \rho_{yx}$ is the transverse (*Hall*) resistivity. This tensor *changes sign* under a rotation by $\frac{\pi}{2}$ but is invariant under a rotation by π. A similar analysis can be done in terms of the dielectric tensor, which is useful in the context of light scattering experiments.

On the other hand, when looking at the spin polarization properties of a system other measures of nematic order are available. For instance, in a neutron scattering experiment, the anisotropy under a rotation $\mathcal{R}$ (say, by $\pi/2$) of the structure factor $S(\mathbf{k})$

$$Q \sim S(\mathbf{k}) - S(\mathcal{R}\mathbf{k}) \tag{2.12}$$

is a measure of the nematic order parameter Q [4, 13].

Other, more complex, yet quite interesting phases are possible. One should keep in mind that the nematic order parameter (as defined above) corresponds to a field that transforms under the lowest (angular momentum $\ell = 2$) irreducible representation of the rotations group, compatible with inversion symmetry. The nematic phase thus defined has d-wave symmetry, the symmetry of a quadrupole. Higher symmetries are also possible, e.g. hexatic ($\ell = 6$). However it is also possible to have states that break both rotational invariance and 2D inversion (mirror reflection), as in the $\ell = 3$ channel. Such states break (although mildly) time-reversal invariance [14, 15]. Other complex phases arise by combining the nematic order in real space with those of some internal symmetry, e.g. spin or orbital degeneracies. Thus one can consider nematic order parameters in the *spin-triplet* channels, which give rise to a host of (as yet undetected) phases with fascinating behaviors in the spin channel or under time-reversal, such as the dynamical generation of spin orbit coupling or the spontaneous breaking of time-reversal invariance [14, 16, 17].

2.1.3 Electronic Liquid Crystal Phases and Strong Correlation Physics

One of the central problems in condensed matter physics is the understanding of doped Mott insulators. Most of the interesting systems in condensed matter, notably high temperature superconductors, are doped Mott insulators [18]. A Mott insulator is a phase of an electronic system in which there is a gap in the single particle spectrum due to the effects of electronic correlations and not to features of the band structure. Thus, Mott insulators have an odd number of electrons in the unit cell.

For a system like this, band theory would predict that such systems must be metallic, not insulating, and be described by the Landau theory of the Fermi liquid. Electronic systems that become insulating due to the effects of strong correlation are states of matter with non-trivial correlations.

Most known Mott insulating states are ordered phases, associated with the spontaneous breaking of some global symmetry of the electronic system, and have a clearly defined order parameter. Typically the Mott state is an antiferromagnetic state (or generalizations thereof). However there has been a sustained interest in possible non-magnetic Mott phases, e.g. dimerized, various sorts of conjectured spin liquids, etc., some of which do not admit an order parameter description (as in the case of the topological phases).

We will not concern ourselves on these questions here. What will matter to us is that doping this insulator by holes disrupts the correlations that define the insulating state. Consequently doped holes are more costly (energetically) if they are apart than if they are together. The net effect is that the disruption of the correlations of the Mott state results in an effective strong attractive interaction between the doped holes. This effect was early on mistaken for a sign of pairing in models of high temperature superconductors (such as the Hubbard and *t-J* models). Further analysis revealed that this effective attraction meant instead the existence of a *generic* instability of strongly correlated systems to *phase separation* [19]. This feature of strong correlation has been amply documented in numerical simulations (see, for instance, Ref. [20]).

Due to the inherent tendency to phase separation of Hubbard-type models (and its descendants), the *insulating* nature of a Mott insulator cannot be ignored and, in particular, its inability to screen the longer range Coulomb interactions. Thus, quite generally, one can expect that the combined effects of the kinetic energy of the doped holes and the repulsive Coulomb interactions should in effect *frustrate* the tendency to phase separation of short-ranged models of strong correlation [21].

The existence of strong short range attractive forces and long range repulsion is a recipe for the formation of phases with complex spatial structure. As noted above, this is what happens in classical liquid crystals. It is also the general mechanism giving rise to generally inhomogeneous phases in classical complex fluids such as ferrofluids and heteropolymers [22], as well as in astrophysical problems such as the crusts of neutron stars [23].

The point of view that we take in these lectures is that the behavior observed in the underdoped regime of high temperature superconductors and in other strongly correlated systems is due to the strong tendency that these systems have to form generally inhomogeneous and anisotropic phases, "stripes". In the following lectures we will go over the experimental evidence for these phases and for their theoretical underpinning.[5]

---

[5]   Ref. [24] is a recent, complementary, review of the phenomenology of nematic phases in strongly correlated systems.

Fig. 2.3 Schematic phase diagram of the cuprate superconductors. The full lines are the phase boundaries for the antiferromagnetic and superconducting phases. The broken line is the phase diagram for a system with static stripe order and a pronounced 1/8 anomaly. The dotted line marks the crossover between the bad metal and pseudogap regimes

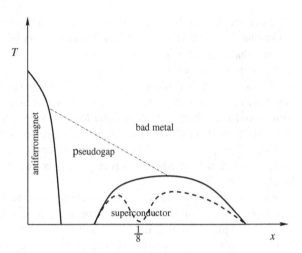

2.2 Experimental Evidence in Strongly Correlated Systems

During the past decade or so experimental evidence has been mounting of the existence of electronic liquid crystal phases in a variety of strongly correlated (as well as not as strongly correlated) electronic systems. We will be particularly interested in the experiments in the copper oxide high temperature superconductors, in the ruthenate materials (notably $Sr_3Ru_2O_7$), and in two-dimensional electron gases (2DEG) in large magnetic fields. However, as we will discuss below, these concepts are also relevant to more conventional CDW systems Fig. 2.3.

2.2.1 Nematic Phases in the 2DEG in High Magnetic Fields

To this date, the best documented electron nematic state is the anisotropic compressible state observed in 2DEGs in large magnetic fields near the middle of a Landau level, with Landau index $N \geq 2$ [25–28] (Figs. 2.4, 2.5). In ultra high mobility samples of a 2DEG in AlAs-GaAs heterostructures, transport experiments in the second Landau level (and above) near the center of the Landau level show a pronounced anisotropy of the longitudinal resistance rising sharply below $T \simeq 80$ mK, with an anisotropy that increases by orders of magnitude as the temperature is lowered. This effect is only seen in ultra-clean samples, with nominal mean free paths of about 0.5 mm (!) and nominal mobilities of $10 - 30 \times 10^6$.[6]

A nematic order parameter can be constructed phenomenologically from the measured resistivity tensor, by taking the symmetric traceless piece of it. This was done in Ref. [9] where a fit of the data of Lilly et al. [25, 26] was shown to be

---

[6] The anisotropy is strongly suppressed by disorder.

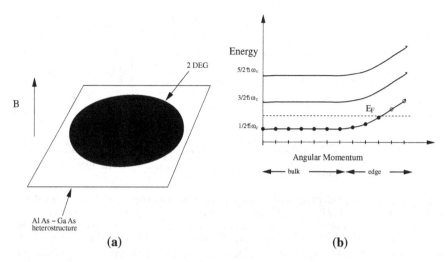

Fig. 2.4 a 2DEG in a magnetic field. **b** Landau levels

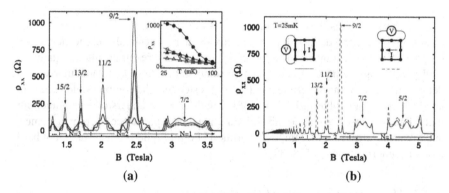

Fig. 2.5 Spontaneous magneto-transport anisotropy in the 2DEG: **a** peaks in ρ_{xx} developing at low T in high LLs (*dotted line*: T = 100 mK; *thick line*: 65 mK; *thin line*: 25 mK). *Inset*: temperature dependence of peak height at $\nu = 9/2$ (*closed circles*), 11/2 (*open circles*), 13/2 (*closed triangles*) and 15/2 (*open triangles*). **b** Anisotropy of ρ_{xx} at T = 25 mK. From Lilly et al. [25], reprinted with permission from APS

consistent with a classical 2D XY model (in a weak symmetry breaking field). A 2D XY symmetry is expected for a planar nematic order provided the weak lattice symmetry breaking is ignored. Presumably lattice anisotropy is responsible for the saturation shown at low temperatures in Fig. 2.6 (left panel).

These experiments were originally interpreted as evidence for a quantum Hall smectic (stripe) phase [29–33]. However, further experiments ([10, 34, 35]) (Fig. 2.6, right panel) did not show any evidence of pinning of this putative unidirectional CDW as the *I-V* curves were found to be strictly linear at low bias. In addition, the observation of broadband noise in the current, which is a character-

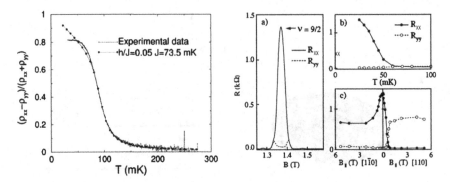

Fig. 2.6 *Left*: Nematic order in the 2DEG; fit of the resistance anisotropy to a 2D XY model Monte carlo simulation. From Fradkin et al [9], reprinted with permission from APS. *Right*: **a** Longitudinal resistance anisotropy around $\nu = 9/2$ at $T = 25$ mK. Solid trace: R_{xx}; average current flow along [110]. **b** Temperature dependence of resistances at $\nu = 9/2$. **c** R_{xx} and R_{yy} at $\nu = 9/2$ at $T = 25$ mK vs in-plane magnetic field along [110] and [1$\bar{1}$0]. From Cooper et al [10], reprinted with permission from APS

istic of CDW systems, has not been detected in the regime where this remarkable anisotropy is observed. In contrast, extremely sharp threshold electric fields and broadband noise in transport was observed in a nearby reentrant integer quantum Hall phase, suggesting a crystallized electronic state. These facts, together with a detailed analysis of the experimental data, suggested that the compressible state is in an electron nematic phase [9, 31, 36–38], which is better understood as a quantum melted stripe phase.[7] An alternative picture, a nematic phase accessed by a Pomeranchuk instability from a "composite fermion" Fermi liquid is conceivable but hard to justify microscopically [38, 39].

2.2.2 The Nematic Phase of $Sr_3Ru_2O_7$

Recent magneto-transport experiments in the quasi-two-dimensional bilayer ruthenate $Sr_3Ru_2O_7$ by the St. Andrews group [12] have given strong evidence of a strong temperature-dependent in-plane transport anisotropy in strongly correlated materials at low temperatures $T \lesssim 800$ mK and for a window of perpendicular magnetic fields around 7.5 Tesla (see Fig. 2.7). $Sr_3Ru_2O_7$ is a quasi-2D bilayer material known to have a metamagnetic transition as a function of applied perpendicular magnetic field and temperature. Contrary to the case of the 2DEG in AlAs-GaAs heterostructures and quantum wells, the magnetic fields applied to $Sr_3Ru_2O_7$ are too weak to produce Landau quantization. However, as in the case of the 2DEG of the previous section, the transport anisotropy appears at very low temperatures and only in the cleanest samples. The observed transport anisotropy has a strong temperature

---

[7] The 2DEG in a strong magnetic field is inherently a strongly correlated system as the interaction is always much bigger than the (vanishing) kinetic energy.

Fig. 2.7 Phase diagram of $Sr_3Ru_2O_7$ in the temperature-magnetic field plane. The nematic phase is the region comprised between $\sim 7.5T$ and $\sim 8.1T$. N transport anisotropy is detected outside this region where the system behaves as an isotropic (metamagnetic) metal. From Grigera et al. [40], reprinted with permission from AAAS

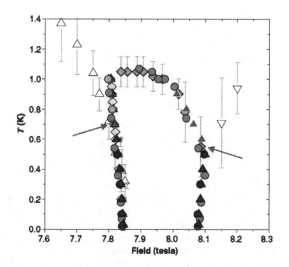

and field dependence (although not as pronounced as in the case of the 2DEG) is shown in Fig. 2.8. These experiments provide strong evidence that the system is in an electronic nematic phase in that range of magnetic fields [12, 41]. The electronic nematic phase appears to have preempted a metamagnetic QCP in the same range of magnetic fields [42–45]. This suggests that proximity to phase-separation may be a possible microscopic mechanism to trigger such quantum phase transitions, consistent with recent ideas on the role of Coulomb-frustrated phase separation in 2DEGs [46, 47].

2.2.3 Stripe Phases and Nematic Phases in the Cuprates

In addition to high temperature superconductivity, the copper oxide materials display a strong tendency to have charge-ordered states, such as stripes. The relation between charge ordered states [48], as well as other proposed ordered states [15, 49], and the mechanism(s) of high temperature superconductivity is a subject of intense current research. It is not, however, the main focus of these lectures.   Stripe phases have been extensively investigated in high temperature superconductors and detailed and recent reviews are available on this subject [4, 52]. Stripe phases in high temperature superconductors have unidirectional order in both spin and charge (although not always) which are typically incommensurate. In general the detected stripe order (by low energy inelastic neutron scattering) in $La_{2-x}Sr_xCuO_4$, $La_{2-x}Ba_xCuO_4$ and $YBa_2Cu_3O_{6+x}$ (see Refs. [4, 52] and references therein) is not static but "fluctuating". As emphasized in Ref. [4], "fluctuating order" means that there is no true long range unidirectional order. Instead, the system is in a (quantum) disordered phase, very close to a quantum phase transition to such an ordered phase, with very low energy fluctuations that reveal the character of the proximate ordered state. On the other hand,

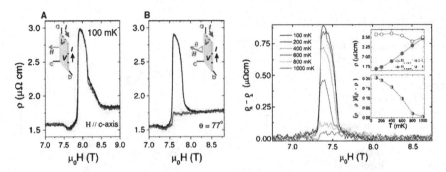

Fig. 2.8 *Left*: ρ_{aa} and ρ_{bb} of the in-plane magnetoresistivity tensor of a high-purity single crystal of $Sr_3Ru_2O_7$. (**A**) For an applied c-axis field, ρ_{aa} (*upper curve*) and ρ_{bb} (*lower curve*). (**B**) Tilted field (13° from c). *Right*: The temperature dependence difference of ρ_{aa} and ρ_{bb} for fields applied at $\theta = 72°$ such that the in-plane field component lies along a (*upper inset*). Temperature dependence of ρ_{aa} (*open symbols*) and ρ_{bb} (*filled symbols*) for $\mu_0 H = 7.4$ T applied in the direction specified above. (*Lower inset*) The temperature dependence of the difference between the two magnetoresistivities shown in the upper inset, normalized by their sum. From Borzi et al. [12], reprinted with permission from AAAS

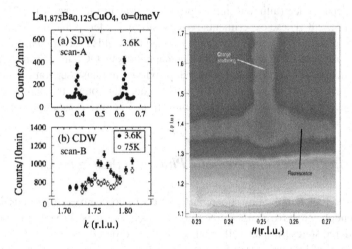

Fig. 2.9 *Left*: Static spin and charge stripe order in $La_{2-x}Ba_xCuO_4$ in neutron scattering. From Fujita et al. [50], reprinted with permission from APS. *Right*: resonant X-ray scattering. From Abbamonte et al. [51], reprinted with permission from Nature Physics

in $La_{2-x}Ba_xCuO_4$ near $x = 1/8$ (and in $La_{1.6-x}Nd_{0.4}Sr_xCuo_4$ also near $x = 1/8$ where they were discovered first [53]), the order detected by elastic neutron scattering [54], and resonant X-ray scattering in $La_{2-x}Ba_xCuO_4$ [51] also near $x = 1/8$, becomes true long range static order (see Fig. 2.9).

In the case of $La_{2-x}Sr_xCuO_4$, away from $x = 1/8$, and particularly on the more underdoped side, the in-plane resistivity has a considerable temperature-dependent anisotropy, which has been interpreted as an indication of electronic nematic

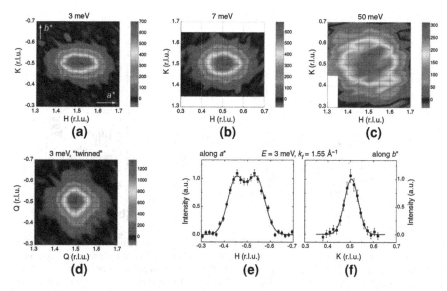

Fig. 2.10 Nematic order in underdoped $YBa_2Cu_3O_{6+x}$ ($y = 6.45$). (**a** to **c**) Intensity maps of the spin-excitation spectrum at 3, 7, and 50 meV, respectively. The a^* and b^* directions are indicated in (**a**). (**d**) Colormap of the intensity at 3 meV, as it would be observed in a crystal consisting of two perpendicular twin domains with equal population. (**e** and **f**) Scans along a^* and b^* through Q_{AF}. From Hinkov et al. [13], reprinted with permission from AAAS

order [11]. The same series of experiments also showed that very underdoped $YBa_2Cu_3O_{6+x}$ is an electron nematic as well.

The most striking evidence for electronic nematic order in high temperature superconductors are the recent neutron scattering experiments in $YBa_2Cu_3O_{6+x}$ at $y = 6.45$ [56] (see Figs. 2.10, 2.11). In particular, the temperature-dependent anisotropy of the inelastic neutron scattering in $YBa_2Cu_3O_{6+x}$ shows that there is a critical temperature for nematic order (with $T_c \sim 150$ K) where the inelastic neutron peaks also become incommensurate. Similar effects were reported by the same group [57] at higher doping levels ($y \sim 6.6$) who observed that the nematic signal was decreasing in strength suggesting the existence of a nematic-isotropic quantum phase transition closer to optimal doping. Fluctuating stripe order in underdoped $YBa_2Cu_3O_{6+x}$ has been detected earlier on in inelastic neutron scattering experiments [58, 59] which, in hindsight, can be reinterpreted as evidence for nematic order. However, as doping increases past a $y \sim 6.6a$ spin gap appears and magnetic scattering is strongly suppressed at low energies (in the absence of magnetic fields) making inelastic neutron scattering experiments less effective in this regime.

In a series of particularly interesting experiments, the Nernst coefficient was measured in $YBa_2Cu_3O_{6+x}$ ranging from the very underdoped regime, where inelastic neutron scattering detects nematic order, to a slightly overdoped regime [60]. The Nernst coefficient is defined as follows. Let j_e and j_Q be the charge and heat currents established in a 2D sample by an electric field E and a temperature

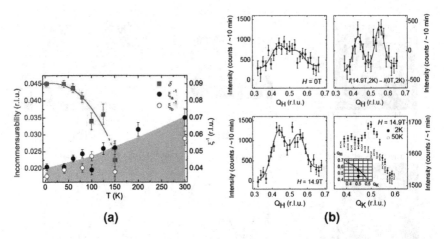

Fig. 2.11 **a** Incommensurability δ (*squares*), half-width-at-half-maximum of the incommensurate peaks along a^* (ξ_a^{-1}, *dark circles*) and along b^* (ξ_b^{-1}, *white circles*) in reciprocal lattice units. From Hinkov et al. [13], reprinted with permission from AAAS. **b** Static stripe order induced by an external magnetic field in YBa$_2$Cu$_3$O$_{6+x}$ at $y = 6.45$. From Haug et al. [55], reprinted with permission from APS

gradient ∇T :

$$\begin{pmatrix} j_e \\ j_Q \end{pmatrix} = \begin{pmatrix} \sigma & \alpha \\ T\alpha & \kappa \end{pmatrix} \begin{pmatrix} E \\ -\nabla T \end{pmatrix} \tag{2.13}$$

where σ, α and κ are 2×2 tensors for the conductivity, the thermoelectric conductivity and the thermal conductivity respectively. The Nernst coefficient, also a 2×2 tensor θ is measured (see Ref. [61]) say by applying a temperature gradient in the x direction and measuring the voltage along the y direction:

$$E = -\theta \nabla T \tag{2.14}$$

Since no current flows through the system, the Nernst tensor is

$$\theta = -\sigma^{-1}\alpha \tag{2.15}$$

These experiments revealed that the Nernst (tensor) coefficient has an anisotropic component whose onset coincides (within the error bars) with the conventionally defined value of the pseudogap temperature T^*, and essentially tracks its evolution as a function of doping. Thus, it appears that, at least in YBa$_2$Cu$_3$O$_{6+x}$, the pseudogap is a regime with nematic order (see Fig. 2.12). The same group had shown earlier than the Nernst coefficient is a sensitive indicator of the onset of stripe charge order in La$_{1.6-x}$Nd$_{0.4}$Sr$_x$CuO$_4$ [62].

Inelastic neutron scattering experiments have found nematic order also in La$_{2-x}$Sr$_x$CuO$_4$ materials where fluctuating stripes where in fact first discovered

Fig. 2.12 The pseudogap as a regime with of charge nematic order in $YBa_2Cu_3O_{6+x}$: measured from the Nernst effect, the onset (data) coincides with the T^* of the pseudogap PG (broken line). Here T_c is the superconducting (SC) critical temperature of $YBa_2Cu_3O_{6+x}$ plotted as a function of p (the hole concentration). From Ref. [60], reprinted with permission from Nature

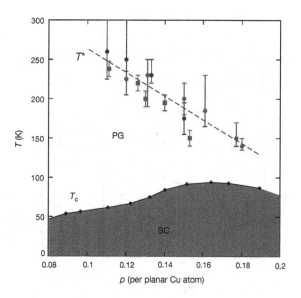

[53]. Matsuda et al. [63] have found in underdoped $La_{2-x}Sr_xCuO_4$ ($x = 0.05$), a material that was known to have "fluctuating diagonal stripes", evidence for nematic order similar to what Hinkov et al. [56] found in underdoped $YBa_2Cu_3O_{6+x}$. Earlier experiments in $La_{2-x}Sr_xCuO_4$ in moderate magnetic fields had also shown that a spin stripe state became static over some critical value of the field [64]. These experiments strongly suggest that the experiments that had previously identified the high temperature superconductors as having "fluctuating stripe order" (both inside and outside the superconducting phase) were most likely detecting an electronic nematic phase, quite close to a state with long range stripe (smectic) order. In all cases the background anisotropy (due to the orthorhombic distortion of the crystal structure) acts as a symmetry breaking field that couples linearly to the nematic order, thus rounding the putative thermodynamic transition to a state with spontaneously broken point group symmetry. These effects are much more apparent at low doping where the crystal orthorhombicity is significantly weaker.

In $La_{2-x}Ba_xCuO_4$ at $x = 1/8$ there is strong evidence for a complex stripe ordered state that intertwines charge, spin and superconducting order [5, 65] (shown in Fig. 2.13). In fact $La_{2-x}Ba_xCuO_4$ at $x = 1/8$ appears to have some rather fascinating properties. As summarized in Fig. 2.14, $La_{2-x}Ba_xCuO_4$ at $x = 1/8$ has a very low critical superconducting $T_c \sim 4$ K (where the Meissner state sets in). However it is known from angle-resolved photoemission (ARPES) experiments that the anti-nodal gap (which roughly gives the pairing scale) is actually largest at $x = 1/8$ [66] (or unsuppressed by the 1/8 anomaly according to Ref. [67].) Static charge stripe order sets in at 54 K (where there is a structural transition from the LTO to the LTT lattice structure), but static spin stripe order only exists below 42 K. As soon as static spin order sets in, the in-plane resistivity begins to decrease very rapidly with decreasing

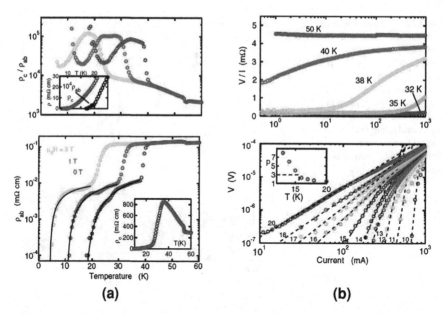

Fig. 2.13 a Dynamical layer decoupling in $La_{2-x}Ba_xCuO_4$ at $x=1/8$ from transport data.
b Kosterlitz-Thouless transition in $La_{2-x}Ba_xCuO_4$ at $x=1/8$. From Q. Li et al. [65], reprinted
with permission from APS

temperature, while the c-axis resistivity increases (see Fig. 2.13, left panel). Below
35 K strong 2D superconducting fluctuations are observed and at 16 K the in-plane
resistivity vanishes at what appears to be a Kosterlitz–Thouless transition (shown
in Fig. 2.13, right panel). However, the full 3D resistive transition is only reached
at 10 K (where $\rho_c \to 0$) although the Meissner state is only established below 4 K!
This dazzling set of phenomena shows clearly that spin, charge and superconducting
order are forming a novel sort of *intertwined* state, rather than *compete*. We have
conjectured that a pair density wave is stabilized in this intermediate temperature
regime [5, 6, 68]. Similar phenomenology, i.e. a dynamical layer decoupling, has
been seen in $La_{2-x}Sr_xCuO_4$ at moderate fields where the stripe order is static [69].
We will return below on how a novel state, the pair-density wave, explains these
phenomena.

An important caveat to the analysis we presented here is that in doped systems
there is always quenched disorder, which has different degrees of short range "orga-
nization" in different high temperature superconductors. Since disorder also couples
linearly to the charge order parameters it ultimately also rounds the transitions and
renders the system to a glassy state (as noted already in Refs. [1, 4]). Such effects are
evident in scanning tunneling microscopy (STM) experiments in $Bi_2Sr_2CaCu_2O_{8+\delta}$
which revealed that the high energy (local) behavior of the high temperature superconductors has charge order and it is glassy [4, 70–73]. This is most remarkable as
the STM data on $Bi_2Sr_2CaCu_2O_{8+\delta}$ at low bias shows quasiparticle propagation in

Fig. 2.14 Summary of the behavior of the stripe-ordered superconductor $La_{2-x}Ba_xCuO_4$ near $x = 1/8$: T_{co} is the charge ordering temperature, T_{spin} the spin ordering temperature, T^{**} marks the beginning of layer decoupling behavior, T_{KT} is the 2D superconducting temperature ("KT"), T_{3D} is the 3D resistive transition, and T_c is the 3D Meissner transition

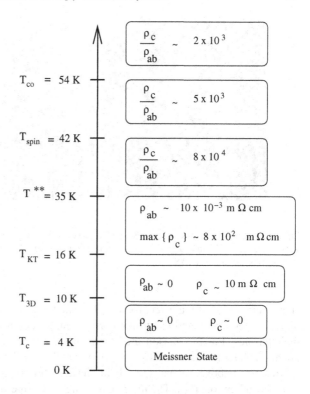

the superconducting state (but not above T_c). Yet, at high bias (i.e. high energies) there are no propagating "quasiparticles" but, instead, provides a vivid image of electronic inhomogeneity with short range charge order. This behavior is contrary to what is commonly the case in conventional superconductors where STM at high energies shows Fermi-liquid like electronic quasiparticles. Similarly, the high energy spectrum of ARPES has never resembled that of a conventional metal. We note that a recent analysis of this data by Lawler and coworkers has revealed the existence of nematic order over much longer length scales than the broken positional order [74] (Fig. 2.15).

Finally, we note that in the recently discovered iron pnictides based family of high temperature superconductors, such as La $(O_{1-x}F_x)FeAs$ and $Ca(Fe_{1-x}Co_x)_2As_2$ [75, 76], a unidirectional spin-density-wave has been found. It has been suggested [77] that the undoped system LaOFeAs and $CaFe_2As_2$ may have a high-temperature nematic phase and that quantum phase transitions also occur as a function of fluorine doping [78, 79]. This suggests that many of the ideas and results that we present here may be relevant to these still poorly understood materials.

The existence of stripe-ordered phases is well established in other complex oxide materials, particularly the manganites and the nickelates. In general, these materials tend to be "less quantum mechanical" than the cuprates in that they are typically insulating (although with interesting magnetic properties) and the observed

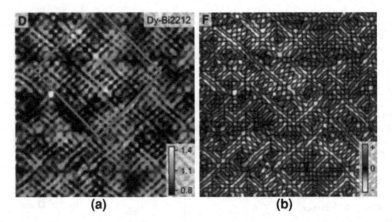

Fig. 2.15 Short range stripe order in $Bi_2Sr_2CaCu_2O_{8+\delta}$ (Dy-Bi2212) as seen in STM experiments. **a** R maps taken at 150 mV; **b** $\bigtriangledown^2 R$ shows the local nematic order. From Kohsaka et al. [70], reprinted with permission from AAAS

charge-ordered phases are very robust. These materials typically have larger electron-phonon interactions and electronic correlation are comparatively less dominant in their physics. For this reason they tend to be "more classical" and less prone to quantum phase transitions. However, at least at the classical level, many of the issues we discussed above, such as the role of phase separation and Coulomb interactions, also play a key role [80]. The thermal melting of a stripe state to a nematic has been seen in the manganite material $Bi_xCa_xMnO_3$ [81].

2.2.4 Conventional CDW Materials

CDWs have been extensively studied since the mid-seventies and there are extensive reviews on their properties [82, 83]. From the symmetry point of view there is no difference between a CDW and a stripe (or electron smectic). CDW states are usually observed in systems which are not particularly strongly correlated, such as the quasi-one-dimensional and quasi-two-dimensional dichalcogenides, and the more recently studied tritellurides. These CDW states are reasonably well described as Fermi liquids (FL) which undergo a CDW transition, commensurate or incommensurate, triggered by a nesting condition of the FS [84, 85]. As a result, a part or all of the FS is gapped in which case the CDW may or may not retain metallic properties. Instead, in a strongly correlated stripe state, which has the same symmetry breaking pattern, at high energy has Luttinger liquid behavior [1, 86, 87].

What will interest us here is that conventional quasi-2D dichalcogenides, the also quasi-2D tritellurides and other similar CDW systems can quantum melt as a function of pressure in $TiSe_2$ [88], or by chemical intercalation as in Cu_xTiSe_2 [89, 90] and Nb_xTaS_2 [91]. Thus, CDW phases in chalcogenides can serve as a weak-coupling version of the problem of quantum melting of a quantum smectic. Interestingly, there is strong experimental evidence that both $TiSe_2$ [88] and $Nb_x TaS_2$ [91] do not melt directly to an isotropic Fermi fluid but go instead through an intermediate, possibly hexatic, phase.[8] Whether or not the intermediate phases are anisotropic is not known as no transport data is yet available in the relevant regime.

The case of the CDWs in tritellurides is more directly relevant to the theory we will present here. Tritellurides are quasi-2D materials which for a broad range of temperatures exhibit a unidirectional CDW (i.e. an electronic smectic phase) and whose anisotropic behavior appears to be primarily of electronic origin [93–96]. However, the quantum melting of this phase has not been observed yet. Theoretical studies have also suggested that it may be possible to have a quantum phase transition to a state with more than one CDW in these materials [97].

2.3 Theories of Stripe Phases

2.3.1 Stripe Phases in Microscopic Models

Of all the electronic liquid crystal phases, stripe states have been studied most. There are in fact a number of excellent reviews on this topic [4, 87, 98] covering both the phenomenology and microscopic mechanisms. I will only give a brief summary of important results and refer to the literature for details.

As we noted in Sect. 2.1.2, stripe and CDW (and SDW) phases have the same order parameter as they correspond to the same broken symmetry state, and therefore the same order parameter ρ_K (and S_Q).[9] There is however a conceptual difference. CDW and SDW are normally regarded as weak coupling instabilities of a Fermi liquid (or free fermion state) typically triggered by a nesting condition satisfied by the ordering wave vector [82, 84]. In this context, the quasiparticle spectrum is modified by the partial opening of gaps and a change in the topology of the original Fermi surface (or, equivalently, by the formation of "pockets"). Because of this inherently weak coupling physics, the ordering wave vector is rigidly tied to the Fermi wave vector k_F.

In one-dimensional systems, non-linearities lead to a more complex form of density wave order, a lattice of solitons, known in this context as discommensurations [85, 99] whose ordering wavevector is no longer necessarily tied to k_F. A stripe state is essentially a two-dimensional generally incommensurate ordered state which is an analog of this strong coupling one-dimensional lattice of discommensurations [100]. Thus, in this picture, the spin stripe seen in neutron scattering [53] is pictured as regions of antiferromagnetic (commensurate) order separated by anti-phase domain walls (the discommensurations) where the majority of the doped

---

[8] $Cu_x TiSe_2$ is known to become superconducting [89]. The temperature-pressure phase diagram of $TiSe_2$ exhibits a superconducting dome enclosing the quantum critical point at which the CDW state melts [92].

[9]   In principle the order parameter of the stripe state may not be pure sinusoidal and will have higher harmonics of the fundamental order parameter.

holes reside. This picture is quite hard to achieve by any weak coupling approximation such as Hartree–Fock.

Stripe phases were first found in Hartree–Fock studies of Hubbard and t-J
models in two dimensions [101–105]. In this picture stripe phases are unidirectional
charge density waves with or without an associated spin-density-wave (SDW) order.
As such they are characterized by a CDW and/or SDW order parameters, ρ_K and
S_Q respectively.[10] A Hartree–Fock theory of stripe phases was also developed in the
context of the 2DEG in large magnetic fields [29, 30, 32] to describe the observed
and very large transport anisotropy we discussed above.

As it is usually the case, a serious limitation of the Hartree–Fock approach is that
it is inherently reliable only at weak coupling, and hence away from the regime of
strong correlation of main interest. In particular, all Hartree–Fock treatments of the
stripe ground state typically produce an "empty stripe state", an insulating crystal and
therefore not a metallic phase. Thus, in this approach a conducting (metallic) stripe
phase can only arise from some sort of quantum melting of the insulting crystal and
hence not describable in mean-field theory. The phenomenological significance of
stripe phases was emphasized by several authors, particularly by Emery and Kivelson
[21, 106, 107].

Mean field theory predicts that, at a fixed value of the electron (or hole) density
(doping), the generally incommensurate ordering wave vectors satisfy the relation
$K = 2Q$. That this result should generally hold follows from a simple Landau–
Ginzburg (LG) analysis of stripe phases (see, e.g. [108, 109] where it is easy to see
that a trilinear term of the form $\rho_K^* S_Q \cdot S_Q$ (and its complex conjugate) is generally
allowed in the LG free energy. In an ordered state of this type the antiferromagnetic
spin order is "deformed" by anti-phase domain walls whose periodic pattern coincides with that of the charge order, as suggested by the observed magnetic structure
factor of the stripe state first discovered in the cuprate $La_{1.6-x}Nd_{0.4}Sr_xCuo_4$ [53].

This pattern of CDW and SDW orders has suggested the popular cartoon of
stripe phases as antiferromagnetic regions separated by narrow "rivers of charge" at
antiphase domain walls. The picture of the stripe phase as an array of rivers suggests
a description of stripe phases as a quasi-one-dimensional system. As we will see in
the next subsection, this picture turned out to be quite useful for the construction
of a strong coupling theory of the physics of the stripe phase. On the other hand, it
should not be taken literally in the sense that these rivers always have a finite width
which does not have to be small compared with the stripe period and in many cases
they may well be of similar magnitude. Thus, one may regard this phase as being
described by narrow 1D regions with significant transversal quantum fluctuations in
shape (as it was presented in Ref.[1]) or, equivalently, as quasi-1D regions with a
significant transversal width.

An alternative picture of the stripe phases can be gleaned from the t-J model, the
strong coupling limit of the Hubbard model. Since in the resulting effective model
there is no small parameter, the only (known) way to treat it is to extend the $SU(2)$
symmetry of the Hubbard (and Heisenberg) model to either $SU(N)$ or $Sp(N)$ and to

---

[10] I will ignore here physically correct but more complex orders such as helical phases.

use the large N expansion to study its properties [98, 110–112]. In this (large N) limit the undoped antiferromagnet typically has a dimerized ground state, a periodic (crystalline) pattern of valence bond spin singlets. Since all degrees of freedom are bound into essentially local singlets this state is a quantum paramagnet. However, it is also "striped" in the sense that the valence bond crystalline state breaks at least the point group symmetry C_4 of the square lattice as well as translation invariance: it is usually a period 2 columnar state.[11] In the doped system the valence bond crystal typically becomes a non-magnetic incommensurate insulating system. Mean-field analyses of these models [112] also suggest the existence of superconducting states, some of which are "striped". Similar results are suggested by variational wave functions based of the RVB state [114–116]. We should note, however, that mean-field states are no longer controlled by a small parameter, such as $1/N$, and hence it is unclear how reliable they may be at the physically relevant case $N = 2$. For a detailed (and up-to-date) review of this approach see Ref. [98].

There are also extensive numerical studies of stripe phases in Hubbard type models. The best numerical data to date is the density matrix renormalization group (DMRG) work of White and Scalapino (and their collaborators) on Hubbard and t-J ladders of various widths (up to 5) and varying particle densities [117–120] and by Jackelmann et al. in fairly wide ladders (up to 7) [121]. An excellent summary and discussion on the results from various numerical results (as well as other insights) can be found in Ref. [87]. The upshot of all the DMRG work is that there are strong stripe correlations in Hubbard and t-J models which may well be the ground state.[12]

Much of the work on microscopic mechanisms of stripe formation has been done in models with short range interactions such as the Hubbard and t-J models. As it is known [19, 20, 122], models of this type have a strong tendency to *electronic phase separation*. As we noted in the introduction, the physics of phase separation is essentially the disruption of the correlations of the Mott (antiferromagnetic) state by the doped holes which leads to an effective *attractive* interaction among the charge carriers. When these effects overwhelm the stabilizing effects of the Fermi pressure (i.e. the fermion kinetic energy), phase separation follows. In more realistic models, however, longer range (and even Coulomb) interactions must be taken into account which tend to frustrate this tendency to phase separation [21], as well as a more complex electronic structure [123]. The structure of actual stripe phases in high T_c materials results from a combination of these effects. One of the (largely) unsolved questions is the relation between the stripe period and the filling fraction of each stripe at a given density. Most simple minded calculation yield simple commensurate filling fractions for each stripe leading to insulating states. At present time, except for results from DMRG studies on wide ladders [121], there are no controlled calculations that reproduce these effects, although suggestive variational estimates have been published [124].

---

[11]   This state is a close relative of the resonating valence bond (RVB) state originally proposed as a model system for a high T_c superconducting state [18, 113], i.e. a (non-resonating) valence bond (VB) state.

[12]   A difficulty in interpreting the DMRG results lies in the boundary conditions that are used that tend to enhance inhomogeneous, stripe-like, phases.

2.3.2 Phases of Stripe States

We will now discuss the strong coupling picture of the stripe phases [48, 86, 87, 112, 125, 126]. We will assume that a stripe phase exists with a fixed (generally incommensurate) wave vector K and a fixed filling fraction (or density) on each stripe. In this picture a stripe phase is equivalent to an array of ladders of certain width. In what follows we will assume that each stripe has a finite *spin gap*: a *Luther–Emery liquid* [127, 128].

2.3.2.1 Physics of the 2-Leg Ladder

The assumption of the existence of a finite spin gap in ladders can be justified in several ways. In DMRG studies of Hubbard and t-J ladders in a rather broad density range, $0 < x < 0.3$, it is found that the ground state has a finite spin gap [129]. Similar results were found analytically in the weak coupling regime [130–133].

Why there is a spin gap? There is actually a very simple argument for it [134]. In the non-interacting limit, $U = V = 0$, the two-leg ladder has two bands with two different Fermi wave vectors, $p_{F1} \neq p_{F2}$. Let us consider the effects of interactions in this weak coupling regime. The only allowed processes involve an *even* number of electrons. In this limit is is easy to see that the coupling of CDW fluctuations with $Q_1 = 2p_{F1} \neq Q_2 = 2p_{F2}$ is suppressed due to the mismatch of their ordering wave vectors. In this case, scattering of electron pairs with zero center of mass momentum from one system to the other is a peturbatively (marginally) relevant interaction. The spin gap arises since the electrons can gain zero-point energy by delocalizing between the two bands. To do that, the electrons need to pair, which may cost some energy. When the energy gained by delocalizing between the two bands exceeds the energy cost of pairing, the system is driven to a spin-gap phase.

This physics is borne out by detailed numerical (DMRG) calculations, even in systems with only repulsive interactions. Indeed, at $x=0$ (the undoped ladder) the system is in a Mott insulating state, with a unique fully gapped ground state ("*C0S0*" in the language of Ref. [130]). In the strong coupling limit (in which the rungs of the ladder are spin singlet valence bonds), $U \gg t$, the spin gap is large: $\Delta_s \sim J/2$ [135].

At low doping, $0 < x < x_c \sim 0.3$, the doped ladder is in a Luther–Emery liquid: there is no charge gap and large spin gap ("*C1S0*"). In fact, in this regime the spin gap is found to decrease monotonically as doping increases, $\Delta_s \downarrow$ as $x \uparrow$, and vanishes at a critical value x_c : $\Delta_s \to 0$ as $x \to x_c$.

The most straightforward way to describe this system is to use bosonization. Although the ladder system has several bands of electrons that have charge and spin degrees of freedom, in the low energy regime the effective description is considerably simplified. Indeed, in this regime it is sufficient to consider only one effective bosonized charge field and one bosonized spin field. Since there is a spin gap, $\Delta_s \neq 0$, the spin sector is massive. In contrast, the charge sector is only massive at $x=0$, where there is a finite Mott gap Δ_M.

The effective Hamiltonian for the charge degrees of freedom in this (Luther–Emery) phase is

$$H = \int dy \frac{v_c}{2} \left[\frac{1}{K} \left(\partial_y \theta \right)^2 + K \left(\partial_x \phi \right)^2 \right] + \ldots \qquad (2.16)$$

where ϕ is the CDW phase field, and θ is the SC phase field. They satisfy canonical commutation relations

$$[\phi(y'), \partial_y \theta(y)] = i\delta(y - y'). \qquad (2.17)$$

The parameters of this effective theory, the spin gap Δ_s, the charge Luttinger parameter K, the charge velocity v_c, and the chemical potential μ, have non-universal but smooth dependences on the doping x and on the parameters of the microscopic Hamiltonian, the hopping matrix elements t'/t and the Hubbard interaction U/t. The ellipsis ... in the effective Hamiltonian represent cosine potentials responsible for the Mott gap Δ_M in the undoped system ($x=0$). It can be shown that the spectrum in the low doping regime, $x \to 0$, consists of gapless and spinless charge $2e$ fermionic solitons.

The charge Luttinger parameter is found to approach $K \to 1/2$ as $x \to 0$. As x increases, so does K reaching the value $K \sim 1$ for $x \sim 0.1$. On the other hand $K \sim 2$ for $x \sim x_c$ where the pin gap vanishes. The temperature dependence of the superconducting and CDW susceptibilities have the scaling behavior

$$\chi_{SC} \sim \frac{\Delta_s}{T^{2-K}} \qquad (2.18)$$

$$\chi_{CDW} \sim \frac{\Delta_s}{T^{2-K^{-1}}}. \qquad (2.19)$$

Thus, both susceptibilities diverge $\chi_{CDW}(T) \to \infty$ and $\chi_{SC}(T) \to \infty$ for $0 < x < x_c$ as $T \to 0$. However, for $x \lesssim 0.1$, the SC susceptibility is more divergent: $\chi_{SC} \gg \chi_{CDW}$. Hence, the doped ladder in the Luther–Emery regime is effectively a 1D superconductor even for a system with nominally repulsive interactions (i.e. without "pairing").

2.3.2.2 The spin-Gap Stripe State

Now consider a system with N stripes, each labeled by an integer $a = 1, \ldots, N$. We will consider first the phase in which there is a spin gap. Here, the spin fluctuations are effectively frozen out at low energies. Nevertheless each stripe a has two degrees of freedom [1]: a transverse displacement field which describes the local dynamics of the configuration of each stripe, and the phase field ϕ_a for the charge fluctuations on each stripe. The action of the generalized Luttinger liquid which describes the smectic charged fluid of the stripe state is obtained by integrating out the local shape fluctuations associated with the displacement fields. These fluctuations give rise to a finite renormalization of the Luttinger parameter and velocity of each stripe. More importantly, the shape fluctuations, combined with the long-wavelength inter-stripe Coulomb interactions, induce inter-stripe density-density and current-current interactions, leading to an imaginary time Lagrangian density of the form

$$\mathcal{L}_{\text{smectic}} = \frac{1}{2} \sum_{a,a',\mu} j_\mu^a(x) \, \widetilde{W}_\mu(a - a') \, j_\mu^{a'}(x). \tag{2.20}$$

where the current operators on stripe a are $j_\mu^a(x) = \frac{1}{\sqrt{\pi}} \epsilon_{\mu\nu} \partial_\nu \phi^a(x)$; here $\mu = t, x$. These operators are *marginal*, i.e., have scaling dimension 2, and preserve the *smectic symmetry* $\phi_a \rightarrow \phi_a + \alpha_a$ (where α_a is constant on each stripe) of the decoupled Luttinger fluids. Notice that the current (and density) operators *of each stripe* are invariant under these transformations. Whenever this symmetry is exact, the charge-density-wave order parameters of the individual stripes do not lock with each other, and the charge density profiles on each stripe can slide relative to each other without an energy cost. In other words, there is no rigidity to shear deformations of the charge configuration on nearby stripes. This is the *smectic metal* phase [1], a *sliding Luttinger liquid* [136].

The fixed point action for a generic smectic metal phase thus has the form (in Fourier space)

$$\begin{aligned} S &= \sum_Q \frac{K(Q)}{2} \left\{ \frac{\omega^2}{v(Q)} + v(Q)k^2 \right\} |\phi(Q)|^2 \\ &= \sum_Q \frac{1}{2K(Q)} \left\{ \frac{\omega^2}{v(Q)} + v(Q)k^2 \right\} |\theta(Q)|^2 \end{aligned} \tag{2.21}$$

where $Q = (\omega, k, k_\perp)$. Here θ_a is the field *dual* to ϕ_a and obey the canonical (equal-time) commutation relations

$$[\phi_a(x'), \partial_x \theta_b(x)] = i\delta(x' - x)\delta_{ab} \tag{2.22}$$

In Eq. 2.21 k is the momentum *along* the stripe and $k_\perp$ perpendicular to the stripes. The kernels $K(Q)$ and $v(Q)$ are analytic functions of Q whose form depends on microscopic details, e.g. at weak coupling they are functions of the inter-stripe Fourier transforms of the forward and backward scattering amplitudes $g_2(k_\perp)$ and $g_4(k_\perp)$, respectively. In practice, up to irrelevant operators, it is sufficient to keep the dependence of the kernels only on the transverse momentum $k_\perp$. Thus, the smectic fixed point is characterized by the effective Luttinger parameter and velocity (functions), $K(k_\perp)$ and $v(k_\perp)$. Much like the ordinary 1D Luttinger liquid, this "fixed point" is characterized by power-law decay of correlations functions. This effective field theory also yields the correct low energy description of the quantum Hall stripe phase of the 2DEG in large magnetic fields [31–33, 137, 138].

In the presence of a spin gap, single electron tunneling is irrelevant [134], and the only potentially relevant interactions involving pairs of stripes a, a' are singlet pair (Josephson) tunneling, and the coupling between the CDW order parameters. These interactions have the form $\mathcal{H}_{int} = \sum_n (\mathcal{H}_{SC}^n + \mathcal{H}_{CDW}^n)$ for $a' - a = n$, where

$$\mathcal{H}_{SC}^n = \left(\frac{\Lambda}{2\pi}\right)^2 \sum_a \mathcal{J}_n \cos[\sqrt{2\pi}(\theta_a - \theta_{a+n})]$$

$$\mathcal{H}_{CDW}^n = \left(\frac{\Lambda}{2\pi}\right)^2 \sum_a \mathcal{V}_n \cos[\sqrt{2\pi}(\phi_a - \phi_{a+n})]. \tag{2.23}$$

Here $\mathcal{J}_n$ are the inter-stripe Josephson couplings (SC), $\mathcal{V}_n$ are the $2k_F$ component of the inter-stripe density-density (CDW) interactions, and Λ is an ultra-violet cutoff, $\Lambda \sim 1/a$ where a is a lattice constant. A straightforward calculation, yields the scaling dimensions $\Delta_{1,n} \equiv \Delta_{SC,n}$ and $\Delta_{-1,n} \equiv \Delta_{CDW,n}$ of $\mathcal{H}_{SC}^n$ and $\mathcal{H}_{CDW}^n$:

$$\Delta_{\pm 1,n} = \int_{-\pi}^{\pi} \frac{dk_\perp}{2\pi} \, [\kappa(k_\perp)]^{\pm 1} \, (1 - \cos nk_\perp), \tag{2.24}$$

where $\kappa(k_\perp) \equiv K(0, 0, k_\perp)$. Since $\kappa(k_\perp)$ is a periodic function of $k_\perp$ with period 2π, $\kappa(k_\perp)$ has a convergent Fourier expansion of the form $\kappa(k_\perp) = \sum_n \kappa_n \cos nk_\perp$. We will parametrize the fixed point theory by the coefficients κ_n, which are smooth non-universal functions. In what follows we shall discuss the behavior of the simplified model with $\kappa(k_\perp) = \kappa_0 + \kappa_1 \cos k_\perp$. Here, κ_0 can be thought of as the intra-stripe inverse Luttinger parameter, and κ_1 is a measure of the nearest neighbor inter-stripe coupling. For stability we require $\kappa_0 > \kappa_1$.

Since it is unphysical to consider longer range interactions in H_{int} than are present in the fixed point Hamiltonian, we treat only perturbations with $n = 1$, whose dimensions are

$$\Delta_{SC,1} \equiv \Delta_{SC} = \kappa_0 - \frac{\kappa_1}{2} \tag{2.25}$$

and

$$\Delta_{CDW,1} \equiv \Delta_{CDW} = \frac{2}{\left(\kappa_0 - \kappa_1 + \sqrt{\kappa_0^2 - \kappa_1^2}\right)} \tag{2.26}$$

For a more general function $\kappa(k_\perp)$, operators with larger n must also be considered, but the results are qualitatively unchanged [136, 139].[13]

In Fig. 2.16 we present the phase diagram of this model. The dark AB curve is the set of points where $\Delta_{CDW} = \Delta_{SC}$, and it is a line of first order transitions.

---

[13] $\Delta_{SC,2}$ is the most relevant operator. For a model with $\kappa(k_\perp) = [\kappa_0 + \kappa_1 \cos(k_\perp)]^2$, all perturbations are irrelevant for large κ_0 and small $|\kappa_0 - \kappa_1|$.

Fig. 2.16 Phase diagram for
a stripe state with a spin gap

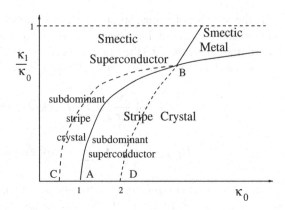

To the right of this line the inter-stripe CDW coupling is the most relevant perturbation, indicating an instability of the system to the formation of a 2D stripe crystal [1]. To the left, Josephson tunneling (which still preserves the smectic symmetry) is the most relevant, so this phase is a 2D smectic superconductor. (Here we have neglected the possibility of coexistence since a first order transition seems more likely). Note that there is a region of $\kappa_0 \geq 1$, and large enough κ_1, where the global order is superconducting although, in the absence of inter-stripe interactions (which roughly corresponds to $\kappa_1 = 0$), the superconducting fluctuations are subdominant. There is also a (strong coupling) regime above the curve CB where *both* Josephson tunneling *and* the CDW coupling *are irrelevant* at low energies. Thus, in this regime *the smectic metal state is stable*. This phase is a 2D smectic non-Fermi liquid in which there is coherent transport *only* along the stripes.

To go beyond this description we need to construct an effective theory of the *two-dimensional* ordered phase. For instance, the superconducting state is a 2D striped superconductor, whereas the crystal is a bidirectional charge density wave. A theory of these 2D ordered phases can be developed by combining the quasi-one-dimensional renormalization group with an effective inter-stripe mean field theory, as in Ref. [140], which in turn can be fed into a 2D renormalization group theory [141]. One advantage of this approach is that the inter-stripe mean field theory has the same analytic structure as the dimensional crossover RG (see Ref. [126]).

Let us consider the superconducting state, a *striped* superconductor. In the way we constructed this state all ladders are *equivalent*. Hence this is a period 2 stripe (columnar) SC phase, similar to the one discussed by Vojta [98]. Let us use inter-stripe mean field theory to estimate the critical temperature of the 2D state. For the isolated ladder, $T_c = 0$ as required by the Mermin-Wagner theorem. If the inter-stripe Josephson and CDW couplings are non-zero, $\mathcal{J} \neq 0$ and $\mathcal{V} \neq 0$, the system will now have a finite SC critical temperature, $T_c > 0$. Now, for $x \lesssim 0.1$, CDW couplings are irrelevant as in this range $1/2 < K < 1$. Hence, in the same range, the inter-ladder Josephson coupling are relevant and lead to a SC state in a small x with a somewhat low T_c which, in inter-stripe (or 'chain') mean field theory can be estimated by

$$2\mathcal{J}\chi_{SC}(T_c) = 1. \tag{2.27}$$

In this regime, however, $T_c \propto \delta t \; x$ and it is low due to the low carrier density. Conversely, for larger x, $K > 1$ and χ_{CDW} is more strongly divergent than χ_{SC}. Thus, for $x \gtrsim 0.1$ the CDW couplings become more relevant. This leads to an insulating incommensurate CDW state with ordering wave number $P = 2\pi x$.

In the scenario we just outlined [48, 126] in the 2D regime the system has a *first order* transition from a superconducting state to a non-superconducting phase with charge order. However at large enough inter-stripe forward scattering interactions both couplings become irrelevant and there is a quantum *bicritical* point separating both phases from a smectic metal (as depicted in Fig. 2.16). However, an alternative possibility is that instead of a bicritical point, we may have a quantum *tetracritical* point and a phase in which SC and CDW orders coexist.

2.4 Is Inhomogeneity Good or Bad for Superconductivity?

The analysis we just did raises the question of whether stripe order (that is, some form of spatial charge inhomogeneity) is good or bad for superconductivity. This question was addressed in some detail in Refs. [48, 126] where it was concluded that (a) there is an *optimal degree of inhomogeneity at which T_c reaches a maximum*, and (b) that charge order in a system with a spin gap can provide a mechanism of "high temperature superconductivity" (the meaning of which we will specify below).

The argument goes as follows. Consider a system with a *period* 4 stripe phase, consisting of an alternating array of inequivalent A and B type ladders in the Luther–Emery regime.[14] The inter-stripe mean field theory estimate for the superconducting and CDW critical temperatures now takes the somewhat more complex form:

$$(2\mathcal{J})^2 \chi_{SC}^A(T_c)\chi_{SC}^B(T_c) = 1 \tag{2.28}$$

for the superconducting T_c, and

$$(2\mathcal{V})^2 \chi_{CDW}^A(P, T_c)\chi_{CDW}^B(P, T_c) = 1 \tag{2.29}$$

for the CDW T_c. In particular, the 2D CDW order is greatly suppressed due to the mismatch between ordering vectors, P_A and P_B, on neighboring ladders.

For inequivalent A and B ladders SC beats CDW if the corresponding Luttinger parameters satisfy the inequalities

$$2 > K_A^{-1} + K_B^{-1} - K_A; \quad 2 > K_A^{-1} + K_B^{-1} - K_B. \tag{2.30}$$

---

[14]   In Ref. [125] a similar pattern was also considered except that the (say) 'B' stripes do no have a spin gap. This patterns was used to show how a crude model with nodal quasiparticles can arise in an inhomogeneous state.

Fig. 2.17 Model of a period
4 stripe phase

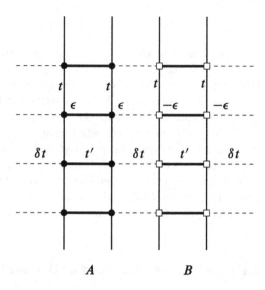

The SC critical temperature is then found to obey a *power law* scaling form (instead of the essential singularity of the BCS theory of superconductivity):

$$T_c \sim \Delta_s \left(\frac{\mathcal{J}}{\widetilde{W}} \right)^\alpha \; ; \alpha = \frac{2K_A K_B}{[4K_A K_B - K_A - K_B]}. \tag{2.31}$$

A simple estimate of the effective inter-stripe Josephson coupling,[15] $\mathcal{J} \sim \delta t^2 / J$ and of the high energy scale $\widetilde{W} \sim J$, implies that the superconducting critical temperature T_c is (power law) small for small $\mathcal{J}$!, with an exponent that typically is $\alpha \sim 1$ (Fig. 2.17).

These arguments can be used to sketch a phase diagram of the type presented in Fig. 2.18 which shows the qualitative dependence of the SC T_c with doping x. The broken line shown is the spin gap $\Delta_s(x)$ as a function of doping x and, within this analysis, it must be an upper bound on T_c. Our arguments then showed that a period 4 structure can have a substantially larger T_c than a period 2 stripe. Consequently, the critical dopings, $x_c(2)$ and $x_c(4)$, for the SC-CDW quantum phase transition must move to higher values of x for period 4 compared with period 2. On the other hand, for $x \gtrsim x_c$ the isolated ladders do not have a spin gap, and this strong coupling mechanism is no longer operative.

How reliable are these estimates? What we have are mean-field estimates for T_c and it is an upper bound to the actual T_c. As it is usually the case, T_c should be suppressed by phase fluctuations by up to a factor of 2. On the other hand, perturbative RG studies for small $\mathcal{J}$ yield the *same power law dependence*. This result

---

[15] Josephson coupling is due to pair tunneling from one stripe to a neighboring one. Josephson processes arise in second order in perturbation theory and involve intermediate states with excitations energies of order J and have an amplitude controlled by δt.

Fig. 2.18 Evolution of the superconducting critical temperature with doping

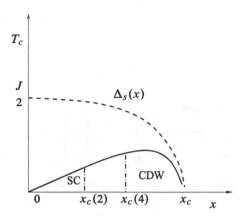

is asymptotically exact for $\mathcal{J} << \widetilde{W}$. Since T_c is a smooth function of $\delta t / \mathcal{J}$, it is reasonable to extrapolate for $\delta t \sim \mathcal{J}$. Hence, $T_c^{max} \propto \Delta_s$ and we have a "high T_c". This is in contrast to the exponentially small T_c obtained in a BCS-like mechanism.

Now, having convinced ourselves that a period 4 stripe will have a larger SC T_c than a period 2 stripe one may wonder if an even longer period stripe state would do better. It is easy to see that there will be a problem with this proposal. Clearly, although the argument we just presented would suggest that the exponents will also be of order 1 for longer periods, the problem now is that the effective couplings become very small very quickly as the Josephson coupling has an *exponential* dependence on distance (tunneling!). Thus, there must be an optimal period for this mechanism and it is likely to be a number larger than 2 but smaller than (say) 6.

In summary, we have shown that in systems with strong repulsive interactions (and without attractive interactions), an (inhomogeneous) stripe-ordered state can support a 2D superconducting state with a high critical temperature, in the sense that it is not exponentially suppressed, with a high paring scale (the spin gap). This state is an inhomogeneous version of the RVB mechanism [18, 113, 142]. The arguments suggest that there is an optimal degree of inhomogeneity. There is suggestive evidence in ARPES data in $La_{2-x}Ba_xCuO_4$ that show a large pairing scale in the stripe-ordered state which support this picture [66, 67].

2.5 The Striped Superconductor: A Pair Density Wave State

We now turn to a novel type of striped superconductor, the pair density wave state. Berg et al. [5, 109, 143] have recently proposed this state as a symmetry-based explanation of the spectacular dynamical layer decoupling seen in stripe-ordered $La_{2-x}Ba_xCuO_4$ (and $La_{1.6-x}Nd_{0.4}Sr_xCuO_4$) [65, 144, 145], and in $La_{2-x}Sr_xCuO_4$ in magnetic fields [69].

Summary of experimental facts for $La_{2-x}Ba_xCuO_4$ near $x = 1/8$:

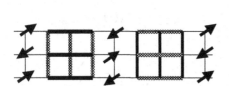 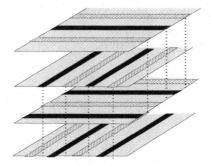

Fig. 2.19 Period 4 striped superconducting state

- ARPES finds an anti-nodal d-wave SC gap that is large and unsuppressed at $x = 1/8$. Hence, there is a large pairing scale in the stripe-ordered state.
- Resonant X-Ray scattering finds static charge stripe order for $T < T_{charge} = 54$ K.
- Neutron Scattering finds static Stripe Spin order $T < T_{spin} = 42$ K.
- The in-plane resistivity ρ_{ab} drops rapidly to zero from T_{spin} to T_{KT} (the Kosterlitz–Thouless (KT) transition).
- ρ_{ab} shows KT behavior for $T_{spin} > T > T_{KT}$.
- ρ_c increases as T decreases for $T > T^{**} \approx 35$ K.
- $\rho_c \to 0$ as $T \to T_{3D} = 10$ K (the bulk 3D resistive transition).
- $\rho_c/\rho_{ab} \to \infty$ for $T_{KT} > T > T_{3D}$.
- Theres is a Meissner state only below $T_c = 4$ K.

How do we understand these remarkable effects that can be summarized as follows: There is a broad temperature range, $T_{3D} < T < T_{2D}$ with 2D superconductivity but not in 3D, as if there is no interlayer Josephson coupling. In this regime there is both striped charge and spin order. This can only happen if there is a special symmetry of the superconductor in the striped state that leads to an almost complete cancellation of the c-axis Josephson coupling.

What else do we know? The stripe state in the LTT ("low temperature tetragonal") crystal structure of $La_{2-x}Ba_xCuO_4$ has two planes in the unit cell. Stripes in the 2nd neighbor planes are shifted by half a period to minimize the Coulomb interaction: 4 planes per unit cell. The anti-ferromagnetic spin order suffers a π phase shift accross the charge stripe which has period 4. Berg et al. [5] proposed that the superconducting order is also striped and also suffers a π phase shift. The superconductivity resides in the spin gap regions and there is a π phase shift in the SC order across the anti-ferromagnetic regions (Fig. 2.19).

The PDW SC state has *intertwined* striped charge, spin and superconducting orders.[16]

---

[16] While there is some numerical evidence for a state of this type in variational Monte Carlo calculations [115] and in slave particle mean field theory [114, 146] (see, however, Ref.[147, 148]), a consistent and controlled microscopic theory is yet to be developed. Since the difference between

How does this state solve the puzzle? If this order is perfect, the Josephson coupling between neighboring planes cancels exactly due to the symmetry of the periodic array of π textures, i.e. the spatial average of the SC order parameter is exactly zero. The Josephson couplings J_1 and J_2 between planes two and three layers apart also cancel by symmetry. The first non-vanishing coupling J_3 occurs at four spacings. It is quite small and it is responsible for the non-zero but very low T_c. Defects and/or discommensurations give rise to small Josephson coupling J_0 between neighboring planes.

Are there other interactions? It is possible to have an inter-plane biquadratic coupling involving the product of the SC order parameters between neighboring planes $\Delta_1 \Delta_2$ and the product of spin stripe order parameters also on neighboring planes $M_1 \cdot M_2$. However in the LTT structure $M_1 \cdot M_2 = 0$ and there is no such coupling. In a large enough perpendicular magnetic field it is possible (spin flop transition) to induce such a term and hence an effective Josephson coupling. Thus in this state there should be a strong suppression of the 3D SC T_c. but not of the 2D SC T_c.

On the other hand, away from $x = 1/8$ there is no perfect commensuration. Discommensurations are defects that induce a finite Josephson coupling between neighboring planes $J_1 |x - 1/8|^2$, leading to an increase of the 3D SC T_c. away from $x = 1/8$. Similar effects arise from disorder which also lead to a rise in the 3D SC T_c.

2.5.1 Landau–Ginzburg Theory of the Pair Density Wave

In what follows we will rely heavily on the results of Refs. [68, 109, 143]. We begin with a description of the order parameters:

1. PDW (Striped) SC:

$$\Delta(r) = \Delta_Q(r)e^{i\,Q \cdot r} + \Delta_{-Q}(r)e^{-i\,Q \cdot r} \tag{2.32}$$

   complex charge 2e singlet pair condensate with wave vector Q, (i.e. an FFLO type state at zero magnetic field)[17]
2. Nematic: detects breaking of rotational symmetry: N, a real neutral pseudo-scalar order parameter
3. Charge stripe: ρ_K, unidirectional charge stripe with wave vector K
4. Spin stripe order parameter: S_Q, a neutral complex spin vector order parameter.

These order parameters have the following transformation properties under rotations by $\pi/2$, $\mathcal{R}_{\pi/2}$:

(Footnote 16 continued)
the energies of the competing states seen numerically is quite small one must conclude that they are all reasonably likely.

[17] A state that is usually described as a pair crystal is commonly known as a pair density wave [149, 150]. However that state cannot be distinguished by symmetry from a (two) CDWs coexisting with a uniform SC.

1. The nematic order parameter changes sign: $N \to -N$
2. The CDW ordering wave vector rotates: $\rho_K \to \rho_{\mathcal{R}_{\pi/2}K}$
3. The SDW ordering wave vector also rotates: $S_Q \to S_{\mathcal{R}_{\pi/2}Q}$
4. The striped SC (s or d wave) order parameter: $\Delta_{\pm Q} \to \pm \Delta_{\pm \mathcal{R}_{\pi/2}Q}$ (+ for s-wave, $-$ for d- wave)

and by translations by R

$$N \to N, \quad \rho_K \to e^{iK \cdot R} \rho_K, \quad S_Q \to e^{iQ \cdot R} S_Q \qquad (2.33)$$

The Landau–Ginzburg free energy functional is, as usual, a sum of terms of the form

$$\mathcal{F} = \mathcal{F}_2 + \mathcal{F}_3 + \mathcal{F}_4 + \dots \qquad (2.34)$$

where $\mathcal{F}_2$, the quadratic term, is simply a sum of decoupled terms for each order parameter. There exist a number of trilinear terms mixing several of the order parameters described above. They are

$$\begin{aligned}
\mathcal{F}_3 =& \gamma_s [\rho_{-K} S_Q \cdot S_Q + \rho_{-\bar{K}} S_{\bar{Q}} \cdot S_{\bar{Q}} + \text{c.c}] \\
&+ \gamma_\Delta [\rho_{-K} \Delta^\star_{-Q} \Delta_Q + \rho_{-\bar{K}} \Delta^\star_{-\bar{Q}} \Delta_{\bar{Q}} + \text{c.c.}] \\
&+ g_\Delta N [\Delta^\star_Q \Delta_Q + \Delta^\star_{-Q} \Delta_{-Q} - \Delta^\star_{\bar{Q}} \Delta_{\bar{Q}} - \Delta^\star_{-\bar{Q}} \Delta_{-\bar{Q}}] \\
&+ g_s N [S_{-Q} \cdot S_Q - S_{-\bar{Q}} \cdot S_{\bar{Q}}] \\
&+ g_c N [\rho_{-K} \rho_K - \rho_{-\bar{K}} \rho_{\bar{K}}],
\end{aligned} \qquad (2.35)$$

where $\bar{Q} = \mathcal{R}_{\pi/2} Q$, and $\bar{K} = \mathcal{R}_{\pi/2} K$. The fourth order term, which is more or less standard, is shown explicitly below.

Several consequences follow directly from the form of the trilinear terms, Eq. 2.35. One is that, at least in a fully translationally invariant system, the first two terms of Eq. 2.35 imply a relation between the ordering wave vectors: $K = 2Q$. Also, as we will see below, these terms imply the existence of vortices of the SC order with *half* the flux quantum.

Another important feature of the PDW SC is that it implies the existence of a nonzero charge $4e$ *uniform* SC state. Indeed, if we denote by Δ_4 the (uniform) charge $4e$ SC order parameter, then the following term in the LG expansion is allowed

$$\mathcal{F}'_3 = g_4 \left[\Delta^*_4 \left(\Delta_Q \Delta_{-Q} + \text{rotation by} \frac{\pi}{2} \right) + \text{c.c.} \right] \qquad (2.36)$$

Hence, the existence of striped SC order (PDW) implies the existence of uniform charge $4e$ SC order![18]

---

[18] A charge $4e$ SC order parameter is an expectation value of a four fermion operator.

We should also consider a different phase in which there are coexisting uniform and striped SC orders, as it presumably happens at low temperatures in $La_{2-x}Ba_x CuO_4$. If this is so, there is a non-zero PDW SC order parameter Δ_Q as well as an uniform (d-wave) SC order parameter Δ_0 which are coupled by new (also trilinear) terms in the LG free energy of the form

$$\mathcal{F}_{3,u} = \gamma_\Delta \Delta_0^*(\rho_Q \Delta_{-Q} + \rho_{-Q} \Delta_Q) + g_\rho \rho_{-2Q} \rho_Q^2 + \frac{\pi}{2} \text{ rotation} + \text{c.c.} \quad (2.37)$$

If $\Delta_0 \neq 0$ and $\Delta_Q \neq 0$, there is a new ρ_Q component of the charge order!. Also, the small uniform component Δ_0 removes the sensitivity to quenched disorder of the PDW SC state.

2.5.2 Charge 4e SC Order and the Topological Excitations of the PDW SC State

If there is a uniform charge 4e SC order, its vortices must be quantized in units of $hc/4e$ instead of the conventional $hc/2e$ flux quantum. Hence, half- vortices are natural in this state. To see how they arise let us consider a system deep in the PDW SC state so that the magnitude of all the order parameters is essentially constant, but their phases may vary. Thus we can write the PDW SC order parameter as

$$\Delta(r) = |\Delta_Q| e^{i Q \cdot r + i\theta_+ Q(r)} + |\Delta_{-Q}| e^{-i Q \cdot r + i\theta_- Q(r)} \quad (2.38)$$

where (by inversion symmetry) $|\Delta_Q| = |\Delta_{-Q}| = \text{const}$. It will be convenient to define the new phase fields $\theta_\pm(r)$ by

$$\theta_{\pm Q}(r) = \frac{1}{2} (\theta_+(r) \pm \theta_-(r)). \quad (2.39)$$

Likewise, in the same regime the CDW order parameter can be written as

$$\rho(r) = |\rho_K| \cos(K \cdot r + \phi(r)) \quad (2.40)$$

(and a similar expression for the SDW order parameter.) In this notation, the second trilinear term shown in Eq. 2.35 takes the form

$$\mathcal{F}_{3,\gamma} = 2\gamma_\Delta |\rho_K \Delta_Q \Delta_{-Q}| \cos(2\theta_-(r) - \phi(r)) \quad (2.41)$$

Hence, the *relative phase* θ_- is locked to ϕ, the Goldstone boson of the CDW (the phason), and they are not independently fluctuating fields. Furthermore, the phase fields $\theta_{\pm Q}$ are defined modulo 2π while θ_+ is defined only modulo π.

This analysis implies that the allowed topological excitations of the PDW SC are

1. A conventional SC vortex with $\Delta\theta_+ = 2\pi$ and $\Delta\phi = 0$, with topological charges $(1,0)$.

Fig. 2.20 Schematic phase diagram of the thermal melting of the PDW state

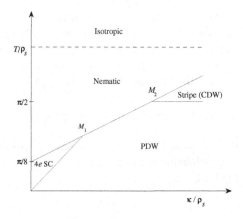

2. A bound state of a 1/2 vortex and a CDW dislocation, $\Delta\theta_+ = \pi$ and $\Delta\phi = 2\pi$, with topological charges $(\pm 1/2, \pm 1/2)$ (any such combination is allowed).
3. A double dislocation, $\Delta\theta_+ = 0$ and $\Delta\phi = 4\pi$, with topological charge $(0, 1)$.

All three topological defects have logarithmic interactions.

There are now three different pathways to melt the PDW SC [68], depending which one of these topological excitations becomes relevant (in the Kosterlitz–Thouless RG sense [151]) first. To determine the relevance or irrelevance of an operator $\mathcal{O}$ one must first compute its scaling dimension $\Delta_{\mathcal{O}}$ given by the exponent of its correlation function

$$\langle \mathcal{O}(x)\mathcal{O}(y)\rangle = \frac{1}{|X - y|^{2\Delta_{\mathcal{O}}}} \tag{2.42}$$

For the case of a topological excitation (vortices, dislocations, etc) this amounts to the computation of the ratio of the partition function of a system without topological excitations with two operators that create these excitations inserted at X and y (respectively) with the free partition function without these insertions. It is the straightforward to show [3, 152, 153] that at temperature T, the scaling dimension of a topological excitation of topological charge (p, q) is

$$\Delta_{p,q} = \frac{\pi}{T}\left(\rho_{sc}p^2 + \kappa_{CDW}q^2\right) \tag{2.43}$$

where ρ_{sc} is the superfluid density (the stiffness of the θ_+ phase field) and κ_{CDW} is the CDW stiffness (that is, of the ϕ phase field).

As usual the criterion of relevance is that an operator that creates an excitation is relevant if its scaling dimension is equal to the space dimension (for details see, for instance, Ref.[153]) which in this case is 2. This condition, $\Delta_{p,q} = 2$ for each one of the topological excitations listed above, leads to the phase thermal phase diagram shown in Fig. 2.20.[19]

Thus, the PDW state may thermally melt in three possible ways:

1. First into a CDW phase by proliferating conventional SC vortices, a $(1, 0)$ topological excitation, followed by a subsequent melting of the CDW into the normal (Ising nematic) high temperature phase. This scenario corresponds to the right side of the phase diagram and, presumably, is what happens in $La_{2-x}Ba_xCuO_4$.
2. A direct melting into the normal (Ising nematic) phase by proliferation of fractional vortices, with topological charge $(\pm 1/2, \pm 1/2)$.
3. Melting into a charge $4e$ uniform SC phase by proliferation of double dislocations, with topological charge $(0, 1)$.

The prediction that the PDW state should effectively have a uniform charge $4e$ SC order with an anomalous $hc/4e$ flux quantization leads to a direct test of this state. this can be done by searching for fractional vortices, and similarly of fractional periodicity in the Josephson effect (and Shapiro steps). Similarly, the prediction that in the phase in which an uniform (d-wave) SC is present there should be a charge-ordered state with period equal to that of the SC (and of the SDW) is another direct test of this theory.

2.6 Nematic Phases in Fermi Systems

We now turn to the theory of the nematic phases. The nematic phase is the simplest of the liquid crystal states. In this state the system is electronically uniform but anisotropic. There are two ways to access this phase. One is by a direct transition from the isotropic electronic fluid. The other is by melting (thermal or quantum mechanical) the stripe phase. We will consider both cases. We will begin with the first scenario in its simplest description as a *Pomeranchuk instability* of a Fermi liquid.

2.6.1 The Pomeranchuk Instability

The central concept of the Landau theory of the Fermi liquid [157] is the quasi-particle. A Landau quasiparticle is the elementary excitation of a Fermi liquid with the same quantum numbers as a non-interacting electron. A necessary condition for the Landau theory to work is the condition that the quasiparticle becomes sharp (or well defined) at asymptotically low energies, i.e. as the Fermi surface is

---

[19]   A more elaborate version of this phase diagram, based on a one-loop Kosterlitz RG calculation for physically very different systems with the same RG structure system, was given in Refs. [154–156].

approached. For the Landau quasiparticle to be well defined it is necessary that the quasiparticle width, i.e. the quasiparticle scattering rate, to be small on the scale of the quasiparticle energy. The quasiparticle scattering rate, the imaginary part of the electron self energy, $\Sigma''(\omega, p)$, is determined by the quasiparticle interactions, which in the Landau theory of the Fermi liquid are parametrized by the *Landau parameters*. Except for the BCS channel, the forward scattering interactions (with or without spin flip) are the only residual interactions among the quasiparticles that survive at low energies [158, 159].

The Landau "parameters" are actually functions $F^{S,A}(p, p')$ that quantifying the strength of the forward scattering interactions among quasiparticles at low energies with momenta p and p' close to the Fermi surface in the singlet (charge) channel (S) or the triplet (spin) channel (A). For a translationally invariant system it depends only on the difference of the two momenta, $F(p, p') = F(p - p')$. Furthermore, if the system is also rotationally invariant, the Landau parameters can be expressed in an angular momentum basis. In 3D they take the form $F_{\ell,m}^{S,A}$ (with $\ell = 0, 1, 2, \ldots$ and $|m| \leq \ell$), while in 2D they are simply $F_m^{S,A}$ (where $m \in \mathbb{Z}$). We will see below that in some cases of interest we will also need to keep the dependence on a small momentum transfer in the Landau parameters (i.e. p and p' will not be precisely at the FS) even though it amounts to keeping a technically irrelevant interaction. On the other hand, for a lattice model rotational invariance is always broken down to the point (or space) group symmetry of the lattice. In that case the Landau parameters are classified according to the irreducible representations of the point (or space) group of the lattice, e.g. the $\mathcal{C}_4$ group of the square lattice.

It is well known in the Landau theory of the Fermi liquid that the thermodynamic stability of the Fermi liquid state requires that the Landau parameters cannot be too negative. This argument, due to Pomeranchuk [160], implies that if in one channel the forward scattering interaction becomes sufficiently negative (attractive) to overcome the stabilizing effects of the Pauli pressure, the Fermi liquid becomes unstable to a distortion of the FS with the symmetry of the unstable channel.[20]

Oganesyan et al. [39] showed that in a 2D system of interacting fermions, the Pomeranchuk instability in fact marks a quantum phase transition to a *nematic Fermi fluid*. We will discuss this theory below in some detail. While the theory of Oganesyan and coworkers applied to a system in the continuum, Kee and coworkers [161, 162] considered a lattice model. Hints of nematic order in specific models had in fact been discovered independently (but not recognized as such originally), notably by the work of Metzner and coworkers [163–167].[21]

There is by now a growing literature on the nematic instability. Typically the models, both in the continuum [39] or on different lattices [161, 171, 172], are

---

[20] Although the Pomeranchuk argument is standard and reproduced in all the textbooks on Fermi liquid theory (see, e.g. Ref.[157]) the consequences of this instability were not pursued until quite recently.

[21] In fact, perturbative renormalization group calculations [168, 169] have found a runaway flow in the $d_{x^2-y^2}$ particle-hole channel, which is a nematic instability, but it was not recognized as such. See, however, Ref.[170].

solved within a Hartree–Fock type approximation (with all the limitations that such an approach has), or in special situations such as vicinity to Van Hove singularities [162, 163] and certain degenerate band crossings [173] (where the theory is better controlled), or using uncontrolled approximations to strong coupling systems such as slave fermion/boson methods [174, 175]. A strong coupling limit of the Emery model of the cuprates was shown to have a nematic state in Ref. [176] (we will review this work below). Finally some non-perturbative work on the nematic quantum phase transition has been done using higher dimensional bosonization in Refs. [177, 178] and by RG methods [179].

Extensions of these ideas have been applied to the problem of the nematic phase seen in the metamagnetic bilayer ruthenate $Sr_3Ru_2O_7$ relying either on the van Hove mechanism [180–183] or on an orbital ordering mechanism [184, 185], and in the new iron-based superconducting compounds [77, 78]. More recently nematic phases of different types have been argued to occur in dipolar Fermi gases of ultra-cold atoms [186, 187].[22]

2.6.2 The Nematic Fermi Fluid

Here I will follow the work of Oganesyan, Kivelson and Fradkin [39] and consider first the instability in the charge (symmetric) channel. Oganesyan et al. defined a charge nematic order parameter for a two-dimensional Fermi fluid as the 2×2 symmetric traceless tensor of the form

$$\hat{Q}(x) \equiv -\frac{1}{k_F^2} \Psi^\dagger(r) \begin{pmatrix} \partial_x^2 - \partial_y^2 & 2\partial_x \partial_y \\ 2\partial_x \partial_y & \partial_y^2 - \partial_x^2 \end{pmatrix} \Psi(r), \qquad (2.44)$$

It can also be represented by a complex valued field $Q_2(x)$ whose expectation value in the nematic phase is

$$\langle Q_2 \rangle \equiv \langle \Psi^\dagger (\partial_x + i\partial_y)^2 \Psi \rangle = |Q_2| e^{2i\theta_2} = Q_{11} + i Q_{12} \neq 0 \qquad (2.45)$$

Q_2 transforms under rotations in the representation of angular momentum $\ell = 2$. Oganesyan et al. showed that if $\langle Q_2 \rangle \neq 0$ then the Fermi surface *spontaneously* distorts and becomes an ellipse with eccentricity $\propto Q$. This state breaks rotational invariance mod π.

More complex forms of order can be considered by looking at particle-hole condensates with angular momenta $\ell > 2$ (see Ref. [14])

$$\langle Q_\ell \rangle = \langle \Psi^\dagger (\partial_x + i\partial_y)^\ell \Psi \rangle \qquad (2.46)$$

---

[22] Another class of nematic state can occur inside a $d_{x^2-y^2}$ superconductor. This quantum phase transition involves primarily the nodal quasiparticles of the superconductor and it is tractable within large N type approximations [188, 189].

For ℓ *odd*, this condensate breaks rotational invariance (mod $2\pi/\ell$). It also breaks parity $\mathcal{P}$ and time reversal $\mathcal{T}$ but $\mathcal{PT}$ is invariant. For example the condensate with $\ell = 3$ is effectively equivalent to the "Varma loop state" [15, 190]. The states with ℓ *even* are also interesting, e.g. a hexatic state is described by a particle-hole condensate with $\ell = 6$ [191].

In a 3D system, the anisotropic state is described by an order parameter $\mathcal{Q}_{ij}$ which is a traceless symmetric tensor (as in conventional liquid crystals [2, 3]). More generally, we can define an order parameter that transforms under the (ℓ, m) representation of the group of $SO(3)$ spatial rotations.

Oganesyan et al. considered in detail a Fermi liquid type model of the nematic transition and developed a (Landau) theory of the transition ("Landau on Landau"). The Hamiltonian of this model describes (spinless) fermions in the continuum with a two-body interaction corresponding to the $\ell = 2$ particle- hole angular momentum channel. The Hamiltonian is

$$H = \int d\boldsymbol{r}\, \Psi^\dagger(\boldsymbol{r})\epsilon(\boldsymbol{\nabla})\Psi(\boldsymbol{r}) + \frac{1}{4}\int d\boldsymbol{r}\int d\boldsymbol{r}'\, F_2(\boldsymbol{r} - \boldsymbol{r}')\mathrm{Tr}[\widehat{\mathcal{Q}}(\boldsymbol{r})\widehat{\mathcal{Q}}(\boldsymbol{r}')] \quad (2.47)$$

where the free-fermion dispersion (near the FS) is $\epsilon(\boldsymbol{k}) = v_F q[1 + a(\frac{q}{k_F})^2]$ (here $q \equiv |\boldsymbol{k}| - k_F$), and the interaction is given in terms of the coupling

$$F_2(\boldsymbol{r}) = (2\pi)^{-2}\int d\boldsymbol{q}\, e^{i\boldsymbol{q}\cdot\boldsymbol{r}}\frac{F_2}{1 + \kappa F_2 q^2} \quad (2.48)$$

where F_2 is the $\ell = 2$ Landau parameter, and κ measures the range of these interactions. Notice that we have kept a cubic momentum dependence in the dispersion, which is strongly irrelevant in the Landau Fermi liquid phase (but it is needed to insure stability in the broken symmetry state).

The Landau energy density functional for this model has the form (which can be derived by Hartree–Fock methods or, equivalently, using a Hubbard–Stratonovich decoupling)

$$\mathcal{U}[\mathcal{Q}] = \mathcal{E}(\mathcal{Q}) - \frac{\widetilde{\kappa}}{4}\mathrm{Tr}[\mathcal{Q}D\mathcal{Q}] - \frac{\widetilde{\kappa}'}{4}\mathrm{Tr}[\mathcal{Q}^2 D\mathcal{Q}] + \ldots \quad (2.49)$$

Here we have use the 2×2 symmetric tensor $D_{ij} \equiv \partial_i\partial_j$, and $\widetilde{\kappa}$ and $\widetilde{\kappa}'$ are the two effective Franck constants (see Ref. [3, 39]). The uniform part of the energy functional, $\mathcal{E}(\mathcal{Q})$, is given by

$$\mathcal{E}(\mathcal{Q}) = \mathcal{E}(\boldsymbol{0}) + \frac{A}{4}\mathrm{Tr}[\mathcal{Q}^2] + \frac{B}{8}\mathrm{Tr}[\mathcal{Q}^4] + \ldots \quad (2.50)$$

where

$$A = \frac{1}{2N_F} + F_2 \quad (2.51)$$

N_F is the density of states at the Fermi surface, and the coefficient of the quartic term is

$$B = \frac{3aN_F|F_2|^3}{8E_F^2} \tag{2.52}$$

$E_F \equiv v_F k_F$ is the Fermi energy [39]. The (normal) Landau Fermi liquid phase is stable provided $A > 0$, or, what is the same, if $2N_F F_2 > -1$ which is the Pomeranchuk condition (in this notation). On the other hand, thermodynamic stability also requires that $B > 0$, which implies that the coefficient of the cubic correction in the dispersion be positive, $a > 0$. If this condition is not satisfied, as it is the case in simple lattice models [161], higher order terms must be kept to insure stability.

However, in this case the transition is typically first order.

This model has two phases:

- an isotropic Fermi liquid phase, $A > 0$
- a nematic (non-Fermi liquid) phase, $A < 0$

separated by a quantum critical point at the Pomeranchuk value, $2N_F F_2 = -1$.

Let us discuss the quantum critical behavior. We will parametrize the distance to the Pomeranchuk QCP by

$$\delta = \frac{1}{2N_F} + F_2 \tag{2.53}$$

and define $s = \omega/qv_F$. The transverse collective nematic modes have Landau damping at the QCP [39]. Their effective action has the form

$$S_\perp = \int d\omega dq \left(\kappa q^2 + \delta - i\frac{|\omega|}{qv_F}\right) |Q_\perp(\omega, q)|^2 \tag{2.54}$$

which implies that the dynamic critical exponent is $z=3$.[23]

According to the standard perturbative criterion of Hertz [193] and Millis [194], the quantum critical behavior is that of an equivalent ϕ^4 type field theory in dimensions $D = d + z$ which in this case is $D=5$. Since the upper critical (total) dimension is 4, the Hertz-Millis analysis would predict that mean field theory is asymptotically exact and that the quartic (and higher) powers of the order parameter field are *irrelevant* at the quantum critical point (for an extensive discussion see Ref. [195]). However we will see below that while this analysis is correct for the bosonic sector of the theory, i.e. the behavior of the bosonic collective modes such as the order parameter itself, the situation is far less clear in the fermionic sector. We will come back to this question below.

Let us discuss now the physics of the nematic phase. In the nematic phase the FS is spontaneously distorted along the direction of the (director) order parameter

---

[23] There are other collective modes at higher energies. In particular there is an underdamped longitudinal collective mode with $z=2$ [39]. These higher energy modes contribute to various crossover effects [192], but decouple in the asymptotic quantum critical regime.

Fig. 2.21 Spontaneous
distortion of the Fermi
surface in the nematic phase
of a 2D Fermi fluid

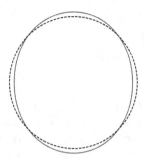

(see Fig. 2.21) and exhibits a quadrupolar (d-wave) pattern, i.e. the Fermi wave vector
has an angular dependence $k_F(\theta) \propto \cos 2\theta$ (in 2D). Indeed, in the nematic phase the
Hartree–Fock wave function is a Slater determinant whose variational parameters
determine the shape of the FS.

In principle, a wave function with a similar structure can be used to suggest
(as it was done in Ref. [39]) that it should also apply to the theory of the electronic
nematic state observed in the 2DEG in large magnetic fields. In that framework one
thinks of the 2DEG in a half-filled Landau level as an equivalent system of "composite
fermions" [196], fermions coupled to a Chern-Simons gauge field [197, 198]. It has
been argued [199] that this state can be well described by a Slater determinant wave
function, projected onto the Landau level. The same procedure can be applied to
the nematic wave function, and some work has been done along this direction [200].
A problem that needs to be solved first is the determination of the Landau parameters
of the composite fermions of which very little (that makes sense) is known.

A simple (Drude) calculation then shows that the transport is anisotropic. The
resistivity in the nematic phase, due to scattering from structureless isotropic impurities, yields the result that the resistivity tensor is anisotropic with an anisotropy
controlled by the strength of the nematic order parameter:

$$\frac{\rho_{xx} - \rho_{yy}}{\rho_{xx} + \rho_{yy}} = \frac{1}{2} \frac{m_y - m_x}{m_y + m_x} = \frac{\text{Re } Q}{E_F} + O(Q^3) \tag{2.55}$$

where m_x and m_y are the (anisotropic) effective masses of the quasiparticles in the
nematic state. In general it is a more complex odd function of the order parameter.

In the nematic phase the transverse Goldstone boson is generically *overdamped*
(Landau damping) except for a finite set of symmetry directions, $\phi = 0$, $\pm\pi/4$,
$\pm\pi/2$, where it is *underdamped*. Thus, $z=3$ scaling also applies to the nematic
phase for general directions. Naturally, in a lattice system the rotational symmetry
is not continuous and the transverse Goldstone modes are gapped. However, the
continuum prediction still applies if the lattice symmetry breaking is weak and if
either the energy or the temperature is larger that the lattice anisotropy scale.

On the other hand, the behavior of the fermionic correlators is much more strongly
affected. To one loop order, the quasiparticle scattering rate, $\Sigma''(\omega, \boldsymbol{p})$ is found to
have the behavior

$$\Sigma''(\omega, k) = \frac{\pi}{\sqrt{3}} \frac{(\kappa k_F^2)^{1/3}}{\kappa N_F} \left| \frac{k_x k_y}{k_F^2} \right|^{4/3} \left| \frac{\omega}{2 v_F k_F} \right|^{2/3} + \dots \qquad (2.56)$$

for k along a general direction. On the other hand, along a symmetry direction

$$\Sigma''(\omega) = \frac{\pi}{3 N_F \kappa} \frac{1}{(\kappa k_F^2)^{1/4}} \left| \frac{\omega}{v_F k_F} \right|^{3/2} + \dots \qquad (2.57)$$

Hence, the entire nematic phase is a non-Fermi liquid (again, with the caveat on lattice symmetry breaking effects).

At the Pomeranchuk quantum critical point the quasiparticle scattering rate obeys the same (one loop) scaling shown in Eq. 2.56, $\Sigma''(\omega) \propto |\omega|^{2/3}$, both in continuum [39] and lattice models [167], but it is isotropic. In the quantum critical regime the electrical resistivity obeys a $T^{4/3}$ law [201]. Also, both in the nematic phase (without lattice anisotropy) and in the quantum critical regime, the strong nematic fluctuations yield an electronic contribution to the specific heat that scales as $T^{2/3}$ (consistent with the general scaling form $T^{d/z}$ [195]) which dominates over the standard Fermi liquid linear T dependence at low temperatures [157].

Since $\Sigma''(\omega) \gg \Sigma'(\omega)$ (as $\omega \to 0$), we need to asses the validity of these results as they signal a *failure of perturbation theory*. To this end we have used higher dimensional bosonization as a non-perturbative tool [202–206]. Higher dimensional bosonization reproduces the collective modes found in Hartree–Fock+ RPA and is consistent with the Hertz-Millis analysis of quantum criticality: $d_{\text{eff}} = d + z = 5$ [177, 178]. Within the higher bosonization approach, the fermion propagator takes the form

$$G_F(x, t) = G_0(x, t) Z(x, t) \qquad (2.58)$$

where $G_0(x, t)$ is the free fermion propagator. In the Fermi liquid phase the quantity $Z(x,t)$ approaches a finite constant at long distances and at long times leading to a reduction of the quasiparticle residue Z (see, e.g. [178] and references therein). However, at the Nematic-FL QCP, $Z(x, t)$ becomes singular, and the full quasiparticle propagator now has the form

$$G_F(x, 0) = G_0(x, 0) \, e^{-\text{const.} |x|^{1/3}} \qquad (2.59)$$

at equal times, and

$$G_F(0, t) = G_0(0, t) \, e^{-\text{const.} |t|^{-2/3} \ln t} \qquad (2.60)$$

at equal positions. Notice that these expressions are consistent with the expected $z = 3$ scaling even though the time and space dependence is not a power law. The quasiparticle residue is then seen to vanish at the QCP:

$$Z = \lim_{x \to \infty} Z(x, 0) = 0 \qquad (2.61)$$

However, the single particle density of states, $N(\omega) = -\frac{1}{\pi} \mathrm{Im} G(\omega, 0)$, turns out to have a milder behavior:

$$N(\omega) = N(0) \left(1 - \mathrm{const}'.|\omega|^{2/3} \ln \omega\right) \tag{2.62}$$

Let us now turn to the behavior near the QCP. For $T=0$ and $\delta \ll 1$ (on the Fermi Liquid side) the quasiparticle residue is now finite (see Fig. 2.22)

$$Z \propto e^{-\mathrm{const.}/\sqrt{\delta}} \tag{2.63}$$

but its dependence on the distance to the nematic QCP is an essential singularity. On the other hand, right at the QCP ($\delta = 0$), and for a temperature range $T_F \gg T \gg T_\kappa$, the equal-time fermion propagator is found to vanish exactly

$$Z(x, 0) \propto e^{-\mathrm{const.}Tx^2 \ln(L/x)} \to 0 \qquad \text{as } L \to \infty \tag{2.64}$$

but, the equal-position propagator $Z(0, t)$ remains *finite* in the thermodynamic limit, $L \to \infty$! This behavior has been dubbed "Local quantum criticality".[24] On the other hand, irrelevant quartic interactions of strength u lead to a renormalization of δ that smears the QCP at $T > 0$ [194]

$$\delta \to \delta(T) = -uT \ln\left(uT^{1/3}\right) \tag{2.65}$$

leading to a milder behavior at equal-times

$$Z(x, 0) \propto e^{-\mathrm{const.}Tx^2 \ln(\xi/x)} \qquad \text{where } \xi = \delta(T)^{-1/2} \tag{2.66}$$

These results are far from being universally accepted. Indeed Chubukov and coworkers [208–210] have argued that the perturbative non-Fermi liquid behavior, $\Sigma''(\omega) \sim \omega^{2/3}$, which is also found at a ferromagnetic metallic QCP, persists to all orders in perturbation theory and can recover the results of higher dimensional bosonization only by taking into account the most infrared divergent diagrams. The same non-Fermi liquid one-loop perturbative scaling has been found in other QCPs such as in the problem of fermions (relativistic or not) at finite density coupled to dynamical gauge fields. This problem has been discussed in various settings ranging from hot and dense QED and QCD [211–213], to the gauge-spinon system in RVB approaches to high T_c superconductors [214– 219] to the compressible states of the 2DEG in large magnetic fields [198]. In all cases these authors have also argued that the one-loop scaling persists to all orders. In a recent paper Metlitski and Sachdev [179] found a different scaling behavior.

We end with a brief discussion on the results in lattice models of the nematic quantum phase transition. This is important since, with the possible exception of the 2DEG in large magnetic fields and in ultra-cold atomic systems, all strongly

---

[24] A similar behavior was found in the quantum Lifshitz model at its QCP [207].

Fig. 2.22 The discontinuity of the quasiparticle momentum distribution function in a Fermi liquid. Z is the quasiparticle residue at the Fermi surface

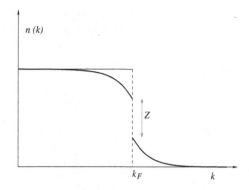

correlated systems of interest have very strong lattice effects. The main difference between the results in lattice models and in the continuum is that in the former the quantum phase transition (at the mean field, Hartree–Fock, level) has a strong tendency to be first order. Although fluctuations can soften the quantum transition and turn the system quantum critical (as emphasized in Ref. [220]), nevertheless there are good reasons for the transition to be first order more or less generically. One is that if the stabilizing quartic terms are negative (e.g. say due to the band structure), this also results, in the case of a lattice system, in a Lifshitz transition at which the topology of the FS changes from closed to open. This cannot happen in a continuous way.

2.7 Generalizations: Unconventional Magnetism and Time Reversal Symmetry Breaking

We will now consider briefly the extension of these ideas to the spin-triplet channel [16, 17]. In addition to particle-hole condensates in the singlet (charge) channel we will be interested in particle-hole condensates in the spin (triplet) channel. In 2D the order parameters for particle-hole condensates in the spin triplet channel are (here $\alpha, \beta = \uparrow, \downarrow$)

$$Q_\ell^a(r) = \langle \Psi_\alpha^\dagger(r) \sigma_{\alpha\beta}^a \left(\partial_x + i \partial_y \right)^\ell \Psi_\beta(r) \rangle \equiv n_1^a + i n_2^a \qquad (2.67)$$

These order parameters transform under both $SO(2)$ spatial rotations and under the internal $SU(2)$ symmetry of spin. If $\ell \neq 0$ the state has a broken rotational invariance in space and in spin space. These states are a particle-hole condensate analog of the unconventional superconductors and superfluids, such as He_3A and He_3B. Indeed one may call these states "unconventional magnetism" as the $\ell = 0$ (isotropic) state is just a ferromagnet. In 2D these states are then given in terms of two order parameters, each in the vector (adjoint) representation of the $SU(2)$ spin symmetry.[25] We will discuss only the 2D case. The order parameters obey the following transformation laws:

1. Time reversal:

$$\mathcal{T} \mathcal{Q}_\ell^a \mathcal{T}^{-1} = (-1)^{\ell+1} \mathcal{Q}_\ell^a \tag{2.68}$$

2. Parity:

$$\mathcal{P} \mathcal{Q}_\ell^a \mathcal{P}^{-1} = (-1)^\ell \mathcal{Q}_\ell^a \tag{2.69}$$

3. $\mathcal{Q}_\ell^a$ rotates under an $SO_{\mathrm{spin}}(3)$ transformation, and transforms as $\mathcal{Q}_\ell^a \to \mathcal{Q}_\ell^a e^{i\ell\theta}$ under a rotation in space by an angle θ.
4. $\mathcal{Q}_\ell^a$ is invariant under a rotation by π/ℓ followed by a spin flip.

Wu and collaborators [16, 17] have shown that these phases can also be accessed by a Pomeranchuk instability in the spin (triplet) channel.[26] They showed that the Landau-Ginzburg free energy takes the simple form

$$F[n] = r(|\boldsymbol{n}_1|^2 + |\boldsymbol{n}_2|^2) + v_1(|\boldsymbol{n}_1|^2 + |\boldsymbol{n}_2|^2)^2 + v_2|\boldsymbol{n}_1 \times \boldsymbol{n}_2|^2 \tag{2.70}$$

The Pomeranchuk instability occurs at $r=0$, i.e., for $N_F F_\ell^A = -2$ (with $\ell \geq 1$), where F_ℓ^A are the Landau parameters in the spin- triplet channel. Notice that this free energy is invariant only by global $SO(3)$ rotations involving both vector order parameters, $\boldsymbol{n}_1$ and $\boldsymbol{n}_2$. Although at this level the $SO(3)$ invariance is seemingly an internal symmetry, there are gradient terms that lock the internal $SO(3)$ spin rotations to the "orbital" spatial rotations (see Ref. [17]). A similar situation also occurs in classical liquid crystals [3].

At the level of the Landau–Ginzburg theory the system has two phases with broken $SO(3)$ invariance:

1. If $v_2 > 0$, then the two $SO(3)$ spin vector order parameters must be parallel to each other, $\boldsymbol{n}_1 \times \boldsymbol{n}_2 = 0$. They dubbed this the "$\alpha$" phase. In the α phases the up and down Fermi surfaces are distorted (with a pattern determined by ℓ) but are rotated from each other by π/ℓ. One case of special interest is the α phase with $\ell = 2$. This is the "nematic-spin-nematic" discussed briefly in Ref. [4].[27] In this phase the spin up and spin down FS have an $\ell = 2$ quadrupolar (nematic) distortion but are rotated by $\pi/2$ (see Fig. 2.23).
2. Conversely, if $v_2 < 0$, then the two $SO(3)$ spin vector order parameters must be orthogonal to each other, $\boldsymbol{n}_1 \cdot \boldsymbol{n}_2 = 0$ and $|\boldsymbol{n}_1| = |\boldsymbol{n}_2|$. Wu et al. dubbed these

[25] In 3D the situation is more complex and the possible are more subtle. In particular, in 3D there are three vector order parameters involved [17].

[26] The $\ell = 0$ case is, of course, just the conventional Stoner ferromagnetic instability.

[27] The term "nematic-spin-nematic" is a poor terminology. A spin nematic is a state with a magnetic order parameter that is a traceless symmetric tensor, which this state does not.

Fig. 2.23 The α-phases in the $\ell = 1$ and $\ell = 2$ spin triplet channels. The Fermi surfaces exhibit the p and d-wave distortions, respectively

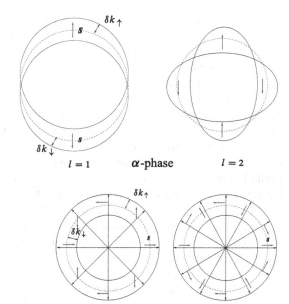

$l = 1$ α-phase $l = 2$

Fig. 2.24 The β-phases in the $\ell = 2$ triplet channel. Spin configurations exhibit the vortex structure with winding number $w = \pm 2$. These two configurations can be transformed to each other by performing a rotation around the x-axis with the angle of π

$w = 2$ $w = -2$

the "β" phases. In the β phases there are two isotropic FS but spin is not a good quantum number. In fact, the electron self energy in the β-phases acquires a spin-orbit type form with a strength given by the magnitude of the order parameter.

The mean-field electronic structure thus resembles that of a system with a strong and tunable spin-orbit coupling (i.e. not of $O((v_F/c)^2)$ as it is normally the case). We can define now a d vector:

$$d(k) = (\cos(\ell\theta_k), \sin(\ell\theta_k), 0) \qquad (2.71)$$

In the β phases the d vector winds about the undistorted FS. For the special case of $\ell = 1$, the windings are $w = 1$ (corresponding to a "Rashba" type state) and $w = -1$ (corresponding to a "Dresselhaus" type state). For the d-wave case, the winding numbers are $w = \pm 2$ (see Fig. 2.24).

These phases have a rich phenomenology of collective modes and topological excitations which we will not elaborate on here. See Ref. [17] for a detailed discussion.[28]

Fermionic systems with dipolar magnetic interactions may be a good candidate for phases similar to the ones we just described (See Refs. [186, 187], and it is quite possible that these systems may be realized in ultra-cold atomic gases. In that context the anisotropic form of the dipolar interaction provides for a mechanism to access some of this physics. Indeed, in the case of a fully polarized (3D) dipolar Fermi gas, the FS will have an uniaxial distortion. If the polarization is spontaneous (in the

---

[28]   The p-wave ($\ell = 1$) β phase has the same physics as the 'spin-split' metal of Ref. [221]. A similar state was proposed in Ref. [222] as an explanation of the "hidden order" phase of URu$_2$Si$_2$.

absence of a polarizing external field) this phase is actually a ferro-nematic state, a state with coexisting ferromagnetism and nematic order. If the system is partially polarized then the phase is a mix of nematic order and ferromagnetism coexisting with a phase with a non-trivial "spin texture" in momentum space.

It turns out that generalizations of the Pomeranchuk picture of the nematic state to multi-band electronic systems can describe metallic states with a spontaneous breaking of time reversal invariance. This was done in Ref. [14] where it was shown that in a two-band system (i.e. a system with two Fermi surfaces) it is possible to have a metallic state which breaks time reversal invariance and exhibits a spontaneous anomalous Hall effect. While the treatment of this problem has a superficial formal similarity with the triplet (spin) case, i.e. regarding the band index as a "pseudo- spin" (or flavor), the physics differs considerably. At the free fermion level the fermion number on each band is separately conserved, leading to a formal $SU(2)$ symmetry. However, the interacting system has either a smaller $U(1) \times U(1)$ invariance or, more generally, $\mathbb{Z}_2 \times \mathbb{Z}_2$ invariance, as the more general interactions preserve only the *parity* of the band fermion number [14]. At any rate it turns out that analogs of the "α" and "β" phases of the triplet channel exist in multi-band systems. The "α" phases break time reversal and parity (but not the product). An example of such metallic (gapless) states is the "Varma loop state" [15, 223]. The "β" states break time reversal invariance (and chirality). In the "β" phases there is a spontaneous anomalous Hall effect, i.e. a zero field Hall effect with a Hall conductivity that is not quantized as this state is a metal,[29] whereas the "α" phases there is not.

2.8 Nematic Order in the Strong Correlation Regime

We will now discuss how a nematic state arises as the exact ground state in the strong coupling limit of the Emery model [176]. The Emery model is a simplified microscopic model of the important electronic degrees of freedom of the copper oxides [123]. In this model, the CuO plane is described as a square lattice with the Cu atoms residing on the sites and the O atoms on the links (the medial lattice of the square lattice). On each site of the square lattice there is a single $d_{x^2-y^2}$ Cu orbital, and a p_x (p_y) O orbital on each site of the medial along the x (y) direction. We will denote by $d_\sigma^\dagger(r)$ the operator that creates a hole on the Cu site r and by $p_{x,\sigma}^\dagger(r + \frac{e_x}{2})$ and $p_{y,\sigma}^\dagger(r + \frac{e_y}{2})$ the operators the create a hole on the O sites $r + \frac{e_x}{2}$ and $r + \frac{e_y}{2}$ respectively.

The Hamiltonian of the Emery model is the sum of kinetic energy and interaction terms. The kinetic energy terms consist of the hopping of a hole from a Cu site to its nearest O sites (with amplitude t_{pd}), an on-site energy $\varepsilon > 0$ on the O sites (accounting for the difference in "affinity" between Cu and O), and a (small) direct hopping between nearest-neighboring O sites, t_{pp}. The interaction terms are just the on-site Hubbard repulsion U_d (on the Cu sites) and U_p (on the O sites) as well as

---

[29] This is consistent with the general arguments of Ref. [224].

Fig. 2.25 The Emery model
of the CuO lattice

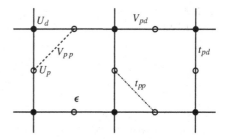

nearest neighbor ("Coulomb") repulsive interactions of strength V_{pd} (between Cu
and O) and V_{pp} (between two nearest O) (see Fig. 2.25). It is commonly believed that
this model is equivalent to its simpler cousin, the one band Hubbard model. However,
while this equivalency is approximately correct in the weak coupling limit, it is known
to fail already at moderate couplings. We will see that in the strong coupling limit,
no such reduction (to a "Zhang-Rice singlet") is possible.

Let us look at the energetics of the 2D Cu O model in the strong coupling limit. By
strong coupling we will mean the regime in which the following inequalities hold:

$$\frac{t_{pd}}{U_p}, \frac{t_{pd}}{U_d}, \frac{t_{pd}}{V_{pd}}, \frac{t_{pd}}{V_{pp}} \to 0, \qquad U_d > U_p \gg V_{pd} > V_{pp}, \qquad \text{and} \qquad \frac{t_{pp}}{t_{pd}} \to 0 \tag{2.72}$$

as a function of hole doping $x > 0$, where x is the number of doped holes per Cu.
In this regime, neither Cu nor O sites can be doubly occupied. At half filling, $x=0$, the
holes occupy all the Cu sites and all O sites are empty. At half-filling and in this strong
coupling regime the Emery model (much as the Hubbard model) is equivalent to a
quantum Heisenberg antiferromagnet with a small exchange coupling $J_H \approx \frac{8t_{pd}^4}{U_p V_{pd}^2}$.
This is the double-exchange mechanism (It turns out that in this model the four-spin
ring exchange interactions can be of the same order of magnitude as the Heisenberg
exchange J_H [176]).

Let us consider now the very low doping regime, $x \to 0$. Any additional hole
will have to be on an O site. The energy to add one hole (i.e. the chemical potential
μ of a hole) is $\mu \equiv 2V_{pd} + \epsilon$. Similarly, the energy of two holes on nearby O sites is
$2\mu + V_{pp}$. It turns out that in this strong coupling regime, with $t_{pp} = 0$, the dynamic
of the doped holes is strongly constrained and effectively becomes one-dimensional.
The simplest allowed move for a hole on an O site, which takes two steps, is shown
in Fig. 2.26. The final and initial states are degenerate, and their energy is $E_0 + \mu$,
where E_0 is the ground state energy of the undoped system. If this was the only
allowed process, the system would behave as a collection of 1D fermionic systems.

To assess if this is correct let us examine other processes to the same (lowest)
order in perturbation theory (in powers of the kinetic energy). One possibility is a
process in which in the final state the hole "turned the corner" (went from being
on an x oxygen to a near y oxygen). However for that to happen it will have to go
through an intermediate state such as the one shown in Fig. 2.27a. This intermediate

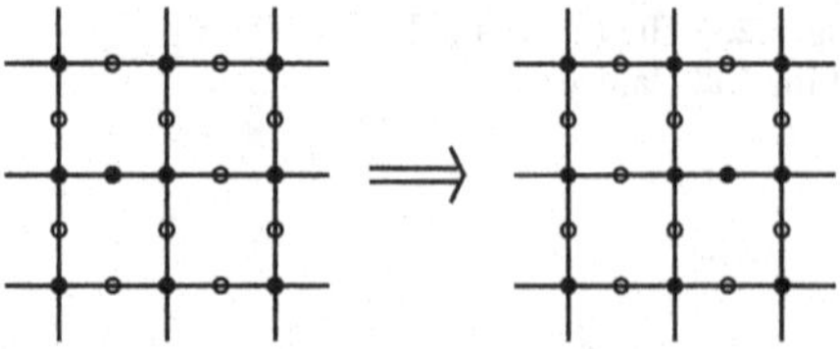

Fig. 2.26 An allowed
two-step move

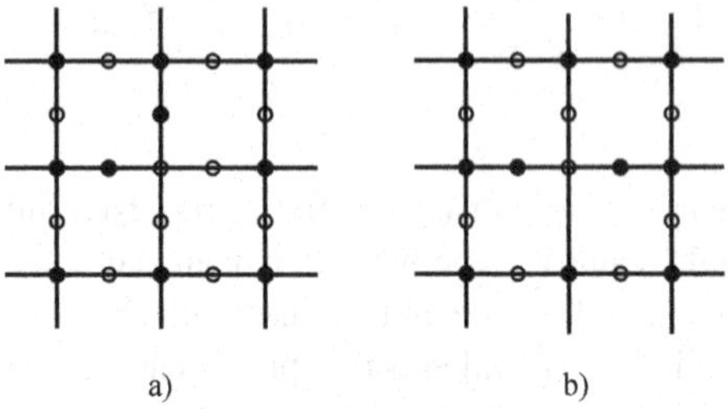

Fig. 2.27 Intermediate states
for processes in which a hole
a turns a corner and **b**
continues on the same row

a)   b)

state has an energy $E_0 + \mu + V_{pp}$. Hence, the effective hopping matrix element to turn the corner is $t_{\text{eff}} = \frac{t_{pd}^2}{V_{pp}} \ll t_{pd}$, which is strongly suppressed by the Coulomb effects of V_{pp}. In contrast, the intermediate state for the hole to continue on the same row (see Fig. 2.27) is $E_0 + \mu + \epsilon$. Thus the effective hopping amplitude instead becomes $t_{\text{eff}} = \frac{t_{pd}^2}{\epsilon}$, which is not suppressed by Coulomb effects of V_{pp}. All sorts of other processes have large energy denominators and are similarly suppressed

(for a detailed analysis see Ref. [176].) The upshot of this analysis is that in the strong coupling limit the doped holes behave like a set of one-dimensional fermions (one per row and column). The argument for one-dimensional effective dynamics can in fact be made more rigorous. In Ref. [176] it is shown that to leading order in the strong coupling expansion the system is a generalized t-J type model (with J effectively set to zero). In this limit there are an infinite number of conserved charges, exactly one per row and one per column.

The next step is to inquire if, at fixed but very small doping $x \to 0$, the rows and columns are equally populated or not. Consider then two cases; a) all rows and columns have the same fermion density, and b) the columns (or the rows) are empty. Case a) is isotropic while case b) is nematic. It turns out that due to the effects of the repulsive Coulomb interaction V_{pp}, the nematic configuration has lower energy at low enough doping. The argument is as follows. For the nematic state (in which all rows are equally populated but all columns are empty), the ground state energy has an expansion in powers of the doping x of the form:

$$E_{\text{nematic}} = E(x = 0) + \Delta_c \, x + W \, x^3 + O(x^5) \qquad (2.73)$$

where $\Delta_c = 2V_{pd} + \epsilon + \ldots$ and $W = \pi^2 \hbar^2 / 6m^*$. The energy of the isotropic state (at the same doping x) is

$$E_{\text{isotropic}} = E(x = 0) + \Delta_c \, x + (1/4)W \, x^3 + V_{\text{eff}} \, x^2 + \dots \qquad (2.74)$$

where $V_{\text{eff}} \propto V_{pp}$ is an effective coupling for holes on intersecting rows and columns. Clearly, for x small enough $E_{\text{isotropic}} > E_{\text{nematic}}$. Therefore, at low enough doping the ground state of the Emery model in the strong coupling limit is a nematic, in the sense that it breaks spontaneously the point symmetry group of the square lattice, C_4.[30] What happens for larger values of x is presently not known. Presumably a complex set of stripe phases exist. How this analysis is modified by the spin degrees of freedom is an open important problem which may illuminate our understanding of the physics of high temperature superconductors.

The nematic state we have found to be the exact ground state is actually maximally nematic: the nematic order parameter is 1. To obtain this result we relied on the fact that we have set $t_{pp} = 0$ as this is by far the smallest energy scale. The state that we have found is reminiscent of the nematic state found in 2D mean field theories of the Pomeranchuk transition [161] in which it is found that the nematic has an open Fermi surface, as we have also found. Presumably, for the more physical case of $t_{pp} \neq 0$, the strong 1D-like nematic state we found will show a crossover to a 2D (Ising) Nematic Fermi liquid state.

2.9 The Quantum Nematic-Smectic QCP and the Melting of the Stripe Phase

We will now turn to the problem of the quantum phase transition between electron stripe and nematic phases. For simplicity we will consider only the simpler case of the charge stripe and the charge nematic, and we will not discuss here the relation with antiferromagnetic stripes and superconductivity. Even this simpler problem is not well understood at present.

In classical liquid crystals there are two well established ways to describe this transition, known as the smectic A-nematic transition. One approach is the McMillan-de Gennes theory, and it is a generalization of the Landau–Ginzburg theory of phase transitions to this problem (see Ref. [2].) The other approach regards this phase transition as a melting of the smectic by proliferation of its topological excitations, the dislocations of the smectic order [3, 225, 226].

There are however important (and profound) differences between the problem of the quantum melting of a stripe phase into a quantum nematic and its classical smectic/nematic counterpart. The classical problem refers to three-dimensional liquid crystals whereas here we are interested in a two- dimensional quantum system. One may think that the time coordinate, i.e. the existence of a time evolution dictated by quantum mechanics, provides for the third dimension and that two

---

[30] It also turns out that in the (so far physically unrealizable) case of $x=1$, the ground state is a nematic insulator as each row is now full. However, for $x \to 1$ the ground state is again a nematic metal.

problems may indeed be closely related. Although to a large extent this is correct (this is apparent in the imaginary time form of the path integral representation) some important details are different.The problem we are interested in involves a metal and hence has dynamical fermionic degrees of freedom which do not have a counterpart in its classical cousin. One could develop a theory of quantum melting of the stripe phase by a defect proliferation mechanism by considering only the collective modes of the stripes. Theories of this type can be found in Refs. [227, 228] and in Ref. [156] (in the context of a system of cold atoms) which lead to several interesting predictions. Theories of this type would describe correctly an insulating stripe state in which the fermionic degrees of freedom are gapped and hence not part of the low energy physics. The problem of how to develop a non-perturbative theory of this transition with dynamical fermions is an open and unsolved problem.[31]

Another important difference is that in most cases of interest the quantum version of this transition takes place in a lattice system. thus, even if the stripe state may be incommensurate, and hence to a first approximation be allowed to "slide" over the lattice background, there is no continuous rotational invariance but only a point group symmetry leftover. Thus, at least at the lowest energies, the nematic Goldstone mode which plays a key role in the classical problem, is gapped and drops out of the critical behavior. However one should not be a purist about this issue as there may be significant crossovers that become observable at low frequencies and temperatures if the lattice effects are weak enough. Thus it is meaningful to consider a system in which the lattice symmetry breaking are ignored at the beginning and considered afterwards.

In Ref. [229] a theory of the quantum melting of a charge stripe phase is developed using an analogy with the McMillan-deGennes theory. The main (and important) difference is the role that the fermionic degrees of freedom play in the dynamics. Thus we will describe the stripe (which at this level is equivalent to a CDW) by its order parameter, the complex scalar field $\Phi(r, t)$, representing the Fourier component of the charge density operator near the ordering wavevector Q.[32]

We will assume that the phase transition occurs in a regime in which the nematic order is well developed and has a large amplitude $|\mathcal{N}|$. In this regime the fluctuations of the amplitude of the nematic order parameter N are part of the high energy physics, and can be integrated out. In contrast we will assume that the phase mode of the nematic order, the Goldstone mode, is either gapless (as in a continuum system) or of low enough energy that it matters to the physics (as if the lattice symmetry breaking is sufficiently weak). In this case we will only need to consider the nematic Goldstone (or 'pseudo-Goldstone') field which we will denote by $\varphi(r, t)$.

We should note that there is another way to think about this problem, in which one considers the competition between the CDW order parameters (two in this case), the nematic order and the normal Fermi liquid near a suitable multi-critical point. This problem was considered briefly in Ref. [229] and revisited in more detail in

---

[31]   Important work with a similar approach has been done on the problem of the quantum melting of the stripe state in quantum Hall systems [36, 37].

[32]   For a different perspective see Ref. [230].

Ref. [231]. The main conclusion is that for a (square) lattice system, the multicritical point is inaccessible as it is replaced by a direct (and weak) fluctuation-induced first-order transition from the FL to the CDW state. Thus, the theory that we discuss here applies far from this putative multicritical point in a regime in which, as we stated above, the nematic order is already well developed and large.

Following Ref. [229] we will think of this quantum phase transition in the spirit of a Hertz-Millis type approach and postulate the existence of an order parameter theory coupled to the fermionic degrees of freedom. The quantum mechanical action of the order parameter theory $S_{op}[\Phi, \varphi_N]$, has the McMillan-deGennes form

$$S_{op} = |N|^2 \int drdt \left((\partial_t \varphi_N)^2 - K_1 (\partial_x \varphi_N)^2 - K_2 (\partial_y \varphi_N)^2 \right)$$
$$+ \int drdt \left(|\partial_t \Phi|^2 - C_y |\partial_y \Phi|^2 - C_x \left| \left(\partial_x - i\frac{Q}{2} \varphi_N \right) \right|^2 \right.$$
$$\left. - \Delta_{CDW} |\Phi|^2 - u_{CDW} |\Phi|^4 \right) \tag{2.75}$$

where $|N|$ is the amplitude of the nematic order parameter, K_1 and K_2 are the two Franck constants (which were discussed before), C_x and C_y are the stiffnesses of the CDW order parameter along the x and y directions, Q is the modulus of the CDW ordering wavevector, Δ_{CDW} and u_{CDW} are parameters of the Landau theory that control the location of the CDW transition ($\Delta_{CDW} = 0$) and stability. Here we have assumed a stripe state, a unidirectional CDW, with its ordering wavevector along the x direction. We have also assumed $z = 1$ ("relativistic") quantum dynamics which would be natural for an insulating system.

The fermionic contribution has two parts. One part of the fermionic action, $S_{FL}[\psi]$, where ψ is the quasiparticle Fermi field (we are omitting the spin indices), describes a conventional Fermi liquid, i.e. the quasiparticle spectrum with a Fermi surface of characteristic Fermi wavevector k_F, and the quasiparticle interactions given in terms of Landau parameters. What will matter to us is the coupling between the fermionic quasiparticles and the nematic order parameter (the complex director field N), and the CDW order parameter Φ,

$$S_{int} = g_N \int drdt \left(\mathcal{Q}_2(r, t) N^\dagger(r, t) + \text{h.c.} \right)$$
$$+ g_{CDW} \int drdt \left(n_{CDW}(r, t) \Phi^\dagger(r, t) + \text{h.c.} \right) \tag{2.76}$$

where g_N and g_{CDW} are two coupling constants and, as before,

$$\mathcal{Q}_2(r, t) = \psi^\dagger(r, t)(\partial_x + i\partial_y)^2 \psi(r, t) \tag{2.77}$$

is the nematic order parameter (in terms of quasiparticle Fermi fields), and

$$n_{CDW}(q, \omega) = \int dk d\Omega \, \psi^\dagger(k + q + Q, \omega + \Omega) \psi(k, \Omega) \tag{2.78}$$

is the CDW order parameter (also in terms of the quasiparticle Fermi fields.)

This theory has two phases: a) a nematic phase, for $\Delta_{CDW} > 0$, where $\langle \Phi \rangle = 0$, and b) a CDW phase, for $\Delta_{CDW} < 0$, where $\langle \Phi \rangle \neq 0$, separated by a QCP at $\Delta_{CDW} = 0$. In the nematic ("normal") phase the only low energy degrees of freedom are the (overdamped) fluctuations of the nematic Goldstone mode φ_N, and nematic susceptibility in the absence of lattice effects (which render it gapped otherwise)

$$\chi_\perp^N(\boldsymbol{q}, \omega) = \frac{1}{g_N^2 N(0)} \frac{1}{\left(i\frac{\omega}{q} \sin^2(2\phi_q) - K_1 q_x^2 - K_2 q_y^2\right)} \tag{2.79}$$

where $\sin(2\phi_q) = 2q_x q_y/q^2$ and $N(0)$ is the quasiparticle density of states at the FS.

We will consider here the simpler case in which the CDW ordering wavevector obeys $Q < 2k_F$ (see the general discussion in Ref. [229].) In this case one can see that the main effect of the coupling to the quasiparticles (aside from some finite renormalizations of parameters) is to change the dynamics of the CDW order parameter due to the effects of Landau damping. The total effective action in this case becomes

$$S = \int \frac{dq d\omega}{(2\pi)^3} C_0 i |\omega| |\Phi(\boldsymbol{q}, \omega)|^2$$

$$- \int dr dt \left(C_y |\partial_y \Phi|^2 + C_x \left| \left(\partial_x - i\frac{Q}{2} \varphi_N \right) \Phi \right|^2 + \Delta_{CDW} |\Phi|^2 + u_{CDW} |\Phi|^4 \right)$$

$$+ \int \frac{dq d\omega}{(2\pi)^3} \left(\tilde{K}_0 \frac{i|\omega|}{q} \sin^2(2\phi_q) - \tilde{K}_1 q_x^2 - \tilde{K}_2 q_y^2 \right) |\varphi_N(\boldsymbol{q}, \omega)|^2 \tag{2.80}$$

where $C_0 \sim g_{CDW}^2$, $\tilde{K}_0 = g_N^2 |N|^2 N(0)$ and $K_{1,2} = g_N^2 |N|^2 N(0) K_{1,2}$.

By inspecting Eq. 2.80 one sees that at $\Delta_{CDW} = 0$, as before the nematic Goldstone fluctuations have $z=3$ (provided they remain gapless), and the CDW fluctuations have $z=2$. Thus the nematic Goldstone modes dominate the dynamics at the nematic-CDW QCP. Even if the nematic Goldstone modes were to become gapped (by the lattice anisotropy), the QCP now will have $z=2$ (due to Landau damping) instead of $z=1$ as in the "pure" order parameter theory. In both cases, the nematic Goldstone mode and the CDW order parameter fluctuations effectively decouple in the nematic phase. The result is that the nematic phase has relatively low energy CDW fluctuations with a dynamical susceptibility

$$\chi^{CDW}(\boldsymbol{q}, \omega) = -i\langle \Phi^\dagger(\boldsymbol{q}, \omega) \Phi(\boldsymbol{q}, \omega) \rangle_{\text{ret}} = \frac{1}{iC_0|\omega| - C_x q_x^2 - C_y q_y^2 - \Delta_{CDW}} \tag{2.81}$$

In other terms, as the QCP is approached, the nematic phase exhibits low energy CDW fluctuations that would show up in low energy inelastic scattering experiments much in the same way as the observed *fluctuating stripes* do in inelastic neutron scattering experiments in the high temperature superconductors [4]. As we saw before, a regime with "fluctuating" CDW (stripe) order is a nematic.

A simple scaling analysis of the effective action of Eq. 2.80 shows that, since $z > 1$ at this QCP, the coupling between the nematic Goldstone mode φ_N and the CDW order parameter Φ is actually *irrelevant*. In contrast, in the (classical) case it is a marginally relevant perturbation leading to a fluctuation induced first order transition [3, 232]. Thus, this "generalized McMillan-de Gennes" theory has a continuous (quantum) phase transition which, possibly, may become weakly first order at finite temperature.

This is not to say, however, that the stripe-nematic quantum phase transition is necessarily continuous. In Ref. [229] it is shown that the nature of the quantum phase transition depends on the relation between the ordering wave vector $\mathbf{Q}$ and the Fermi wave vector k_F. For $|\mathbf{Q}| < 2k_F$ the transition is continuous and has dynamical scaling $z = 2$. Instead, for $|\mathbf{Q}| = 2k_F$ it depends on whether $|\mathbf{Q}|$ is commensurate or incommensurate with the underlying lattice: for the incommensurate case the transition is (fluctuation induced) first order (consistent with the results of Ref. [233]) but it is continuous for the commensurate case with $z = 2$ and anisotropic scaling in space.

As in the case of the Pomeranchuk transition, the quasiparticles are effectively destroyed at the stripe-nematic QCP as well. Indeed, already to order one loop it is found [229] that the quasiparticle scattering rate scales with frequency as $\Sigma''(\omega) \propto \sqrt{|\omega|}$, signaling a breakdown of Fermi liquid theory. As in our discussion of the nematic-FL QCP, this behavior must be taken as an indication of a breakdown of perturbation theory and not as the putative ultimate quantum critical behavior, which remains to be understood.

In the quasiparticle picture we are using, the stripe state is similar to a CDW. Indeed, in the broken symmetry state the Fermi surface of the nematic is reconstructed leading to the formation of fermion pockets. As we noted above, we have not however assumed a rigid connection between the ordering wave vector and the Fermi surface and, in this sense, this is not a weak coupling state. Aside from that, in the presence of lattice pinning of the nematic Goldstone mode, the asymptotic low-energy properties of the stripe state are similar to those of a CDW (for details, see Ref. [229]).

2.10 Outlook

In these lectures we have covered a wide range of material on the theory of electronic liquid crystal phases and on the experimental evidence for them. As it is clear these lectures have a particular viewpoint, developed during the past decade in close collaboration with Steven Kivelson. I have tried, primarily at the level of citations as well an on numerous caveats, to make it clear that there are many important unsolved and still poorly understood questions that (at present) allow for more than one answer. It is a problem that requires the development of many points of view which eventually complement each other.

Several major problems remain open. One of them, in my view the most pressing one, is to establish the relation between the existence of these phases (stripes, nematics, etc.) and the mechanism(s) of high temperature superconductivity. In my opinion there is mounting experimental evidence, some of which I discussed here, that strongly suggests the existence of a close and probably unavoidable connection. A question that deserves more consideration is the particular connection between nematic order and superconductivity. Superficially these two issues would seem quite orthogonal to each other. Indeed, it is hard to see any connection within the context of a weak coupling theory. However if the nematic order arises from melting a stripe state which has a spin gap (such as the pair density wave state we discussed in these lectures) it is quite likely that a close connection may actually exist and be related. The current experimental evidence suggests such a relation.

Another key theoretical question that is wide open is to develop a more microscopic theory of the pair density wave state. In spite with the formal analogy with the Larkin-Ovchinnikov state, it seems very unlikely that a a "straight BCS approach" would be successful in this case. This state seems to have a strong coupling character.

As it must be apparent from the presentation of these lectures, the theory that has been done (and that is being done now) is for the most part quite phenomenological in character. There are very few "rigorous" results on the existence of these phases in strongly correlated systems. The notable exception are the arguments we presented for the existence of nematic order in the strong coupling limit of the Emery model. Clearly more work is needed.

Acknowledgments I am deeply indebted to Steve Kivelson with whom we developed many of the ideas that are presented here. Many of these results were obtained also in collaboration with my former students Michael Lawler and Kai Sun, as well as to John Tranquada, Vadim Oganesyan, Erez Berg, Daniel Barci, Congjun Wu, Benjamin Fregoso, Siddhartha Lal and Akbar Jaefari, and many other collaborators. I would like to thank Daniel Cabra, Andreas Honecker and Pierre Pujol for inviting me to this very stimulating Les Houches Summer School on "Modern theories of correlated electron systems" (Les Houches, May 2009). This work was supported in part by the National Science Foundation, under grant DMR 0758462 at the University of Illinois, and by the Office of Science, U.S. Department of Energy, under Contracts DE-FG02-91ER45439 and DE-FG02-07ER46453 through the Frederick Seitz Materials Research Laboratory of the University of Illinois.

References

1. Kivelson, S.A., Fradkin, E., Emery, V.J.: Electronic liquid-crystal phases of a doped Mott insulator. Nature **393**, 550 (1998)
2. de Gennes, P.G., Prost, J.: The Physics of Liquid Crystals. Oxford Science Publications/ Clarendon Press, Oxford, UK (1993)
3. Chaikin, P.M., Lubensky, T.C.: Principles of Condensed Matter Physics. Cambridge University Press, Cambridge, UK (1995)
4. Kivelson, S.A., Fradkin, E., Oganesyan, V., Bindloss, I., Tranquada, J., Kapitulnik, A., Howald, C.: How to detect fluctuating stripes in high tempertature superconductors. Rev. Mod. Phys. **75**, 1201 (2003)
5. Berg, E., Fradkin, E., Kim, E.-A., Kivelson, S., Oganesyan, V., Tranquada, J.M., Zhang, S.: Dynamical layer decoupling in a stripe-ordered high T_c superconductor. Phys. Rev. Lett. **99**, 127003 (2007)

6. Berg, E., Chen, C.-C., Kivelson, S.A.: Stability of nodal quasiparticles in superconductors with coexisting orders. Phys. Rev. Lett. **100**, 027003 (2008)
7. Fulde, P., Ferrell, R.A.: Superconductivity in a strong spin-exchange field. Phys. Rev. **135**, A550 (1964)
8. Larkin, A.I., Ovchinnikov, Y.N.: Nonuniform state of superconductors. Zh. Eksp. Teor. Fiz. **47**, 1136 (1964). (Sov. Phys. JETP. **20**, 762 (1965))
9. Fradkin, E., Kivelson, S.A., Manousakis, E., Nho, K.: Nematic phase of the two-dimensional electron gas in a magnetic field. Phys. Rev. Lett. **84**, 1982 (2000)
10. Cooper, K.B., Lilly, M.P., Eisenstein, J.P., Pfeiffer, L.N., West, K.W.: Onset of anisotropic transport of two-dimensional electrons in high Landau levels: possible isotropic-to-nematic liquid-crystal phase transition. Phys. Rev. B **65**, 241313 (2002)
11. Ando, Y., Segawa, K., Komiya, S., Lavrov, A.N.: Electrical resistivity anisotropy from self-organized one-dimensionality in high-temperature superconductors. Phys. Rev. Lett. **88**, 137005 (2002)
12. Borzi, R.A., Grigera, S.A., Farrell, J., Perry, R.S., Lister, S.J.S., Lee, S.L., Tennant, D.A., Maeno, Y., Mackenzie, A.P.: Formation of a nematic fluid at high fields in $Sr_3Ru_2O_7$. Science **315**, 214 (2007)
13. Hinkov, V., Haug, D., Fauqué, B., Bourges, P., Sidis, Y., Ivanov, A., Bernhard, C., Lin, C.T., Keimer, B.: Electronic liquid crystal state in superconducting $YBa_2Cu_3O_{6.45}$. Science **319**, 597 (2008)
14. Sun, K., Fradkin, E.: Time-reversal symmetry breaking and spontaneous anomalous Hall effect in Fermi fluids. Phys. Rev. B **78**, 245122 (2008)
15. Varma, C.M.: A theory of the pseudogap state of the cuprates. Philos. Mag. **85**, 1657 (2005)
16. Wu, C., Zhang, S.-C.: Dynamic generation of spin-orbit coupling. Phys. Rev. Lett. **93**, 036403 (2004)
17. Wu, C.J., Sun, K., Fradkin, E., Zhang, S.-C.: Fermi liquid instabilities in the spin channel. Phys. Rev. B **75**, 115103 (2007)
18. Anderson, P.W.: The resonating valence bond state of La_2CuO_4 and superconductivity. Science **235**, 1196 (1987)
19. Emery, V.J., Kivelson, S.A., Lin, H.Q.: Phase separation in the t-J model. Phys. Rev. Lett. **64**, 475 (1990)
20. Dagotto, E.: Correlated electrons in high-temperature superconductors. Rev. Mod. Phys. **66**, 763 (1994)
21. Emery, V.J., Kivelson, S.A.: Frustrated electronic phase separation and high-temperature superconductors. Physica C **209**, 597 (1993)
22. Seul, M., Andelman, D.: Domain shapes and patterns: the phenomenology of modulated phases. Science **267**, 476 (1995)
23. Lorenz, C.P., Ravenhall, D.G., Pethick, C.J.: Neutron star crusts. Phys. Rev. Lett. **70**, 379 (1993)
24. Fradkin, E., Kivelson, S.A., Lawler, M.J., Eisenstein, J.P., Mackenzie, A.P.: Nematic Fermi fluids in condensed matter physics. Annu. Rev. Condens. Matter Phys. **1**, 71 (2010)
25. Lilly, M.P., Cooper, K.B., Eisenstein, J.P., Pfeiffer, L.N., West, K.W.: Evidence for an anisotropic state of two-dimensional electrons in high Landau levels. Phys. Rev. Lett. **82**, 394 (1999)
26. Lilly, M.P., Cooper, K.B., Eisenstein, J.P., Pfeiffer, L.N., West, K.W.: Anisotropic states of two-dimensional electron systems in high Landau levels: effect of an in-plane magnetic field. Phys. Rev. Lett. **83**, 824 (1999)
27. Du, R.R., Tsui, D.C., Störmer, H.L., Pfeiffer, L.N., Baldwin, K.W., West, K.W.: Strongly anisotropic transport in higher two-dimensional Landau levels. Solid State Comm. **109**, 389 (1999)
28. Pan, W., Du, R.R., Störmer, H.L., Tsui, D.C., Pfeiffer, L.N., Baldwin, K.W., West, K.W.: Strongly anisotropic electronic transport at Landau level filling factor $\nu = 9/2$ and $\nu = 5/2$ under tilted magnetic field. Phys. Rev. Lett. **83**, 820 (1999)

29. Koulakov, A.A., Fogler, M.M., Shklovskii, B.I.: Charge density wave in two-dimensional electron liquid in weak magnetic field. Phys. Rev. Lett. **76**, 499 (1996)

30. Moessner, R., Chalker, J.T.: Exact results for interacting electrons in high Landau levels. Phys. Rev. B **54**, 5006 (1996)

31. Fradkin, E., Kivelson, S.A.: Liquid crystal phases of quantum Hall systems. Phys. Rev. B **59**, 8065 (1999)

32. MacDonald, A.H., Fisher, M.P.A.: Quantum theory of quantum Hall smectics. Phys. Rev. B **61**, 5724 (2000)

33. Barci, D.G., Fradkin, E., Kivelson, S.A., Oganesyan, V.: Theory of the quantum Hall smectic phase. I. Low-energy properties of the quantum Hall smectic fixed point. Phys. Rev. B **65**, 245319 (2002)

34. Cooper, K.B., Lilly, M.P., Eisenstein, J.P., Jungwirth, T., Pfeiffer, L.N., West, K.W.: An investigation of orientational symmetry-breaking mechanisms in high Landau levels. Sol. State Commun. **119**, 89 (2001)

35. Cooper, K.B., Eisenstein, J.P., Pfeiffer, L.N., West, K.W.: Observation of narrow-band noise accompanying the breakdown of insulating states in high Landau levels. Phys. Rev. Lett. **90**, 226803 (2003)

36. Wexler, C., Dorsey, A.T.: Disclination unbinding transition in quantum Hall liquid crystals. Phys. Rev. B **64**, 115312 (2001)

37. Radzihovsky, L., Dorsey, A.T.: Theory of quantum Hall nematics. Phys. Rev. Lett. **88**, 216802 (2002)

38. Doan, Q.M., Manousakis, E.: Quantum nematic as ground state of a two-dimensional electron gas in a magnetic field. Phys. Rev. B **75**, 195433 (2007)

39. Oganesyan, V., Kivelson, S.A., Fradkin, E.: Quantum theory of a nematic Fermi fluid. Phys. Rev. B **64**, 195109 (2001)

40. Grigera, S.A., Gegenwart, P., Borzi, R.A., Weickert, F., Schofield, A.J., Perry, R.S., Tayama, T., Sakakibara, T., Maeno, Y., Green, A.G. et al.: Disorder-sensitive phase formation linked to metamagnetic quantum criticality. Science **306**, 1154 (2004)

41. Fradkin, E., Kivelson, S.A., Oganesyan, V.: Discovery of a nematic electron fluid in a transition metal oxide. Science **315**, 196 (2007)

42. Grigera, S.A., Perry, R.S., Schofield, A.J., Chiao, M., Julian, S.R., Lonzarich, G.G., Ikeda, S.I., Maeno, Y., Millis, A.J., Mackenzie, A.P.: Magnetic field-tuned quantum criticality in the metallic ruthenate $Sr_3Ru_2O_7$. Science **294**, 329 (2001)

43. Millis, A.J., Schofield, A.J., Lonzarich, G.G., Grigera, S.A.: Metamagnetic quantum criticality. Phys. Rev. Lett. **88**, 217204 (2002)

44. Perry, R.S., Kitagawa, K., Grigera, S.A., Borzi, R.A., Mackenzie, A.P., Ishida, K., Maeno, Y.: Multiple first-order metamagnetic transitions and quantum oscillations in ultrapure $Sr_3Ru_2O_7$. Phys. Rev. Lett. **92**, 166602 (2004)

45. Green, A.G., Grigera, S.A., Borzi, R.A., Mackenzie, A.P., Perry, R.S., Simons, B.D.: Phase bifurcation and quantum fluctuations in $Sr_3Ru_2O_7$. Phys. Rev. Lett. **95**, 086402 (2005)

46. Jamei, R., Kivelson. Spivak, B.: Universal aspects of Coulomb-frustrated phase separation. Phys. Rev. Lett. **94**, 056805 (2005)

47. Lorenzana, J., Castellani, C., Di Castro, C.: Mesoscopic frustrated phase separation in electronic systems. Euro. Phys. Lett. **57**, 704 (2002)

48. Kivelson, S.A., Fradkin, E.: In: Schrieffer, J.R., Brooks, J. (eds.) Handbook of High Temperature Superconductivity, pp. 569–595. Springer-Verlag, New York (2007)

49. Chakravarty, S., Laughlin, R.B., Morr, D.K., Nayak, C.: Hidden order in the cuprates. Phys. Rev. B **63**, 094503 (2001)

50. Fujita, M., Goka, H., Yamada, K., Tranquada, J.M., Regnault, L.P.: Stripe order depinning and fluctuations in $La_{1.875}Ba_{0.125}CuO_4$ and $La_{1.875}Ba_{0.075}Sr_{0.050}CuO_4$. Phys. Rev. B **70**, 104517 (2004)

51. Abbamonte, P., Rusydi, A., Smadici, S., Gu, G.D., Sawatzky, G.A., Feng, D.L.: Spatially modulated 'Mottness' in $La_{2-x}Ba_xCuO_4$. Nature Phys. **1**, 155 (2005)

52. Tranquada, J.M.: In: Schrieffer, J.R., Brooks, J. (ed.) Treatise of High Temperature Supercon-ductivity, pp. 257–298. Springer-Verlag, New York (2007)

53. Tranquada, J.M., Sternlieb, B.J., Axe, J.D., Nakamura, Y., Uchida, S.: Evidence for stripe correlations of spins and holes in copper oxide superconductors. Nature **375**, 561 (1995)

54. Tranquada, J.M., Woo, H., Perring, T.G., Goka, H., Gu, G.D., Xu, G., Fujita, M., Yamada, K.: Quantum magnetic excitations from stripes in copper-oxide superconductors. Nature **429**, 534 (2004)

55. Haug, D., Hinkov, V., Suchaneck, A., Inosov, D.S., Christensen, N.B., Niedermayer, C., Bourges, P., Sidis, Y., Park, J.T., Ivanov, A. et al.: Magnetic-field-enhanced incommensurate magnetic order in the underdoped high-temperature superconductor $YBa_2Cu_3O_{6.45}$. Phys. Rev. Lett. **103**, 017001 (2009)

56. Hinkov, V., Bourges, P., Pailhés, S., Sidis, Y., Ivanov, A., Frost, C.D., Perring, T.G., Lin, C.T., Chen, D.P., Keimer, B.: Spin dynamics in the pseudogap state of a high-temperature superconductor. Nature Phys. **3**, 780 (2007)

57. Hinkov, V., Bourges, P., Pailhés, S., Sidis, Y., Ivanov, A., Lin, C., Chen, D., Keimer, B.: In-plane anisotropy of spin excitations in the normal and superconducting states of underdoped $YBa_2Cu_3O_{6+x}$. Nature Phys. **3**, 780 (2007)

58. Mook, H.A., Dai, P., Doğan, F., Hunt, R.D.: One-dimensional nature of the magnetic fluctuations in $YBa_2Cu_3O_{6.6}$. Nature **404**, 729 (2000)

59. Stock, C., Buyers, W.J.L., Liang, R., Peets, D., Tun, Z., Bonn, D., Hardy, W.N., Birgeneau, R.J.: Dynamic stripes and resonance in the superconducting and normal phases of $YBa_2Cu_3O_{6.5}$ ortho-II superconductor. Phys. Rev. B **69**, 014502 (2004)

60. Daou, R., Chang, J., LeBoeuf, D., Cyr-Choinière, O., Laliberté, F., Doiron-Leyraud, N., Ramshaw, B.J., Liang, R., Bonn, D.A., Hardy, W.N. et al.: Broken rotational symmetry in the pseudogap phase of a high-T_c superconductor. Nature **463**, 519 (2010)

61. Li, L., Wang, Y., Naughton, M.J., Komiya, S., Ono, S., Ando, Y., Ong, N.P.: Magnetization, nernst effect and vorticity in the cuprates. J. Magn. Magn. Mater. **310**, 460 (2007)

62. Cyr-Choinière, O., Daou, R., Laliberté, F., LeBoeuf, D., Doiron-Leyraud, N., Chang, J., Yan, J.-Q., Cheng, J.-G., Zhou, J.-S., Goodenough, J.B. et al.: Enhancement of the nernst effect by stripe order in a high-T_c superconductor. Nature **458**, 743 (2009)

63. Matsuda, M., Fujita, M., Wakimoto, S., Fernandez-Baca, J.A., Tranquada, J.M., Yamada, K.: Magnetic excitations of the diagonal incommensurate phase in lightly-doped $La_{2-x}Sr_xCuO_4$. Phys. Rev. Lett. **101**, 197001 (2008)

64. Lake, B., Rønnow, H.M., Christensen, N.B., Aeppli, G., Lefmann, K., McMorrow, D.F., Vorderwisch, P., Smeibidl, P., Mangkorntong, N., Sasagawa, T., Nohara, M., Takagi, H., Mason, T.E.: Antiferromagnetic order induced by an applied magnetic field in a high temperature superconductor. Nature **415**, 299 (2002)

65. Li, Q., Hücker, M., Gu, G.D., Tsvelik, A.M., Tranquada, J.M.: Two-dimensional supercon-ducting fluctuations in stripe-ordered $La_{1.875}Ba_{0.125}CuO_4$. Phys. Rev. Lett. **99**, 067001 (2007)

66. Valla, T., Fedorov, A.V., Lee, J., Davis, J.C., Gu, G.D.: The ground state of the pseudogap in cuprate superconductors. Science **314**, 1914 (2006)

67. He, R.-H., Tanaka, K., Mo, S.-K., Sasagawa, T., Fujita, M., Adachi, T., Mannella, N., Yamada, K., Koike, Y., Hussain, Z. et al.: Energy gaps in the failed high-T_c superconductor $La_{1.875}Ba_{0.125}CuO_4$. Nat. Phys. **5**, 119 (2008)

68. Berg, E., Fradkin, E., Kivelson, S.A.: Charge 4e superconductivity from pair density wave order in certain high temperature superconductors. Nature Phys. **5**, 830 (2009)

69. Schafgans, A.A., LaForge, A.D., Dordevic, S.V., Qazilbash, M.M., Komiya, S., Ando, Y., Basov, D.N.: Towards two-dimensional superconductivity in $La_{2-x}Sr_xCuO_4$ in a moderate magnetic field. Phys. Rev. Lett. **104**, 157002 (2010)

70. Kohsaka, Y., Taylor, C., Fujita, K., Schmidt, A., Lupien, C., Hanaguri, T., Azuma, M., Takano, M., Eisaki, H., Takagi, H. et al.: An intrinsic bond-centered electronic glass with unidirectional domains in underdoped cuprates. Science **315**, 1380 (2007)

71. Howald, C., Eisaki, H., Kaneko, N., Kapitulnik, A.: Coexistence of charged stripes and super-conductivity in $Bi_2Sr_2CaCu_2O_{8+\delta}$. Proc. Natl. Acad. Sci. U.S.A. **100**, 9705 (2003)

72. Hanaguri, T., Lupien, C., Kohsaka, Y., Lee, D.H., Azuma, M., Takano, M., Takagi, H., Davis, J.C.: A 'checkerboard' electronic crystal state in lightly hole-doped $Ca_{2-x}Na_xCuO_2Cl_2$. Nature **430**, 1001 (2004)

73. Vershinin, M., Misra, S., Ono, S., Abe, Y., Ando, Y., Yazdani, A.: Local ordering in the pseudogap state of the high-T_c superconductor $Bi_2Sr_2CaCu_2O_{8+\delta}$. Science **303**, 1005 (2004)

74. Lawler, M.J., Fujita, K., Lee, J.W., Schmidt, A.R., Kohsaka, Y., Kim, C.K., Eisaki, H., Uchida, S., Davis, J.C., Sethna, J.P. et al.: Electronic nematic ordering of the intra-unit-cell pseudogap states in underdoped $Bi_2Sr_2CaCu_2O_{8+\delta}$. Nature **466**, 347 (2009)

75. Kamihara, Y., Watanabe, T., Hirano, M., Hosono, H.: Iron-based layered superconductor $La[O_{1-x}F_x]$.FeAs (x = 0.05–0.12) with $T_c = 26$ K. J. Am. Chem. Soc. **130**, 3296 (2008)

76. Mu, G., Zhu, X., Fang, L., Shan, L., Ren, C., Wen, H.H.: Nodal gap in Fe-based layered superconductor $LaO_{0.9}F_{0.1-\delta}$FeAs probed by specific heat measurements. Chin. Phys. Lett. **25**, 2221 (2008)

77. Fang, C., Yao, H., Tsai, W.-F., Hu, J.P., Kivelson, S.A.: Theory of electron nematic order in LaOFeAs. Phys. Rev. B **77**, 224509 (2008)

78. Xu, C., Müller, M., Sachdev, S.: Ising and spin orders in Iron-based superconductors. Phys. Rev. B **78**, 020501 (R) (2008)

79. Chuang, T.-M., Allan, M., Lee, J., Xie, Y., Ni, N., Bud'ko, S., Boebinger, G.S., Canfield, P.C., Davis, J.C.: Nematic electronic structure in the 'parent' state of the iron-based superconductor $Ca(Fe_{1-x}Co_x)_2As_2$. Science **327**, 181 (2010)

80. Dagotto,, E., Hotta,, T., Moreo, A.: Colossal magnetoresistant materials: the key role of phase separation. Phys. Rep. **344**, 1 (2001)

81. Rübhausen, M., Yoon, S., Cooper, S.L., Kim, K.H., Cheong, S.W.: Anisotropic optical signatures of orbital and charge ordering in $Bi_{1-x}Ca_xMnO_3$. Phys. Rev. B **62**, R4782 (2000)

82. Grüner, G.: The dynamics of charge-density-waves. Rev. Mod. Phys. **60**, 1129 (1988)

83. Grüner, G.: The dynamics of spin-density-waves. Rev. Mod. Phys. **66**, 1 (1994)

84. McMillan, W.L.: Landau theory of charge density waves in transition-metal dichalcogenides. Phys. Rev. B **12**, 1187 (1975)

85. McMillan, W.L.: Theory of discommensurations and the commensurate-incommensurate charge-density-wave phase transition. Phys. Rev. B **14**, 1496 (1976)

86. Emery, V.J., Fradkin, E., Kivelson, S.A., Lubensky, T.C.: Quantum theory of the smectic metal state in stripe shases. Phys. Rev. Lett. **85**, 2160 (2000)

87. Carlson, E.W., Emery, V.J., Kivelson, S.A., Orgad, D.: In: Bennemann, K.H., Ketterson, J.B. (ed.) The Physics of Conventional and Unconventional Superconductors, vol. II, Springer-Verlag, Berlin (2004)

88. Snow, C.S., Karpus, J.F., Cooper, S.L., Kidd, T.E., Chiang, T.-C.: Quantum melting of the charge-density-wave state in 1T-TiSe$_2$. Phys. Rev. Lett. **91**, 136402 (2003)

89. Morosan, E., Zandbergen, H.W., Dennis, B.S., Bos, J.W., Onose, Y., Klimczuk, T., Ramirez, A.P., Ong, N.P., Cava, R.J.: Superconductivity in Cu_xTiSe_2. Nature Phys. **2**, 44 (2006)

90. Barath, H., Kim, M., Karpus, J.F., Cooper, S.L., Abbamonte, P., Fradkin, E., Morosan, E., Cava, R.J.: Quantum and classical mode softening near the charge-density-wave/superconductor transition of Cu_xTiSe_2 : Raman spectroscopic studies. Phys. Rev. Lett. **100**, 106402 (2008)

91. Dai, H., Chen, H., Lieber, C.M.: Weak pinning and hexatic order in a doped two-dimensional charge-density-wave system. Phys. Rev. Lett. **66**, 3183 (1991)

92. Kusmartseva, A.F., Sipos, B., Berker, H., Forró, L., Tutiš, E.: Pressure induced superconductivity in pristine 1T-TiSe$_2$. Phys. Rev. Lett. **103**, 236401 (2009)

93. Brouet, V., Yang, W.L., Zhou, X.J., Hussain, Z., Ru, N., Shin, K.Y., Fisher, I.R., Shen, Z.X.: Fermi surface reconstruction in the CDW state of CeTe$_3$ observed by photoemission. Phys. Rev. Lett. **93**, 126405 (2004)

94. Laverock, J., Dugdale, S.B., Major, Z., Alam, M.A., Ru, N., Fisher, I.R., Santi, G., Bruno, E.: Fermi surface nesting and charge-density wave formation in rare-earth tritellurides. Phys. Rev. B **71**, 085114 (2005)

95. Sacchetti, A., Degiorgi, L., Giamarchi, T., Ru, N., Fisher, I.R.: Chemical pressure and hidden one-dimensional behavior in rare-earth tri-telluride charge-density-wave compounds. Phys. Rev. B **74**, 125115 (2006)

96. Fang, A., Ru, N., Fisher, I.R., Kapitulnik, A.: STM studies of Tb Te$_3$: evidence for a fully incommensurate charge density wave. Phys. Rev. Lett. **99**, 046401 (2007)

97. Yao, H., Robertson, J.A., Kim, E.-A., Kivelson, S.A.: Theory of stripes in quasi-two-dimensional rare-earth tritellurides. Phys. Rev. B **74**, 245126 (2006)

98. Vojta, M.: Lattice symmetry breaking in cuprate superconductors: stripes, nematics, and superconductivity. Adv. Phys. **58**, 564 (2009)

99. Brazovskii, S., Kirova, N.: Electron self-localization and superstructures in quasi one-dimensional dielectrics. Sov. Sci. Rev. A **5**, 99 (1984)

100. Kivelson, S.A., Emery, V.J.: In: Bedell, K., Wang, Z., Meltzer, D.E., Balatsky, A.V., Abrahams, E. (ed.) Strongly Correlated Electron Materials: The Los Alamos Symposium 1993, pp. 619–650. Addison-Wesley, Redwood City (1994)

101. Zaanen, J., Gunnarsson, O.: Charged magnetic domain lines and the magnetism of high T_c oxides. Phys. Rev. B **40**, 7391 (1989)

102. Schulz, H.J.: Incommensurate antiferromagnetism in the 2-dimensional Hubbard model. Phys. Rev. Lett. **64**, 1445 (1990)

103. Poilblanc, D., Rice, T.M.: Charged solitons in the hartree–fock approximation to the large-U Hubbard model. Phys. Rev. B **39**, 9749 (1989)

104. Machida, K.: Magnetism in La$_2$CuO$_4$ based compounds. Physica C **158**, 192 (1989)

105. Kato, M., Machida, K., Nakanishi, H., Fujita, M.: Soliton lattice modulation of incommensurate spin density wave in two dimensional Hubbard model —a mean field study. J. Phys. Soc. Jpn. **59**, 1047 (1990)

106. Kivelson, S.A., Emery, V.J.: Topological doping. Synth. Met. **80**, 151 (1996)

107. Emery, V.J., Kivelson, S.A., Tranquada, J.M.: Stripe phases in high-temperature superconductors. Proc. Natl. Acad. Sci. USA **96**, 8814 (1999)

108. Pryadko, L.P., Kivelson, S.A., Emery, V.J., Bazaliy, Y.B., Demler, E.A.: Topological doping and the stability of stripe phases. Phys. Rev. B **60**, 7541 (1999)

109. Berg, E., Fradkin, E., Kivelson, S.A., Tranquada, J.M.: Striped superconductors: how the cuprates intertwine spin, charge and superconducting orders. New J. Phys. **11**, 115004 (2009)

110. Read, N., Sachdev, S.: Valence-bond and Spin-Peierls ground states in low-dimensional quantum antiferromagnets. Phys. Rev. Lett. **62**, 1694 (1989)

111. Vojta, M., Sachdev, S.: Charge order, superconductivity, and a global phase diagram of doped antiferromagnets. Phys. Rev. Lett. **83**, 3916 (1999)

112. Vojta, M., Zhang, Y., Sachdev, S.: Competing orders and quantum criticality in doped antiferromagnets. Phys. Rev. B **62**, 6721 (2000)

113. Kivelson, S.A., Rokhsar, D., Sethna, J.P.: Topology of the resonating valence-bond state: solitons and high T_c superconductivity. Phys. Rev. B **35**, 865 (1987)

114. Capello, M., Raczkowski, M., Poilblanc, D.: Stability of RVB hole stripes in high temperature superconductors. Phys. Rev. B **77**, 224502 (2008)

115. Himeda, A., Kato, T., Ogata, M.: Stripe states with spatially oscillating d-wave superconductivity in the two-dimensional $t - t' - J$ model. Phys. Rev. Lett. **88**, 117001 (2002)

116. Yamase, H., Metzner, W.: Competition of Fermi surface symmetry breaking and superconductivity. Phys. Rev. B **75**, 155117 (2007)

117. White, S.R., Scalapino, D.J.: Ground states of the doped four-leg t-J ladder. Phys. Rev. B **55**, 14701 (R) (1997)

118. White, S.R., Scalapino, D.J.: Density matrix renormalization group study of the striped phase in the 2D t-J model. Phys. Rev. Lett. **80**, 1272 (1998)

119. White, S.R., Scalapino, D.J.: Ground-state properties of the doped three-leg t-J ladder. Phys. Rev. B **57**, 3031 (1998)
120. White, S.R., Scalapino, D.J.: Phase separation and stripe formation in the two-dimensional t-J model: a comparison of numerical results. Phys. Rev. B **61**, 6320 (2000)
121. Hager, G., Wellein, G., Jackelmann, E., Fehske, H.: Stripe formation in doped Hubbard ladders. Phys. Rev. B **71**, 075108 (2005)
122. Kivelson, S.A., Emery, V.J., Lin, H.Q.: Doped antiferromagnets in the small t limit. Phys. Rev. B **42**, 6523 (1990)
123. Emery, V.J.: Theory of high T_c superconductivity in oxides. Phys. Rev. Lett. **58**, 2794 (1987)
124. Lorenzana, J., Seibold, G.: Metallic mean-field stripes, incommensurability, and chemical potential in cuprates. Phys. Rev. Lett. **89**, 136401 (2002)
125. Granath, M., Oganesyan, V., Kivelson, S.A., Fradkin, E., Emery, V.J.: Nodal quasi-particles and coexisting orders in striped superconductors. Phys. Rev. Lett. **87**, 167011 (2001)
126. Arrigoni, E., Fradkin, E., Kivelson, S.A.: Mechanism of high temperature superconductivity in a striped Hubbard model. Phys. Rev. B. **69**, 214519 (2004)
127. Emery, V.J.: In: Devreese, J.T., Evrard, R.P., van Doren, V.E. (ed.) Highly Conducting One-Dimensional Solids, p. 327. Plenum Press, New York (1979)
128. Luther, A., Emery, V.J.: Backward scattering in the one-dimensional electron gas. Phys. Rev. Lett. **33**, 589 (1974)
129. Noack, R.M., Bulut, N., Scalapino, D.J., Zacher, M.G.: Enhanced $d_{x^2-y^2}$ pairing correlations in the two-leg Hubbard ladder. Phys. Rev. B **56**, 7162 (1997)
130. Balents, L., Fisher, M.P.A.: Weak-coupling phase diagram of the two-chain Hubbard model. Phys. Rev. B **53**, 12133 (1996)
131. Lin, H.H., Balents, L., Fisher, M.P.A.: N-chain Hubbard model in weak coupling. Phys. Rev. B **56**, 6569 (1997)
132. Lin, H.H., Balents, L., Fisher, M.P.A.: Exact SO(8) symmetry in the weakly-interacting two-leg ladder. Phys. Rev. B **58**, 1794 (1998)
133. Emery, V.J., Kivelson, S.A., Zachar, O.: Classification and stability of phases of the multi-component one-dimensional electron gas. Phys. Rev. B **59**, 15641 (1999)
134. Emery, V.J., Kivelson, S.A., Zachar, O.: Spin-gap proximity effect mechanism of high temperature superconductivity. Phys. Rev. B **56**, 6120 (1997)
135. Tsunetsugu, H., Troyer, M., Rice, T.M.: Pairing and excitation spectrum in doped t-J ladders. Phys. Rev. B **51**, 16456 (1995)
136. Vishwanath, A., Carpentier, D.: Two-dimensional anisotropic non-Fermi-liquid phase of coupled luttinger liquids. Phys. Rev. Lett. **86**, 676 (2001)
137. Fertig, H.A.: Unlocking transition for modulated surfaces and quantum Hall stripes. Phys. Rev. Lett. **82**, 3693 (1999)
138. Lawler, M.J., Fradkin, E.: Quantum Hall smectics, sliding symmetry and the renormalization group. Phys. Rev. B **70**, 165310 (2004)
139. O'Hern, C.S., Lubensky, T.C., Toner, J.: Sliding phases in XY-models, crystals, and cationic lipid-DNA complexes. Phys. Rev. Lett. **83**, 2746 (1999)
140. Carlson, E.W., Orgad, D., Kivelson, S.A., Emery, V.J.: Dimensional crossover in quasi one-dimensional and high T_c superconductors. Phys. Rev. B **62**, 3422 (2000)
141. Affleck, I., Halperin, B.I.: On a renormalization group approach to dimensional crossover. J. Phys. A.: Math. Gen. **29**, 2627 (1996)
142. Lee, P.A., Nagaosa, N., Wen, X.-G.: Doping a mott insulator: physics of high temperature superconductivity. Rev. Mod. Phys. **78**, 17 (2006)
143. Berg, E., Fradkin, E., Kivelson, S.A.: Theory of the striped superconductor. Phys. Rev. B **79**, 064515 (2009)
144. Tranquada, J.M., Gu, G.D., Hücker, M., Kang, H.J., Klingerer, R., Li, Q., Wen, J.S., Xu, G.Y., Zimmermann, M.v.: Evidence for unusual superconducting correlations coexisting with stripe order in La$_{1.875}$Ba$_{0.125}$CuO$_4$. Phys. Rev. B **78**, 174529 (2008)

145. Hücker, M., Zimmermann, M.V., Debessai, M., Schilling, J.S., Tranquada, J.M., Gu, G.D.: Spontaneous symmetry breaking by charge stripes in the high-pressure phase of superconducting $La_{1.875}Ba_{0.125}CuO_4$. Phys. Rev. Lett. **104**, 057004 (2010)

146. Raczkowski, M., Capello, M., Poilblanc, D., Frésard, R., Oleś, A.M.: Unidirectional d-wave superconducting domains in the two-dimensional t-J model. Phys. Rev. B **76**, 140505 (R) (2007)

147. Yang, K.-Y., Chen, W.-Q., Rice, T.M., Sigrist, M., Zhang, F.-C.: Nature of stripes in the generalized t-J model applied to the cuprate superconductors. New J. Phys. **11**, 055053 (2009)

148. Loder, F., Kampf, A.P., Kopp, T.: Superconductivity with finite-momentum pairing in zero magnetic field. Phys. Rev. B **81**, 020511 (2010)

149. Chen, H.D., Vafek, O., Yazdani, A., Zhang, S.-C.: Pair density wave in the pseudogap state of high temperature superconductors. Phys. Rev. Lett. **93**, 187002 (2004)

150. Melikyan, A., Tešanović, Z.: A model of phase fluctuations in a lattice d-wave superconductor: application to the Cooper pair charge-density-wave in underdoped cuprates. Phys. Rev. B **71**, 214511 (2005)

151. Kosterlitz, J.M., Thouless, D.J.: Order metastability and phase transitions in two-dimensional systems. J. Phys. C: Solid State Phys. **6**, 1181 (1973)

152. José, J.V., Kadanoff, L.P., Kirkpatrick, S., Nelson, D.R.: Renormalization, vortices, and symmetry-breaking perturbations in the wto-dimensional planar model. Phys. Rev. B **16**, 1217 (1977)

153. Cardy, J.: Scaling and Renormalization in Statistical Physics. Cambridge University Press, Cambridge, UK (1996) Chapter 8

154. Krüger, F., Scheidl, S.: Non-universal ordering of spin and charge in stripe phases. Phys. Rev. Lett. **89**, 095701 (2002)

155. Podolsky, D., Chandrasekharan, S., Vishwanath, A.: Phase transitions of $S = 1$ spinor condensates in an optical lattice. Phys. Rev. B **80**, 214513 (2009)

156. Radzihovsky, L., Vishwanath, A.: Quantum liquid crystals in imbalanced Fermi gas: fluctuations and fractional vortices in Larkin-Ovchinnikov states. Phys. Rev. Lett. **103**, 010404 (2009)

157. Baym, G., Pethick, C.: Landau Fermi Liquid Theory. Wiley , New York, NY (1991)

158. Polchinski, J.: In: Harvey, J., Polchinski, J. (ed.) Recent directions in particle theory: from superstrings and black holes to the Standard Model (TASI - 92). Theoretical Advanced Study Institute in High Elementary Particle Physics (TASI 92), Boulder, Colorado, USA, 1–26 Jun, 1992. (World Scientific, Singapore, 1993).

159. Shankar, R.: Renormalization-group approach to interacting fermions. Rev. Mod. Phys. **66**, 129 (1994)

160. Pomeranchuk, I.I.: On the stability of a Fermi liquid. Sov. Phys. JETP **8**, 361 (1958)

161. Kee, H.-Y., Kim, E.H., Chung, C.-H.: Signatures of an electronic nematic phase at the isotropic-nematic phase transition. Phys. Rev. B **68**, 245109 (2003)

162. Khavkine, I., Chung, C.-H., Oganesyan, V., Kee, H.-Y.: Formation of an electronic nematic phase in interacting fermion systems. Phys. Rev. B **70**, 155110 (2004)

163. Yamase, H., Oganesyan, V., Metzner, W.: Mean-field theory for symmetry-breaking Fermi surface deformations on a square lattice. Phys. Rev. B **72**, 035114 (2005)

164. Halboth, C.J., Metzner, W.: D-wave superconductivity and pomeranchuk instability in the two-dimensional Hubbard model. Phys. Rev. Lett. **85**, 5162 (2000)

165. Metzner, W., Rohe, D., Andergassen, S.: Soft Fermi surfaces and breakdown of Fermi-liquid behavior. Phys. Rev. Lett. **91**, 066402 (2003)

166. Neumayr, A., Metzner, W.: Renormalized perturbation theory for Fermi systems: Fermi surface deformation and superconductivity in the two-dimensional Hubbard model. Phys. Rev. B **67**, 035112 (2003)

167. Dell'Anna, L., Metzner, W.: Fermi surface fluctuations and single electron excitations near pomeranchuk instability in two dimensions. Phys. Rev. B **73**, 45127 (2006)

168. Honerkamp, C., Salmhofer, M., Furukawa, N., Rice, T.M.: Breakdown of the Landau-Fermi liquid in two dimensions due to umklapp scattering. Phys. Rev. B **63**, 035109 (2001)

169. Honerkamp, C., Salmhofer, M., Rice, T.M.: Flow to strong coupling in the two-dimensional Hubbard model. Euro. Phys. J. B **27**, 127 (2002)

170. Hankevych, V., Grote, I., Wegner, F.: Pomeranchuk and other instabilities in the t-t' Hubbard model at the van hove filling. Phys. Rev. B **66**, 094516 (2002)

171. Lamas, C.A., Cabra, D.C., Grandi, N.: Fermi liquid instabilities in two-dimensional lattice models. Phys. Rev. B **78**, 115104 (2008)

172. Quintanilla, J., Haque, M., Schofield, A.J.: Symmetry-breaking Fermi surface deformations from central interactions in two dimensions. Phys. Rev. B **78**, 035131 (2008)

173. Sun, K., Yao, H., Fradkin, E., Kivelson, S.A.: Topological insulators and nematic phases from spontaneous symmetry breaking in 2D Fermi systems with a quadratic band crossing. Phys. Rev. Lett. **103**, 046811 (2009)

174. Yamase, H., Kohno, H.: Possible quasi-one-dimensional Fermi surface in $La_{2-x}Sr_xCuO_4$. J. Phys. Soc. Jpn. **69**, 2151 (2000)

175. Miyanaga, A., Yamase, H.: Orientational symmetry-breaking correlations in square lattice t-J model. Phys. Rev. B **73**, 174513 (2006)

176. Kivelson, S.A., Fradkin, E., Geballe, T.H.: Quasi-1D dynamics and the Nematic phase of the 2D emery model. Phys. Rev. B **69**, 144505 (2004)

177. Lawler, M.J., Barci, D.G., Fernández, V., Fradkin, E., Oxman, L.: Nonperturbative behavior of the quantum phase transition to a nematic Fermi fluid. Phys. Rev. B **73**, 085101 (2006)

178. Lawler, M.J., Fradkin, E.: Local quantum criticality in the nematic quantum phase transition of a Fermi fluid. Phys. Rev. B **75**, 033304 (2007)

179. Metlitski, M.A., Sachdev, S.: Quantum phase transitions of metals in two spatial dimensions:I. Ising-nematic order. Phys. Rev. B **82**, 075127 (2010)

180. Kee, H.Y., Kim, Y.B.: Itinerant metamagnetism induced by electronic nematic order. Phys. Rev. B **71**, 184402 (2005)

181. Yamase, H., Katanin, A.A.: Van Hove singularity and spontaneous Fermi surface symmetry breaking in $Sr_3Ru_2O_7$. J. Phys. Soc. Jpn. **76**, 073706 (2007)

182. Puetter, C.M., Doh, H., Kee, H.-Y.: Meta-nematic transitions in a bilayer system: application to the bilayer ruthenate. Phys. Rev. B **76**, 235112 (2007)

183. Puetter, C.M., Rau, J.G., Kee, H.-Y.: Microscopic route to nematicity in $Sr_3Ru_2O_7$. Phys. Rev. B **81**, 081105 (2010)

184. Raghu, S., Paramekanti, A., Kim, E.-A., Borzi, R.A., Grigera, S., Mackenzie, A.P., Kivelson, S.A.: Microscopic theory of the nematic phase in $Sr_3Ru_2O_7$. Phys. Rev. B **79**, 214402 (2009)

185. Lee, W.C., Wu, C.: Nematic electron states enhanced by orbital band hybridization. Phys. Rev. B **80**, 104438 (2009)

186. Fregoso, B.M., Sun, K., Fradkin, E., Lev, B.L.: Biaxial nematic phases in ultracold dipolar Fermi gases. New J. Phys. **11**, 103003 (2009)

187. Fregoso, B.M., Fradkin, E.: Ferro-Nematic ground state of the dilute dipolar Fermi gas. Phys. Rev. Lett. **103**, 205301 (2009)

188. Kim, E.A., Lawler, M.J., Oreto, P., Sachdev, S., Fradkin, E., Kivelson, S.A.: Theory of the nodal nematic quantum phase transition in superconductors. Phys. Rev. B **77**, 184514 (2008)

189. Huh, Y., Sachdev, S.: Renormalization group theory of nematic ordering in d-wave superconductors. Phys. Rev. B **78**, 064512 (2008)

190. Varma, C.M.: Non-Fermi-liquid states and pairing instability of a general model of copper oxide metals. Phys. Rev. B **55**, 4554 (1997)

191. Barci, D.G., Oxman, L.E.: Strongly correlated fermions with nonlinear energy dispersion and spontaneous generation of anisotropic phases. Phys. Rev. B **67**, 205108 (2003)

192. Zacharias, M., Wölfle, P., Garst, M.: Multiscale quantum criticality: Pomeranchuk instability in isotropic metals. Phys. Rev. B **80**, 165116 (2009)

193. Hertz, J.A.: Quantum critical phenomena. Phys. Rev. B **14**, 1165 (1976)

194. Millis, A.J.: Effect of a nonzero temperature on quantum critical points in itinerant Fermion systems. Phys. Rev. B **48**, 7183 (1993)
195. Sachdev, S.: Quantum Phase Transitions. Cambridge University Press, Cambridge, UK (1999)
196. Jain, J.K.: Composite-fermion approach for the fractional quantum Hall effect. Phys. Rev. Lett. **63**, 199 (1989)
197. Lopez, A., Fradkin, E.: Fractional quantum Hall effect and Chern-Simons gauge theories. Phys. Rev. B **44**, 5246 (1991)
198. Halperin, B.I., Lee, P.A., Read, N.: Theory of the half-filled Landau level. Phys. Rev. B **47**, 7312 (1993)
199. Rezayi, E., Read, N.: Fermi-liquid-like state in a half-filled Landau level. Phys. Rev. Lett. **72**, 900 (1994)
200. Doan, Q.M., Manousakis, E.: Variational Monte Carlo calculation of the nematic state of the two-dimensional electron gas in a magnetic field. Phys. Rev. B **78**, 075314 (2008)
201. Dell'Anna, L., Metzner, W.: Electrical resistivity near pomeranchuk instability in two dimensions. Phys. Rev. Lett. **98**, 136402 (2007). Erratum: Phys. Rev. Lett. **103**, 220602 (2009)
202. Haldane, F.D.M.: In: Schrieffer, J.R., Broglia, R. (ed.) Proceedings of the International School of Physics Enrico Fermi, course 121, Varenna, 1992. North-Holland, New York (1994)
203. Castro Neto, A.H., Fradkin, E.: Bosonization of the low energy excitations of Fermi liquids. Phys. Rev. Lett. **72**, 1393 (1994)
204. Castro Neto, A.H., Fradkin, E.H.: Exact solution of the Landau fixed point via bosonization. Phys. Rev. B **51**, 4084 (1995)
205. Houghton, A., Marston, J.B.: Bosonization and fermion liquids in dimensions greater than one. Phys. Rev. B **48**, 7790 (1993)
206. Houghton, A., Kwon, H.J., Marston, J.B.: Multidimensional bosonization. Adv. Phys. **49**, 141 (2000)
207. Ghaemi, P., Vishwanath, A., Senthil, T.: Finite temperature properties of quantum Lifshitz transitions between valence-bond solid phases: an example of local quantum criticality. Phys. Rev. B **72**, 024420 (2005)
208. Chubukov, A.V., Pépin, C., Rech, J.: Instability of the quantum critical point of itinerant ferromagnets. Phys. Rev. Lett. **92**, 147003 (2004)
209. Chubukov, A.V.: Self-generated locality near a ferromagnetic quantum critical point. Phys. Rev. B **71**, 245123 (2005)
210. Rech, J., Pépin, C., Chubukov, A.V.: Quantum critical behavior in itinerant electron systems–Eliashberg theory and instability of a ferromagnetic quantum critical point. Phys. Rev. B **74**, 195126 (2006)
211. Holstein, T., Norton, R.E., Pincus, P.: de Haas-van Alphen effect and the specific heat of an electron gas. Phys. Rev. B **8**, 2649 (1973)
212. Baym, G., Monien, H., Pethick, C.J., Ravenhall, D.G.: Transverse interactions and transport in relativistic quark-gluon and electromagnetic plasmas. Phys. Rev. Lett. **64**, 1867 (1990)
213. Boyanovsky, D., de Vega, H.J.: Non-Fermi-liquid aspects of cold and dense QED and QCD: equilibrium and non-equilibrium. Phys. Rev. D **63**, 034016 (2001)
214. Reizer, M.Y.: Relativistic effects in the electron density of states, specific heat, and the electron spectrum of normal metals. Phys. Rev. B **40**, 11571 (1989)
215. Ioffe, L.B., Wiegmann, P.B.: Linear temperature dependence of resistivity as evidence of gauge interaction. Phys. Rev. Lett. **65**, 653 (1990)
216. Nagaosa, N., Lee, P.A.: Experimental consequences of the uniform resonating-valence-bond state. Phys. Rev. B **43**, 1233 (1991)
217. Polchinski, J.: Low-energy dynamics of the spinon-gauge system. Nucl. Phys. B **422**, 617 (1994)
218. Chakravarty, S., Norton, R.E., Syljuasen, O.F.: Transverse gauge interactions and the vanquished Frmi liquid. Phys. Rev. Lett. **74**, 1423 (1995)
219. Lee, S.S.: Low-energy effective theory of Fermi surface coupled with U(1) gauge field in $2+1$ dimensions. Phys. Rev. B **80**, 165102 (2009)

220. Jakubczyk, P., Metzner, W., Yamase, H.: Turning a first order quantum phase transition contin-
     uous by fluctuations. Phys. Rev. Lett. **103**, 220602 (2009)
221. Hirsch, J.E.: Spin-split states in metals. Phys. Rev. B **41**, 6820 (1990)
222. Varma, C.M., Zhu, L.: Helicity order: hidden order parameter in URu_2Si_2. Phys. Rev. Lett.
     96, 036405 (2006)
223. Simon, M.E., Varma, C.M.: Detection and implications of a time-reversal breaking state in
     underdoped cuprates. Phys. Rev. Lett. **89**, 247003 (2002)
224. Haldane, F.D.M.: Berry Curvature on the Fermi surface: anomalous Hall effect as a topological
     Fermi-liquid property. Phys. Rev. Lett. **93**, 206602 (2004)
225. Nelson, D.R., Toner, J.: Bond-orientational order, dislocation loops, and melting of solids and
     smectic-A liquid crystals. Phys. Rev. B **24**, 363 (1981)
226. Toner, J., Nelson, D.R.: Smectic, cholesteric, and Rayleigh-Benard order in two dimensions.
     Phys. Rev. B **23**, 316 (1981)
227. Zaanen, J., Nussinov, Z., Mukhin, S.I.: Duality in $2+1$ D quantum elasticity: superconductivity
     and quantum nematic order. Ann. Phys. **310**, 181 (2004)
228. Cvetkovic, V., Nussinov, Z., Zaanen, J.: Topological kinematical constraints: quantum
     dislocations and glide principle. Phil. Mag. **86**, 2995 (2006)
229. Sun, K., Fregoso B.M., Lawler M.J., Fradkin E.: Fluctuating stripes in strongly correlated
     electron systems and the nematic-smectic quantum phase transition. Phys. Rev. B **78**, 085124
     (2008). Erratum: Phys. Rev. B **80**, 039901(E) (2008).
230. Kirkpatrick, T.R., Belitz, D.: Soft modes in electronic stripe phases and their consequences
     for thermodynamics and transport Phys. Rev. B **80**, 075121 (2009)
231. Millis, A.J.: Fluctuation-driven first order behavior near the $T=0$ two dimensional stripe to
     Fermi liquid transition. Phys. Rev. B **81**, 035117 (2010)
232. Halperin, B.I., Lubensky, T.C., Ma, S.-K.: First-order phase transitions in superconductors
     and smectic—a liquid crystals. Phys. Rev. Lett. **32**, 292 (1974)
233. Altshuler, B.L., Ioffe, L.B., Millis, A.J.: Critical behavior of the $T=0$, $2k_F$, density-wave
     phase transition in a two-dimensional Fermi liquid. Phys. Rev. B **52**, 5563 (1995)

Chapter 3
Selected Topics in Graphene Physics

Antonio H. Castro Neto

Abstract Graphene research is currently one of the largest fields in condensed matter. Due to its unusual electronic spectrum with Dirac-like quasiparticles, and the fact that it is a unique example of a metallic membrane, graphene has properties that have no match in standard solid-state textbooks. In these lecture notes, I discuss some of these properties that are not covered in detail in recent reviews (Castro Neto AH, Guinea F, Peres NMR, Novoselov KS, Geim AK (2009) Rev Mod Phys 81:109). We study the particular aspects of the physics/chemistry of carbon that influence the properties of graphene; the basic features of graphene's band structure including the π and σ bands; the phonon spectra in free floating graphene; the effects of a substrate on the structural properties of graphene; and the effect of deformations in the propagation of electrons. The objective of these notes is not to provide an unabridged theoretical description of graphene but to point out some of the peculiar aspects of this material.

3.1 Introduction

Graphene, a two-dimensional allotrope of carbon, has created an immense interest in the condensed matter community and in the media since its isolation in 2004 [1]. On the one hand, graphene has unique properties that derive from its honeycomb-like lattice structure such as the Dirac-like spectrum (that mimics effects of matter under extreme conditions), its low dimensionality (that leads to enhanced quantum and thermal fluctuations), and its membrane-like nature (that mixes aspects of soft and hard condensed matter). On the other hand, because of the strength and specificity of its covalent bonds, graphene is one of the strongest materials in nature

A. H. Castro Neto (✉)
Department of Physics, Boston University,
590 Commonwealth Avenue, Boston, MA 02215, USA
e-mail:neto@bu.edu

D. C. Cabra et al. (eds.), *Modern Theories of Many-Particle Systems in Condensed Matter Physics*, Lecture Notes in Physics 843, DOI: 10.1007/978-3-642-10449-7_3,
© Springer-Verlag Berlin Heidelberg 2012

(albeit one of the softest), with literally none extrinsic substitutional impurities, leading to the highest electronic mobilities among metals and semiconductors [2]. Therefore, graphene is being considered for a plethora of applications that range from conducting paints, and flexible displays, to high speed electronics. In fact, it can be said that perhaps, not since the invention of the transistor out of germanium in the 1950s, a material has had this kind of impact in the solid-state literature. However, unlike ordinary semiconductors such as germanium, gallium–arsenide, and silicon, graphene's unusual properties have to be understood before it can really have an impact in technological applications.

Any material has a hierarchy of energy scales that range from the atomic physics (~ 10eV), to many-body effects ($\sim 10^{-3}$ eV). To understand the behavior of a material one needs to understand how these different energy scales affect its macroscopic properties. While structural properties such as strength against strain, shear and bending may depend on the covalent bonds formed by the atoms, magnetism and superconductivity are governed by the particular way electrons interact with each other through Coulomb forces. Furthermore, while the properties of metals and semiconductors depend on the physics close to the Fermi energy (a direct consequence of Pauli's exclusion principle), the nature of the vibrational spectrum depends on the particular way ions interact among themselves and how the electrons screen these interactions.

One of the great accomplishments of the application of quantum mechanics to the theory of metals is the understanding that while different materials can be structurally very different from each other, their long wavelength and low-energy physics is essentially identical and depend on very few parameters. This so-called renormalization towards the Fermi energy [3] is one of the greatest theoretical accomplishments of the twentieth century and is the basis of Landau's theory of the Fermi liquid [4]. In systems where the low-energy physics is goverened by *Galilean invariance* the most significant parameter is the "effective" mass of the carriers that acts as to generate a *scale* from which is possible to compute most of the important physical quantities such as specific heat, magnetic susceptibility, electronic compressibility, and so on. The most basic difference between graphene and other materials is that its low-energy physics is not Galilean invariant, but instead *Lorentz invariant*, just like systems in particle and high-energy physics with Dirac particles as elementary excitations [5]. In this case, the renormalization towards the Fermi energy is different from other materials [6] because, in the absence of a "mass" (which in graphene means the absence of a gap in the electronic spectrum), all physical quantities depend on a characteristic "effective" velocity that plays an analogous role as the speed of light plays in relativistic quantum mechanics.[1] However, unlike true relativistic fermionic systems [7], the Dirac quasiparticles in graphene still propagate with a velocity that is much smaller than the speed of light, the speed that Coulomb interactions propagate. Therefore, the Coulomb field can be considered instantaneous in first approximation,

---

[1] It should be noted that, even if a gap can be opened in graphene (as if, for instance, one finds a way to break the symmetry between different sublattices), its dispersion would be hyperbolic, not parabolic, because the very basic Lorentz invariance is preserved.

rendering the electrodynamics of graphene electrons a mix between a relativistic and a non-relativist problem. Clearly this unusual situation requires a re-evaluation of the Fermi liquid theory for this material.

In these notes, however, we are going to focus on the more basic aspects of graphene's properties. In Sect. 3.2 we are going to discuss the s-p hybridization theory and how it leads to the basic energy scales of the graphene problem. We are going to show that the problem of hybridization is controlled by the angle between the s-p hybridized orbitals. Using the Slater–Koster theory [8], we compute the main matrix elements of the problems in terms of this angle. In Sect. 3.3 we move from molecular orbitals to the crystal and discuss the simplest tight-binding Hamiltonian that describes the *full* band structure, that is, that includes both π and σ bands. We show that even this simple band-structure reproduces quite well the results of more sophisticated *ab initio* methods. This description becomes particularly good close to the Fermi energy where the Dirac particles emerge naturally. Phonons in free floating graphene are discussed in Sect. 3.4. We show how the flexural modes result from the bending energy of a soft membrane and how those modes can be quantized. We show that as a result of the presence of flexural modes the linear phonon theory predicts an instability of the graphene sheet towards crumpling. The effect of a substrate on the flexural mode spectrum is discussed in Sect. 3.5. We show that the presence of a substrate, that breaks rotational and translational symmetry, allows for new terms in the phonon Hamiltonian that change considerably the energy dispersion of the flexural modes. We also show that within the linear theory, graphene follows the substrate in a smooth way with the characteristic length scales that are dependent on the details of the interaction with the substrate. When graphene is deformed in some way, either by bending or strain, the electronic motion is affected directly. In Sect. 3.6 we show that at long wavelengths deformations lead to new terms in the Dirac equation that couple to the electrons as vector and scalar potentials. There are cases, however, where graphene is not slightly deformed but strongly deformed in which case the Dirac theory has to be completely reconsidered. In these notes we discuss briefly the case of a graphene scroll which results from the competition between the bending energy (that favors flat graphene) and the van der Waals interaction of graphene with itself (that wants it to have maximum area overlap). This is only one example where structural deformations can have a strong effect on many of the electronic properties of this material. Our conclusions are given in Sect. 3.7.

3.2 The Chemistry

The electronic configuration of atomic carbon is $1s^2 2s^2 2p^2$. In a solid, however, carbon forms s-p hybridized orbitals. The 1s electrons form a deep valence band and essentially all properties of carbon-based materials can be described in terms of the $2s$ and $2p_x$, $2p_y$ and $2p_z$ orbitals that can be written as [9]:

Fig. 3.1 Absolute value of
the wave-function for a
hybridized s-p state, Eq. 3.2,
for $A = 0.5$

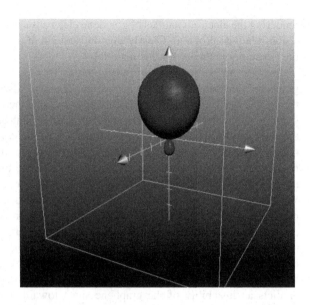

$$\langle r | s \rangle = R_s(r) \times 1,$$
$$\langle r | p_x \rangle = R_p(r) \times \sqrt{3} \sin\theta \cos\phi,$$
$$\langle r | p_y \rangle = R_p(r) \times \sqrt{3} \sin\theta \sin\phi,$$
$$\langle r | p_z \rangle = R_p(r) \times \sqrt{3} \cos\theta, \tag{3.1}$$

where $R_s(r) = (2 - r/a_0)e^{-r/(2a_0)}$, and $R_p(r) = (r/a_0)e^{-r/(2a_0)}$ are the radial
wave-functions. One particular way to parametrize a hybridized s-p state is given
by Pauling [10]:

$$|0\rangle = A|s\rangle + \sqrt{1 - A^2}|p_z\rangle, \tag{3.2}$$

where A is a parameter that describes the degree of hybridization between s and p
states. This basic orbital is shown in Fig. 3.1. The energy associated with this orbital
can be obtained from the hydrogen atom spectrum:

$$E_0 = \epsilon_\pi = \langle 0|H_0|0\rangle = A^2 E_s + (1 - A^2) E_p \tag{3.3}$$

where H_0 is the hydrogen atom Hamiltonian where $E_s \approx -19.38\,\text{eV}$ is the energy
of the 2s-state and $E_p \approx -11.07\,\text{eV}$ is the energy of the 2p-state.

All the other 3 orthogonal orbitals can be constructed starting from (3.2). A particularly simple parametrization is the following [11]:

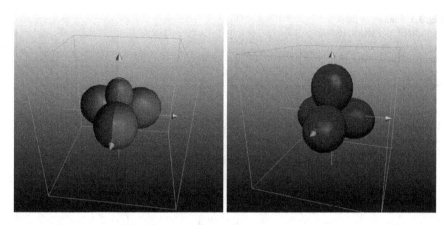

Fig. 3.2 sp^2 (*left*) and sp^3 (*right*) hybridized orbitals

$$|1\rangle = \sqrt{(1-A^2)/3}|s\rangle + \sqrt{2/3}|p_x\rangle - (A/\sqrt{3})|p_z\rangle,$$

$$|2\rangle = \sqrt{(1-A^2)/3}|s\rangle - \sqrt{1/6}|p_x\rangle - \sqrt{1/2}|p_y\rangle - (A/\sqrt{3})|p_z\rangle,$$

$$|3\rangle = \sqrt{(1-A^2)/3}|s\rangle - \sqrt{1/6}|p_x\rangle + \sqrt{1/2}|p_y\rangle - (A/\sqrt{3})|p_z\rangle. \qquad (3.4)$$

Notice that A controls the angle between the z axis and these states. We can clearly see that the direction of largest amplitude for one of these orbitals (say, $\langle r|1\rangle$) is given by:

$$\frac{\partial}{\partial\theta}\langle r|1\rangle(\phi=0) = \sqrt{2}\cos\theta_m + A\sin\theta_m = 0,$$

$$\theta_m = -\arctan(\sqrt{2}/A). \qquad (3.5)$$

Hence, for $A=0$ the hybridized state $|1\rangle$ is perpendicular to the other orbitals that remain in the x-y plane. This is the so-called sp^2 hybridization (see Fig. 3.2). For $A=1/2$ the orbitals have tetragonal structure making an angle of 109.47 degrees with the z axis. This is the sp^3 hybridization (see Fig. 3.2).

In free space, the states $|1\rangle$, $|2\rangle$, and $|3\rangle$ are clearly degenerate and their energy is given by:

$$E_1 = E_2 = E_3 = \langle 1|H_0|1\rangle = \epsilon_\sigma = \frac{1-A^2}{3}E_s + \frac{2+A^2}{3}E_p. \qquad (3.6)$$

The energy of these states is shown in Fig. 3.3. Notice that in the sp^3 case ($A=1/2$) all orbitals are degenerated while in the sp^2 case the orbitals are separated by an energy of approximately 2.77 eV.

The presence of another carbon atom induces a hybridization between the different orbitals. This hybridization depends on the distance ℓ between the atoms and also

Fig. 3.3 Energy of the
hybridized s-p states, in *blue*
(*top curve*) we show the
energy of the π state, (3.3),
and in *red* (*bottom curve*) the
energy of the σ state, (3.6),
as a function of the
hybridization parameter A.
$A = 0$ corresponds to the sp^2
and $A = 1/2$ corresponds to
the sp^3 configuration

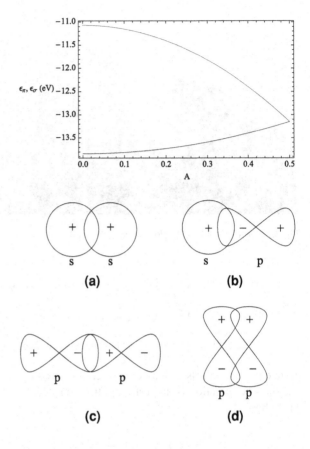

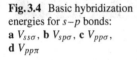

Fig. 3.4 Basic hybridization
energies for s–p bonds:
a $V_{ss\sigma}$, **b** $V_{sp\sigma}$, **c** $V_{pp\sigma}$,
d $V_{pp\pi}$

on the orientation of the orbitals relative to each other. The distance dependence is
usually well described by an exponential behavior:

$$V_\alpha(\ell) \approx V_\alpha^0 e^{-\kappa_\alpha \ell}, \tag{3.7}$$

where $\kappa_\alpha = d \ln(V_\alpha)/d\ell$ and α labels the different orientations of the orbitals.
In terms of orientation, there are four different types of elementary hybridization
between different orbitals (shown in Fig. 3.4): $V_{ss\sigma}$ ($\approx -5\,\text{eV}$ for $\ell = 1.42\,\text{Å}$);
$V_{sp\sigma}$ ($\approx +5.4\,\text{eV}$ for $\ell = 1.42\,\text{Å}$); $V_{pp\sigma}$ ($\approx +8.4\,\text{eV}$ for $\ell = 1.42\,\text{Å}$); $V_{pp\pi}$
($\approx -2.4\,\text{eV}$ for $\ell = 1.42\,\text{Å}$) [12].

Any hybridization energy can be obtained from those basic hybridizations shown
in Fig. 3.4 as linear combinations. For instance, let us consider the intra-atomic
hybridization between between $|2\rangle$ and $|3\rangle$. This is given by the matrix element:

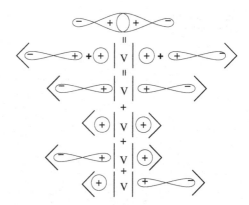

Fig. 3.5 Calculation of the hybridization of two s-p orbitals in terms of the basic hybridization energies shown in Fig. 3.4

$$V_{\text{intra}} = \langle 2|H_0|3\rangle = \langle 2| \left(\sqrt{\frac{1-A^2}{3}} E_s|s\rangle - \frac{E_p}{\sqrt{6}}|p_x\rangle + \frac{E_p}{\sqrt{2}}|p_y\rangle - \frac{A}{\sqrt{3}} E_p|p_z\rangle \right)$$

$$= \frac{1-A^2}{3}(E_s - E_p). \tag{3.8}$$

We can also compute inter-atomic hybridization energies such as the hybridization between two $|2\rangle$ states oriented as in Fig. 3.5:

$$V_\sigma = -\frac{2}{3}V_{pp\sigma} + \frac{1-A^2}{3}V_{ss\sigma} - \frac{2}{3}\sqrt{2(1-A^2)}V_{sp\sigma}. \tag{3.9}$$

3.3 The Crystal and Band Structure

In this section we focus on the flat graphene (sp^2 only) lattice. The geometry of the lattice is shown in Fig. 3.6a and it can be easily seen that the honeycomb lattice can be considered a crystal lattice with two atoms per unit cell. These two sublattices, A and B, form two interpenetrating triangular lattices. The lattice vectors are:

$$a_1 = \frac{a}{2}(3, \sqrt{3}),$$
$$a_2 = \frac{a}{2}(3, -\sqrt{3}), \tag{3.10}$$

and the nearest neighbor vectors are (see Fig. 3.7):

$$d_1 = a(1, 0),$$
$$d_2 = \frac{a}{2}(-1, \sqrt{3}),$$
$$d_3 = \frac{a}{2}(-1, -\sqrt{3}), \tag{3.11}$$

Fig. 3.6 a Lattice structure,
b First Brillouin zone

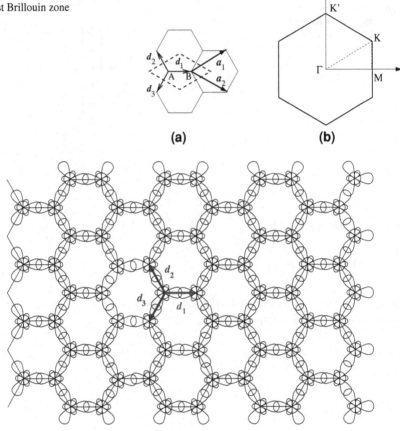

Fig. 3.7 Honeycomb lattice and electronic orbitals

where $a = 1.42\,\text{Å}$ is the carbon–carbon distance. The first Brillouin zone is shown in Fig. 3.6b and the reciprocal lattice is spanned by the vectors:

$$\boldsymbol{b}_1 = \frac{2\pi}{3a}(1, \sqrt{3}),$$
$$\boldsymbol{b}_2 = \frac{2\pi}{3}(1, -\sqrt{3}). \tag{3.12}$$

The K point, located at

$$\boldsymbol{k} = \frac{2\pi}{3a}\left(1, \frac{1}{\sqrt{3}}\right) \tag{3.13}$$

is of particular importance here.

Let us label the state at each carbon atom as: $|n, a, i\rangle$ where $n = 1, \ldots, N$ labels the unit cell, $a = A, B$ labels the sub-lattice (see Fig. 3.6a), and $i = 0,1,2,3$ labels the states defined in (3.2) and (3.4). In the simplest approximation we can consider a tight-binding description where the electrons hop between different orbitals (intraband) in the same atom with characteristic energy V_{intra}, between the two different sublattices (interband) along the planar orbitals with energy V_σ, and between two different sublattices with orbitals perpendicular to the plane with energy $V_{pp\pi}$. In this case the Hamiltonian can be written as:

$$
\begin{aligned}
H = {} & \epsilon_\pi \sum_{n,a} c^\dagger_{n,a,0} c_{n,a,0} + \epsilon_\sigma \sum_{n,a,i\neq 0} c^\dagger_{n,a,i} c_{n,a,i} \\
& + V_{\text{intra}} \sum_{n,a,i\neq j\neq 0} (c^\dagger_{n,a,i} c_{n,a,j} + \text{h.c.}) + V_{pp\pi} \sum_{\langle n,m\rangle} (c^\dagger_{n,A,0} c_{m,B,0} + \text{h.c.}) \\
& + V_\sigma \sum_{\langle n,m\rangle, i\neq 0} (c^\dagger_{n,A,i} c_{m,B,i} + \text{h.c.}).
\end{aligned}
\tag{3.14}
$$

The first step for the solution of this problem is to use the fact that the system is invariant under discrete translations and to Fourier transform the operators:

$$
c_{n,a,i} = \frac{1}{\sqrt{N}} \sum_k e^{i k \cdot R_n} c_{k,a,i},
\tag{3.15}
$$

which leads to:

$$
\begin{aligned}
H = \sum_k \Bigg\{ & \epsilon_\pi \sum_a c^\dagger_{k,a,0} c_{k,a,0} + \epsilon_\sigma \sum_{a,i\neq 0} c^\dagger_{k,a,i} c_{k,a,i} \\
& + V_{\text{intra}} \sum_{a,i\neq j\neq 0} (c^\dagger_{k,a,i} c_{k,a,j} + \text{h.c.}) + V_{pp\pi} \gamma_k (c^\dagger_{k,A,0} c_{k,B,0} + \text{h.c.}) \\
& + V_\sigma \left[e^{i k \cdot d_1} c^\dagger_{k,A,1} c_{k,B,1} + e^{i k \cdot d_2} c^\dagger_{k,A,2} c_{k,B,2} + e^{i k \cdot d_3} c^\dagger_{k,A,3} c_{k,B,3} + \text{h.c.} \right] \Bigg\}
\end{aligned}
\tag{3.16}
$$

where

$$
\gamma_k = \sum_{i=1,2,3} e^{i k \cdot d_i},
$$

$$
|\gamma_k| = \sqrt{3 + 2\cos(k \cdot (d_1 - d_2)) + 2\cos(k \cdot (d_1 - d_3)) + 2\cos(k \cdot (d_2 - d_3))}.
\tag{3.17}
$$

In this case the Hamiltonian can be written as:

$$
H = \sum_k \Psi^\dagger_k \cdot [H] \cdot \Psi_k,
\tag{3.18}
$$

where $\Psi^\dagger_k = (c^\dagger_{k,A,0}, c^\dagger_{k,B,0}, c^\dagger_{k,A,1}, c^\dagger_{k,B,1}, c^\dagger_{k,A,2}, c^\dagger_{k,B,2}, c^\dagger_{k,A,3}, c^\dagger_{k,B,3})$ and

$$[H] = \begin{bmatrix} \epsilon_\pi & V_{pp\pi}\gamma_k & 0 & 0 & 0 & 0 & 0 & 0 \\ V_{pp\pi}^*\gamma_k^* & \epsilon_\pi & 0 & 0 & 0 & 0 & 0 & 0 \\ 0 & 0 & \epsilon_\sigma & V_\sigma e^{ik\cdot d_1} & V_{intra} & 0 & V_{intra} & 0 \\ 0 & 0 & V_\sigma e^{-ik\cdot d_1} & \epsilon_\sigma & 0 & V_{intra} & 0 & V_{intra} \\ 0 & 0 & V_{intra} & 0 & \epsilon_\sigma & V_\sigma e^{ik\cdot d_2} & V_{intra} & 0 \\ 0 & 0 & 0 & V_{intra} & V_\sigma e^{-ik\cdot d_2} & \epsilon_\sigma & 0 & V_{intra} \\ 0 & 0 & V_{intra} & 0 & V_{intra} & 0 & \epsilon_\sigma & V_\sigma e^{ik\cdot d_3} \\ 0 & 0 & 0 & V_{intra} & 0 & V_{intra} & V_\sigma e^{-ik\cdot d_3} & \epsilon_\sigma \end{bmatrix}.$$

(3.19)

It is immediately obvious that the Hamiltonian for the π band completely decouples from the σ band and can be treated separately.

Although (3.19) is a 8×8 matrix (a 2×2 block plus a 6×6 block), it can be diagonalized analytically to produce 8 energy bands that read:

$$E_{\pi,\pm}(k) = \epsilon_\pi \pm |V_{pp\pi}||\gamma_k|,$$

$$E_{\sigma,1,\pm}(k) = \epsilon_\sigma - V_{intra} \pm V_\sigma,$$

$$E_{\sigma,2,\pm}(k) = \epsilon_\sigma + \frac{V_{intra}}{2} + \sqrt{\left(\frac{3V_{intra}}{2}\right)^2 + V_\sigma^2 \pm |V_{intra}V_\sigma||\gamma_k|},$$

$$E_{\sigma,3,\pm}(k) = \epsilon_\sigma + \frac{V_{intra}}{2} - \sqrt{\left(\frac{3V_{intra}}{2}\right)^2 + V_\sigma^2 \pm |V_{intra}V_\sigma||\gamma_k|}, \quad (3.20)$$

which represent two π bands and six σ bands, respectively. The π-bands are shown in Fig. 3.8a and the σ-bands are shown in Fig. 3.8b. In Fig. 3.9 we compare the results of this simple tight-binding with more sophisticated calculations [13]. One can clearly see that the tight-binding approach produces a fairly good description of the band structure, especially the π bands, although more hopping parameters have to be introduced in order to reproduce the details of the σ bands. The fact that this full band structure can be obtained analytically makes this particular parametrization of the bands rather attractive as a first step towards the electronic description of the graphene electrons.

As we said in Sect. 4.2, the carbon $1s$ states are completely filled, leaving 4 electrons per carbon atom (8 per unit cell) to fill these 8 bands. Hence, the first 4 bands are fully occupied meaning that the Fermi energy crosses exactly at the "conical" points of the dispersion at the K point of the Brillouin zone in Fig. 3.9 (and similarly at the K' point). Notice that there are 6 of these points in the Brillouin zone with vectors given by:

$$Q_1 = 4\pi/(3\sqrt{3}a)(0, 1),$$

$$Q_2 = 4\pi/(3\sqrt{3}a)(\sqrt{3}/2, 1/2),$$

$$Q_3 = 4\pi/(3\sqrt{3}a)(\sqrt{3}/2, -1/2), \quad (3.21)$$

and also at $-Q_i$ with $i = 1,2,3$. Close to these particular points, that is, close to the Fermi energy, we can find a simple expression for the electronic dispersion by

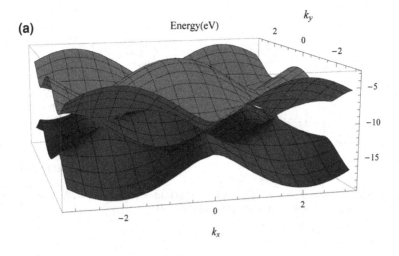

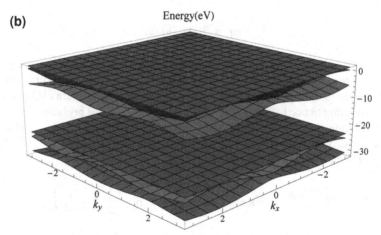

Fig. 3.8 Band structure of graphene (energy in eV). **a** π-bands, **b** σ-bands

expanding γ_k as:

$$\gamma_{Q_1+k} \approx \frac{3a}{2}(k_x + ik_y), \tag{3.22}$$

for $k \ll Q$. From (3.20) we see that the spectrum becomes:

$$E_\pm(k_x, k_y) = \pm v_F k = \pm v_F \sqrt{k_x^2 + k_y^2}, \tag{3.23}$$

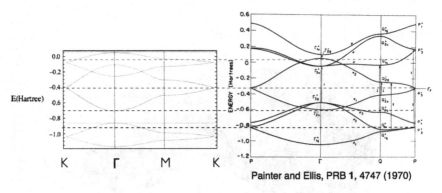

Painter and Ellis, PRB **1**, 4747 (1970)

Fig. 3.9 Band structure of graphene (energy is given in Hartrees – 1 Hartree = 27.21 eV): *left*, result of the model discussed in the text; *right*, result of Ref. [13]

where

$$v_F = 3|V_{pp\pi}|a/2, \tag{3.24}$$

is the Fermi velocity. Thus the electronic spectrum close to the Fermi energy has a conical form, mimicking the dispersion of a relativistic, massless, Dirac particle. Notice that in the first Brillouin zone each corner of the zone contains 1/3 of a Dirac cone but we do not need all the vectors in (3.21) to describe the problem. We can use reciprocal lattice vectors (3.12) in order to translate each piece of the cone to two corners located at $\pm Q_1$ making 2 Dirac cones in the extended zone scheme. Hence, close to Q_1 we can rewrite the Hamiltonian as:

$$H_0 \approx v_F \sum_k (\psi_{A,k}^\dagger, \psi_{B,k}^\dagger) \cdot \begin{bmatrix} 0 & k_x + ik_y \\ k_x - ik_y & 0 \end{bmatrix} \cdot \begin{pmatrix} \psi_{A,k} \\ \psi_{B,k} \end{pmatrix}, \tag{3.25}$$

where $\psi_{A,k} = c_{Q_1+k,A,0}$ and similarly for the B sublattice. By Fourier transforming the Hamiltonian back to real space we obtain the 2D Dirac Hamiltonian:

$$H_0 = \int d^2r \Psi^\dagger(r)(iv_F\sigma \cdot \nabla - \mu)\Psi(r), \tag{3.26}$$

where μ is the chemical potential measured away from the Dirac point, $\sigma = (\sigma_x, \sigma_y)$ are Pauli matrices, and

$$\Psi^\dagger(r) = (\psi_A^\dagger(r), \psi_B^\dagger(r)). \tag{3.27}$$

Fig. 3.10 Lattice displacements associated with some of the long-wavelength phonon modes in graphene

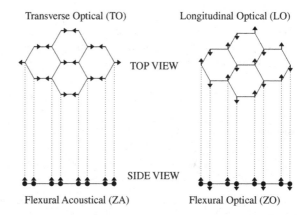

3.4 Phonons in Free Floating Graphene

Since graphene is two-dimensional and has 2 atoms per unit cell there are 4 in-plane lattice degrees of freedom that give rise to 2 acoustic, longitudinal (LA) and transverse (TA), and 2 optical, longitudinal (LO) and transverse (TO), phonon modes. However, graphene is embedded in a three-dimensional space and hence there are two extra degrees of freedom associated with out-of-plane motion: flexural acoustical (ZA) and optical (ZO) phonon modes. The displacements associated with these modes at long wavelengths is shown in Fig. 3.10. In leading order in the displacements the in-plane and out-of-plane modes decouple and we can study them separately. Here we focus in the out-of-plane modes since, as we are going to show, those dominate the low-energy physics of graphene.

We use the Monge representation where a point $\boldsymbol{R}$ in the graphene lattice is described by the vector $\boldsymbol{R} = (\boldsymbol{r}, h(\boldsymbol{r}))$ where $\boldsymbol{r} = (x, y)$ is the 2D coordinate vector and $h(\boldsymbol{r})$ is the height variable. The unit vector normal to the surface is given by:

$$\boldsymbol{N} = (-\nabla h + \boldsymbol{z})/\sqrt{1 + (\nabla h)^2}, \tag{3.28}$$

where $\nabla = (\partial_x, \partial_y)$ is the 2D gradient operator, and $\boldsymbol{z}$ is the unit vector in the third direction. Distortions from the flat configuration cost energy because they rotate the sigma orbitals that want to be aligned in order to get maximum overlap of the wavefunctions that bind the atoms together. Let us consider the triangular lattice centered at sublattice A, say at $\boldsymbol{R}_i$, that covers the entire graphene lattice. The normal at this point is $\boldsymbol{N}_i$. The energy cost for the bending of graphene can then be written as:

$$U_B = -\frac{\kappa_L}{2} \sum_{\langle i,j \rangle} \boldsymbol{N}_i \cdot \boldsymbol{N}_j, \tag{3.29}$$

where κ_L is the lattice bending rigidity of graphene. Notice that (3.29) has the form of the energy of a classical ferromagnet where the normal vectors play the role of spins. Define $R_j = R_i + u$ where u is the next-to-nearest neighbor vector that connects two triangles and let us consider only length scales much larger than the lattice spacing, $|u| \approx \sqrt{3}a$. In this, assuming that N is a smooth function over the lattice scale, we can write:

$$N_j \approx N_i + (u \cdot \nabla)N_i + \frac{1}{2}(u \cdot \nabla)^2 N_i \tag{3.30}$$

and hence:

$$N_i \cdot N_j \approx 1 + N_i \cdot [(u \cdot \nabla)N_i] + \frac{1}{2}N_i \cdot [(u \cdot \nabla)^2 N_i], \tag{3.31}$$

where we used $N_i^2 = 1$. Again, assuming the distortions to be small and smooth, we can rewrite (3.28) as:

$$N \approx z - \nabla h - \frac{z}{2}(\nabla h)^2, \tag{3.32}$$

and hence,

$$\partial_a N \approx -\partial_a \nabla h - z \nabla h \cdot (\partial_a \nabla h), \tag{3.33}$$

where $a = x, y$. Therefore, using (3.32) and (3.33) and the fact that $z \cdot \nabla h = 0$, we find:

$$N \cdot [\partial_a N] \approx 0. \tag{3.34}$$

Once again:

$$\partial_{a,b}^2 N \approx -\partial_{a,b}^2 \nabla h - z(\partial_n \nabla h) \cdot (\partial_a \nabla h) - z \nabla h \cdot (\partial_{a,b}^2 \nabla h), \tag{3.35}$$

and finally:

$$N \cdot [\partial_{a,b}^2 N] \approx -(\partial_b \nabla h) \cdot (\partial_a \nabla h). \tag{3.36}$$

Substituting (3.34) and (3.36) into (3.31) we find:

$$N_i \cdot N_j \approx 1 - \frac{1}{2}[(u \cdot \nabla)\nabla h]^2. \tag{3.37}$$

Substituting (3.37) in (3.29) we find:

$$\delta U_B \approx \frac{\kappa_L}{4} \sum_{i,u} [(u \cdot \nabla)\nabla h(R_i)]^2 \approx \frac{\kappa}{2} \int d^2 r \left[\nabla^2 h(r)\right]^2, \tag{3.38}$$

where κ ($\propto \kappa_L$) is the bending rigidity. Notice that the energy (3.38) is invariant under translations along the z direction and rotations around any axis (the energy is invariant under $h \rightarrow h + C_1 a + C_2 r \times \theta$ where C_1 and C_2 are constants and θ is the anti-clockwise oriented angle of rotation along a given axis). In momentum space we can rewrite (3.38) as:

$$\delta U_B = \frac{\kappa}{2} \int d^2 q q^4 |h(\boldsymbol{q})|^2, \tag{3.39}$$

showing the elastic cost in energy for bending the graphene sheet behaves as q^4 when $q \rightarrow 0$.

The quantum mechanics of bending can be easily obtained by quantization of the field $h(\boldsymbol{r})$. Introducing a momentum operator $P(\boldsymbol{q})$ that is canonically conjugated to $h(\boldsymbol{q})$:

$$\left[h(\boldsymbol{q}), P(\boldsymbol{q'}) \right] = i \delta^2 (\boldsymbol{q} - \boldsymbol{q'}), \tag{3.40}$$

we can write the Hamiltonian for the bending modes as:

$$H = \int d^2 q \left\{ \frac{P(-\boldsymbol{q})P(\boldsymbol{q})}{2\sigma} + \frac{\kappa q^4}{2} h(-\boldsymbol{q})h(\boldsymbol{q}) \right\}, \tag{3.41}$$

where σ is graphene's 2D mass density. From the Heisenberg equations of motion for the operators it is trivial to find that $h(\boldsymbol{q})$ oscillates harmonically with a frequency given by:

$$\omega_{\text{flex}}(\boldsymbol{q}) = \left(\frac{\kappa}{\sigma} \right)^{1/2} q^2, \tag{3.42}$$

which is the dispersion of the acoustical flexural mode.

Notice that the existence of a mode dispersing like q^2 in two dimensions has some strong consequences. At long wavelengths, $q \rightarrow 0$, these modes have lower energy than acoustical (that disperse like $\omega_{\text{acoust.}}(q) \approx v_s q$, where v_s is the sound speed) and optical ($\omega_{\text{opt.}}(q) \approx \omega_0$ is independent of q) phonon modes and thus dominate the thermodynamics of free floating graphene at low temperatures. Consider, for instance, the mean square displacement of the height ($\omega_n = 2\pi n/\beta$ with $\beta = 1/T$ is the Matsubara frequency at temperature T):

$$\langle h^2 \rangle = \frac{1}{\beta} \sum_n \int \frac{d^2 q}{(2\pi)^2} \frac{1}{(\omega_n)^2 + \kappa q^4} = \int \frac{d^2 q}{(2\pi)^2} \frac{\coth \left(\frac{\beta \sqrt{\kappa} q^2}{2\pi} \right)}{2\sqrt{\kappa} q^2}$$

$$= \frac{1}{8\pi \sqrt{\kappa}} \int\limits_{\beta\sqrt{\kappa}/(2\pi L^2)}^{\beta\sqrt{\kappa}\Lambda^2/(2\pi)} du \frac{\coth(u)}{u}, \tag{3.43}$$

where we had to introduce an ultraviolet cut-off ($\Lambda \sim 1/a$) and an infrared cut-off ($1/L$ where L is the system size) because the integral is formally divergent in both limits. The ultraviolet divergence is not of concern since there is a physical cut-off which is the lattice spacing, the problem is the infrared divergence since it indicates an instability in the system. Let us consider two different limits. At low temperatures $(T \to 0)$ we can approximate $\coth(u) \approx 1$ and we get:

$$\langle h^2(T \to 0) \rangle \approx \frac{1}{4\pi \sqrt{\kappa}} \ln(\Lambda L) \tag{3.44}$$

indicating that even at zero temperature the quantum fluctuations are logarithmically divergent. This behavior is a manifestation of the Hohenberg–Mermin–Wagner theorem [14, 15] that states that long-range order is not possible is systems with a broken continous symmetry (in this case, translation in the direction perpendicular to the flat graphene configuration) in two dimensions. However, at high T we can approximate $\coth(u) \approx 1/u$, the ultraviolet cut-off becomes irrelevant, and we find

$$\langle h^2(T \to \infty) \rangle \approx \frac{L^2}{4\kappa\beta} \tag{3.45}$$

and one finds a severe infrared divergence of the thermal fluctuations. Similar arguments can be made about the fluctuations of the normal N, namely,

$$\langle (\delta N)^2 \rangle \approx \langle (\nabla h)^2 \rangle \approx \frac{1}{4\kappa\beta} \ln \left[\frac{\sinh(\beta \sqrt{\kappa} \Lambda^2/(2\pi))}{\sinh(\beta \sqrt{\kappa}/(2\pi L^2))} \right], \tag{3.46}$$

and hence at low temperatures we find $\langle (\delta N)^2 (T \to 0) \rangle \approx \Lambda^2/(16\pi \sqrt{\kappa})$ and the problem is free of infrared divergences. At high temperatures we have $\langle (\delta N)^2 (T \to \infty) \rangle \approx \ln[\Lambda L]/(2\beta\kappa)$ and the fluctuations of the normal are logarithmically divergent. The interpretation of these results are straightforward: at zero temperature the graphene sheet should be rough but mostly flat, as one increases the temperature the thermal fluctuations become large and the graphene crumples (divergence of the normal indicates the formation of folds and creases). These results are modified by non-linear effects that we do not consider here [16]. More important than those is the fact that in most graphene experiments, where the measurement of electric properties depends of physically constraining the samples, the basic symmetries of free floating graphene are explicitly broken. This is what we consider in the next section.

3.5 Constrained Graphene

Since graphene is not floating in free space and since there can be impurities that hybridize with the carbon atoms changing the structural properties, it is important to consider symmetry breaking processes that change the picture presented in the last section. Consider, for instance, if graphene is under tension. In this case, the rotational symmetry along the x and y axis is broken and a new term is allowed in the energy, that reads:

$$U_T = \frac{\gamma}{2} \int d^2r \, (\nabla h(\mathbf{r}))^2 , \qquad (3.47)$$

where γ plays the role of the surface tension. It is easy to see that the dispersion is modified to:

$$\omega(\mathbf{q}) = q\sqrt{\frac{\kappa}{\sigma}q^2 + \frac{\gamma}{\sigma}}, \qquad (3.48)$$

indicating that the dispersion of the flexural modes becomes linear in q, as $q \to 0$, under tension. Notice that this is the case of a 3D solid, that is, the flexural mode becomes an ordinary acoustic phonon mode. This is what happens in graphite where the interaction between layers does not preserve the rotational symmetry around a single plane. It is easy to see that in this case the graphene fluctuations become suppressed: at low temperatures we find $\langle h^2(T \to 0) \rangle$ is constant while $\langle h^2(T \to \infty) \rangle \approx T \ln(\Lambda L)$ is log divergent. Also, $\langle (\delta N)^2(T) \rangle$ is well behaved at all temperatures.

In the presence of a substrate, the translational symmetry along the z direction is broken and a new term is allowed:

$$U_S = \frac{v}{2} \int d^2r (h(\mathbf{r}) - s(\mathbf{r}))^2, \qquad (3.49)$$

where v is the interaction strength with the substrate and $s(\mathbf{r})$ is a reference height. Notice that in this case the flexural mode becomes gapped (by means of the h^2 term) and hence all fluctuations (quantum and thermal) are quenched. Therefore, the most generic energy for small and smooth height distortions of graphene is:

$$U = \frac{1}{2} \int d^2r \left\{ \kappa \left[\nabla^2 h(\mathbf{r}) \right]^2 + \gamma \, [\nabla h(\mathbf{r})]^2 + v(h(\mathbf{r}) - s(\mathbf{r}))^2 \right\}. \qquad (3.50)$$

Let us consider the case of (3.50) in the presence of a substrate [17]. Minimization of the free energy with respect to h leads to the equation:

$$\kappa \nabla^4 h - \gamma \nabla^2 h + vh = vs. \qquad (3.51)$$

A particular solution of the non-homogeneous equation can be obtained by Fourier transform:

$$h(\mathbf{q}) = \frac{s(\mathbf{q})}{1 + \ell_t^2 q^2 + \ell_c^4 q^4}, \qquad (3.52)$$

where

$$\ell_t = \left(\frac{\gamma}{\upsilon}\right)^{1/2}, \qquad \ell_c = \left(\frac{\kappa}{\upsilon}\right)^{1/4}, \tag{3.53}$$

are the length scales associated with tension and curvature, respectively. Notice that (3.52) implies that the surface height more or less follows the substrate landscape, as expected.

Let us consider the case of a random substrate where the probability of a substrate height between s and $s + ds$ is given by:

$$P(s) = \frac{1}{\mathcal{N}} \exp\left\{ -\int d^2r \frac{s^2}{2s_0^4} \right\}, \tag{3.54}$$

where $\mathcal{N}$ is a normalization factor:

$$\mathcal{N} = \int\limits_{-\infty}^{+\infty} ds\, P(s) = \left(\frac{\sqrt{2\pi}\, s_0^2}{a}\right)^N, \tag{3.55}$$

where N is the number of surface cells, a is their lattice spacing, and s_0 is the average height variation. In this case, the height correlation function can be shown to be:

$$\overline{s(r)s(r')} = s_0^4 \delta(r - r'). \tag{3.56}$$

In momentum space (3.56) is written as:

$$\overline{s(q)s(q')} = (2\pi)^2 s_0^4 \delta(q + q'). \tag{3.57}$$

Using (3.57) we can immediately compute the height-height correlation function as a function of the in-plane distance:

$$\overline{h(r)h(0)}$$

$$= \int \frac{d^2q}{(2\pi)^2} \int \frac{d^2q'}{(2\pi)^2} \frac{e^{iq\cdot r}}{(1 + \ell_t^2 q^2 + \ell_c^4 q^4)(1 + \ell_t^2 (q')^2 + \ell_c^4 (q')^4)} \overline{s(q)s(q')}$$

$$= \frac{s_0^4}{2\pi} \int\limits_0^\infty dq \frac{q\, J_0(qr)}{(1 + \ell_t^2 q^2 + \ell_c^4 q^4)^2}$$

$$= \frac{s_0^4}{2\pi \ell_t^8} \int\limits_0^\infty dq \frac{q\, J_0(qr)}{(q^2 + P_+^2)^2 (q^2 + P_-^2)^2}, \tag{3.58}$$

where we have defined:

$$P_\pm^2 = \frac{1}{2\ell^2}\left(1 \pm \sqrt{1 - \zeta^2}\right),$$

$$\ell = \frac{\ell_c^2}{\ell_t} = \sqrt{\frac{\kappa}{\gamma}},$$

$$\zeta = 2\left(\frac{\ell_c}{\ell_t}\right)^2 = 2\sqrt{\frac{\kappa \upsilon}{\gamma^2}}, \qquad (3.59)$$

and the behavior of the integral depends on whether ζ is smaller or bigger than 1. For $r \gg \lambda = 1/\mathrm{Min}(\mathrm{Re}(P_\pm))$ one finds:

$$\overline{h(r)h(0)} \sim \frac{s_0^4}{2\pi \lambda^2} e^{-r/\lambda}, \qquad (3.60)$$

showing that the height fluctuations are short ranged and decay with a characteristic length scale given by λ. The local height variations are given by:

$$\overline{h^2(0)} = \frac{s_0^4}{2\pi \ell_t^2} w_\pm(\zeta), \qquad (3.61)$$

where for $\zeta > 1$ we have:

$$w_+(\zeta) = \frac{1}{(\zeta^2 - 1)^{3/2}}\left[\frac{\pi}{2} - \arctan\left(\frac{1}{\sqrt{\zeta^2 - 1}}\right) - \frac{\sqrt{\zeta^2 - 1}}{\zeta^2}\right], \qquad (3.62)$$

and for $\zeta < 1$:

$$w_-(\zeta) = \frac{1}{(1 - \zeta^2)^{3/2}}\left[\frac{\sqrt{1 - \zeta^2}}{\zeta^2} - \frac{1}{2}\ln\left(\frac{1 + \sqrt{1 - \zeta^2}}{1 - \sqrt{1 - \zeta^2}}\right)\right]. \qquad (3.63)$$

The function $w(\zeta)$ is shown in Fig. 3.11. Notice as $\zeta \gg 1$, that is, when $\ell_c \gg \ell_t$, the height variations are strongly suppressed and the system is essentially flat. This happens because in this limit the interaction with the substrate is strong, and the bending rigidity is large, compared with the tension.

3.6 Deformed Graphene

Graphene is one atom thick and hence very soft. Just as any soft material, graphene can be easily deformed, especially in the out of the plane direction where the restitution forces vanish. These deformations couple directly to the electrons, rendering graphene a unique example of a metallic membrane [18]. Let us consider what happens to the electrons when the graphene is deformed either by bending or strain.

Fig. 3.11 Amplitude of the
local height variations in
graphene supported in a
random substrate, $w(\zeta)$

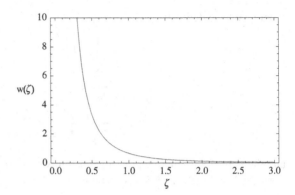

One effect is the change in the distance between the atoms, the other is the change in the overlap between the different orbitals. In both cases, the hopping energy between different carbon atoms is affected. Consider the case where at a certain site $\boldsymbol{R}_i$ the nearest neighbor hopping energy changes from t_0 to $t_0 + \delta t_i(\boldsymbol{\delta})$ in the direction of $\boldsymbol{\delta}$. In this case we have to add a new term to the Hamiltonian (3.16):

$$
\begin{aligned}
\delta H_0 &= -\sum_{k,k'} a_k^\dagger b_{k'} \sum_{i,\delta} \delta t_i(\boldsymbol{\delta}) e^{i(k-k')\cdot\boldsymbol{R}_i - i\boldsymbol{\delta}\cdot k'} \\
&\approx -\sum_{q,q'} \psi_{A,q}^\dagger \psi_{B,q'} \sum_{i,\delta} \delta t_i(\boldsymbol{\delta}) e^{i(q-q')\cdot\boldsymbol{R}_i} e^{-i\boldsymbol{\delta}\cdot\boldsymbol{Q}_1} \\
&\approx \int d^2 r\, \psi_A^\dagger(\boldsymbol{r}) \psi_B(\boldsymbol{r}) \mathcal{A}(\boldsymbol{r}),
\end{aligned}
\tag{3.64}
$$

where

$$
\mathcal{A}(\boldsymbol{r}) = \mathcal{A}_x(\boldsymbol{r}) + i\mathcal{A}_y(\boldsymbol{r}) = -\sum_{\delta} \delta t_i(\boldsymbol{\delta}) e^{-i\boldsymbol{\delta}\cdot\boldsymbol{Q}_1}.
\tag{3.65}
$$

Notice that if we insert (3.64) into (3.26), the full Hamiltonian becomes:

$$
H = \int d^2 r\, \Psi^\dagger(\boldsymbol{r}) \left[\boldsymbol{\sigma} \cdot (i v_F \nabla + \mathcal{A}) - \mu \right] \Psi(\boldsymbol{r}),
\tag{3.66}
$$

where $\mathcal{A} = (\mathcal{A}_x, \mathcal{A}_y)$ and hence the changes in the nearest neighbor hopping couple like a gauge field. One should not worry about a possible broken time reversal symmetry here because there is another nonequivalent point in the Brillouin zone, at K', which is related to the K point by inversion symmetry. Hence, in the K' point the "magnetic field" is reversed and therefore there is no net time reversal symmetry breaking.

Let us consider the problem of the hopping between next-nearest neighbor sites. The Hamiltonian in this case is:

$$H_1 = -t_0' \sum_{\langle i,j \rangle} \left(a_i^\dagger a_j + b_i^\dagger b_j + \text{h.c.} \right), \tag{3.67}$$

which describes essentially hopping on a triangular lattice with lattice spacing $\sqrt{3}a$ with nearest neighbor vectors:

$$\begin{aligned}
\boldsymbol{v}_1 &= \sqrt{3}a(0, 1), \\
\boldsymbol{v}_2 &= \sqrt{3}a(\sqrt{3}/2, -1/2), \\
\boldsymbol{v}_3 &= \sqrt{3}a(-\sqrt{3}/2, -1/2).
\end{aligned} \tag{3.68}$$

By Fourier transforming (3.67) we find:

$$H_1 = \sum_k \epsilon(\boldsymbol{k}) \left(a_k^\dagger a_k + b_k^\dagger b_k \right), \tag{3.69}$$

where

$$\begin{aligned}
\epsilon(\boldsymbol{k}) &= -t_0' \sum_v e^{i\boldsymbol{k}\cdot\boldsymbol{v}} \\
&= 2t_0' \left[\cos(\sqrt{3}ak_y) + 4\cos(3ak_x/2)\cos(\sqrt{3}ak_y/2) \right].
\end{aligned} \tag{3.70}$$

For $\boldsymbol{k} = \boldsymbol{Q}_1 + \boldsymbol{q}$ we find:

$$\epsilon(\boldsymbol{Q}_1 + \boldsymbol{q}) \approx -3t_0' + \frac{9t_0'a^2}{2}q^2 \tag{3.71}$$

for $q \ll Q_1$. Notice that the first term of Eq. 3.71 leads to a shift in the chemical potential and the second term introduces a quadratic term in the dispersion. For $q \ll t_0/(3at_0')$ we can disregard this term and consider only the chemical potential shift. *Ab initio* calculations estimate that $t_0' \approx 0.1t_0$ providing a good range where this approximation is valid [19].

Consider again the deformations of the graphene surface. In this case t_0' in (3.71) has to be replaced by $t' = t_0' + \delta t'$ and, in complete analogy with (3.64), we find:

$$\delta H_1 \approx \int d^2 r \Phi(r) \left(\psi_A^\dagger(r)\psi_A(r) + \psi_B^\dagger(r)\psi_B(r) \right), \tag{3.72}$$

where

$$\Phi(r) = -3 \sum_v \delta t_i'(v) e^{-iv\cdot\boldsymbol{Q}_1} \tag{3.73}$$

where we have used (3.11), (3.21), and (3.68). In this case, changes in the next-to-nearest neighbor hopping couple to the Dirac fermions as a scalar potential. The final Dirac Hamiltonian has the form:

$$H = \int d^2r \Psi^\dagger(\boldsymbol{r}) \left\{ \boldsymbol{\sigma} \cdot (i v_F \nabla + \boldsymbol{\mathcal{A}}) - \mu + \Phi(\boldsymbol{r}) \right\} \Psi(\boldsymbol{r}). \qquad (3.74)$$

Given the local changes in the hopping energies the scalar and vector potentials can be readily computed through (3.65) and (3.73), respectively.

These changes in hopping energies can be connected to changes in the structure of the lattice. Let us consider the case of in-plane distortions. In this case, the only change in the hopping energy is due to the change in the distance between the p_z orbitals. Consider the case where the distance between sublattices changes by a distance $\delta\ell$ in the direction of $\boldsymbol{\delta}$, in first order we have:

$$\delta t \approx (\partial t / \partial a)\delta\ell, \qquad (3.75)$$

$$\delta\ell \approx (\boldsymbol{\delta} \cdot \nabla)\boldsymbol{u}, \qquad (3.76)$$

where $u(\boldsymbol{r})$ is the lattice displacement. Substituting the above expression in (3.65) we get:

$$\begin{aligned}
\mathcal{A}_x^{(u)}(\boldsymbol{r}) &\approx \alpha a(u_{xx} - u_{yy}), \\
\mathcal{A}_y^{(u)}(\boldsymbol{r}) &\approx \alpha a\, u_{xy},
\end{aligned} \qquad (3.77)$$

where α is a constant with dimensions of energy and we have used the standard definition of the strain tensor:

$$u_{ij} = \frac{1}{2}\left(\frac{\partial u_i}{\partial x_j} + \frac{\partial u_j}{\partial x_i}\right). \qquad (3.78)$$

An analogous calculation for the change in the next-to-nearest neighbor hopping energy leads to:

$$\Phi^{(u)}(\boldsymbol{r}) \approx g(u_{xx} + u_{yy}). \qquad (3.79)$$

Equations 3.77 and 3.79 relate the strain tensor to potentials that couple directly to the Dirac particles.

Less trivial is the coupling to the out-of-plane modes since those involve rotations of the orbitals. In Fig. 3.12 we show a rotation of two orbitals by an angle θ. The rotation mixes π and σ states and for small angles the change in the hybridization energy is given by:

$$t \approx t_0 + \delta V \theta^2, \qquad (3.80)$$

where δV is the energy mixing between π and σ states. Notice that $\theta \approx a/R$ where R is the curvature radius. In terms of the height variable we can write, in analogy with (3.37), that the change in the hopping amplitude due to bending in the direction $\boldsymbol{u}$ is given by:

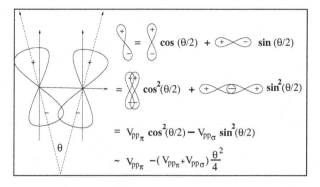

Fig. 3.12 Rotation of p_z orbitals

$$\delta t \approx \delta V \left[(\boldsymbol{u} \cdot \boldsymbol{\nabla}) \boldsymbol{\nabla} h \right]^2 . \tag{3.81}$$

On the one hand, if $\boldsymbol{u}$ is a nearest neighbor vector $\boldsymbol{\delta}$ we get from (3.65):

$$\mathcal{A}_x^{(h)}(\boldsymbol{r}) = -\frac{3 E_{ab} a^2}{8} \left[(\partial_x^2 h)^2 - (\partial_y^2 h)^2 \right],$$
$$\mathcal{A}_y^{(h)}(\boldsymbol{r}) = \frac{3 E_{ab} a^2}{4} \left(\partial_x^2 h + \partial_y^2 h \right) \partial_x h \partial_y h, \tag{3.82}$$

where the coupling constant E_{ab} depends on microscopic details. On the other hand, if $\boldsymbol{u}$ is the next-to-nearest neighbor vector we find, in accordance with (3.73):

$$\Phi^{(h)}(\boldsymbol{r}) \approx -E_{aa} a^2 \left[\nabla^2 h(\boldsymbol{r}) \right]^2 , \tag{3.83}$$

where E_{aa} is an energy scale associated with the mixing between orbitals. The main conclusion is therefore that for smooth distortions of graphene due to strain or bending the Dirac particles are subject to scalar and vector potentials leading to an "electrodynamics" that is purely geometrical (there is no electric charge associated with the "electric" and "magnetic" fields created by structural deformations). This structural "electrodynamics" has strong consequences for the electronic motion in graphene leading to many unusual effects that cannot be found in ordinary materials. In particular, one can manipulate the electrons by "constructing" appropriate deformations of the lattice that mimic electric and magnetic fields. This is the so-called *strain engineering* and is a field of research that is still in its infancy [20, 21].

3.6.1 A Non-Trivial Example: The Scroll

So far we have discussed small distortions of the graphene sheet, but as a soft material graphene can be bent by large angles making completely new structures (Castro

Fig. 3.13 Graphene scroll in the form of an Archimedian spiral

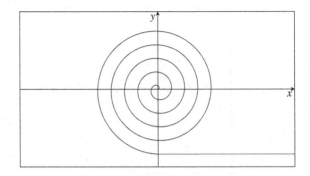

Neto et al., Bending and scrolling of a suspended graphene edge, "unpublished"). The scroll is an example of the softness of graphene that shows up as an interplay between the van der Waals energy, E_{vdW}, that makes graphene stick to itself and the bending energy, E_B, the energy cost to introduce curvature in the graphene sheet. Let us show first that the scroll is a stable configuration of a free-standing graphene sheet (Fig. 3.13).

We assume here that when graphene forms a scroll the equilibrium distance between folded regions is the same that one would find between the planes in graphite, that is, $d = 0.34$ nm. In this case, the scroll is described by the so-called Archimedian spiral:

$$r(\theta) = a\theta, \tag{3.84}$$

where $a = d/(2\pi)$. We assume that the scroll has length L along the scroll axis and consider the two main energy scales in this problem:

$$\frac{E_{vdW}}{L} = -\gamma \int_{\theta_0 + 2\pi}^{\theta_1} ds(\theta), \tag{3.85}$$

$$\frac{E_B}{L} = \frac{\kappa}{2} \int_{\theta_0}^{\theta_1} ds(\theta) \frac{1}{R^2(\theta)}, \tag{3.86}$$

where θ_0 is the angle at the edge and θ_1 is the total rolling angle of the scroll, $\gamma \approx 2.5$ eV/nm is the van der Waals coupling [22], $\kappa \approx 1$eV is the bending energy [19]. The radius of curvature of the Archimedian spiral is given by:

$$R(\theta) = \frac{a(1 + \theta^2)^{3/2}}{2 + \theta^2}, \tag{3.87}$$

Fig. 3.14 Energy of a scroll as a function of the angle of rotation

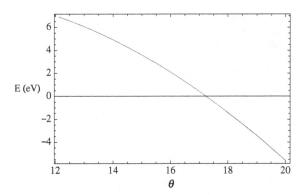

and

$$ds = a\sqrt{1 + \theta^2}d\theta, \tag{3.88}$$

is the infinitesimal arc-length of the spiral. The evaluation of the integrals is straight-forward:

$$\frac{E_{vdW}}{L} = -\frac{\gamma a}{2}\left[\ln\left(\frac{\theta_1 + \sqrt{1 + \theta_1^2}}{(\theta_0 + 2\pi) + \sqrt{1 + (2\pi + \theta_0)^2}}\right) + \frac{\theta_1(9 + 8\theta_1^2)}{(1 + \theta_1^2)^{3/2}}\right.$$
$$\left. -\frac{(\theta_0 + 2\pi)(9 + 8(\theta_0 + 2\pi)^2)}{(1 + (\theta_0 + 2\pi)^2)^{3/2}}\right], \tag{3.89}$$

$$\frac{E_B}{L} = \frac{\kappa a}{2}\left[\ln\left(\frac{\theta_1 + \sqrt{1 + \theta_1^2}}{\theta_0 + \sqrt{1 + \theta_0^2}}\right) + \theta_1(1 + \theta_1^2)^{1/2} - \theta_0(1 + \theta_0^2)^{1/2}\right]. \tag{3.90}$$

The angle θ_0 is determined by the condition that the force on the graphene sheet vanishes at the edge of the scroll:

$$F = -\frac{1}{L}\frac{\partial E}{\partial \theta_0} = -\frac{\kappa}{2a}\frac{(2 + \theta_0^2)^2}{(1 + \theta_0^2)^{5/2}} + \gamma a\sqrt{1 + (\theta_0 + 2\pi)^2} = 0, \tag{3.91}$$

and using the values of the parameters we find $\theta_0 \approx 5.85$. The total energy as a function of θ_1 is given by (3.89) and (3.90) and shown in Fig. 3.14. Notice that the scroll becomes stable for angles of rotation bigger than ≈ 17.23 which is equivalent to approximately 3 turns of the scroll. This result is in agreement with more involved calculations [23].

Consider what happens when a *uniform* magnetic field is applied perpendicular to the graphene plane. Notice that in the region of the scroll the component of the field

Fig. 3.15 Magnetic field perpendicular to the graphene sheet close to the scroll

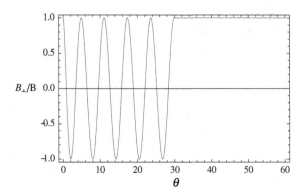

perpendicular to the graphene surface is changing periodically. In fact, the normal component to the graphene sheet at the scroll can be written as:

$$N(\theta) = \frac{1}{\sqrt{1+\theta^2}}(-\sin(\theta) - \theta\cos(\theta), \cos(\theta) - \theta\sin(\theta)), \tag{3.92}$$

and hence the component of the field perpendicular to the graphene sheet is:

$$B_\perp(\theta) = B \cdot N(\theta) = \frac{B}{\sqrt{1+\theta^2}}(\cos(\theta) - \theta\sin(\theta)), \tag{3.93}$$

which is shown in Fig. 3.15. Notice that from the point of view of the electrons, which are constrained to live in a two-dimensional universe, a uniform magnetic field in three dimensions, becomes an oscillating magnetic field in the presence of the scroll. An oscillating magnetic field has unusual effects on the electronic motion because the Lorentz force changes sign in the region where the field changes sign producing a "snake-like" motion around the regions of zero field. These *snake states* are chiral one-dimensional correlated systems with rather unusual properties [24]. While it is very hard to produce a magnetic field that oscillates in a short length scale, it is very easy to roll graphene on a short length scale. Therefore, snake states should be easy to observe in rolled graphene sheets.

3.7 Conclusions

Graphene is an unusual system that shares properties of soft and hard matter and mimics problems in particle physics. These properties have their origin in the nature of its chemical bonds, the s-p hybridization, and the low dimensionality. All these ingredients create a new framework for theoretical exploration which is still very much in its beginning. The possibility of studying electronics of deformed surfaces is quite intriguing because of its exotic "electrodynamics". The field of electronic membranes was born out of the discovery of graphene in 2004 and much has still to be understood.

In these notes I covered some very basic aspects of graphene chemistry and physics. We have studied the basic Hamiltonian that describes the overall band structure of graphene. We have seen that flexural modes, nonexistent in three dimensional solids, are fundamental for the understanding of the structural stability of the material. We have seen that substrates can change considerably the flexural modes and hence control the height fluctuations in supported samples. Finally, we have studied the various ways that graphene deformations, either by strain or bending, can modify the electron propagation in this material. In particular, new structures that can be created out of flat graphene, such as the graphene scroll, can have exotic properties, such as snake states, in the presence of applied uniform magnetic fields. The present notes add a notch to the material already published in Ref. [19]. Nevertheless, the whole subject of graphene electronics and structure is much bigger than that and there is so much more to be understood.

It is quite obvious that a material that is structurally robust, still flexible, and extremely clean, has enormous potential for technological applications. By understanding these unusual properties one can harvest new functionalities that did not exist before. Progress in material science unavoidably leads to new riches. Graphene is considered today one of the most promising candidates for a new era of a carbon-based technology, possibly supplanting the current silicon-based one. However, only progress in the understanding of the basic properties of these systems can actually drive the technological progress.

Acknowledgments It is a pleasure to acknowledge countless conversations with Eva Andrei, Misha Fogler, Andre Geim, Francisco Guinea, Silvia Kusminskiy, Valeri Kotov, Alessandra Lanzara, Caio Lewenkopf, Johan Nilsson, Kostya Novoselov, Eduardo Mucciolo, Vitor Pereira, Nuno Peres, Marcos Pimenta, Tatiana Rappoport, João Lopes dos Santos, and Bruno Uchoa. This work was possible due to the financial support of a Department of Energy grant DE-FG02-08ER46512 and the Office of Naval research grant MURI N00014-09-1-1063.

References

1. Novoselov, K.S., Geim, A.K., Morozov, S.V., Jiang, D., Zhang, Y., Dubonos, S.V., Grigorieva, I.V., Firsov, A.A.: Electric field effect in atomically thin carbon films. Science **22**, 666 (2004)
2. Geim, A.K., Novoselov, K.S.: The rise of graphene. Nat. Mat. **6**, 183 (2007)
3. Shankar, R.: Renormalization group approach to interacting fermions. Rev. Mod. Phys. **66**, 129 (1994)
4. Baym, G., Pethick, C.: Landau Fermi-Liquid Theory. Wiley, New York (1991)
5. Castro Neto, A.H., Guinea, F., Peres, N.M.R.: Drawing conclusions from graphene. Phys. World **19**, 33 (2006)
6. Gonzalez, J., Guinea, F., Vozmediano, M.A.H.: Non-Fermi liquid behavior of electrons in the half-filled honeycomb lattice: a renormalization group approach. Nucl. Phys. B **424**, 595 (1994)
7. Baym, G., Chin, S.A.: Landau theory of relativistic Fermi liquids. Nucl. Phys. A **262**, 527 (1976)

8. Slater, J.C., Koster, G.F.: Simplified LCAO method for the periodic potential problem. Phys. Rev. **94**, 1498 (1954)
9. Gordon, B.: Lectures on Quantum Mechanics. Addison-Wesley, Reading, MA (1990)
10. Pauling, L.: The Nature of the Chemical Bond. Cornell University Press, Ithaca (1960)
11. Castro Neto, A.H., Guinea, F.: Impurity induced spin-orbit coupling in graphene. Phys. Rev. Lett. **103**, 026804 (2009)
12. Harrison, W.A.: Elementary Electronic Structure. World Scientific, Singapore (2005)
13. Painter, G.S., Ellis, D.E.: Electronic structure and optical properties of graphite from a variational approach. Phys. Rev. B **1**, 4747 (1970)
14. Hohenberg, P.C.: Existence of long-range order in one and two dimensions. Phys. Rev. **158**, 383 (1967)
15. Mermin, N.D., Wagner, H.: Absence of ferromagnetism or antiferromagnetism in one- or two-dimensional isotropic Heisenberg models. Phys. Rev. Lett. **17**, 1133 (1966)
16. Chaikin, P., Lubensky, T.C.: Introduction to Condensed Matter Physics. Cambridge University Press, Cambridge (1995)
17. Swain, P.S., Andelman, D.: The influence of substrate structure on membrane adhesion. Langmuir **15**, 8902 (1999)
18. Kim, E.-A, Castro Neto, A.H.: Graphene as an electronic membrane. Europhys. Lett. **84**, 57007 (2008)
19. Castro Neto, A.H., Guinea, F., Peres, N.M.R., Novoselov, K.S., Geim, A.K.: The Electronic Properties of Graphene. Rev. Mod. Phys. **81**, 109 (2009)
20. Pereira, V.M., Castro Neto, A.H.: All-graphene integrated circuits via strain engineering. Phys. Rev. Lett. **103**, 046801 (2009)
21. Guinea, F., Katsnelson, M.I., Geim, A.K.: Energy gaps and a zero-field quantum Hall effect in graphene by strain engineering. Nat. Phys. **6**, 30 (2010)
22. Tománek, D.: Mesoscopic origami with graphite: scrolls, nanotubes, peapods. Phys. B **323**, 86 (2002)
23. Braga, S.F., Coluci, V.R., Legoas, S.B., Giro, R., Galvão, D.S., Baughman, R.H.: Structure and dynamics of carbon nanoscrolls. Nano Lett. **4**, 881 (2004)
24. Müller, J.E.: Effect of a nonuniform magnetic field on a two-dimensional electron gas in the ballistic regime. Phys. Rev. Lett. **68**, 385 (1992)

Chapter 4
Strong Electronic Correlations: Dynamical Mean-Field Theory and Beyond

Hartmut Hafermann, Frank Lechermann, Alexei N. Rubtsov,
Mikhail I. Katsnelson, Antoine Georges and Alexander I. Lichtenstein

Abstract This chapter aims at a pedagogical introduction to theoretical approaches
of strongly correlated materials based on dynamical mean-field theory (DMFT) and
its extensions. The goal of this theoretical construction is to retain the many-body
aspects of local atomic physics within the extended solid. After introducing the
main concept at the level of the Hubbard model, we briefly review the theoretical insights into the Mott metal-insulator transition that DMFT provides. We then
describe realistic extensions of this approach which combine the accuracy of first-
principle Density-Functional Theory with the treatment of local many-body effects
within DMFT. We further provide an elementary discussion of the continuous-time
Quantum Monte Carlo schemes for the numerical solution of the DMFT effective
quantum impurity problem. Finally, the effects of short-range non-local correla-

H. Hafermann (✉)
Centre de Physique Théorique, CNRS,
École Polytechnique, 91128 Palaiseau Cedex, France
e-mail:hartmut.hafermann@cpht.polytechnique.fr

F. Lechermann · A. I. Lichtenstein
I. Institut für Theoretische Physik,
Universität Hamburg, 20355 Hamburg, Germany

A. N. Rubtsov
Department of Physics, Moscow State University,
119992 Moscow, Russia

M. I. Katsnelson
Institute for Molecules and Materials,
Radboud University Nijmegen,
Nijmegen, 6525 AJ, The Netherlands

A. Georges
Centre de Physique Théorique, CNRS,
École Polytechnique, 91128 Palaiseau Cedex, France

Collège de France,
11 place Marcellin Berthelot, 75005 Paris, France

D. C. Cabra et al. (eds.), *Modern Theories of Many-Particle Systems in Condensed
Matter Physics*, Lecture Notes in Physics 843, DOI: 10.1007/978-3-642-10449-7_4,
© Springer-Verlag Berlin Heidelberg 2012

tions within cluster extensions of the DMFT scheme, as well as long-range fluctuations within the fully renormalized dual-fermion perturbation scheme are discussed extensively.

4.1 The Physics of Strongly Correlated Materials

The field of strongly correlated electron systems is at the frontier of modern condensed matter research. The complex interplay of many degrees of freedom in many-body quantum systems with strong interactions leads to very diverse phenomena such as the fractional quantum Hall effect [1], itinerant electron magnetism, heavy-fermion [2, 3] and non-Fermi-liquid behavior, the Kondo effect [4] and high-temperature superconductivity [5, 6].

The term "correlations" is often used as a synonym for interacting, without acknowledging its actual meaning; it emphasizes the fact that electrons, e.g. in a crystal, do not move independently. Whether or not an electron will hop to a neighboring site or orbital (considering only processes that are compatible with the Pauli principle) will strongly depend on whether the site is already occupied, due to the Coulomb repulsion. This real space picture is in strong contrast to the reciprocal space picture of independent, nearly free electrons moving in the background potential of the nuclei as described by Bloch waves. Strong electronic correlations are therefore expected in materials containing atoms with open d- or f-shells, with a small overlap between orbitals and correspondingly narrow bands. Such materials often exhibit remarkable properties. Examples are the transition metals V, Fe, Co, Ni, Cu and their oxides such as the Mott insulator NiO and the possible singlet insulator VO_2 [7]. The CuO_2 planes determine the properties of the high-temperature cuprate superconductors. Recently superconductivity has been observed in iron-based compounds, the iron-pnictide superconductors [8]. Materials containing rare earth or actinide ions with partially filled f-shells can exhibit heavy fermion behavior with effective masses thousand times larger than the electron mass.

The Coulomb repulsion due to the confinement of the electrons in these orbitals is of the order of the kinetic energy. This often results in a delicate balance between localization and delocalization. As a consequence, these materials are very sensitive to externally controllable parameters, such as pressure and doping, and are therefore promising candidates for applications. As of today, applications arising from a deepened understanding of strongly correlated materials are only beginning to be explored [9]. The development of reliable theoretical tools to calculate the material specific properties of strongly correlated materials therefore remains one of the primary concerns of modern theoretical condensed matter physics.

4.1.1 Theoretical Description

The description of correlated materials is theoretically challenging. The framework of density functional theory (DFT) provides a quantitative description with remarkably accurate predictions (apart from some shortcomings, such as the band-gap problem) for simple metals, semiconductors and band-insulating materials. However, the failure of approximations to the DFT, such as the local density approximation (LDA), to capture the physics of strong correlations, renders a realistic description of correlated systems difficult. A classic example is the Mott insulator NiO which is erroneously predicted to be metallic in absence of magnetic long-range order. There have been attempts to improve upon the LDA, such as the LDA+U approach [10] and the GW approximation [11]. While introducing a wavevector dependence into the self-energy, the applicability of the GW approximation is still restricted to systems with relatively weak correlations. The static nature of the LDA+U describes the formation of Hubbard bands for systems with spin and orbital order, but ignores dynamical correlations and fails to reliably predict the temperature dependent phase diagram [12].

Due to such difficulties, a quantitative understanding has only become possible in recent years. This is not primarily due to a strong increase of computational resources, but rather the development of new theoretical concepts and methods. The main difficulty lies in the fact that two (and possibly more) vastly different energy scales play an important role in the redistribution of the spectral weight. The description of the low energy physics requires a very accurate understanding of the high-energy excitations and their feedback on low energy excitations. In addition, the huge degeneracy in the spin sector leads to a large number of possible ground states. Quantum fluctuations and singlet correlations play an important role and might favor exotic states [3]. The interplay between low energy spin fluctuations and the fermionic excitations is believed to ultimately lead to the phenomenon of high-temperature superconductivity [6].

In Sect. 4.2 we will introduce the dynamical mean-field theory (DMFT), which has been a major step forward in understanding correlated materials [13, 14]. As outlined in Sect. 4.3, it considerably improved our insight into the physics of the Mott transition in fermionic systems [15]. For the first time it allowed a consistent description of both the low-energy coherent features—the long-lived quasiparticle excitations—and the incoherent high-energy excitations due to the Coulomb interaction, acting on short timescales. The former give rise to the quasiparticle peak at the Fermi level, while the latter manifest themselves in the broad Hubbard bands at high energies, giving rise to the well-known three peak structure of the spectrum [14].

The success of DMFT to capture the local temporal quantum fluctuations was exploited to merge it with a material specific theory in the LDA + DMFT approach [16, 17], which by now has been applied to a variety of systems [12]. A successful application was the correct prediction of the temperature dependence of the local moment in the transition metals iron and nickel [18]. Section 4.4 discusses further examples.

Despite the fact that DMFT considerably simplifies the calculation of material specific properties of correlated materials, the underlying multiorbital Anderson impurity model still remains a complicated many-body problem, which requires highly accurate and efficient methods for its solution. The versatile applications of the Anderson model in particular in the context of DMFT have brought about a wealth of solution methods [14]. Apart from approximate methods, established methods such as exact diagonalization and the Hirsch-Fye quantum Monte Carlo technique provide reliable solutions, however at the cost of introducing systematical errors, either through a discrete representation of the impurity-bath hybridization function for the former or the discretization of the imaginary time interval into "slices" for the latter.

Recently, a new class of Monte Carlo based impurity solvers has emerged, the so-called continuous-time quantum Monte Carlo algorithms. These constitute a progress as regards efficiency as well as accuracy, since they are free of any systematic errors. They allow to handle impurity problems with as many as seven orbitals required for problems with open f-shells. These solvers exist in mainly two distinct flavors, determined by the expansion parameter of the effective action, i.e. the weak-coupling (or interaction expansion) and strong-coupling (hybridization expansion) approach. These will be described in detail in Sect. 4.5.

The local nature of DMFT restricts its applicability to materials and properties where spatial correlations can be neglected. Attempts to include the effect of short-range correlations have primarily been based on various cluster generalizations of DMFT [19–22], which will be reviewed in Sect. 4.6. The range of the correlations included is determined by the cluster size. While these approaches constitute a systematic way to improve upon DMFT, they become computationally highly demanding for large cluster size. They break translational invariance either in real or momentum space explicitly and might artificially favor states which order at some finite wave-vector. Effects caused by long-wavelength fluctuations and correlations related with a narrow region of reciprocal space, such as the vicinity of Van Hove singularities [23], can hardly be taken into account by cluster approaches. At present, no computationally feasible approaches are available to reliably predict the properties of correlated materials where the correlations are manifestly long-ranged. Attempts to include the effect of long-range spatial correlations beyond the dynamical mean-field description are mainly based on diagrammatic corrections to DMFT [24–26]. In the dual fermion approach a fully renormalized expansion around DMFT is formulated in terms of auxiliary, so-called dual fermions. This approach has several advantages over straightforward diagrammatic expansions and will be introduced in detail in Sect. 4.7.

4.2 The Dynamical Mean-Field Theory Construction: An Embedded Atom

Dynamical mean-field theory is by now a well-established description of strongly correlated electron systems [12, 14]. DMFT maps the lattice problem to a local quantum impurity problem embedded in an electronic bath subject to a self-consistency condition [27]. It becomes exact in the limit of infinite coordination number [28], where quantum fluctuations become purely local. Its applicability is justified when the physics is dominated by strong local interactions and spatial correlations do not play a too important role.

4.2.1 A Reminder on Classical Mean-Field Theory

DMFT may be viewed as a quantum-mechanical generalization of classical mean-field theories, hence a brief reminder of the latter is useful. Consider for example the Ising model:

$$H = \sum_{\langle ij \rangle} J_{ij} S_i S_j - h \sum_i S_i. \tag{4.1}$$

Mean-field theory introduces an effective Hamiltonian of *independent spins*:

$$H_{\text{eff}} = -\sum_i \Delta_i S_i. \tag{4.2}$$

The Weiss effective field Δ_i is chosen in such a way that the local magnetization m_i is exactly reproduced by this Hamiltonian. This requires choosing the Δ_i's such that:

$$m_i = \tanh(\beta \Delta_i), \quad \text{i.e.: } \beta \Delta_i = \tanh^{-1} m_i, \tag{4.3}$$

where β is the inverse temperature. Up to this point, everything is exact: we have just represented the local magnetization as that of an effective Hamiltonian of independent spins (4.2). The mean-field theory *approximation*, first put forward by Pierre Weiss (1907), under the name of "molecular field theory", is that Δ_i can be approximated by the thermal average of the local field seen by the spin at site i, namely:

$$\Delta_i \simeq h + \sum_j J_{ij} m_j. \tag{4.4}$$

This is a *self-consistency condition* which relates the Weiss fields Δ_i to the local observables m_i. Taken together, (4.3) and (4.4) provide a set of self-consistent equations for the local magnetization:

$$m_i = \tanh\left(\beta h + \beta \sum_j J_{ij} m_j\right), \tag{4.5}$$

which, when specialized to a translationally invariant system with only nearest-neighbor couplings on a lattice of connectivity z, reads:

$$m = \tanh(\beta h + z\beta J m). \qquad (4.6)$$

The mean-field approximation becomes *exact* in the limit where the connectivity z of the lattice becomes large, with a proper scaling of the coupling $J = J^\star/z$: in this limit, spatial fluctuations of the local field become negligible, and it converges with probability one to its mean-value (4.4).

4.2.2 Generalization to the Quantum Case: Dynamical Mean-Field Theory

DMFT is a generalization of this construction to quantum many-body systems, following the same two essential steps: (1) a local (on-site) effective problem involving a Weiss field and (2) a self-consistency condition relating this Weiss field to the local observable. The two key differences with the classical case above is that the Weiss field is promoted to a full function of energy (or time) $\Delta(\omega)$ and that the effective problem, although local, is a fully interacting many-body model: an atom embedded in an effective medium.

Let us illustrate this construction for the simplest example of the Hubbard model:

$$H = \sum_{ij,\sigma} t_{ij} c_{i\sigma}^{\dagger} c_{j\sigma} + U \sum_{i} n_{i\uparrow} n_{i\downarrow} - \mu \sum_{i\sigma} n_{i\sigma}. \qquad (4.7)$$

It describes a collection of single-orbital "atoms" placed at the nodes R_i of a periodic lattice. The orbitals overlap from site to site, so that the fermions can hop with an amplitude t_{ij}. In the absence of hopping, each "atom" has four eigenstates: $|0\rangle$, $|\uparrow\rangle$, $|\downarrow\rangle$ and $|\uparrow\downarrow\rangle$ with energies 0, $-\mu$ and $U - 2\mu$, respectively. Clearly, $-\mu \equiv \epsilon_0$ plays the role of the one-electron energy of that single-level "atom".

The key observable on which DMFT focuses is the local Green function at a given lattice site:

$$G_{ii}^{\sigma}(\tau - \tau') \equiv -\langle T_\tau c_{i\sigma}(\tau) c_{i\sigma}^{\dagger}(\tau') \rangle. \qquad (4.8)$$

In order to represent that local observable, DMFT introduces the following effective problem [27]: an atom, with the same on-site level $-\mu$ and interaction U as the original lattice model, embedded into an effective medium with which it can exchange electrons.

The effective action describing this embedded atom is that of a single-orbital Anderson impurity model and reads:

$$S_{\text{eff}} = \int_0^\beta d\tau \left(\sum_\sigma c_\sigma^*(\tau) \frac{\partial}{\partial \tau} c_\sigma(\tau) + H_{\text{at}} \right) + \int_0^\beta d\tau \int_0^\beta d\tau' \sum_\sigma c_\sigma^*(\tau) \Delta_\sigma(\tau - \tau') c_\sigma(\tau').$$

$$(4.9)$$

In this expression, $H_{\text{at}} \equiv U \sum_i n_{i\uparrow} n_{i\downarrow} - \mu \sum_{i\sigma} n_{i\sigma}$ is the atomic Hamiltonian and the *"Weiss function"* $\Delta(\tau - \tau')$ has to be chosen in such a way that the local Green function

$$g(\tau - \tau'; [\Delta]) \equiv -\langle T_\tau c_\sigma(\tau) c_\sigma^\dagger(\tau') \rangle_{S_{\text{eff}}[\Delta]} \qquad (4.10)$$

resulting from S_{eff}, which we always denote by lowercase g, coincides with the on-site Green function of the lattice model (for simplicity we consider a case in which translational and spin symmetry are both unbroken):

$$g(\tau - \tau'; [\Delta]) = G_{ii}(\tau - \tau'). \qquad (4.11)$$

We now need to write the proper generalization of the self-consistency condition, relating the Weiss function to the local Green function on the lattice, hence closing the self-consistency loop: this is where an approximation comes in. One observes that the action S_{eff} can be viewed as the action of an interacting problem with an effective bare Green function:

$$\mathcal{G}^{-1}(i\omega) \equiv i\omega + \mu - \Delta(i\omega). \qquad (4.12)$$

Hence, the self-energy of the embedded effective atom reads:

$$\Sigma(i\omega) \equiv \mathcal{G}^{-1}(i\omega) - g^{-1}(i\omega) = i\omega + \mu - \Delta(i\omega) - g^{-1}(i\omega). \qquad (4.13)$$

The DMFT approximation then consists in *identifying this local self-energy with that of the lattice model*. The lattice model Green function correspondingly reads:

$$G^{-1}(k, i\omega) = i\omega + \mu - \epsilon_k - \Sigma(i\omega), \qquad (4.14)$$

in which ϵ_k is the Fourier transform of the hopping integral, i.e the dispersion relation of the non-interacting tight-binding band:

$$\epsilon_k \equiv \frac{1}{N} \sum_{ij} t_{ij} e^{ik \cdot (R_i - R_j)}. \qquad (4.15)$$

Under the assumption that the lattice self-energy can be identified to that of the embedded atom, Eq. 4.13, the condition (4.11) that the embedded atom Green's function coincides with the on-site Green's function of the lattice model $G_{ii}(i\omega) = 1/N \sum_k G(k, i\omega)$ hence reads:

$$\frac{1}{N} \sum_k \frac{1}{g^{-1}(i\omega) + \Delta(i\omega) - \epsilon_k} = g(i\omega). \qquad (4.16)$$

This equation should be thought of as the generalization of the classical Eq. 4.4 relating the Weiss-field to the local observable. Taken together, Eqs. 4.10 and 4.16 provide two equations which determine both G_{loc} and Δ in a self-consistent manner.

Some remarks are in order:

- The DMFT equations reproduce exactly both the non-interacting limit $U = 0$ and the isolated atom limit $t_{ij} = \epsilon_k = 0$. In the former, the solution of the effective action yields $\Sigma = 0$, and the Weiss function just adapts to reproduce the correct local non-interacting Green's function from (4.16). In the atomic limit, Eq. 4.16 clearly implies $\Delta = 0$, so that S_{eff} is the action of the isolated atom, as expected. The fact that the DMFT construction correctly reproduces both limits is a key to its success.

- DMFT neglects the momentum dependence of the self-energy (at least for the purpose of computing the local Green's function). Hence, all non local components are neglected: $\Sigma_{ij}(\omega) \simeq \Sigma(\omega)\delta_{ij}$. This can be shown to become exact in the limit of infinite lattice connectivity (or infinite dimensions) [28], with proper scaling $t_{ij} \propto 1/\sqrt{z}$ of the hopping amplitude.

4.2.3 The Embedded Atom as an Anderson Impurity Model

It is often quite useful to use a Hamiltonian form of the embedded atom described by the effective action S_{eff}. This can be done by explicitly introducing degrees of freedom describing the electron reservoir with which the atom can exchange electrons. Let us call $f_{l\sigma}^{\dagger}$ the creation operator associated with these degrees of freedom, where l is an index running over the full energy range required to faithfully represent Δ (see below). The Hamiltonian of the Anderson impurity model reads:

$$H_{eff} = \sum_{l\sigma} \tilde{\epsilon}_l f_{l\sigma}^{\dagger} f_{l\sigma} + \sum_{l\sigma} (\tilde{V}_l c_{\sigma}^{\dagger} f_{l\sigma} + \tilde{V}_l^* f_{l\sigma}^{\dagger} c_{\sigma}) + H_{at}[c^{\dagger}, c]. \qquad (4.17)$$

This Hamiltonian exactly reproduces the action (4.9), provided the energy levels $\tilde{\epsilon}_l$ and hybridization amplitudes $\tilde{V}_l$ are chosen in such a way that a faithful spectral representation of $\Delta(i\omega)$ is obtained:

$$\Delta(i\omega) = \sum_l \frac{|\tilde{V}_l|^2}{i\omega - \tilde{\epsilon}_l}. \qquad (4.18)$$

4.2.4 The Self-Consistency Loop in Practice

Since in general neither the Weiss field nor the Green function are known a priori, they have to be determined self-consistently. This is done by performing a self-consistency loop, as depicted in Fig. 4.1. The procedure is as follows: Starting from

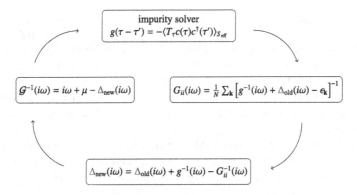

Fig. 4.1 Schematic illustration of the DMFT self-consistency loop

an initial guess for the hybridization function $\Delta_{old}(i\omega)$ and letting $\Sigma(i\omega) \equiv 0$ in $g^{-1}(i\omega) = i\omega + \mu - \Delta(i\omega) - \Sigma(i\omega)$, the lattice Green function is constructed according to

$$G(\boldsymbol{k}, i\omega) = \left[g^{-1}(i\omega) + \Delta_{old} - \epsilon_{\boldsymbol{k}}\right]^{-1}. \tag{4.19}$$

From its local part, $G_{ii}(i\omega) = (1/N) \sum_{\boldsymbol{k}} G(\boldsymbol{k}, i\omega)$, an updated hybridization function can be computed as

$$\Delta_{new}(i\omega) = \Delta_{old}(i\omega) + \left[g^{-1}(i\omega) - G_{ii}^{-1}(i\omega)\right]. \tag{4.20}$$

The hybridization enters the bare Green function of the impurity model, which serves as the input to the impurity solver. In the impurity solver step a new Green function $g(i\omega)$ is computed and inserted into (4.19) to close the self-consistency loop. According to (4.20), self-consistency is obviously reached when the Green function of the impurity model coincides with the on-site Green function of the lattice. In order to avoid oscillations around the fixed point of these equations, the old and new guesses for the hybridization function should be mixed. In practice, a linear mixing proves sufficient, which is easily implemented by multiplying the angular brackets in (4.20) with a parameter $\zeta \in (0, 1)$.

4.3 The Metal to Mott Insulator Transition from a DMFT Perspective

Arguably the most important phenomenon of strong correlation physics is that the Coulomb interaction can drive a material insulating, when a metal would be expected from band-structure (independent-electron) considerations. Outstanding examples are transition-metal oxides (e.g superconducting cuprates), fullerene compounds and organic conductors. For an extensive review, see e.g. Ref. [9]. A limited number of materials are right on the verge of this electronic instability. This is the case, for example, for V_2O_3, $NiS_{2-x}Se_x$ and for quasi two-dimensional organic conductors of the $\kappa-$ BEDT family. These materials are particularly interesting for the fundamental investigation of the Mott transition, since they offer the possibility of going from one phase to the other by varying some external parameter (e.g chemical composition, temperature, pressure,...).

One of the early successes of the DMFT approach has been to formulate a detailed theory of this phenomenon. Here we give a brief summary of the main aspects of this theoretical description, referring to [29] for a more detailed review.

- *Correlated metals: quasiparticles and Hubbard bands.* A strongly correlated metal close to the Mott transition displays two types of excitations. The low-energy excitations are Landau Fermi-liquid quasiparticles. High-energy excitations correspond to atomic-like transitions (broadened by the solid-state environment), e.g. removing or adding an electron in the d-shell. The low-energy quasiparticles carry a fraction Z of the spectral weight, and the high-energy excitations a fraction $1-Z$. Within DMFT, because the self-energy depends on frequency only, the effective mass enhancement of quasiparticles is directly related to Z: $m^\star/m = 1/Z$. The $k-$ integrated spectral function $A(\omega) = \sum_k A(k, \omega)$ (density of states, DOS) of the strongly correlated metal displays a three-peak structure [27], made of a quasiparticle band around the Fermi energy surrounded by lower and upper Hubbard bands (Fig. 4.2). The central quasiparticle peak in the DOS has a reduced width of order $ZW \sim \varepsilon_F^*$, where W is the bare electronic bandwidth. This energy scale is also the quasiparticle coherence scale below which a Fermi-liquid description applies. The lower and upper Hubbard bands are separated by an energy scale $\Delta\epsilon$. It is one of the main strengths of DMFT to be able to describe both types of excitations on an equal footing, as well as transfers of spectral weight between these spectral features as temperature or coupling are varied. These transfers of spectral weight are frequently observed in spectroscopies of strongly correlated materials.
- *The Mott transition: Brinkman-Rice and Mott-Hubbard.* The paramagnetic DMFT solution of the half-filled Hubbard model displays a Mott transition as the coupling U/W is increased. The mean-field solution corresponding to the paramagnetic metal at $T = 0$ disappears at a critical coupling U_{c2} (Fig. 4.3). At this critical value, the quasiparticle weight vanishes ($Z \propto 1 - U/U_{c2}$) as in Brinkman-Rice theory. Hence, the approach to the Mott transition, in this theory, is associated with a divergence of the quasiparticle effective mass (in reality, magnetic correlations cut off this divergency, see below). On the other hand, a mean-field insulating solution is found for $U > U_{c1}$, with the Mott gap $\Delta\epsilon$ opening up at this critical coupling (Mott-Hubbard transition). As a result, $\Delta\epsilon$ is a finite energy scale for $U = U_{c2}$ and the quasiparticle peak in the DOS is well separated from the Hubbard bands in the strongly correlated metal: within DMFT, a separation of energy scales exists between the quasiparticle bandwidth and the distance from the lower to the upper Hubbard band. These two critical couplings extend at finite

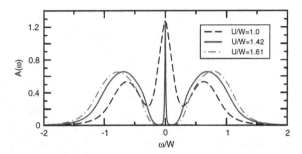

Fig. 4.2 Local spectral function for several values of the interaction strength in DMFT. These results have been obtained using the numerical renormalization group for the half-filled paramagnetic Hubbard model (from Ref. [30]). Close to the transition ($U/W = 1.42$, *solid line*), the separation of scales between the quasiparticle coherence energy (ε_F^*) and the distance between Hubbard bands ($\Delta\epsilon$) is clearly seen

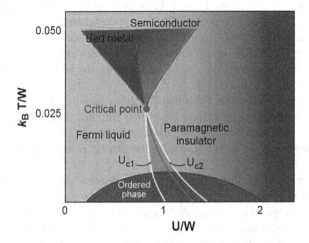

Fig. 4.3 Generic phase diagram of the half-filled Hubbard model within DMFT, as a function of temperature T and coupling U (normalized to the bandwidth W). The figure displays schematically: (1) the spinodal lines ($U_{c1}(T)$ and $U_{c2}(T)$, respectively) of the Mott insulating and metallic mean-field solutions, (2) the first-order transition line ending at the critical endpoint, (3) the crossover lines separating the different transport regimes and (4) the phase boundary corresponding to magnetic long-range order. The figure corresponds to a situation in which the magnetic boundary has been driven below the Mott critical endpoint, e.g. due to frustration. Reproduced from Ref. [31]

temperature into two spinodal lines $U_{c1}(T)$ and $U_{c2}(T)$, which delimit a region of the $(U/W, T/W)$ parameter space in which two mean-field solutions (insulating and metallic) coexist (Fig. 4.3). Hence, within DMFT, a first-order Mott transition occurs at finite temperature even in a purely electronic model. The corresponding critical temperature T_c is of order $T_c \sim \Delta E/\Delta S$, with ΔE and $\Delta S \sim \ln(2S + 1)$ the energy and entropy differences between the metal and the insulator. Because the energy difference is small ($\Delta E \sim (U_{c2} - U_{c1})^2/W$), the critical temperature is much lower than W and U_c (by almost two orders of magnitude). Indeed, in V_2O_3

as well as in the organics, the critical temperature corresponding to the endpoint of the first-order Mott transition line is a factor of 50–100 smaller than the bare electronic bandwith.

- *Magnetism: local moments, long-range order and frustration.* The paramagnetic insulating solution of the DMFT equations displays unscreened local moments, corresponding to a Curie law for the local susceptibility $\sum_q \chi_q \propto 1/T$, and an extensive entropy. Note however that the uniform susceptibility $\chi_{q=0}$ is finite, of order $1/J \sim U/W^2$. As temperature is lowered, these local moments order into an antiferromagnetic phase. Because many DMFT studies have put a strong emphasis on the description of the paramagnetic Mott transition (and rightly so), the possibility of calculating magnetic phase boundaries within DMFT is often under-appreciated. If anything, the domain of existence of magnetic phases is actually *overestimated* within DMFT, as expected from the mean-field nature of the theory and the freezing of spatial fluctuations. However, the inclusion of local fluctuations repair a basic problem of the Hartree–Fock mean field: the Néel temperature T_N within DMFT is indeed proportional to the superexchange ($T_N \propto J \propto W^2/U$) and not to the Mott gap ($\sim U$) as in Hartree–Fock. The Néel temperature is strongly dependent on frustration, e.g. on the ratio t'/t between the next-nearest-neighbor and nearest-neighbor hopping, on the geometry of the lattice and on the orbital degeneracy.

  In the absence of frustration, T_N is usually higher than the electronic Mott transition T_c, so that only the crossovers associated with this transition are visible. T_N is strongly suppressed however, as frustration is increased, so that the Mott critical endpoint can be revealed, leading to the generic phase diagram displayed in Fig. 4.3.

- *Transport crossovers.* The rather rich phase diagram implies several different transport regimes, separated by crossover lines. At low-temperature, conventional behaviors corresponding to a Fermi-liquid metal $\rho_{dc} = AT^2$ with $A \propto 1/Z^2$ or to an insulator with activation energy $\sim \Delta\epsilon$ are found. At intermediate temperatures, away from the Fermi-liquid regime, a bad-metal behavior is found, with a resistivity which can vastly exceed the Mott-Ioffe-Regel criterion. In this regime where $T > \varepsilon_F^*$, no description in terms of long-lived quasiparticle excitations is possible. For a more detailed description of these crossovers, see Ref. [29].

4.4 From Models to Real Materials: Combining DMFT with First-Principle Electronic Structure Methods

The combination of the local density approximation with DMFT allows one to capture the effects of local correlations within a description that maintains the material specific aspects of the problem. We will sketch the essential aspects in the following. For a more complete treatment, see Ref. [12].

In the context of LDA + DMFT, it is important to distinguish between the complete basis set $\{|B_{k\alpha}\rangle\}$ in which the full electronic structure problem on a lattice is formulated (and accordingly the lattice Green function is represented), and local orbitals $|\chi_{km}^R\rangle$ which span a "correlated" subspace C of the total Hilbert space. The index α labels the basis functions for each wave vector $\mathbf{k}$ in the Brillouin zone (BZ). The index R denotes the correlated atom within the primitive unit cell, around which the local orbital $|\chi_{km}^R\rangle$ is centered, and $m = 1, \ldots, M$ is an orbital index within the correlated subset. Many-body corrections will be considered only inside this subspace. Projection onto that subset, for atom type R, is done with the projection operator

$$\hat{P}_R^{(C)} \equiv \sum_{km \in C} |\chi_{km}^R\rangle\langle\chi_{km}^R|. \tag{4.21}$$

The DMFT self-consistency condition, which relates the impurity Green function G_{imp}^R to the Green function of the solid computed locally on atom R then reads [32]:

$$G_{mm'}^{R,\text{loc}}(i\omega_n) = \sum_{k,\alpha\alpha'} \langle\chi_{km}^R|B_{k\alpha}\rangle G_{\alpha\alpha'}(\mathbf{k}, i\omega_n)\langle B_{ka'}|\chi_{km'}^R\rangle,$$

$$G_{\alpha\alpha'}(\mathbf{k}, i\omega_n) = \left\{[i\omega_n + \mu - \mathbf{H}_s(\mathbf{k}) - \Delta\Sigma(\mathbf{k}, i\omega_n)]^{-1}\right\}_{\alpha\alpha'} \tag{4.22}$$

In this expression,

$$|\chi_{km}^R\rangle = \sum_T e^{i\mathbf{k}\cdot(T+R)}|\chi_{Tm}^R\rangle \tag{4.23}$$

denotes the Bloch transform of the local orbitals whereby T denotes the Bravais lattice translation vectors. Note that $G_{\alpha\alpha'}(\mathbf{k}, i\omega_n)$ in Eq. 4.22 is, of course, nothing else than the full lattice Green function in the chosen $\{|B_{k\alpha}\rangle\}$ basis. The single-particle Hamiltonian $\mathbf{H}_s(\mathbf{k})$ can be expressed in the $\{|B_{k\alpha}\rangle\}$ basis set as:

$$\mathbf{H}_{s,\alpha\alpha'}(\mathbf{k}) = \sum_\nu \langle B_{k\alpha}|\Psi_{k\nu}\rangle \varepsilon_{k\nu}\langle\Psi_{k\nu}|B_{ka'}\rangle, \tag{4.24}$$

where $\varepsilon_{k\nu}$ and $|\Psi_{k\nu}\rangle$ are eigenvalues and eigenfunctions in the Bloch description of the lattice problem (ν is the band index). In order to obtain the self-energy for the full solid, one has to promote ("upfold") the DMFT impurity self-energy $\Sigma_{mm'}^{\text{imp}}$ to the lattice via [32, 33]:

$$\Delta\Sigma_{\alpha\alpha'}(\mathbf{k}, i\omega_n) = \sum_{R,mm'} \langle B_{k\alpha}|\chi_{km}^R\rangle \left[\Sigma_{mm'}^{\text{imp}}(i\omega_n) - \Sigma_{mm'}^{\text{dc}}\right]\langle\chi_{km'}^R|B_{ka'}\rangle, \tag{4.25}$$

whereby a double-counting correction $\Sigma_{mm'}^{\text{dc}}$ takes care of correlation effects possibly already accounted for in the single-particle Hamiltonian in an effective way.

The choice of this term itself is a complicated issue because this correction is not well defined. Consequently different schemes which derive from different physical motivations are in use. For details, see e.g. Ref. [33]. A natural simple choice for the general basis set is the Bloch basis itself, i.e., $\{|B_{k\alpha}\rangle\} = \{|\Psi_{k\nu}\rangle\}$. This basis is most conveniently used, since it is, e.g. a direct output of any DFT-LDA calculation and furthermore diagonalizes the single-particle Hamiltonian: $H^s_{\nu\nu'}(k) = \delta_{\nu\nu'}\varepsilon_{k\nu}$.

The basic DMFT equations for a realistic system in the Bloch basis set are easily formulated using the projection matrix elements of the local orbitals onto the Bloch functions, defined as:

$$P^R_{m\nu}(k) \equiv \langle \chi^R_{km} | \Psi_{k\nu} \rangle, \quad P^{R*}_{\nu m}(k) \equiv \langle \Psi_{k\nu} | \chi^R_{km} \rangle. \tag{4.26}$$

Equations 4.22, 4.25 then read:

$$G^{R,\text{loc}}_{mm'}(i\omega_n) = \sum_{k,\nu\nu'} P^R_{m\nu}(k) G^{\text{bl}}_{\nu\nu'}(k, i\omega_n) P^{R*}_{\nu'm'}(k), \tag{4.27}$$

$$\Delta\Sigma^{\text{bl}}_{\nu\nu'}(k, i\omega_n) = \sum_R \sum_{mm'} P^{R*}_{\nu m}(k) \Delta\Sigma^{\text{imp}}_{mm'}(i\omega_n) P^R_{m'\nu'}(k), \tag{4.28}$$

where

$$G^{\text{bl}}_{\nu\nu'}(k, i\omega_n) = \left\{[(i\omega_n + \mu - \varepsilon_{k\nu})\delta_{\nu\nu'} - \Delta\Sigma_{\text{bl}}(k, i\omega_n)]^{-1}\right\}_{\nu\nu'}, \tag{4.29}$$

$$\Delta\Sigma^{\text{imp}}_{mm'}(i\omega_n) = \Sigma^{\text{imp}}_{mm'}(i\omega_n) - \Sigma^{\text{dc}}_{mm'}. \tag{4.30}$$

Starting from an orthonormalized set of local orbitals $\langle \chi^R_{km} | \chi^{R'}_{k'm'} \rangle = \delta_{mm'}\delta_{RR'}\delta_{kk'}$, it is easily checked that the matrix $\langle \Psi_{k\nu} | \chi^R_{km} \rangle$ is unitary (from the completeness of the Bloch basis). Hence the χ_m's can formally be viewed as Wannier functions associated with the complete basis set of all Bloch states,

$$|\chi^R_{km}\rangle \equiv \sum_\nu \langle \Psi_{k\nu} | \chi^R_{km} \rangle | \Psi_{k\nu} \rangle. \tag{4.31}$$

In practice, the Bloch basis needs to be truncated, involving only Bloch functions which span a certain energy window, for which the projections (4.26) are significant. In the vanadates, e.g. V_2O_3 and VO_2, the p bands are well separated from the d bands and the d orbitals are split into two nearly empty e_g and three t_{2g} bands. Restricting the sum in (4.31) to the subset $\mathcal{W}$ of the Bloch bands and defining

$$|\tilde{\chi}^R_{km}\rangle \equiv \sum_{\nu\in\mathcal{W}} \langle \Psi_{k\nu} | \chi^R_{km} \rangle | \Psi_{k\nu} \rangle, \tag{4.32}$$

it is seen that the functions $|\tilde{\chi}^R_{km}\rangle$ associated with the subspace $\mathcal{W}$ are not true Wannier functions since the truncated projection matrix is no longer unitary. However, these functions can be promoted to true Wannier functions $|w^{R}_{\mathbf{k}m}\rangle$ by orthonormalizing this set according to:

$$|w^{R}_{\mathbf{k}m}\rangle = \sum_{R'm'} S^{RR'}_{mm'}(\mathbf{k})|\tilde{\chi}^{R'}_{\mathbf{k}m'}\rangle, \tag{4.33}$$

where $\mathbf{S}^{RR'}(\mathbf{k})$ is given by the inverse square root of the overlap matrix between the Wannier-like orbitals, i.e.,

$$O^{RR'}_{mm'}(\mathbf{k}) \equiv \langle\tilde{\chi}^{R}_{\mathbf{k}m}|\tilde{\chi}^{R'}_{\mathbf{k}m'}\rangle = \sum_{v\in\mathcal{W}} P^{R}_{mv}(\mathbf{k})P^{R'*}_{vm'}(\mathbf{k}), \tag{4.34}$$

$$S^{RR'}_{mm'}(\mathbf{k}) \equiv \left\{(\mathbf{O}(\mathbf{k}))^{-1/2}\right\}^{RR'}_{mm'}. \tag{4.35}$$

Naturally, the functions w^{R}_{m} are *more extended* in space than the original atomic-like functions χ^{R}_{m} since they can be decomposed on a *smaller* number of Bloch functions, spanning a restricted energy range.

In the end, realistic DMFT is implemented by identifying $\mathcal{C}$ with the correlated subset generated by the set of functions $|w^{R}_{\mathbf{k}m}\rangle$. Since those functions have a vanishing overlap with all Bloch functions which do not belong to the set $\mathcal{W}$, the DMFT equations can now be put in a computationally tractable form, involving only an $N_b \times N_b$ matrix inversion within the selected space $\mathcal{W}$ (N_b marks the number of considered Bloch bands). Hence, the equations which are finally implemented read:

$$G^{R,\mathrm{loc}}_{mm'}(i\omega_n) = \sum_{\mathbf{k},(vv')\in\mathcal{W}} \bar{P}^{R}_{mv}(\mathbf{k})G^{\mathrm{bl}}_{vv'}(\mathbf{k},i\omega_n)\bar{P}^{R*}_{v'm'}(\mathbf{k}), \tag{4.36}$$

$$\Delta\Sigma^{\mathrm{bl}}_{vv'}(\mathbf{k},i\omega_n) = \sum_{R,mm'} \bar{P}^{R*}_{vm}(\mathbf{k})\Delta\Sigma^{\mathrm{imp}}_{mm'}(i\omega_n)\bar{P}^{R}_{m'v'}(\mathbf{k}), \tag{4.37}$$

where

$$\bar{P}^{R}_{mv}(\mathbf{k}) = \sum_{R'm'} S^{RR'}_{mm'}(\mathbf{k})P^{R'}_{m'v}(\mathbf{k}), \tag{4.38}$$

$$\bar{P}^{R*}_{vm}(\mathbf{k}) = \sum_{R'm'} S^{RR'*}_{m'm}(\mathbf{k})P^{R'*}_{m'v}(\mathbf{k}). \tag{4.39}$$

Note that the $\mathbf{k}$-dependence of the self-energy in (4.37) solely stems from the $\mathbf{k}$-dependence of the projection matrix elements. The impurity self-energy is purely local. It is important to realize that the truncation to a limited set of Bloch functions was not achieved by simply neglecting matrix elements between the local orbitals and Bloch functions outside this set. Rather a new set of (more extended) local orbitals is constructed such that the desired matrix elements automatically vanish, hence redefining $\mathcal{C}$ accordingly. In this view, the spaces $\mathcal{C}$ and $\mathcal{W}$, although independent in principle, become actually interrelated.

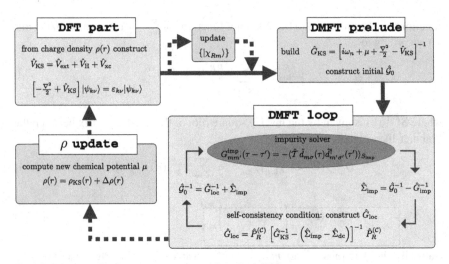

Fig. 4.4 LDA + DMFT flowchart (reproduced from Ref. [32])

We also note that it is not compulsory to insist on forming true Wannier functions out of the (non-orthogonal) set $|\tilde{\chi}_{\mathbf{k}m}^{R}\rangle$. It is perfectly legitimate to formally choose the correlated subspace $\mathcal{C}$ as generated by orbitals having a decomposition in $\mathcal{W}$, but not necessarily unitarily related to Bloch functions spanning $\mathcal{W}$. Although orthogonality of the χ_m's is also not compulsory, several (but not all) impurity solvers used within DMFT do require however that the $\chi_m's$ be orthogonal on a given atomic site. One possibility, for example, is to orthonormalize this set on identical unit-cells only, i.e requiring that $|w_{\mathbf{T}m}^{R}\rangle$ and $|w_{\mathbf{T}'m'}^{R'}\rangle$ in real space are orthogonal for $\mathbf{T} = \mathbf{T}'$, but not in neighboring cells $\mathbf{T} \neq \mathbf{T}'$. This amounts to orthonormalize the $|\tilde{\chi}_{\mathbf{k}m}^{R}\rangle$ set with respect to the $\mathbf{k}$-summed overlap matrix, instead of the one computed at each $\mathbf{k}$-point.

In many actual implementations, the wave-functions spanning the correlated subspace $\mathcal{C}$ are obtained by following the above orthonormalization procedure, starting from atomic-like orbitals χ_m^{R} centered on the atomic site $\mathbf{R}$ in the primitive unit-cell. These local orbitals may, e.g., be chosen as the *all-electron atomic partial waves* in the PAW framework.

The self-consistency loop of the LDA + DMFT scheme is essentially unchanged with respect to Sect. 4.2 The main difference is the computation of the local Green function from the impurity self-energy via Eqs. 4.36, 4.37. The flowchart of the resulting LDA + DMFT computational scheme including the self-consistent determination of the charge density is illustrated in Fig. 4.4.

The local interaction in realistic calculations enters in the impurity solver step and is given by

$$H_{\text{int}} = \frac{1}{2} \sum_{mm'm''m'''} \sum_{\sigma\sigma'} U_{mm'm''m'''} c_{im\sigma}^{\dagger} c_{im'\sigma'}^{\dagger} c_{im'''\sigma'} c_{im''\sigma}. \qquad (4.40)$$

In principle, notwithstanding the sign problem, the continuous-time impurity solvers described in Sect. 4.5 are able to handle the interaction in this general form. The matrix elements

$$U_{mm'm''m'''} = \langle m, m' | V_{ee} | m'', m''' \rangle \tag{4.41}$$

are taken with respect to the *screened* Coulomb interaction $V_{ee} = 1/\varepsilon |x - x'|$. The screening ε can be determined within constrained RPA and the matrix elements may then be evaluated in terms of Wannier functions [34].

A simpler form of the interaction is obtained by assuming that within the solid, the interactions largely retain their atomic nature. Within a spherical approximation, the matrix elements are expressed in terms of effective (i.e. screened) Slater integrals F^k [35],

$$\langle mm' | V_{ee} | m''m''' \rangle = \sum_k a_k(m, m', m'', m''') F^k. \tag{4.42}$$

For d-electrons, the non-zero elements are F^0, F^2 and F^4. Defining the average Coulomb and Stoner parameters U and J, these are linked to the Slater integrals through (4.42) as $U = F^0$ and $J = (F^2 + F^4)/14$ and can be computed within constrained LDA calculations. The three Slater integrals and hence the matrix elements are unambiguously determined by requiring that the ratio F^2/F^4 be equal to its atomic value ~ 0.625 [35].

For nearly degenerate bands, the U-matrix is often approximated by a parameterization restricted to the three parameters

$$
\begin{aligned}
U &= \langle m, m | V_{ee} | m, m \rangle, \\
U' &= \langle m, m' | V_{ee} | m, m' \rangle, \\
J &= \langle m, m' | V_{ee} | m', m \rangle,
\end{aligned}
\tag{4.43}
$$

with the direct and exchange Coulomb matrix elements. In this case, the fully rotationally invariant Hamiltonian (in spin and orbital space) takes the form

$$
\begin{aligned}
H_{\text{loc}} =& \frac{1}{2} \sum_{mm'} \sum_\sigma [U - 2J(1 - \delta_{mm'})] n_{m\sigma} n_{m'\bar{\sigma}} + \frac{1}{2} \sum_{m \neq m'} \sum_\sigma (U' - J) n_{m\sigma} n_{m'\sigma} \\
&+ \frac{1}{2} \sum_{m \neq m'} \sum_\sigma J \left(c^\dagger_{m\sigma} c^\dagger_{m'\bar{\sigma}} c_{m\bar{\sigma}} c_{m'\sigma} + c^\dagger_{m\sigma} c^\dagger_{m\bar{\sigma}} c_{m'\bar{\sigma}} c_{m'\sigma} \right),
\end{aligned}
\tag{4.44}
$$

where $\bar{\sigma} = -\sigma$ and $U' = U - 2J$ is required by symmetry. The interaction terms in the first line are of density-density type, while the two terms in the second line correspond to spin-flip and pair-hopping terms, respectively. Often the latter are neglected. This reduces significantly the computational complexity of the problem, e.g. when using the hybridization expansion solver discussed in Sect. 4.5. For the Hirsch-Fye quantum Monte Carlo solver these terms are known to cause a serious sign-problem. Neglecting these terms is not always legitimate or physically motivated however. Indeed, it has been shown that spin-flip and pair-hopping terms become quite important near the Mott transition [36].

4.4.1 Illustrations

The LDA + DMFT framework has by now been applied to a wealth of different materials, and attempting a review is far beyond the scope of these lecture notes. We shall be content here with just a couple of illustrations.

The first one concerns a semi-metal, ErAs. In this material, the 4f states of Er are very localized and not very close to the Fermi level. Under such circumstances, a simplified impurity solver can be used to solve the DMFT equations: in Ref. [37], the very simple Hubbard-I approximation was used, which consists in replacing the self-energy with that of the self-consistently embedded but *isolated* atom (hence, the self-consistent aspect only comes in through the total charge density and total electron number) [17]. In Fig. 4.5, we reproduce the LDA+U and LDA + DMFT bandstructures and density of states obtained in Ref. [37] for this material. This figure reveals important differences between the two methods, regarding the description of the localized f-states. Most notably: (1) within LDA+U the Hubbard bands have a strong spin polarization (majority-polarized for the lower Hubbard band, minority polarized for the other) and (2) the multiplet structure of these Hubbard bands is properly described within DMFT (because it describes the atom correctly), but not within LDA+U. As explained in detail in Ref. [37], a good description of the internal structure of the Hubbard bands (multiplet structure, spin polarization) turns out to be essential to obtain a proper description of the bandstructure of this semi-metal close to the Fermi level (because of hybridization effects). This low-energy bandstructure is, in this case, known with great precision from quantum oscillation experiments with which LDA + DMFT results are in good agreement.

As a second example, we consider a prototypical strongly correlated metal: the metallic phase of V_2O_3. This material lies just on the verge of the metal to Mott insulator transition, which can be induced by applying (negative) 'chemical pressure' in the form of Cr-substitution on the V-sites. Photoemission experiments conducted by Mo et al. [38] in 2003—and compared to theoretical calculations in the LDA + DMFT framework—clearly revealed a prominent quasiparticle peak in the momentum-integrated DOS, as well as a higher binding energy satellite corresponding to the lower Hubbard band. These are key predictions of DMFT (see discussion of the three-peak structure above). In Fig. 4.6, we reproduce the recent comparison of the photoemission experiment to theory of Panaccione et al. [39], which clearly displays these two spectral features and demonstrates the level of agreement that is possible to achieve with LDA + DMFT calculations for concrete materials.

As a last illustration of great current relevance, let us mention that realistic dynamical mean-field theory has allowed to quantify the degree of correlations of the recently discovered iron-based superconductors. From LDA + DMFT calculations

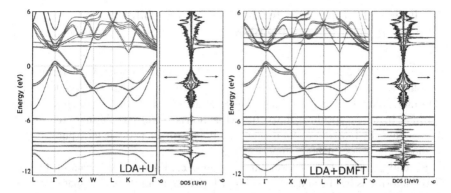

Fig. 4.5 Band structure and density of states (DOS) of ErAs: from LDA+U (*left*), and LDA + DMFT at an applied field of 5 T (*right*). The DOS of the total, Er 5*d*, Er 4*f* and As 4*p* states is displayed by the *black*, *red*, *green*, and *blue* curves respectively. Reproduced from Ref. [37]

Fig. 4.6 Valence band photoemission spectrum (*circles*) and LDA+DMFT calculation (*plain line*) for the paramagnetic phase of V_2O_3. The theoretical spectral function has been broadened with the estimated experimental resolution of 0.25eV. The two features corresponding to the quasiparticle peak and lower Hubbard bands are clearly visible. From Ref. [39]

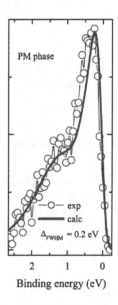

[40–45] combined with first-principle estimates of Coulomb interaction parameters [46–48], it appears that these materials are indeed correlated, but definitely on the metallic side and not very close to the metal-insulator transition. The degree of correlation increases significantly from the 1111 (e.g. LaFeAsO) to the 11 (e.g. FeSe) family [49, 50]. The Hund's rule has been shown to play an essential role in these multi-band materials ([41, 51], see also Ref. [50]).

4.5 Efficient Algorithms for Solving Multiorbital Quantum Impurity Problems

Monte Carlo simulations have been employed for the computation of fermionic field-theoretical models since the advent of determinantal quantum Monte Carlo algorithms due to Blankenbecler, Sugar, Scalapino [52, 53], and Hirsch and Fye [54]. In these algorithms, the partition function is expressed in terms of fermionic determinants by integrating out the fermion fields. Hirsch and Fye later published a method suitable for quantum impurity models [55].

Recently, a new class of fermionic, so-called continuous-time quantum Monte Carlo algorithms has been developed [56–61]. These algorithms also share the fact that classes of diagrams are collected into fermionic determinants.

The two continuous-time algorithms are numerically exact in the sense that, notwithstanding the sign problem, results can in principle be obtained to any desired accuracy. Being based on expansions around different limits, the methods are complementary: both algorithms converge to the same results, but perform differently depending on the parameter regime. A detailed performance analysis of the two approaches in comparison with the Hirsch-Fye method has been conducted in Ref. [62]. In the following we will assume knowledge of the basic principles of Monte Carlo sampling. An introduction can e.g. be found in Ref. [63].

4.5.1 Interaction Expansion Algorithm

The interaction expansion or weak-coupling continuous-time quantum Monte Carlo algorithm for fermions was introduced by Rubtsov et al. [56, 57], following the work of Prokof'ev and coworkers [64] who devised a continuous-time scheme to sample the infinite series generated by the path integral representation of the partition function for bosons.

Here we introduce it in the path integral formulation for the single-orbital Anderson impurity model with Hubbard interaction $U n_\uparrow n_\downarrow$. The generalization to the multiorbital case is straightforward and has been given in [56]. The action (4.9) for the Anderson impurity model is divided into a Gaussian part S_0 and an interaction part S_U as follows:

$$S_0 = \int_0^\beta d\tau \sum_\sigma c_\sigma^*(\tau)[\partial_\tau - \mu + U\alpha_{-\sigma}(\tau)]c_\sigma(\tau)$$

$$+ \int_0^\beta d\tau \int_0^\beta d\tau' \sum_\sigma c_\sigma^*(\tau)\Delta(\tau - \tau')c_\sigma(\tau'), \qquad (4.45)$$

$$S_U = U \int_0^\beta d\tau [c_\uparrow^*(\tau) c_\uparrow(\tau) - \alpha_\uparrow(\tau)][c_\downarrow^*(\tau) c_\downarrow(\tau) - \alpha_\downarrow(\tau)]. \tag{4.46}$$

Here the fields α, or so-called α-parameters have been introduced such that the impurity action $S = S_0 + S_U$ is only changed up to an irrelevant additive constant. They are necessary to control the sign problem, as discussed below. The partition function

$$\mathcal{Z} = \int \mathcal{D}[c^*, c] e^{-S[c^*, c]} \tag{4.47}$$

is written as a functional integral over Grassmann fields. A formal series expansion is obtained by expanding the exponential in the interaction term:

$$\mathcal{Z} = \int \mathcal{D}[c^*, c] e^{-S_0[c^*, c]} \sum_{k=0}^\infty \frac{(-1)^k}{k!} U^k \int_0^\beta d\tau_1 \ldots \int_0^\beta d\tau_k$$
$$\times [c_\uparrow^*(\tau_1) c_\uparrow(\tau_1) - \alpha_\uparrow(\tau_1)][c_\downarrow^*(\tau_1) c_\downarrow(\tau_1) - \alpha_\downarrow(\tau_1)] \ldots [c_\downarrow^*(\tau_k) c_\downarrow(\tau_k) - \alpha_\downarrow(\tau_k)]. \tag{4.48}$$

Using the definition of the average over the noninteracting system, i.e. $\langle \ldots \rangle_0 = (1/\mathcal{Z}_0) \int \mathcal{D}[c^*, c] \ldots \exp(-S_0)$, the partition function can be expressed in the form

$$\mathcal{Z} = \mathcal{Z}_0 \sum_{k=0}^\infty \int_0^\beta d\tau_1 \int_{\tau_1}^\beta d\tau_2 \ldots \int_{\tau_{k-1}}^\beta d\tau_k \, \text{sgn}(\Omega_k) |\Omega_k|, \tag{4.49}$$

where the integrand is given by

$$\Omega_k = (-1)^k U^k \langle [c_\uparrow^*(\tau_1) c_\uparrow(\tau_1) - \alpha_\uparrow(\tau_1)][c_\downarrow^*(\tau_1) c_\downarrow(\tau_1) - \alpha_\downarrow(\tau_1)]$$
$$\ldots [c_\uparrow^*(\tau_k) c_\uparrow(\tau_k) - \alpha_\uparrow(\tau_k)][c_\downarrow^*(\tau_k) c_\downarrow(\tau_k) - \alpha_\downarrow(\tau_k)] \rangle_0. \tag{4.50}$$

Note that the range of time integration in (4.49) has been changed with respect to (4.48) such that time ordering is explicit: $\tau_k > \tau_{k-1} > \ldots > \tau_1$. For a given set of times all $k!$ permutations of this sequence contribute to Eq. 4.48. These can be brought into the standard sequence by permuting quadruples of Grassmann numbers, and hence without gaining an additional sign. Since all terms are subject to time-ordering, their contribution to the integral is identical, so that the factor $1/k!$ in (4.48) cancels. A configuration is hence fully characterized by specifying a perturbation order k and an (unnumbered) set of k times: $C_k = \{\tau_1, \ldots, \tau_k\}$. The entire configuration space comprises the configurations C_k up to all orders:

$$C = \left\{ C_0 = \{\}, C_1 = \{\tau_1\}, \ldots, C_k = \{\tau_1 \ldots \tau_k\}, \ldots \right\}. \tag{4.51}$$

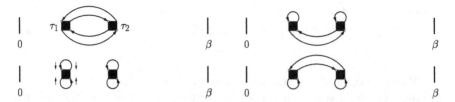

Fig. 4.7 The four contributions to the partition function for $k = 2$. The interaction vertices are depicted by *squares*. Bare Green functions are shown as *lines*. *Vertical arrows* indicate the spin direction. Connecting the vertices by Green functions in all possible ways is the interpretation of the determinant. Here only four out of 4! terms contribute since the Green function is diagonal in spin-space.

The algorithm performs importance sampling over the configuration space, with the weight of a configuration taken to be equal to the modulus of the integrand in Eq. 4.49, $|\Omega_k|$.

Since S_0 is Gaussian, the average over the noninteracting system can be evaluated by application of Wick's theorem. Hence the weight of a configuration is essentially given by a fermionic determinant of a matrix containing the bare Green functions:

$$\Omega_k = (-1)^k U^k \prod_\sigma \det \hat{g}^\sigma, \qquad (\hat{g}^\sigma)_{ij} = g_0^\sigma(\tau_i - \tau_j) - \alpha_\sigma(\tau_i)\delta_{ij}. \qquad (4.52)$$

The determinants for different spin projections factorize since the Green function is diagonal in spin-space. Each configuration to the partition function can be visualized as a collection of Feynman diagrams. The determinant contains all possible contractions of the sequence of Grassmann numbers and corresponds to connecting the Green functions to the vertices in all possible ways. As an example, the four contributions to the partition function for a configuration with $k = 2$ are depicted in Fig. 4.7.

Setting $\alpha_\uparrow = \alpha_\downarrow = 0$ for simplicity, the contribution of this configuration is

$$\Omega_2 = U^2 \langle c_\uparrow^*(\tau_1)c_\uparrow(\tau_1)c_\downarrow^*(\tau_1)c_\downarrow(\tau_1)c_\uparrow^*(\tau_2)c_\uparrow(\tau_2)c_\downarrow^*(\tau_2)c_\downarrow(\tau_2)\rangle_0$$
$$= U^2 \langle c_\uparrow^*(\tau_1)c_\uparrow(\tau_1)c_\uparrow^*(\tau_2)c_\uparrow(\tau_2)\rangle_0 \langle c_\downarrow^*(\tau_1)c_\downarrow(\tau_1)c_\downarrow^*(\tau_2)c_\downarrow(\tau_2)\rangle_0. \qquad (4.53)$$

Using Wick's theorem, the contribution to the configuration can be written in terms of fermionic determinants in the form $\Omega_2 = U^2 \det \hat{g}_\uparrow \det \hat{g}_\downarrow$, where $\hat{g}$ denotes the matrix of Green functions

$$\hat{g}^\sigma = \begin{pmatrix} -\langle c_\sigma(\tau_1)c_\sigma^*(\tau_1)\rangle_0 & -\langle c_\sigma(\tau_1)c_\sigma^*(\tau_2)\rangle_0 \\ -\langle c_\sigma(\tau_2)c_\sigma^*(\tau_1)\rangle_0 & -\langle c_\sigma(\tau_2)c_\sigma^*(\tau_2)\rangle_0 \end{pmatrix}. \qquad (4.54)$$

Different columns correspond to different creator times, while rows correspond to the annihilator times in the particular configuration. Equal time averages yield the noninteracting density. Multiplying out the product of determinants for the two spin projections gives the four contributions of Fig. 4.7. All other out of the 4! possible configurations vanish due to spin conservation.

4.5.2 Monte Carlo Sampling

The algorithm samples configurations by performing a Markovian random walk in the configuration space with the weight of a configuration determined by $|\Omega_k|$. Two kinds of Monte Carlo steps are sufficient to ensure ergodicity (other, e.g. global moves can be implemented to increase the sampling efficiency). A vertex is either inserted or removed from the configuration. The probability to insert a vertex (corresponding to adding a row and column to both $\hat{g}_\uparrow$ and $\hat{g}_\downarrow$) into a domain $d\tau$ is taken to be proportional to the fraction of the domain to the full interval from 0 to β,

$$q^{\text{add}}(k \to k+1) = \frac{d\tau}{\beta}. \tag{4.55}$$

For the inverse move, one needs to remove one out of $k+1$ vertices, each with the same probability:

$$q^{\text{rem}}(k+1 \to k) = \frac{1}{k+1}. \tag{4.56}$$

The probabilities for the old and new configurations are proportional to

$$p^{\text{new}}(k+1) \sim |U|^{k+1} \Pi_\sigma |\det \hat{g}_\sigma^{(k+1)}| d\tau, \quad p^{\text{old}}(k) \sim |U|^k \Pi_\sigma |\det \hat{g}_\sigma^{(k)}| \tag{4.57}$$

and the Metropolis-Hastings acceptance criterion for this step is given by

$$p^{\text{acc}} = \min\left(1, \frac{p^{\text{new}}(k+1)q^{\text{rem}}(k+1 \to k)}{p^{\text{old}}(k)q^{\text{add}}(k \to k+1)}\right) = \min\left(1, \frac{\beta|U|}{k+1}\Pi_\sigma \left|\frac{\det \hat{g}_\sigma^{(k+1)}}{\det \hat{g}_\sigma^{(k)}}\right|\right). \tag{4.58}$$

The acceptance criterion for removing a vertex is given by the inverse of (4.58). In the simulation not the matrix $\hat{g}$, but its inverse M is stored and manipulated. The reason is that M gives faster and more direct access to determinant ratios which appear in acceptance probabilities and Green function measurements (see below).

The algorithm samples diagrams for finite orders of k. Typically the perturbation order distribution has a Gaussian shape with a mean increasing with the interaction U and inverse temperature, as shown in Fig. 4.8.

4.5.3 Fast Update Formulae

The key to an efficient algorithm are the so-called fast-update formulae, which allow a fast update of the matrix M when a configuration is changed and direct access to determinant ratios. Here the fast update formulae are derived using the Sherman–Morrison formula, which states how the *inverse* of a given matrix changes when a row and column is added to the matrix. Inserting a pair of Grassmann numbers into a

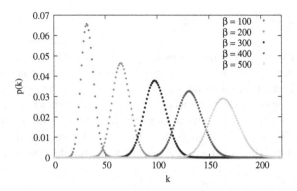

Fig. 4.8 Probability distribution of the perturbation order k for the single-orbital Anderson impurity model with a wide featureless band (corresponding to $\Delta(i\omega) = (-i/2)\mathrm{sgn}(\omega)$) and Hubbard interaction $U = 1$. The histogram shows a shift of the perturbation order to larger values and a broadening of the distribution with increasing inverse temperature β. The average perturbation order is an estimator for the average of the interaction operator $\langle S_U \rangle$

configuration at order k corresponds to adding a column and a row to the matrix $\hat{g}^{(k)}$. In order to apply these formulas in a simple way, the matrix $\hat{g}^{(k)}$ is extended to be a $k + 1 \times k + 1$ matrix with zeros in the $k + 1$th row and column, except for the $k + 1, k + 1$-element, which is 1 and the same for its inverse, $M^{(k)}$. Abbreviate the noninteracting Green function as $g_0(\tau_i - \tau_j') =: g_{ij}$. Adding the column vector $u_{\mathrm{col}} = (g_{1,k+1}, \ldots, g_{k,k+1} | g_{k+1,k+1} - 1)^T$ is expressed in terms of adding the direct product $u_{\mathrm{col}} \otimes e_{k+1}^T$, where e_{k+1}^T is the row vector with entries $(e_{k+1}^T)_i = \delta_{i,l+1}$ (here -1 compensates the $k+1, k+1-$ element of $\hat{g}^{(k)}$ so that $\hat{g}^{(k+1)}$ has the desired form). In a second step, add the row $v_{\mathrm{row}} = (g_{k+1,1}, \ldots, g_{k+1,k} | 0)$ by adding $e_{k+1} \otimes v_{\mathrm{row}}$. Application of the Sherman-Morrison formula for the first step yields the intermediate matrix

$$\bar{M} = M^{(k)} - \frac{M^{(k)} u_{\mathrm{col}} \otimes e_{k+1}^T M^{(k)}}{1 + e_{k+1}^T M^{(k)} u_{\mathrm{col}}}. \tag{4.59}$$

Now define $L_{ij} = \sum_{l=1}^{k} M_{il}^{(k)} g_{lj}$ and $R_{ij} = \sum_{l=1}^{k} g_{il} M_{lj}^{(k)}$. Then $e_{k+1}^T M^{(k)} = e_{k+1}^T$, which implies that $M^{(k)} u_{\mathrm{col}} = (L_{1,k+1}, \ldots L_{k,k+1} | g_{k+1,k+1} - 1)^T$ divided by $1 + e_{k+1}^T M^{(k)} u_{\mathrm{col}} = g_{k+1,k+1}$ is added to the $k + 1$th column of $M^{(k)}$ (the $k + 1, k + 1-$ element is $1 - (g_{k+1,k+1} - 1)/g_{k+1,k+1}$):

$$\bar{M} = \begin{pmatrix} & M^{(k)} & & \begin{array}{c} -L_{1,k+1}/g_{k+1,k+1} \\ \vdots \\ -L_{k,k+1}/g_{k+1,k+1} \end{array} \\ \hline 0 \cdots 0 & 1/g_{k+1,k+1} \end{pmatrix}. \tag{4.60}$$

Likewise, for the second step

$$M^{(k+1)} = \bar{M} - \frac{\bar{M} e_{k+1} \otimes v_{\text{row}} \bar{M}}{1 + v_{\text{row}} \bar{M} e_{k+1}}, \tag{4.61}$$

where now $v_{\text{row}} \bar{M} = (R_{k+1,1}, \ldots R_{k+1,k} | - g_{k+1,l} L_{l,k+1}/g_{k+1,k+1})$. Concatenating the updates using the formula for the determinant, $\det M^{(k)}/\det \bar{M} = 1 + e_{k+1}^T M^{(k)} u_{\text{col}}$, yields

$$r = \frac{\det \hat{g}^{(k+1)}}{\det \hat{g}^{(k)}} = \frac{\det M^{(k)}}{\det \bar{M}} \frac{\det \bar{M}}{\det M^{(k+1)}} = (1 + e_{k+1}^T M^{(k)} u_{\text{col}})(1 + v_{\text{row}} \bar{M} e_{k+1})$$

$$= g_{k+1,k+1} - \sum_{ij=1}^{k} g_{k+1,i} M_{ij}^{(k)} g_{j,k+1}. \tag{4.62}$$

The evaluation of the determinant ratio hence requires $\mathcal{O}(k^2)$ arithmetic operations for adding a vertex (instead of $\mathcal{O}(k^3)$ for a straightforward evaluation). Evaluating the direct product in (4.60) with (4.61) yields an explicit expression for the update of the matrix elements of $M^{(k+1)}$ [56]. Determinant ratios for matrices which differ by two or more rows and columns can be obtained by successive application of the Sherman-Morrison formula, or directly using block matrix manipulation (see below). Note that for the strong coupling solver discussed in Sect. 4.5, the structure of the equations is the same.

4.5.4 Measurement of Green's Function

An expansion similar to that of the partition function can be written for Green's function:

$$g^\sigma(\tau - \tau') = \frac{\mathcal{Z}_0}{\mathcal{Z}} \sum_{k=0}^{\infty} \int_0^\beta d\tau_1 \ldots \int_{\tau_{k-1}}^\beta d\tau_k \, \tilde{g}^\sigma(\tau, \tau'; \tau_1, \ldots, \tau_k) \, \text{sgn}(\Omega_k) |\Omega_k|, \tag{4.63}$$

where $\tilde{g}$ denotes a contribution to Green's function which can be expressed as a ratio of fermionic determinants:

$$\tilde{g}^\sigma(\tau - \tau'; C_k) = \frac{\langle c_\sigma^*(\tau') c_\sigma(\tau) [c_\uparrow^*(\tau_1) c_\uparrow(\tau_1) - \alpha_\uparrow(\tau_1)] \ldots [c_\downarrow^*(\tau_k) c_\downarrow(\tau_k) - \alpha_\downarrow(\tau_k)] \rangle_0}{\langle [c_\uparrow^*(\tau_1) c_\uparrow(\tau_1) - \alpha_\uparrow(\tau_1)] \ldots [c_\downarrow^*(\tau_k) c_\downarrow(\tau_k) - \alpha_\downarrow(\tau_k)] \rangle_0}. \tag{4.64}$$

Diagrammatically, the measurement of Green's function is obtained by removing all lines connecting a given creator and annihilator from the diagrams of the configuration. This corresponds to the removal of a row and a column from the matrix $\hat{g}$. In order to measure a contribution to, say, $-\langle c_\uparrow(\tau_1) c_\uparrow^*(\tau_2) \rangle$ in the configuration depicted in Fig. 4.7, remove row 1 and column 2 from the matrix

Fig. 4.9 The four contributions to Ω_2 after removing all Green functions connected to $c_\uparrow(\tau_1)$ and $c_\uparrow^*(\tau_2)$. The diagrams in the first row have only two unconnected endpoints and contribute to the Green function $g_\uparrow(\tau_1 - \tau_2) = -\langle c_\uparrow(\tau_1) c_\uparrow^*(\tau_2) \rangle$

$$
\begin{array}{c}
\qquad\quad c_\uparrow^*(\tau_1) \qquad\qquad\quad c_\uparrow^*(\tau_2) \\
\begin{array}{c} c_\uparrow(\tau_1) \\ c_\uparrow(\tau_2) \end{array}
\begin{pmatrix}
-\langle c_\uparrow(\tau_1) c_\uparrow^*(\tau_1) \rangle_0 & -\langle c_\uparrow(\tau_1) c_\uparrow^*(\tau_2) \rangle_0 \\
-\langle c_\uparrow(\tau_2) c_\uparrow^*(\tau_1) \rangle_0 & -\langle c_\uparrow(\tau_2) c_\uparrow^*(\tau_2) \rangle_0
\end{pmatrix} .
\end{array}
\tag{4.65}
$$

In doing so, one is left with $-\langle c_\uparrow(\tau_2) c_\uparrow^*(\tau_1) \rangle_0$ and the full matrix for $\sigma = \downarrow$, which gives two possibilities to connect the Green functions to the vertices. This corresponds to the diagrams in the first row of Fig. 4.9. These are the diagrams with exactly two unconnected endpoints. According to (4.62), the measurement for the Green function (4.59) in imaginary time evaluates to

$$
\tilde{g}^\sigma(\tau - \tau'; C_k) = g_0^\sigma(\tau - \tau') - \sum_{ij=1}^{k} g_0^\sigma(\tau - \tau_i) M_{ij}^\sigma g_0^\sigma(\tau_j - \tau'),
\tag{4.66}
$$

where $M^\sigma = M^\sigma[C_k]$ is the inverse of the matrix $\hat{g}^\sigma$ of noninteracting Green functions evaluated at the times corresponding to the particular configuration. Note that the Green function is measured as a correction to a known function (g_0^σ). This is a consequence of the expansion in the interaction. The Monte Carlo average of the Green function is given by

$$
g^\sigma(\tau - \tau') = \sum_{\{k, C_k\}} \tilde{g}^\sigma(\tau - \tau'; C_k) \, \mathrm{sgn}(\Omega_k) \Big/ \sum_{\{k, C_k\}} \mathrm{sgn}(\Omega_k).
\tag{4.67}
$$

Alternatively, the Green function can be measured directly on Matsubara frequencies, by taking the Fourier transform of (4.66):

$$
\tilde{g}^\sigma(\omega; C_k) = g_0^\sigma(\omega) - \frac{g_0^\sigma(\omega)^2}{\beta} \sum_{ij=1}^{k} M_{ij}^\sigma e^{-i\omega(\tau_i - \tau_j)}.
\tag{4.68}
$$

The generalization to measurements for higher-order correlation functions of the impurity is straightforward. Introduce the matrix

$$\Delta = \hat{g}^{(k+n)} - \hat{g}^{(k)} = \begin{pmatrix} 0 & \cdots & 0 & g_{1,k+1} & \cdots & g_{1,k+n} \\ \vdots & \ddots & \vdots & \vdots & & \vdots \\ 0 & \cdots & 0 & g_{k,k+1} & \cdots & g_{k,k+n} \\ g_{k+1,1} & \cdots & g_{k+1,k} & g_{k+1,k+1}-1 & \cdots & g_{k+1,k+n} \\ \vdots & & \vdots & \vdots & & \vdots \\ g_{k+n,1} & \cdots & g_{k+n,k} & g_{k+n,k+1} & \cdots & g_{k+n,k+n}-1 \end{pmatrix},$$

$$(4.69)$$

where the matrix $\hat{g}^{(k)}$ has been extended to be the $k+n \times k+n-$ matrix

$$\hat{g}^{(k)} \rightarrow \begin{pmatrix} \hat{g}^{(k)} & 0 \\ 0 & \mathbb{1}_{(n)} \end{pmatrix}. \qquad (4.70)$$

With this, one has $g^{(k+n)} = g^{(k)}(\mathbb{1}+M^{(k)})$ and the determinant ratio for the n-particle Green function can be written

$$\frac{\det \hat{g}^{(k+n)}}{\det \hat{g}^{(k)}} = \det(\mathbb{1}^{(k)} + \hat{g}_{(k)}^{-1}\Delta) = \det(\mathbb{1}^{(k)} + M^{(k)}\Delta). \qquad (4.71)$$

Then

$$(\mathbb{1} + M^{(k)}\Delta) = \begin{pmatrix} 1 & \cdots & 0 & M_{1,j}g_{j,k+1} & \cdots & M_{1,j}g_{j,k+n} \\ \vdots & \ddots & \vdots & \vdots & & \vdots \\ 0 & \cdots & 1 & M_{k,j}g_{j,k+1} & \cdots & M_{k,j}g_{j,k+n} \\ g_{k+1,1} & \cdots & g_{k+1,k} & g_{k+1,k+1} & \cdots & g_{k+1,k+n} \\ \vdots & & \vdots & \vdots & & \vdots \\ g_{k+n,1} & \cdots & g_{k+n,k} & g_{k+n,k+1} & \cdots & g_{k+n,k+n} \end{pmatrix}. \qquad (4.72)$$

To evaluate the determinant, use the following identity for the determinant of a general 2×2 block matrix:

$$\det \begin{pmatrix} A & B \\ C & D \end{pmatrix} = \det(A)\det(D - CA^{-1}B) \qquad (4.73)$$

for the special case $A = \mathbb{1}$. Application to Eq. 4.72 yields the determinant ratio in the form

$$\frac{\det \hat{g}^{(k+n)}}{\det \hat{g}^{(k)}} = \det \begin{pmatrix} g_{k+1,k+1}-g_{k+1,i}M_{i,j}g_{j,k+1} & \cdots & g_{k+1,k+n}-g_{k+1,i}M_{i,j}g_{j,k+n} \\ \vdots & & \vdots \\ g_{k+n,k+1}-g_{k+n,i}M_{i,j}g_{j,k+1} & \cdots & g_{k+n,k+n}-g_{k+n,i}M_{i,j}g_{j,k+n} \end{pmatrix}$$

$$(4.74)$$

Comparing with (4.62) one sees that measurements for n-particle correlation functions can be expressed in terms of measurements for the single-particle Green function (this is, of course, not true for the Monte Carlo averaged quantities). This property is a consequence of Wick's theorem and provides a mnemonic rule to construct measurements for higher-order moments of the impurity: One (symbolically) enumerates all complete contractions of the Grassmann numbers and replaces each contraction by the corresponding measurement (determinant ratio) of the single-particle Green function. The measurement for the two-particle Green function (Fig. 4.10), $\chi^{\sigma\sigma'}(\omega_1, \omega_2, \omega_3, \omega_4) := \langle c_\sigma(\omega_1) c_\sigma^*(\omega_2) c_{\sigma'}(\omega_3) c_{\sigma'}^*(\omega_4) \rangle$, is obtained in Fourier space as

$$
\begin{aligned}
\langle c_\sigma(\omega_1) c_\sigma^*(\omega_2) c_{\sigma'}(\omega_3) c_{\sigma'}^*(\omega_4) \rangle \longrightarrow\ & c_\sigma(\omega_1) c_\sigma^*(\omega_2)\ c_{\sigma'}(\omega_3) c_{\sigma'}^*(\omega_4) \\
& - c_\sigma(\omega_1) c_{\sigma'}^*(\omega_4)\ c_{\sigma'}(\omega_3) c_\sigma^*(\omega_2) \\
\longrightarrow\ & \tilde{g}^\sigma(\omega_1, \omega_2)\ \tilde{g}^{\sigma'}(\omega_3, \omega_4) \\
& - \delta_{\sigma\sigma'} \tilde{g}^\sigma(\omega_1, \omega_4)\ \tilde{g}^\sigma(\omega_3, \omega_2). \quad (4.75)
\end{aligned}
$$

The functions $\tilde{g}^\sigma(\omega, \omega')$ are measured in analogy to (4.68). The measurement of the two-particle Green function provides the basis of the dual fermion approach introduced in Sect. 4.7.

4.5.5 Sign Problem and α-Parameters

Apparently, the alternating sign in Eq. 4.48 will lead to a severe sign problem (for $U > 0$): the alternating sign causes cancellations in both enumerator and denominator of Eq. 4.67 and error amplification. However, this "trivial" sign problem can be completely suppressed by a suitable choice of the α-parameters. At half-filling the sign problem is suppressed for the choice $\alpha_\downarrow = 1 - \alpha_\uparrow = \alpha$ for any α (by exploiting particle-hole symmetry). Away from half-filling, a sign problem occurs only for $0 < \alpha < 1$. In practice α is chosen randomly and close to this interval (slightly above 1) to minimize the interaction term and hence the perturbation order. While no sign problem exists for the single-orbital problem [65], this is in general not the case for a multiorbital impurity and depends on the local interaction. No general recipe to avoid the sign is available in this case, but a proper choice of the α parameters can significantly reduce the sign problem.

4.5.6 Hybridization Expansion Algorithm

The hybridization expansion or strong-coupling algorithm was initially introduced by P. Werner et al. [58] and has been generalized to multiorbital systems with general interactions [59, 60, 66]. Here the algorithm is discussed in the segment representation, which exploits the possibility of a very fast computation of the trace for interactions of density-density type. The action is regrouped into the atomic part

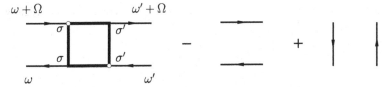

Fig. 4.10 Diagrammatic representation of the local two-particle Green function in terms of scattering of particle-hole pairs, $\chi^{\sigma\sigma'}(\omega_1, \omega_2, \omega_3, \omega_4) := \langle c_{\alpha\sigma}(\omega_1)c^*_{\beta\sigma}(\omega_2)c_{\gamma\sigma'}(\omega_3)c^*_{\delta\sigma'}(\omega_4)\rangle$. Defining the (bosonic) transferred frequency as $\Omega := \omega_1 - \omega_2$, which is conserved in scattering processes, it can be written $\chi^{\sigma\sigma'}(\omega, \omega', \Omega) := \chi^{\sigma\sigma'}(\omega+\Omega, \omega, \omega', \omega'+\Omega)$. The first term contains the vertex part depicted by the square. Lines are fully dressed propagators

$$S_{\mathrm{at}} = \int_0^\beta d\tau \sum_\sigma c^*_\sigma(\tau)[\partial_\tau - \mu]c_\sigma(\tau) + U\int_0^\beta d\tau c^*_\uparrow(\tau)c_\uparrow(\tau)c^*_\downarrow(\tau)c_\downarrow(\tau) \quad (4.76)$$

and the part of the action S_Δ which contains the hybridization term:

$$S_\Delta = -\int_0^\beta d\tau' \int_0^\beta d\tau \sum_\sigma c_\sigma(\tau)\Delta(\tau - \tau')c^*_\sigma(\tau'). \quad (4.77)$$

Here the sign is taken out by reversing the original order of c and c^* to avoid an alternating sign in the expansion. To simplify the notation, consider first the spinless fermion model, which is obtained by disregarding the spin sums and interaction in Eqs. 4.76, 4.77. The series expansion for the partition function is generated by expanding in the hybridization term:

$$\mathcal{Z} = \int \mathcal{D}[c^*, c]e^{-S_{\mathrm{at}}} \sum_k \frac{1}{k!} \int_0^\beta d\tau'_1 \int_0^\beta d\tau_1 \ldots \int_0^\beta d\tau'_k \int_0^\beta d\tau_k$$
$$\times c(\tau_k)c^*(\tau'_k)\ldots c(\tau_1)c^*(\tau'_1)\Delta(\tau_1 - \tau'_1)\ldots\Delta(\tau_k - \tau'_k). \quad (4.78)$$

The important observation now is that, at any order, the diagrams can be collected into a determinant of hybridization functions. In order to see this, one can use a similar reasoning as in the weak-coupling case. The range of integration in (4.73) is changed such that $\tau_k > \tau'_k > \ldots > \tau_1 > \tau'_1$. Note that time ordering is not explicitly indicated since it is implicit in the construction of the path integral. For any given time-ordered sequence of $2k$ times $\tau_k, \tau'_k, \ldots \tau_1, \tau'_1$, the integration in (4.78) generates exactly $2k!$ terms with a different order of times for which the Grassmann numbers can be brought into the same order. These are the $k!$ permutations with times τ_i permuted among themselves and correspondingly for the times τ'_i. The Grassmann numbers can always be brought into the original sequence by permuting first *pairs* of Grassmann numbers, which does not yield an additional sign, and in the second step permuting the annihilators among themselves. The latter operation is associated with an eventual sign depending on the number of permutations required. Specifically, for $k = 2$, one has the $2k! = 4$ terms

$$c(\tau_2)c^*(\tau_2')c(\tau_1)c^*(\tau_1')\Delta(\tau_1 - \tau_1')\Delta(\tau_2 - \tau_2')$$
$$c(\tau_1)c^*(\tau_1')c(\tau_2)c^*(\tau_2')\Delta(\tau_2 - \tau_2')\Delta(\tau_1 - \tau_1')$$
$$c(\tau_2)c^*(\tau_1')c(\tau_1)c^*(\tau_2')\Delta(\tau_2 - \tau_1')\Delta(\tau_1 - \tau_2')$$
$$c(\tau_1)c^*(\tau_2')c(\tau_2)c^*(\tau_1')\Delta(\tau_2 - \tau_1')\Delta(\tau_1 - \tau_2'). \tag{4.79}$$

Upon time ordering, the Grassmann numbers in the last three lines are brought into the same order as in the first line. It is easy to check that only the last two lines acquire a minus sign. Therefore, these terms can be collected to give

$$c(\tau_2)c^*(\tau_2')c(\tau_1)c^*(\tau_1')2\det\begin{pmatrix} \Delta(\tau_1 - \tau_1') & \Delta(\tau_1 - \tau_2') \\ \Delta(\tau_2 - \tau_1') & \Delta(\tau_2 - \tau_2') \end{pmatrix}. \tag{4.80}$$

There are always $k!$ terms which can be brought into the same order by permuting pairs of Grassmann numbers and hence the factor $1/k!$ in (4.78) cancels. The partition function then takes the form

$$\mathcal{Z} = \mathcal{Z}_{\text{at}}\sum_k\int_0^\beta d\tau_1'\int_{\tau_1'}^\beta d\tau_1\ldots\int_{\tau_{k-1}}^\beta d\tau_k'\int_{\tau_k'}^{\circ\tau_k'}d\tau_k$$
$$\times\,\langle c(\tau_k)c^*(\tau_k')\ldots c(\tau_1)c^*(\tau_1')\rangle_{\text{at}}\det\hat\Delta^{(k)}, \tag{4.81}$$

where the average is over the states of the atomic problem described by S_{at}. Here $\det\hat\Delta^{(k)}$ denotes the determinant of the matrix of hybridizations $\hat\Delta_{ij} = \Delta(\tau_i - \tau_j')$. The diagrams contributing to the partition function for $k = 3$ are shown in Fig. 4.11. A diagram is depicted by a collection of segments, where a segment is symbolic for the time interval where the impurity is occupied. The collection of diagrams obtained by connecting the hybridization lines in all possible ways corresponds to the determinant. Collecting the diagrams into a determinant is essential to alleviate or completely suppress the sign problem. Note that the imaginary time interval in (4.81) is viewed as a circle denoted by $\circ\tau_k'$. The trajectories in the path integral are subject to antiperiodic boundary conditions which is accommodated by an additional sign if a segment winds around the circle.

The diagrams of the partition function are sampled by randomly inserting or removing segments of varying length (and further moves to increase sampling efficiency). Empty and fully occupied states also have to be sampled. The weight of a configuration is proportional to

$$w^{(k)} \sim \left|\langle c(\tau_k)c^*(\tau_k')\ldots c(\tau_1)c^*(\tau_1')\rangle_{\text{at}}\det\hat\Delta^{(k)}\right|. \tag{4.82}$$

The insertion of c^* is attempted anywhere in the interval from 0 to β. If the place is already occupied, the move is rejected. Otherwise the c can be inserted at a later time

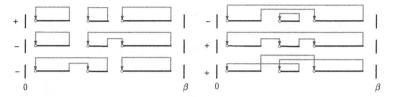

Fig. 4.11 Diagrammatic representation of the six contributions to the partition function for spinless fermions at $k = 3$. An electron is inserted at the start of a segment (marked by an *open circle*) and removed at the segment endpoint. The hybridization function lines $\Delta(\tau_i - \tau'_j)$ (*bold red lines*) are connected to the segments in all possible ways. The sign of each diagram is given on its left. The diagrams collect into a determinant. Reproduced from Ref. [58]

in the interval $[\tau'_{k+1}, \tau'_{k+1} + l_{max})$, where l_{max} is the distance to the starting point of the next segment. The probability to add a segment is hence given by $q^{add} = d\tau^2/\beta l_{max}$. The probability to remove a pair is $q^{rem} = 1/(k + 1)$. The probabilities of the configuration before and after the move are proportional to $p^{new} \sim w^{(k+1)} d\tau^2/\beta^2$ and $p^{old} \sim w^{(k)}$. This yields the acceptance criterion

$$p^{add} = \min\left(1, \frac{1}{k+1} \frac{l_{max}}{\beta} \frac{w^{(k+1)}}{w^{(k)}}\right), \tag{4.83}$$

which involves the computation of ratios of the determinants and the average over the atomic states. The latter is evaluated using exact diagonalization of the atomic problem. For the spinless fermion model, this is particularly simple: $H_{at} = -\mu \int_0^\beta d\tau n(\tau)$ is already diagonal in the occupation number basis $\{|0\rangle, |1\rangle\}$ and, for $k = 2$,

$$\text{Tr}[T_\tau e^{\mu \int_0^\beta d\tau n(\tau)} c(\tau_2) c^\dagger(\tau'_2) c(\tau_1) c^\dagger(\tau'_1)] =$$
$$\text{Tr}[e^{\mu(\beta-\tau_2)n} c(\tau_2) e^{\mu(\tau_2-\tau'_2)n} c^\dagger(\tau'_2) e^{\mu(\tau'_2-\tau_1)n} c(\tau_1) e^{\mu(\tau_1-\tau'_1)n} c^\dagger(\tau'_1) e^{\mu(\tau_1)n}] = e^{\mu l}. \tag{4.84}$$

where l is the length of the time interval on which the impurity is occupied. Here only a single "path" through alternating states $|0\rangle$ and $|1\rangle$ contributes, since the operators have only a single nonzero matrix element.

It is straightforward to generalize the method to general density-density interactions, for which Grassmann numbers of a given flavor (spin, orbitals) appear in alternating order. For general interactions the segment picture has to be abandoned and the trace over a number of matrices that scales with the perturbation order has to be computed explicitly. For d- or f-systems, the number of states becomes very large (2^{10} and 2^{14}, respectively) and several paths leading through different matrix elements contribute. For this case, different optimizations for the efficient computation of the trace have been proposed, such as storing the operators in a binary tree [66], an adjustable base [60], or a Krylov implementation [67].

For the single-orbital Anderson impurity model with Hubbard interaction the segment picture still holds and gives a very intuitive picture of the imaginary time

Fig. 4.12 Typical configuration for the simulation of a single-orbital Anderson impurity model with Hubbard interaction. A separate timeline is drawn for each spin. The shaded regions correspond to the intervals where the impurity is doubly occupied. The segments on each timeline are connected by hybridization lines in all possible ways as in Fig. 4.11 (not shown). The hybridization does not connect segments across timelines, because it is spin-diagonal

dynamics. A configuration is visualized by two separate timelines, one for each spin. The additional sum over spins, $\sum_{\sigma_1 \ldots \sigma_k}$, which enters in the first line of Eq. 4.81, generates contributions such as the one shown in Fig. 4.12. The only difference to the spinless fermion model is that in case the impurity is doubly occupied, the energy U has to be paid and the trace is $e^{\mu(l_\uparrow + l_\downarrow)} e^{-U l_d}$, where l_σ is the time spent on the impurity for an electron with spin σ and l_d is the time the impurity is doubly occupied. The acceptance criterion is modified accordingly.

4.5.7 Measurement of Green's Function

In the strong-coupling formalism, the expansion for Green's function is given by

$$
g(\tau - \tau') = -\frac{\mathcal{Z}_{at}}{\mathcal{Z}} \sum_k \int_0^\beta d\tau_1' \int_{\tau_1'}^\beta d\tau_1 \ldots \int_{\tau_{k-1}}^\beta d\tau_k' \int_{\tau_k'}^\beta d\tau_k \, \tilde{g}(\tau - \tau'; C_k)
$$

$$
\times \langle c(\tau_k) c^*(\tau_k') \ldots c(\tau_1) c^*(\tau_1') \rangle_{at} \det \hat{\Delta}(\tau_i - \tau_j'), \qquad (4.85)
$$

where now

$$
\tilde{g}(\tau - \tau'; C_k) = \frac{\langle c(\tau) c^*(\tau') c(\tau_k) c^*(\tau_k') \ldots c(\tau_1) c^*(\tau_1') \rangle_{at}}{\langle c(\tau_k) c^*(\tau_k') \ldots c(\tau_1) c^*(\tau_1') \rangle_{at}} \qquad (4.86)
$$

is a measurement for Green's function. Hence $\tilde{g}$ is obtained by computing a ratio of traces. Denoting the trace over a product of $k + 1$ pairs of operators in the numerator in (4.86) in the short-hand notation form $\mathrm{Tr}^{(k+1)}$, the integrand for the Green function at some perturbation order k can be written

$$
\mathrm{Tr}^{(k+1)} \det \hat{\Delta}^{(k)} = \frac{\mathrm{Tr}^{(k+1)}}{\mathrm{Tr}^{(k)}} \mathrm{Tr}^{(k)} \det \hat{\Delta}^{(k)}
$$

$$
= \frac{\det \hat{\Delta}^{(k)}}{\det \hat{\Delta}^{(k+1)}} \mathrm{Tr}^{(k+1)} \det \hat{\Delta}^{(k+1)}. \qquad (4.87)
$$

The weight of the partition function appears to the right. Therefore the Green function can either be measured by adding a pair to the trace and compute the Monte Carlo average of the ratio of traces (according to the first line), or remove a column and a row from the matrix $\hat{\Delta}$ and calculate the Monte Carlo average of the ratio of determinants (second line). Diagrammatically, the first possibility corresponds to adding a pair of operators into one of the timelines in Fig. 4.11 (not a segment) and the second to cutting all hybridization lines connecting a given pair. In each case, two unconnected operators $c(\tau)$, $c^\dagger(\tau')$ contribute to $g(\tau - \tau')$. It is clear how to compute the ratio of traces. The ratio of determinants for the strong-coupling case can be obtained using the following identity from linear algebra (A^{-1} is the inverse of the matrix of hybridization functions, i.e. the M-matrix):

$$(A^{-1})_{ji} = \frac{1}{\det(A)} C_{ij}, \tag{4.88}$$

where C_{ji} is the j, i cofactor of A, i.e. the determinant of the j, i minor of A times the factor $(-1)^{i+j}$, which takes care of the sign acquired in the row/column permutations. Note the order of indices, j, i. After cutting the hybridization lines connecting an arbitrary pair of operators, it can be commuted through to the beginning of the trace yielding a factor $(-1)^{i+j}$, so that the right-hand side of (4.88) is the desired determinant ratio. Hence

$$g(\tau) = \left\langle \sum_{i,j=1}^{k} M_{ji} \bar{\delta}(\tau, \tau_i^e - \tau_j^s) \right\rangle_{MC}, \quad \bar{\delta}(\tau, \tau') := \begin{cases} \delta(\tau - \tau') & \tau' > 0 \\ -\delta(\tau - (\tau' + \beta)) & \tau' < 0 \end{cases}.$$
$$\tag{4.89}$$

The $\bar{\delta}$ function arises from the β-antiperiodic definition of the Green function. The density is more accurately determined from the average length of the segments and the average perturbation order serves as an estimator for the kinetic energy, $E_{\text{kin}} = \langle k \rangle_0 / \beta$. The formula for the measurement of the two-particle Green function can be obtained by generalization of (4.83).

4.6   Cluster Extensions of DMFT: Short-Range Non-Local Correlations

DMFT in its single-site formulation neglects spatial correlations at all length scales. The generalization of DMFT to capture the effect of strong short-range correlations leads to the cluster approaches [19, 20, 22]. Short range correlations are relevant in particular for low-dimensional systems and those with strong dimerization. An early application of cluster DMFT to a real material was the case of Ti_2O_3 [68] where single-site DMFT does not reproduce the insulating state stabilized by nonlocal Coulomb interactions and strong chemical bonding between Ti-Ti pairs. Intersite correlations between titanium atoms were also found to be important in cluster calculations of the low-dimensional quantum spin system TiOCl [69]. The metal-insulator

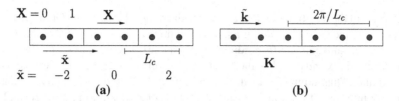

Fig. 4.13 **a** Division of a one-dimensional chain into real-space clusters of size $L_c = 2$. A vector of the original lattice is decomposed as $x = \tilde{x} + X$, where $\tilde{x}$ points to the origin of the clusters in the superlattice and X labels sites within a cluster. All vectors are in units of the (original) lattice spacing. **b** Corresponding division of the reciprocal space into cells of width $2\pi/L_c$. Points within a cell are connected by superlattice momentum $\tilde{k}$ and cells are connected by the cluster momentum K

transition in VO_2 was attributed to the formation of dynamical singlets between neighboring Vanadium atoms [7].

In quantum cluster approaches (for a review, see Ref. [22]), the lattice problem with infinite degrees of freedom is reduced to a cluster problem with less degrees of freedom. The cluster need not be a physical subsystem of the original lattice [21, 70, 71]. Here the cellular DMFT (CDMFT), the dynamical cluster approximation (DCA) and the variational cluster approximation (VCA) will be shortly outlined. The introduction of CDMFT provides a basis for the discussion of the cluster dual fermion approach discussed in Sect. 4.7.

The notation is adapted from Ref. [22]. The d-dimensional lattice containing N sites is grouped into clusters of linear dimension L_c containing $N_c = L_c^d$ sites. As depicted in Fig. 4.13 (a) for $d = 1$, the lattice vectors are decomposed as $x = X + \tilde{x}$, where the vector $\tilde{x}$ denotes the position of a cluster within the superlattice and X labels sites within the cluster. The reciprocal space is split into cells accordingly, as shown in (b). A wave vector in the original lattice is given by $k = \tilde{k} + K$, where $\tilde{k}$ is a superlattice wavevector and K is the cluster momentum. The number of clusters or superlattice momenta is N/N_c. Tiling the lattice into clusters breaks translational invariance of the original lattice. While superlattice momentum is still conserved, the lattice momentum is conserved up to K only. The cluster momentum K has components $K_i = n_i(2\pi/L_c)$ and becomes a reciprocal lattice vector, i.e. $\exp(i K \tilde{x}) \equiv 1$. The Fourier transform and its inverse of a quantity G are thus given by

$$G(K, \tilde{k}) = \sum_{X\tilde{x}} G(X, \tilde{x}) e^{-i[(K+\tilde{k})X + \tilde{k}\tilde{x}]},$$

$$G(X, \tilde{x}) = \frac{1}{N} \sum_{K\tilde{k}} G(K, \tilde{k}) e^{i[(K+\tilde{k})X + \tilde{k}\tilde{x}]}. \qquad (4.90)$$

The sum over $\tilde{x}$ gives the superlattice transform and summing over $\tilde{X}$ corresponds to the intracluster transform.

Following Ref. [22], CDMFT can be obtained from a locator expansion, which is an expansion in real space around the finite cluster. To this end, the hopping and the self-energy are split into intercluster and intracluster parts,

$$t(\tilde{x} - \tilde{x}') = \delta t(\tilde{x} - \tilde{x}) + t_c \delta_{\tilde{x}, \tilde{x}'},$$
$$\tilde{\Sigma}(\tilde{x} - \tilde{x}', i\omega) = \delta \Sigma(\tilde{x} - \tilde{x}', i\omega) + \Sigma_c(i\omega)\delta_{\tilde{x}, \tilde{x}'}, \tag{4.91}$$

where all quantities are matrices in the cluster sites. The locator expansion around the cluster described by the cluster Green function $g_c(i\omega) = [(i\omega + \mu)\mathbb{1} - t_c - \Sigma_c(i\omega)]^{-1}$ in δt and $\delta \Sigma$ reads

$$G(\tilde{x} - \tilde{x}', i\omega) = g_c(i\omega)\delta_{\tilde{x}, \tilde{x}'} + g_c(i\omega) \sum_i [\delta t(\tilde{x} - \tilde{x}_i) + \delta \Sigma(\tilde{x} - \tilde{x}_i, i\omega)] G(\tilde{x}_i - \tilde{x}', i\omega). \tag{4.92}$$

In the cluster approaches, correlations beyond the extension of the cluster are neglected and the remainder of the system is assumed to be uncorrelated. This corresponds to neglecting the intercluster self-energy $\delta \Sigma$ and allows one to map the lattice problem to a cluster embedded in an uncorrelated host. Due to the translational invariance of the superlattice, Eq. 4.92 is diagonalized with respect to superlattice momenta by Fourier transform and with the approximation $\delta \Sigma = 0$ reads

$$G(\tilde{k}, i\omega) = g_c(i\omega) + g_c(i\omega)\delta t(\tilde{k})G(\tilde{k}, i\omega) = [g_c^{-1}(i\omega) - \delta t(\tilde{k})]^{-1}. \tag{4.93}$$

A corresponding relation restricted to the cluster is obtained by averaging this equation over all superlattice momenta,

$$\bar{G}(i\omega) = \frac{N_c}{N} \sum_{\tilde{k}} G(\tilde{k}, i\omega) = \frac{N_c}{N} \sum_{\tilde{k}} [g_c^{-1}(i\omega) - \delta t(\tilde{k})]^{-1}. \tag{4.94}$$

This step is referred to as coarse-graining. It corresponds to neglecting the phase factors $e^{i\tilde{k}\tilde{x}}$ on the vertices of self-energy diagrams which are associated with the position of the cluster in the original superlattice. The cluster Green function contains the self-energy $\Sigma_c(i\omega)$, which can therefore be determined as a functional of the coarse-grained Green function $\bar{G}(i\omega)$ from the solution of an impurity model. Since $g_c(i\omega)$ is independent of $\tilde{k}$, it is possible to determine a local function $\Delta(i\omega)$ such that

$$\bar{G}^{-1}(i\omega) = g_c^{-1}(i\omega) - \Delta(i\omega), \tag{4.95}$$

which defines the hybridization function. In order to establish the relation to an impurity model, one defines the excluded cluster Green function $\mathcal{G}^{-1}(i\omega) = \bar{G}^{-1}(i\omega) + \Sigma_c(i\omega)$ as the bare Green function to $\bar{G}$. By comparison with (4.95) it follows that

$$\mathcal{G}^{-1}(i\omega) = (i\omega + \mu)\mathbb{1} - t_c - \Delta(i\omega). \tag{4.96}$$

The interacting Green function of the impurity model $g(i\omega)$ is hence related to the cluster Green function $g_c(i\omega)$ by

$$g^{-1}(i\omega) = \mathcal{G}^{-1}(i\omega) - \Sigma_c = g_c^{-1}(i\omega) - \Delta(i\omega),\qquad(4.97)$$

so that Eq. 4.95 is seen to be equivalent to requiring that the coarse-grained Green function $\bar{G}$ be equal to the Green function g of the effective cluster impurity model. In particular, Eq. 4.94 takes the form

$$\bar{G}(i\omega) = \frac{N_c}{N} \sum_{\tilde{k}} \left\{ g^{-1}(i\omega) + [\Delta(i\omega) - \delta t(\tilde{k})] \right\}^{-1}.\qquad(4.98)$$

These equations clearly resemble those in Sect. 4.2. Using the local degrees of freedom to label sites within a cluster, instead of (or in addition to) orbitals and identifying the intra- and inter-cluster parts t_c and δt of the hopping with ϵ_{loc} and $\epsilon_{\mathbf{k}}$, the self-consistency loop in Sect. 4.2 is seen to be general enough to accommodate an implementation of the CDMFT equations.[1]

As mentioned previously, CDMFT neglects the phase factors $e^{i\tilde{k}\tilde{x}}$. In CDMFT, the Laue function, which describes momentum conservation at the vertices, is thus approximated by

$$\Delta_{\text{CDMFT}}^{\text{L}} = \sum_{X} e^{iX(K_1+\tilde{k}_1+K_2+\tilde{k}_2-K_3-\tilde{k}_3-K_4-\tilde{k}_4)}\qquad(4.99)$$

for a two-particle vertex. Clearly, cluster momentum K is not conserved due to the presence of the phase factors e^{ikx}. Therefore CDMFT violates translational invariance with respect to the cluster sites which hence are not equivalent. This is obvious for clusters with $L_c \geq 3$, where bulk and surface sites of a cluster may be distinguished, but also applies for $L_c = 2$. CDMFT calculations are carried out in the cluster real-space representation (i.e. all quantities are matrices in the cluster sites), since there is no benefit in changing to the cluster k-space representation, which is not diagonal.

Since translational invariance is broken, the lattice quantities are functions of two independent momenta k and k'. They can differ by a reciprocal lattice vector Q, where $Q_i = 0, \ldots, (L_c - 1)2\pi/L_c$. For example, the self-energy is expressed in terms of the cluster self-energy as

$$\Sigma(k, k', i\omega) = \frac{1}{N_c} \sum_{Q} \sum_{X,X'} e^{ikX} \Sigma_c(X, X', i\omega) e^{-ik'X'} \delta(k - k' - Q),\quad(4.100)$$

where the dependence on cluster sites is written explicitly. A translationally invariant solution is obtained by approximating the lattice quantities by the $Q = 0$ contribution,

---

[1] In the notation of Sect. 4.2 N should then be replaced the number of clusters N/N_c and the momentum $\mathbf{k}$ should be identified with the superlattice momentum $\tilde{k}$.

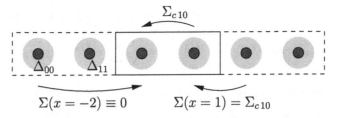

Fig. 4.14 Illustration of the CDMFT lattice self-energy. The original lattice is replaced by a collection of clusters embedded in a self-consistent bath. The intercluster self-energy $\Sigma(x = 1)$ is approximated by the intracluster self-energy Σ_{c10} for this distance not exceeding the maximal distance between cluster sites and zero otherwise

$$\Sigma(k, i\omega) = \frac{1}{N_c} \sum_{X,X'} = e^{ik(X-X')} \Sigma_c(X, X', i\omega). \tag{4.101}$$

Transforming back to real space shows that the lattice quantities for a given distance $x - x'$ are obtained as an average over the cluster quantities for the same distance,

$$\Sigma(x - x', i\omega) = \frac{1}{N_c} \sum_{X,X'} \Sigma_c(X, X', i\omega) \delta_{X-X', x-x'}. \tag{4.102}$$

Spatial correlations are hence included up to a length determined by the extension of the cluster. Note that (4.102) underestimates the nonlocal contributions, in particular for small clusters. Using the shorthand notation $\Sigma_{X,X'} = \Sigma(X, X')$, one sees that the local self-energy is averaged correctly, $\Sigma(x = 0) = (\Sigma_{c00} + \Sigma_{c11})/2$, while the nearest-neighbor self-energy contribution according to (4.102) would read $\Sigma(x = 1) = (1/2)\Sigma_{c10}$, since Σ_{c01} contributes to $\Sigma(x = -1)$. It was therefore suggested to reweigh the terms in the sum [72]. For the above example, $\Sigma(x = 1) = \Sigma_{c10}$.

When translational invariance is recovered in this way, the solution of the lattice problem may be viewed as shown in Fig. 4.14: The lattice is replaced by a lattice of clusters all of which are embedded in a self-consistent bath. The self-energy on a cluster is obtained from the self-consistent solution of the local problem and the intercluster self-energy between sites on neighboring clusters at a distance $x - x'$ is artificially set equal to the average of the intracluster self-energy for the same distance. The self-energy for distances exceeding the maximum distance between sites within the cluster is zero. The cluster dual fermion approach discussed in Sect. 4.7 is concerned with reintroducing the neglected intercluster self-energy perturbatively (Fig. 4.16).

The idea of the DCA is to restore momentum conservation within the cluster by a different choice of the intracluster Fourier transform. In CDMFT, the intracluster transform of the dispersion reads

$$[t(\tilde{k})]_{X,X'} = \frac{1}{N_c} \sum_{K} e^{i(K+\tilde{k})(X-X')} \epsilon_{K+\tilde{k}}, \tag{4.103}$$

while in the DCA, the phase factors $e^{i\tilde{k}X}$ are excluded using the transform

$$[t_{\text{DCA}}(\tilde{k})]_{X,X'} = [t(\tilde{k})]_{X,X'} e^{-i\tilde{k}(X-X')} = \frac{1}{N_c} \sum_K e^{iK(X-X')} \epsilon_{K+\tilde{k}}. \qquad (4.104)$$

Correspondingly, the Laue function reads

$$\Delta_{\text{DCA}}^L = \sum_X e^{iX(K_1+K_2-K_3-K_4)} = N_c \delta_{K_1+K_2,K_1'+K_2'}, \qquad (4.105)$$

which shows that cluster momentum is conserved. The intracluster hopping in DCA is therefore given by the intracluster Fourier transform of the dispersion, which is obvious by coarse-graining Eq. 4.104. Denoting hopping restricted to the cluster as $\bar{\epsilon}_K$, one has $\delta t(K + \tilde{k}) = \epsilon_{K+\tilde{k}} - \bar{\epsilon}_K$. The analog of (4.93) is hence diagonal in cluster Fourier space:

$$G(K + \tilde{k}, i\omega) = g_c(K, i\omega) + g_c(K, i\omega)\delta t(K + \tilde{k})G(K + \tilde{k}, i\omega)$$
$$= \frac{1}{1/g_c(K, i\omega) - \delta t(K + \tilde{k})}. \qquad (4.106)$$

By viewing these quantities as diagonal matrices with respect to cluster momenta, the Eqs. 4.95–4.98 also apply for the DCA. The self-energy becomes a piecewise constant function in k-space [22].

Finally, the variational cluster approach (VCA) is based on the self-energy functional theory (SFT) [73]. In SFT, a functional $\Omega[\Sigma]$ of the self-energy is constructed, which can be shown to be stationary at the physical self-energy. This functional is in general unknown. A numerically solvable reference system is introduced, which shares the interaction part with the original system. The crucial point is that the functional of the original system can be evaluated *exactly* at the trial self-energies of the reference system and the stationary point can be found on the restricted subset of trial self-energies. The approximation lies in the choice of the reference system. The DMFT and CDMFT can be shown to be special cases of this unifying approach [21].

4.7 Long-Range Correlations and the Dual-Fermion Perturbation Expansion

The shortcomings of DMFT on the one hand and its ability to describe the frequently dominant local temporal correlations on the other have triggered many efforts to go beyond the mean-field description, while maintaining DMFT as a starting point. DMFT becomes exact in the limit of infinite coordination number z. An expansion in $1/z$, however, leads to difficulties as the action depends in a non-analytic way on the coordination number [74]. In an empirical approach, DMFT has been supplemented

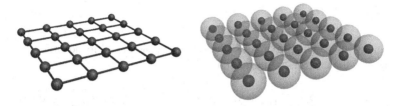

Fig. 4.15 Construction of the dual fermion approximation: In a first step, the original lattice problem (*left*) with bonds (*blue lines*) is replaced by a collection of decoupled impurities exerted to an electronic bath, as indicated by the *transparent (blue) spheres* (*right*)

Fig. 4.16 Illustration of the dual fermion approach. Spatial correlations in the original lattice problem are mediated between the impurities of Fig. 4.15 through dual fermions, which in turn interact via n-particle interactions. Illustration by I. Labuhn

with a momentum dependent self-energy by including spin and charge density wave-like fluctuations through a Gaussian random field [75]. Building on earlier work on strong-coupling expansions for the Hubbard model [76–78], a general framework to perform a systematic cumulant expansion around DMFT even considering non-local Coulomb interaction was developed in Ref. [79].

While cluster extensions to DMFT become computationally highly demanding for large clusters, the combination of numerical and analytic (diagrammatic) methods is a promising route for including the effects of long-range correlations. Recent developments have led to approaches which include long-range correlations via straight-forward diagrammatic corrections to DMFT [24–26]. Based on earlier suggestions for bosonic fields [80, 81], it was recognized that a systematic, fully renormalized expansion around DMFT can be formulated in terms of auxiliary fermions [82]. We will introduce this approach in detail in the following. Additional details can be found in Ref. [83].

An example of a material for which such corrections are relevant is elementary iron: DMFT grossly overestimates the Curie temperatures since it completely neglects spatial correlations and cannot capture the reduction of the critical temperature due to the presence of long-wavelength magnons. Furthermore, evidence was found for a quasiparticle mass enhancement due to electron-magnon coupling in angular resolved photoemission spectroscopy (ARPES) experiments [84]. The same renormalization effects are expected to play a crucial role in the CuO_2 planes of the cuprates [85, 86]. Long-range correlations are expected to be essential in low-dimensional systems such as nanostructures, or quasi-one-dimensional organic conductors, such as TTF-TCNQ [87].

4.7.1 Formalism

The goal is to find an (approximate) solution to the general multiband lattice problem
described by the imaginary time action

$$S[c^*, c] = - \sum_{\omega k \sigma m m'} c^*_{\omega k \sigma m} [(i\omega + \mu)\mathbb{1} - h_{k\sigma}]_{mm'} c_{\omega k \sigma m'} + \sum_i S_{\mathrm{loc}}[c^*_i, c_i].$$

(4.107)

Here $h_{k\sigma}$ is the one-electron part of the Hamiltonian, $\omega_n = (2n + 1)\pi/\beta, n = 0, \pm 1, \ldots$ are the Matsubara frequencies, β and μ are the inverse temperature and chemical potential, respectively, $\sigma = \uparrow, \downarrow$ labels the spin projection, m, m' are orbital indices and c^*, c are Grassmann variables. The index i labels the lattice sites and **k**-vectors are quasimomenta. In order to keep the notation simple, it is useful to introduce the combined index $\alpha \equiv \{m\sigma\}$. Translational invariance is assumed for simplicity in the following. For applications it is important to note that the local part of the action, S_{loc}, may contain *any* type of local interaction. The only requirement is that it is local within the multiorbital atom or cluster. For realistic calculations in combination with density functional theory, one can use the Hamiltonian (4.40).

In order to formulate a perturbation expansion around DMFT, a local quantum impurity problem (Anderson impurity model) is introduced in its spirit, in the form

$$S_{\mathrm{imp}}[c^*, c] = - \sum_{\omega \alpha \beta} c^*_{\omega \alpha} [(i\omega + \mu)\mathbb{1} - \Delta_\omega]_{\alpha\beta} c_{\omega\beta} + S_{\mathrm{loc}}[c^*, c],$$

(4.108)

where Δ_ω is the hybridization matrix describing the coupling of the (cluster) impurity to an electronic bath. Apart from the connection to DMFT, another motivation for rewriting the lattice action in this form is to express it in terms of a reference problem that can be solved accurately for an arbitrary hybridization function using the methods of Sect. 4.5. Exploiting the locality of the hybridization function Δ_ω, the lattice action (4.107) is rewritten *exactly* by adding and subtracting Δ_ω at each lattice site:

$$S[c^*, c] = \sum_i S_{\mathrm{imp}}[c^*_i, c_i] - \sum_{\omega k \alpha\beta} c^*_{\omega k \alpha} (\Delta_\omega - h_k)_{\alpha\beta} c_{\omega k \beta}.$$

(4.109)

Note that this step leaves the hybridization function unspecified. This will be used later to optimize the approach. The lattice may now be viewed as a collection of impurities, which are coupled through the bilinear term to the right of this equation. The effect of spatial correlations enters here and renders an exact solution (of large systems) impossible. A perturbative treatment is desirable, but not straightforward as the impurity action is non-Gaussian and hence there is no Wick theorem. Therefore, dual fermions are introduced in the path integral representation of the partition function

$$\mathcal{Z} = \int \mathcal{D}[c^*, c] \exp(-S[c^*, c]),$$

(4.110)

through the continuous Hubbard-Stratonovich transformation (HST)

$$
\exp\left(c_\alpha^* b_{\alpha\beta} (a^{-1})_{\beta\gamma} b_{\gamma\delta} c_\delta\right) =
$$
$$
\frac{1}{\det a} \int \exp\left(-f_\alpha^* a_{\alpha\beta} f_\beta - f_\alpha^* b_{\alpha\beta} c_\beta - c_\alpha^* b_{\alpha\beta} f_\beta\right) \prod_\gamma df_\gamma^* df_\gamma. \tag{4.111}
$$

In order to transform the exponential of the bilinear term in (4.109), we choose the matrices a, b in accordance with Refs. [82, 88] as

$$
a = g_\omega^{-1} (\Delta_\omega - h_k)^{-1} g_\omega^{-1}, \qquad b = g_\omega^{-1}, \tag{4.112}
$$

where g_ω is the local, interacting Green function of the impurity problem. With this choice, the lattice action transforms to

$$
S[c^*, c, f^*, f] = \sum_i S_{\text{site},i} + \sum_{\omega k \alpha\beta} f_{\omega k \alpha}^* [g_\omega^{-1} (\Delta_\omega - h_k)^{-1} g_\omega^{-1}]_{\alpha\beta} f_{\omega k \beta}. \tag{4.113}
$$

Hence the coupling between sites is transferred to a local coupling to the auxiliary fermions:

$$
S_{\text{site},i}[c_i^*, c_i, f_i^*, f_i] = S_{\text{imp}}[c_i^*, c_i] + \sum_{\alpha\beta} f_{\omega i \alpha}^* g_{\omega\alpha\beta}^{-1} c_{\omega i \beta} + c_{\omega i \alpha}^* g_{\omega\alpha\beta}^{-1} f_{\omega i \beta}. \tag{4.114}
$$

Since g_ω is local, the sum over all states labeled by k could be replaced by the equivalent summation over all sites by a change of basis in the second term. The crucial point is that the coupling to the auxiliary fermions is purely local and S_{site} decomposes into a sum of local terms. The lattice fermions can therefore be integrated out from S_{site} for each site i separately. This completes the change of variables:

$$
\int \exp\left(-S_{\text{site}}[c_i^*, c_i, f_i^*, f_i]\right) \mathcal{D}[c_i^*, c_i] =
$$
$$
\mathcal{Z}_{\text{imp}} \exp\left(-\sum_{\omega\alpha\beta} f_{\omega i \alpha}^* g_{\omega\alpha\beta}^{-1} f_{\omega i \beta} - V_i[f_i^*, f_i]\right). \tag{4.115}
$$

The above equation may be viewed as the defining equation for the dual potential $V[f^*, f]$. The choice of matrices (4.112) ensures a particularly simple form of this potential. An explicit expression is found by expanding both sides of Eq. 4.115 and equating the resulting expressions by order. Formally this can be done up to all orders and in this sense the transformation to the dual fermions is exact. The nth-order term in the infinite series for the dual potential is given by $[(-1)^{n-1}/(n!)^2]\gamma_{12...2n}^{(2n)} f_1^* f_2 \cdots f_{2n}$ for $n \geq 2$. The dual potential involves the impurity correlation functions at all orders. In practice the series has to be terminated. For most applications, the dual potential is approximated as

$$
V[f^*, f] = -\frac{1}{4}\gamma_{1234}^{(4)} f_1^* f_2 f_3^* f_4, \tag{4.116}
$$

where the combined index $1 \equiv \{\omega\alpha\}$ comprises frequency, spin and orbital degrees of freedom. $\gamma^{(4)}$ is the exact, fully antisymmetric, reducible vertex of the local quantum impurity problem. It is given by

$$\gamma^{(4)}_{1234} = g^{-1}_{11'}g^{-1}_{33'}\left[\chi^{imp}_{1'2'3'4'} - \chi^{imp,0}_{1'2'3'4'}\right]g^{-1}_{2'2}g^{-1}_{4'4}, \qquad (4.117)$$

with the two-particle Green function of the impurity being defined as

$$\chi^{imp}_{1234} := \langle c_1 c^*_2 c_3 c^*_4 \rangle_{imp} = \frac{1}{\mathcal{Z}_{imp}}\int c_1 c^*_2 c_3 c^*_4 \exp\left(-S_{imp}[c^*, c]\right)\mathcal{D}[c^*, c]. \qquad (4.118)$$

The disconnected part reads

$$\chi^{imp,0}_{1234} = g_{12}g_{34} - g_{14}g_{32}. \qquad (4.119)$$

The single- and two-particle Green functions can be calculated using the Monte Carlo algorithms introduced in Sect. 4.5. After integrating out the lattice fermions, the dual action depends on the new variables only and reads

$$S_d[f^*, f] = -\sum_{\omega k \, \alpha\beta} f^*_{\omega k \alpha}[G^{d,0}_\omega(\mathbf{k})]^{-1}_{\alpha\beta}f_{\omega k \beta} + \sum_i V_i[f^*_i, f_i]. \qquad (4.120)$$

and the bare dual Green function is found to be

$$G^{d,0}_\omega(\mathbf{k}) = -g_\omega\left[g_\omega + (\Delta_\omega - h_k)^{-1}\right]^{-1}g_\omega, \qquad (4.121)$$

which involves the local Green function g_ω of the impurity model.

Up to now, Eqs. 4.120, 4.121 are merely a reformulation of the original problem. In practice, approximate solutions are constructed by treating the dual problem perturbatively. Several diagrams that contribute to the dual self-energy are shown in Fig. 4.17. These are constructed from the impurity vertices and dual Green functions as lines. The first diagram (a) is purely local, while higher orders contain nonlocal contributions, e.g. diagram (b). Inserting the renormalized Green function into diagram (a) includes contributions such as the one in (a'). In practice, approximations to the self-energy are constructed in terms of skeleton diagrams. The lines shown in Fig. 4.17 are therefore understood to be fully dressed propagators. The use of skeleton diagrams is necessary to ensure the resulting theory to be conserving in the Baym-Kadanoff sense [89], i.e. it fulfills the basic conservation laws for energy, momentum, spin and particle number. It is an important consequence of the exact transformation (4.111) that for a theory which is conserving in terms of dual fermions, the result is also conserving in terms of lattice fermions [82, 88]. This allows to construct general conserving approximations within the dual fermion approach. Numerically, the self-energy is obtained in terms of skeleton diagrams by performing a self-consistent renormalization as described below.

Once an approximate dual self-energy is found, the result may be transformed back to a physical result in terms of lattice fermions using exact relations. Due to the

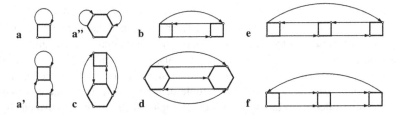

Fig. 4.17 Diagrams contributing to the dual self-energy Σ^d. Diagrams **a**, **a'**, **a"** and **c** give local, the other ones nonlocal contributions. The three diagrams labeled by (**a**) do not contribute in case the condition (4.130) is fulfilled

abstract nature of the dual variables, the truncation of the series for the dual potential and the perturbation series may appear as arbitrary and uncontrolled approximations. As will be shown in the following, these can be well justified and in fact, approximations may be constructed based on physical considerations much like in the usual way.

4.7.2 Dual Perturbation Theory

The action (4.121) allows for a Feynman-type diagrammatic expansion in powers of the dual potential V. The rules are similar to those of the antisymmetrized diagrammatic technique [90]. Extension of these rules to include generic n-particle interaction vertices is straightforward. In summary, the rules for the perturbation theory of the dual self-energy in momentum space are the following:

- Draw all topologically distinct, connected diagrams involving any of the n-body interactions $\gamma^{(2n)}$ depicted by regular polygons with $2n$ endpoints, whereof n are outgoing (incoming) endpoints, where a directed line originates (terminates)
- connect the vertex endpoints with directed lines, compliant with the designation of ingoing and outgoing endpoints
- with each line associate a dual Green function G^d
- Assign a frequency, momentum, orbital and spin label to each endpoint, taking into account energy-, momentum-, and spin-conservation at each vertex
- Sum/integrate over all internal variables
- For each tuple of n equivalent lines (such as equally directed lines connecting the same two vertices), associate a factor $1/n!$
- Multiply the resulting expression by a factor $(T/N)^m S^{-1}$, where m is the number of independent frequency/momentum summations and S is the symmetry factor (see e.g. Ref. [63]).
- For diagrams containing two-particle interactions only, determine the sign by replacing the square by a vertical interaction line and counting the number n_l of resulting closed loops. The sign is $(-1)^{n_l}$.

Note that energy conservation is associated with a δ-function $\delta(\omega_1 + \omega_3 + \cdots + \omega_{2n-1} - \omega_2 - \omega_4 - \cdots - \omega_{2n})$ at each vertex. The number m of independent frequency and momentum summations is the number of summations remaining after eliminating these δ-functions. $T = \beta^{-1}$ is the temperature and N the number of k-points in the first Brillouin zone. Due to the use of an antisymmetrized interaction, the diagrams acquire a combinatorial prefactor. For a tuple of n equivalent lines, the expression has to be multiplied by a factor $1/n!$. For example, there are two equivalent lines pointing from right to left and three equivalent lines from left to right in diagram d) of Fig. 4.17. The prefactor is hence $1/2! \cdot 1/3! = 1/12$. For certain diagrams, additional symmetry factors have to be taken into account. For example, after numbering the vertices in the generic ring diagrams in Fig. 4.35, the $2n$ cyclic permutations of the sequences $(1, 2, \ldots, n)$ and $(n, \ldots, 2, 1)$ yield the same diagram so that $S = 2n$. The symmetry factor for self-energy diagrams is unity.

Two examples follow. Writing spins explicitly for clarity, the self-energy correction of the Hartree–Fock-like diagram (a) in Fig. 4.17 contains a single closed loop and reads

$$[\Sigma_{\omega\sigma}^{d(a)}]_{\alpha\beta,} = -T \sum_{\omega'} \sum_{\sigma'} \gamma_{\alpha\beta\gamma\delta}^{\sigma\sigma\sigma'\sigma'}(\omega, \omega, \omega', \omega')[G_{\omega'\sigma'}^{d\,loc}]_{\delta\gamma}, \qquad (4.122)$$

where summation over repeated indices is implied and $G^{d\,loc} = (1/N) \sum_k G^d(k)$ denotes the local part of the dual Green function. The second-order contribution represented by diagram b) contains two equivalent lines and one closed loop and hence evaluates to

$$[\Sigma_{\omega\sigma}^{d(b)}(k)]_{\alpha\nu} = -\frac{1}{2}\left(\frac{T}{N}\right)^2 \sum_{k_1 k_2} \sum_{\omega_1 \omega_2} \sum_{\sigma'} \gamma_{\alpha\beta\gamma\delta}^{\sigma\sigma\sigma'\sigma'}(\omega, \omega_1, \omega_2, \omega_3)[G_{\omega_1\sigma}^d(k_1)]_{\beta\mu}$$

$$\times [G_{\omega_2\sigma'}^d(k_2)]_{\lambda\gamma}[G_{\omega_3\sigma'}^d(k_3)]_{\delta\kappa}\gamma_{\kappa\lambda\mu\nu}^{\sigma'\sigma'\sigma\sigma}(\omega_3, \omega_2, \omega_1, \omega)$$

$$-\frac{1}{2}\left(\frac{T}{N}\right)^2 \sum_{k_1 k_2} \sum_{\omega_1 \omega_2} \gamma_{\alpha\beta\gamma\delta}^{\sigma\bar{\sigma}\bar{\sigma}\sigma}(\omega, \omega_1, \omega_2, \omega_3)[G_{\omega_1\bar{\sigma}}^d(k_1)]_{\beta\mu}$$

$$\times [G_{\omega_2\bar{\sigma}}^d(k_2)]_{\lambda\gamma}[G_{\omega_3\sigma}^d(k_3)]_{\delta\kappa}\gamma_{\kappa\lambda\mu\nu}^{\sigma\bar{\sigma}\bar{\sigma}\sigma}(\omega_3, \omega_2, \omega_1, \omega), \qquad (4.123)$$

where $\omega_3 = \omega + \omega_2 - \omega_1$ and $k_3 = k + k_2 - k_1$ by energy conservation. Here $\bar{\sigma} := -\sigma$ and the sum over spins runs over $\sigma' = \uparrow, \downarrow$. In practice, it is more efficient to evaluate the lowest order diagrams in real space and transform back to reciprocal space using the fast Fourier transform.

4.7.3 Exact Relations

After an approximate result for the dual self-energy or the dual Green function has been obtained, it has to be transformed back to the corresponding physical quantities in terms of lattice fermions. The fact that dual fermions are introduced through the exact HST (4.111) allows to establish exact identities between dual and lattice quantities. Hence the transformation does not involve any additional approximations.

The relations between the n-particle cumulants of dual and lattice fermions can be established using the cumulant (linked cluster) technique. To this end, one may consider two different, equivalent representations of the following generating functional:

$$F[J^*, J; K^*, K] := \ln \mathcal{Z}_f \int \mathcal{D}[c^*, c; f^*, f]$$

$$\exp\Big(- S[c^*, c; f^*, f] + J_1^* c_1 + c_2^* J_2 + K_1^* f_1 + f_2^* K_2\Big).$$
(4.124)

Integrating out the lattice fermions from this functional similar to (4.115) (this can be done with the sources J and J^* set to zero) yields

$$F[K^*, K] = \ln \tilde{\mathcal{Z}}_f \int \mathcal{D}[f^*, f] \exp\left(-S_d[f^*, f] + K_1^* f_1 + f_2^* K_2\right). \quad (4.125)$$

with $\tilde{\mathcal{Z}}_f = \mathcal{Z}/\mathcal{Z}_d$. The dual Green function and higher order cumulants are obtained from (4.125) by suitable functional derivatives, e.g.

$$G_{12}^d = -\frac{\delta^2 F}{\delta K_2 \delta K_1^*}\bigg|_{K^*=K=0}, \ \left[\chi^d - G^d \otimes G^d\right]_{1234} = \frac{\delta^4 F}{\delta K_4 \delta K_3^* \delta K_2 \delta K_1^*}\bigg|_{K^*=K=0},$$
(4.126)

where $G \otimes G$ is the antisymmetrized direct product of Green functions, so that the term in the angular brackets is the connected part of the dual two-particle Green function. Conversely, integrating out the dual fermions from (4.124) using the HST, one obtains an alternative representation, which more clearly reveals a connection of the functional derivatives with respect to the sources J, J^* and K, K^*. The result is

$$F[J^*, J; K^*, K] = K_1^* [g(\Delta - h)g]_{12} K_2 + \ln \int \mathcal{D}[c^*, c] \exp\Big(- S[c^*, c]$$

$$+ J_1^* c_1 + c_2^* J_2 + K_1^* [g(\Delta - h)]_{12} c_2 + c_1^* [(\Delta - h)g]_{12} K_2\Big).$$
(4.127)

In analogy to (4.126), the cumulants in terms of lattice fermions are obviously obtained by functional derivative with respect to the sources J and J^* with K and K^* set to zero. Applying the derivatives with respect to K, K^* to (4.127) with $J = J^* = 0$ and comparing to (4.126), e.g. yields the following identity:

$$G_{12}^d = -[g(\Delta - h)g]_{12} + [g(\Delta - h)]_{11'} G_{1'2'} [(\Delta - h)g]_{2'2}. \quad (4.128)$$

Solving for G provides the rule how to transform the dual Green function to the physical quantity in terms of lattice fermions. For higher-order cumulants the additive term in (4.127) does not contribute and the relation between the two-particle cumulants evaluates to

$$\left[\chi^d - G^d \otimes G^d \right]_{1234} =$$

$$[g(\Delta - h)]_{11'} [g(\Delta - h)]_{33'} [\chi - G \otimes G]_{1'2'3'4'} [(\Delta - h)g]_{2'2} [(\Delta - h)g]_{4'4} ,$$
$$(4.129)$$

It is apparent that similar relations hold for higher-order cumulants. Note that the transformation only involves single-particle functions. Hence one may conclude that n-particle collective excitations (and corresponding instabilities) are the same for dual and lattice fermions.

4.7.4 Self-Consistency Condition and Relation to DMFT

The hybridization function Δ, which so far has not been specified, allows to optimize the starting point of the perturbation theory and should be chosen in an optimal way. The condition of the first diagram (Fig. 4.17a) in the expansion of the dual self-energy to be equal to zero at all frequencies fixes the hybridization. This eliminates the leading order diagrammatic correction to the self-energy and establishes a connection to DMFT, which can be seen as follows: Since $\gamma^{(4)}$ is local, this condition amounts to demanding that the local part of the dual Green function be zero:

$$\sum_{\mathbf{k}} G^d_\omega(\mathbf{k}) = 0. \qquad (4.130)$$

The simplest nontrivial approximation is obtained by taking the leading-order correction, diagram (a), evaluated with the bare dual propagator (4.121). Recalling the expression for the DMFT Green function, Eq. 4.19, it is readily verified that

$$G^{DMFT}_\omega(\mathbf{k}) - g_\omega = \left[g_\omega^{-1} + \Delta_\omega - h_{\mathbf{k}} \right]^{-1} - g_\omega$$

$$= -g_\omega \left[g_\omega + (\Delta_\omega - h_{\mathbf{k}})^{-1} \right]^{-1} g_\omega = G^{d,0}_\omega(\mathbf{k}). \qquad (4.131)$$

It immediately follows that (4.130) evaluated with the bare dual Green function is exactly equivalent to the DMFT self-consistency condition (4.16):

$$\frac{1}{N} \sum_{\mathbf{k}} G^{d,0}_\omega(\mathbf{k}) = 0 \quad \Longleftrightarrow \quad \frac{1}{N} \sum_{\mathbf{k}} G^{DMFT}_\omega(\mathbf{k}) = g_\omega. \qquad (4.132)$$

Hence DMFT appears as the zero-order approximation in this approach and corrections to DMFT are included perturbatively. A formal relation to DMFT can be established using the Feynman variational functional approach. In this context, DMFT appears as the optimal approximation to a Gaussian ensemble of dual fermions [88].

When diagrammatic corrections are taken into account and the first diagram is evaluated with the dressed propagator G^d, the condition (4.130) will in general be violated. It can be reinforced by adjusting the hybridization function iteratively. This

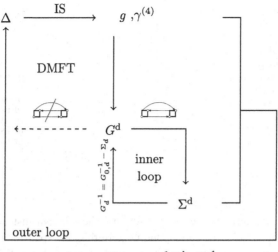

$$\Delta_{\text{new}} = \Delta_{\text{old}} + g^{-1} G^{\text{d}}_{\text{loc}} G^{-1}_{\text{loc}}$$

Fig. 4.18 Illustration of the dual fermion calculation procedure. For DMFT calculations no diagrammatic corrections are evaluated and the new guess for the hybridization function is constructed from the bare dual Green function (indicated by the *dashed line*). In dual fermion calculations, the diagrammatic corrections are included into the dual Green function via a self-consistent renormalization procedure (*inner loop*). A new guess for the hybridization function is computed from the local part of the renormalized dual Green function (*outer loop*)

corresponds to eliminating an infinite partial series starting from the diagrams labeled by (a) in Fig. 4.17. These contributions are effectively absorbed into the impurity problem. Note that such an expansion is not one around DMFT, but rather around an optimized impurity problem. For practical calculations, the hybridization function is updated in successive iterations using the rule

$$\Delta^{\text{new}}_\omega = \Delta^{\text{old}}_\omega + \left[g^{-1}_\omega G^{\text{d,loc}}_\omega (G^{\text{loc}}_\omega)^{-1} \right]. \tag{4.133}$$

The dual fermion calculation procedure is depicted in Fig. 4.18 . If no diagrammatic corrections are taken into account (this is indicated by the dashed line), the self-consistency condition (4.133) is equivalent to (4.20).

The only difference between a DMFT and a DF calculation are the diagrammatic corrections which are included into the dual Green function. To this end, the local impurity vertex $\gamma^{(4)}$ has to be calculated in addition to the Green function in the impurity solver step. The self-energy is calculated within a self-consistent renormalization procedure (inner loop).

Since the choice of the hybridization function is not unique, one may replace it by a discrete version $\Delta^{(n)} = \sum_{k=1}^{n} |V_k|^2 / (i\omega - \epsilon_k)$ for a small number n of bath degrees of freedom, for which the impurity problem can be solved efficiently using exact diagonalization. In this case, the condition (4.130) cannot be fulfilled

Fig. 4.19 Imaginary part of the local Green function for the one-dimensional Hubbard model with nearest-neighbor hopping. Curves from successive outer loop iterations are shown in comparison with DMFT and a result from a zero-temperature DMRG calculation. Parameters are $U/t = 6$, $T/t = 0.1$

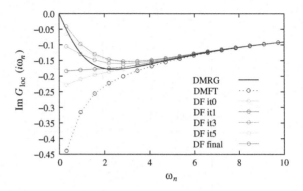

4.7.5 Numerical Results

In the following, we show some illustrative results for the Hubbard model, which is governed by the Hamiltonian (4.7). Unless otherwise stated, only the two lowest-order diagrams (a) and (b) of Fig. 4.17 have been used. Figure 4.19 shows results for the local Green function of the one-dimensional model. It may be considered as a benchmark system for the approach, because the importance of nonlocal correlations is expected to increase by reducing the dimensionality. This is clearly an unfavorable situation for DMFT, which completely neglects spatial correlations. The quality of the approximate solution is assessed by comparison to a numerically accurate Density Matrix Renormalization Group (DMRG) calculation.

For these parameters, DMFT fails to describe the physics of the model even qualitatively, as it predicts a metallic phase. The first correction ("it0") corresponds to a diagrammatic correction of DMFT obtained by self-consistent renormalization of the dual Green function with diagrams (a) and (b). The correction is significant, but the solution remains metallic. This is not unexpected, as the Mott transition cannot be described perturbatively [15]. By adjusting the hybridization function in successive outer loop iterations to reinforce the condition (4.130)—this corresponds to the summation of an infinite partial series—the Green function becomes closer to the DMRG result and the system becomes insulating. This illustrates the importance of the self-consistency condition and the modification of the Weiss dynamical field. It is particularly important for low-dimensional systems, where DMFT gives a rather poor description of the local environment of the embedded atom (e.g. whether it is metallic or insulating) and an unfavorable starting point of the perturbation theory.

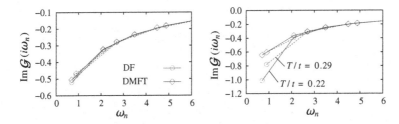

Fig. 4.20 Imaginary part of the bare effective DMFT Green function $\mathcal{G}(i\omega) = [i\omega + \mu - \Delta(i\omega)]^{-1}$ and its modification after a fully self-consistent dual fermion calculation, for $U/t = 4$ (*left*) and $U/t = 8$. Short-range antiferromagnetic correlations lead to a temperature dependent modification of the local environment of the embedded atom

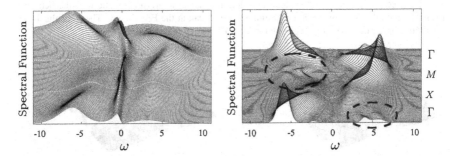

Fig. 4.21 Spectral function $A(k, \omega)$ for the 2D Hubbard model at half-filling obtained within DMFT (*left*) and dual fermion calculations (*right*) for $U = 8t$ and $T/t = 0.235$. From bottom to top, the curves are plotted along the high-symmetry lines $\Gamma \rightarrow X \rightarrow M \rightarrow \Gamma$. The high-symmetry points $X = (0, \pi)$ and $M = (\pi, \pi)$ are marked by dashed lines. *The encircled structures* can be attributed to dynamical short-range antiferromagnetic correlations

The modification of the bare effective DMFT Green function, which is related to the Weiss field as $\mathcal{G}(i\omega) = [i\omega + \mu - \Delta(i\omega)]^{-1}$, is illustrated within self-consistent dual fermion calculations for the two-dimensional Hubbard model in Fig. 4.20. For an on-site interaction $U/W = 1/2$ ($W = 8t$ is the bandwidth), the modification is negligible and does not depend on temperature. For $U/W = 1$ on the contrary, there is a strong, temperature dependent modification of the Weiss field in the dual fermion calculation, while its DMFT value remains temperature independent. This suggests nonlocal correlations as the origin of this modification. Support of this conjecture is provided by the k-resolved spectral function $A(k, \omega)$ obtained from maximum-entropy analytical continuation shown in Fig. 4.21. The DMFT spectral function displays a quasiparticle band, while in the DF calculation, spectral weight is transferred away from the Fermi level. Recalling the nesting condition $\epsilon_{k+Q} = -\epsilon_k$ for the antiferromagnetic wavevector $Q = (\pi, \pi)$, the locus of these features allows to interpret them as shadow bands due to dynamical short-range antiferromagnetic correlations. The strength of these correlations increases as the temperature is lowered.

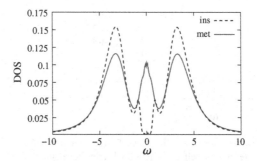

Fig. 4.22 Metallic and insulating local density of states obtained in the coexistence region of the Mott transition for $U/t = 6.5$ and $T/t = 0.08$. The insulating solution exhibits characteristic peaks at the gap edge. The antiferromagnetic correlations lead to antiferromagnetic-gap-like behavior [92]. The metallic solution exhibits shoulders on the peak at the Fermi level. These results are in qualitative agreement with the ones obtained from a cluster calculation [91]

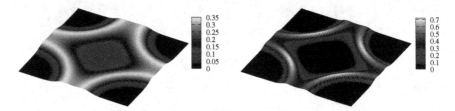

Fig. 4.23 Spectral function $A(k, \omega = 0)$ (Γ-centered) obtained by polynomial extrapolation of the Matsubara lattice Green function $G(k, \omega_n)$ for finite doping $\delta = 7\%$, $U/t = 10$, $T/t = 0.05$ and next-nearest-neighbor hopping $t'/t = -0.3$. While the DMFT spectral function (*left*) is constant along the Fermi surface (lines of highest intensity) as expected, the dual fermion result (*right*) shows the destruction of quasiparticles in the antinodal region

A detailed analysis of the phase diagram shows that these correlations lead to a drastic reduction of the critical U from $U_c/t \sim 9.35$ in DMFT down to $U_c/t \sim 6.5$ within the dual fermion calculation. This, as well as the density of states in the coexistence region (Fig. 4.22) and the slope of the transition lines in the $U - T$ phase diagram below the critical point, which are modified from negative within DMFT to positive [83], is in qualitative agreement with cluster DMFT results [92]. We emphasize that these results cannot be obtained from a straightforward diagrammatic expansion around DMFT as the modification of the Weiss field is essential. This distinguishes the present method from related approaches [25, 26].

As a final illustration, consider the case of finite doping. Figure 4.23 shows results for the spectral function $A(k, \omega = 0)$ at the Fermi level of the hole-doped system for $\delta = 7\%$ obtained by polynomial extrapolation from Matsubara frequencies. A next-nearest-neighbor hopping $t'/t = -0.3$ relevant to the real experimental situation in the cuprates has been included. In DMFT, the spectral function must be flat everywhere along the Fermi surface, as the self-energy does not depend on momentum. Including nonlocal corrections to the self-energy, the high momentum resolution of this approach allows to easily resolve a k-selective destruction of quasiparticles in the antinodal region (in the proximity of the X-point) and the formation of Fermi arcs. A concomitant feature appears in the effective quasiparticle dispersion law $h_k^{\text{eff}} = \text{Re}\{[h_k - \mu + \Sigma(k, \omega = 0)]/[1 + i\partial/\partial_\omega \Sigma(k, \omega = 0)]\}$, which is flattened in the antinodal region. This behavior was predicted earlier as being due to non-Fermi-liquid behavior as the Van-Hove singularity crosses the Fermi level, see Ref. [88] and references therein. As noted in Ref. [93], this can be attributed to short-range correlations, which is confirmed in the present calculations. In recent DCA calculations this phenomenon manifests itself as an orbital selective Mott transition in momentum space [94].

4.7.6 Generalizations of the Dual Fermion Approach

The dual fermion approach has been generalized in various respects. The case of commensurate antiferromagnetic order has been considered in Ref. [88]. Broken symmetry phases can also be studied using susceptibilities. This has been introduced in Refs. [95, 96] and has been applied to the Hubbard model on the triangular lattice [97] and used to calculate the d-wave pairing susceptibility [98]. The method has been improved upon by the extension of the formalism to clusters [99] and through the construction of an infinite-order ladder diagram summation [100]. An expansion in terms of dual fermions has further been employed to construct an efficient impurity solver based on a reference problem with few bath degrees of freedom which can be solved by exact diagonalization [101]. Work to combine the dual fermion approach with realistic band structure calculations is currently in progress. In the following, we briefly discuss the generalization to clusters, the calculation of susceptibilities and the ladder approximation.

4.7.7 Cluster Dual Fermion Approach

The cluster dual fermion approach (CDFA) is a straightforward generalization of the single site approach. The expansion is performed around a cluster impurity as the reference system. The construction is based on a tiling of the lattice into clusters of equal size. A cluster impurity problem is introduced at the position of each cluster. After introducing the auxiliary fermions, the lattice degrees of freedom are integrated out locally on the cluster. One may readily verify that the self-consistency condition (4.130) is exactly equivalent to the respective condition in cellular DMFT (cf. Sect. 4.6). We note at this point that it is also possible to formulate a diagrammatic extension of the dynamical cluster approximation (DCA).

The dual fermion approach employs a two-level self-consistency (Fig. 4.18). Since the CDMFT solution captures the usually dominant short-range correlations, the adjustment of the hybridization function in the outer self-consistency loop is expected to be much less important compared to the single site case. In the CDFA, the full vertex of the cluster impurity problem is determined from CTQMC without approximations.

The CDFA breaks translational invariance with respect to the cluster sites, as CDMFT does. One can imagine the CDMFT solution as a superlattice of clusters, each embedded in a selfconsistently determined host. The intercluster self-energy is zero, but is formally set equal to the (averaged) intracluster self-energy to regain translational invariance. Self-energy contributions at distances exceeding the maximum distance of sites inside the cluster are neglected. In the CDFA, the intercluster self-energy is reintroduced perturbatively.

Using the notation of Sect. 4.6, a translationally invariant self-energy can be constructed as follows. Unlike CDMFT, the CDF self-energy additionally depends on the distance $\tilde{x} - \tilde{x}'$ between clusters in the superlattice. Due to translational invariance of the superlattice, all quantities are diagonal with respect to superlattice momenta $\tilde{k}$. The momentum representation of the CDF self-energy is related to the self-energy in real space representation by the Fourier transform

$$\Sigma(K, K', \tilde{k}) = \sum_{\tilde{x}} \sum_{XX'} e^{-i\tilde{k}(\tilde{x}-\tilde{x}')} e^{-i(K+\tilde{k})X} \Sigma(X, X', \tilde{x} - \tilde{x}') e^{i(K'+\tilde{k})X'}, \quad (4.134)$$

where the frequency dependence is omitted for brevity. Since translational invariance with respect to the cluster sites is broken within the CDFA, the lattice self-energy in general depends on two independent momenta k and k'. These differ only through the difference between K and K', which in turn can differ by a reciprocal lattice vector Q. The lattice self-energy is hence given by

$$\Sigma(k, k') = \frac{1}{N_c} \sum_{Q} \sum_{XX'} e^{-ikX} \Sigma(X, X', k) e^{ik'X'} \delta(k - k' - Q). \quad (4.135)$$

Here the superlattice Fourier transform has been carried out and $\Sigma(X, X', \tilde{k})$ has been replaced by $\Sigma(X, X', k)$. The latter can be done because K is a reciprocal lattice vector so that the phase $\tilde{k}(\tilde{x} - \tilde{x}')$ can be replaced by $k(\tilde{x} - \tilde{x}')$ in (4.134). As in CDMFT, it is natural to approximate the lattice self-energy by the homogeneous $Q = 0$ component. This yields the translationally invariant self-energy

$$\Sigma(k) = \frac{1}{N_c} \sum_{XX'} e^{-ik(X-X')} \Sigma(X, X', k)$$

$$= \frac{1}{N_c} \sum_{XX'} \sum_{\tilde{x}} e^{-ik(X-X')} e^{-ik\tilde{x}} \Sigma(X, X', \tilde{x}). \quad (4.136)$$

The meaning of this equation is more obvious after transforming back to real space,

$$\Sigma(x - x') = \frac{1}{N} \sum_k \Sigma(k) e^{ik(x-x')}$$

$$= \frac{1}{NN_c} \sum_{\tilde{x}} \sum_{XX'} \sum_k e^{-i[k(X-X')+k\tilde{x}-k(x-x')]} \Sigma(X, X', \tilde{x})$$

$$= \frac{1}{N_c} \sum_{\tilde{x}} \sum_{XX'} \Sigma(X, X', \tilde{x}) \delta_{(X-X'+\tilde{x}),x-x'}. \tag{4.137}$$

The construction of the CDF self-energy for the $N_c = 2$ cluster in a 1D system is exemplified by the following table and illustrated in Fig. 4.24.

Coordinates					$\Sigma(x - x')$
$x - x'$	$\tilde{x}$	X	X'	$X - X'$	
0	0	0	0	0	$[\Sigma_{00}(0) + \Sigma_{11}(0)]/2$
0	0	1	1	0	
1	0	1	0	1	$[\Sigma_{10}(0) + \Sigma_{01}(2)]/2$
1	2	0	1	−1	
2	2	0	0	0	$[\Sigma_{00}(2) + \Sigma_{11}(2)]/2$
2	2	1	1	0	
3	2	1	0	1	$[\Sigma_{10}(2) + \Sigma_{01}(4)]/2$
3	4	0	1	−1	

Here the matrix notation $\Sigma_{X,X'}(\tilde{x} - x') := \Sigma(X, X', \tilde{x})$ is used. The translationally invariant self-energy in real space at a given distance is obtained by averaging all contributions $\Sigma(X, X', \tilde{x})$ corresponding to that distance. Note that the $N_c = 2$ cluster constitutes a special case. The local cluster quantities are averaged over all phases and hence are symmetric, i.e. $\Sigma_{c00} = \Sigma_{c11}$. Consequently, the $Q = (\pi)$ contribution $(\Sigma_{00} - \Sigma_{11})/2$ is exactly zero. For an $N_c = 3$ cluster, contributions to the local and nonlocal self-energy from bulk and surface sites are averaged by choosing the $Q = (0)$ component. Apart from averaging the intracluster and intercluster components, in practice, the symmetry-equivalent quantities on the impurity should be averaged to reduce the Monte Carlo error, e.g. $\Sigma_{10}(0)$ and $\Sigma_{01}(0)$ for the $N_c = 2$ cluster.

The number of contributions to the self-energy at a given distance is of course exactly equal to N_c. Hence there is no underestimation of the nonlocal contributions and no need for reweighing as in CDMFT (see Sect. 4.6).

As a benchmark, the CDF approach is applied to the 1D Hubbard model using an $N_c = 2$ cluster. Periodic boundary conditions are assumed with respect to the chain. The dispersion is readily obtained by applying the cluster Fourier transform (4.103) to the dispersion $h_k = -2t \cos(k)$, which yields

$$h_{\tilde{k}} = -\begin{pmatrix} 0 & t + t' e^{i2\tilde{k}} \\ t + t' e^{-i2\tilde{k}} & 0 \end{pmatrix}. \tag{4.138}$$

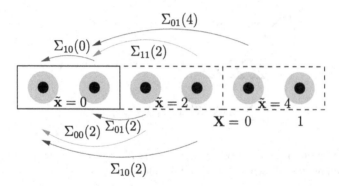

Fig. 4.24 Construction of the translationally invariant CDF self-energy for an $N_c = 2$ cluster. Only nonlocal contributions are shown. Self-energy contributions that are averaged according to (4.137) are indicated by arrows connecting sites at the same distance. Arrows are plotted from X to X'. The nearest-neighbor contribution involves averaging of intra- and intercluster components. All other contributions are obtained perturbatively

Note that the sign of the phases $2\tilde{k}$ in (4.138) is immaterial for CDMFT calculations, since local quantities are averaged over all phases. In the CDFA, the sign of the phase must be compatible with the definition of the superlattice Fourier transform. Otherwise, for example the two self-energies $\Sigma_{10}(\tilde{x} = 2)$ and $\Sigma_{01}(\tilde{x} = 2)$ shown in Fig. 4.24 will be interchanged.

The CDFA results shown in the following have been obtained from a fully self-consistent calculation within second order dual perturbation theory. The DMFT hybridization function remains nearly unchanged in the outer loop iterations and the results from different iterations are essentially the same as expected. This is important because the calculation of the full vertex for the cluster impurity problem requires a sizeable numerical effort. Figure 4.25 shows the local Green function obtained within the CDFA and other approximations in comparison to results from a density matrix renormalization group (DMRG) calculation (for a review on DMRG, see Ref. [103]). Already the cluster approximations for $N_c = 2$ cluster sites, i.e. the cellular DMFT (CDMFT) and the variational cluster approach (VCA) considerably improve the result compared to the single-site DF calculation. This emphasizes the importance of nearest-neighbor correlations in the model. The $N_c = 2$ CDF Green function nevertheless considerably improves upon the CDMFT result and appears to be the best approximation that can be achieved for $N_c = 2$. Note that the CDMFT solution for $N_c = 2$ corresponds to the zero-order approximation of the CDFA. On the other hand, as shown in the inset, it is also apparent that the VCA solution for $N_c = 4$ is superior compared to CDF for $N_c = 2$. This is an indication that correlations beyond the $N_c = 2$ cluster are not negligible and are insufficiently captured by the second-order CDF perturbation theory, which predominantly takes short-range correlations into account. Note that although the VCA and DMRG data correspond to $T = 0$, finite temperature effects are found to be negligible here. Indeed, the VCA and the finite temperature CDMFT calculation give very similar results for $N_{c=2}$.

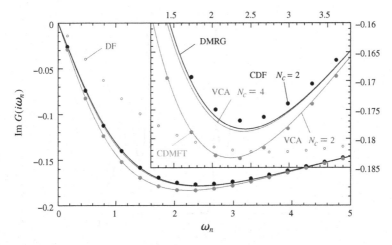

Fig. 4.25 Imaginary part of the local $N_c = 2$ cluster dual fermion (CDF) Green function $G(i\omega_n)$ for the one-dimensional Hubbard model at $U/t = 6$ and $T/t = 0.1$ in comparison with density matrix renormalization group (DMRG) results and various other approximations: cellular DMFT (CDMFT) for $N_c = 2$, second-order dual perturbation theory (DF), and variational cluster approach (VCA) data from Ref. [102] for the indicated number of cluster sites in the reference system. VCA and DMRG results are for $T = 0$

For CDMFT and the CDFA essentially the same results are obtained by reducing the temperature down to $T/t = 0.05$.

The local density of states obtained by analytical continuation of the Matsubara Green functions from Fig. 4.25 are compared to the DMRG result in Fig. 4.26. The DF solution considerably underestimates the width of the gap. It is still somewhat underestimated in the cluster approaches. CDMFT agrees qualitatively, but the position of the Hubbard bands is not correctly captured and the width of the peaks at the gap edge is overestimated. The CDF result already resembles the DMRG solution rather well and provides a satisfactory description of the local properties. The height of the peaks however appears to be too small. This is an artifact of the analytical continuation procedure, which has problems to resolve such sharp features. Here the method of Ref. [104] was used, which for the present case is superior to the maximum entropy method. The peaks at the gap edge are overlooked completely in the maximum entropy density of states (not shown).

Results for the nearest-neighbor Green function are shown in Fig. 4.27. The CDMFT solution strongly overestimates this quantity. In the CDFA, the intra-cluster component of the nearest-neighbor Green function is improved compared to CDMFT. The intercluster component, which is not present in CDMFT, is obtained perturbatively in the CDFA. It appears to be underestimated and decays quickly at high energy. This is not a consequence of the finite frequency cutoff for the vertex and the dual self-energy, which is considerably larger ($\sim 40t$). The nearest-neighbor component of the translationally invariant Green function is also shown. It is obtained by averaging the inter- and intracluster components according to (4.137):

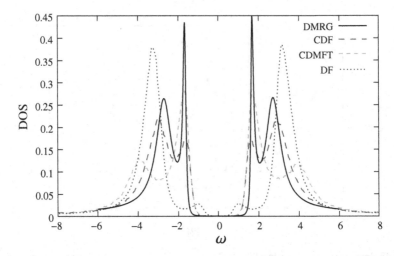

Fig. 4.26 Comparison of the local density of states. For CDF, DF and CDMFT, the spectral function has been obtained by analytical continuation of the Matsubara Green functions from the previous figure using the method of Ref. [104]. Parameters are the same as in Fig. 4.25

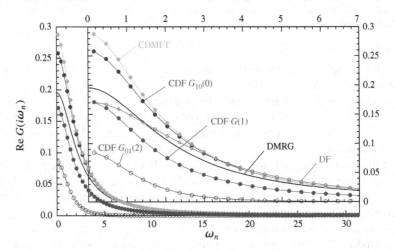

Fig. 4.27 Real part of the nearest-neighbor Green function as a function of Matsubara frequencies in comparison to DMRG. The intra and intercluster Green functions $G_{10}(0)$ and $G_{01}(2)$ are averaged to give the translationally invariant solution $G(1)$

$G(1) = [G_{10}(0) + G_{01}(2)]/2$. For small frequencies, it approximates the DMRG solution rather well. However, due to the fast decay of the intercluster component, it does not have the correct high-frequency behavior. It is interesting to note that while the single-site dual fermion approach is clearly insufficient for the description of the local properties of the 1D system, it seems to capture the nearest-neighbor correlations rather well. This may be attributed to the fact that the single-site approach does

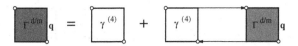

Fig. 4.28 Bethe-Salpeter equation for the dual vertex in the electron-hole channel with a local approximation $\Gamma_{irr} = \gamma^{(4)}$ to the irreducible vertex. The solution Γ contains the sum of all ladder diagrams up to infinite order in γ

not break translational invariance. The CDFA shows a clear tendency to restore the translational invariance. This is however insufficient at this level of approximation to the dual self-energy, which is reflected in the large difference between the intra and intercluster components $G_{10}(0)$ and $G_{01}(2)$.

4.7.8 Calculation of Susceptibilities

For the calculation of the dual susceptibility, the dual vertex function is first calculated by means of a Bethe-Salpeter equation [105, 106] (in the following, the index "d" on dual Green functions is omitted and we write the equations for a single-orbital model for simplicity)

$$\Gamma^{\alpha}_{\omega\omega'\Omega}(q) = \gamma^{\alpha}_{\omega\omega'\Omega} - \frac{T}{N}\sum_{\omega''}\sum_{k}\gamma^{\alpha}_{\omega\omega''\Omega}G_{\omega''}(k)G_{\omega''+\Omega}(k+q)\Gamma^{\alpha}_{\omega''\omega'\Omega}(q). \quad (4.139)$$

This equation is depicted diagrammatically in Fig. 4.28. Here the irreducible vertex is approximated by the local irreducible interaction of dual fermions to lowest order and is hence given by the *reducible* vertex of the impurity model γ (the index '(4)' is omitted in what follows). Here $\alpha = d, m$ stands for the density (d) and magnetic (m) electron-hole channels: $\Gamma^d = \Gamma^{\uparrow\uparrow\uparrow\uparrow} + \Gamma^{\uparrow\uparrow\downarrow\downarrow}$, $\Gamma^m = \Gamma^{\uparrow\uparrow\uparrow\uparrow} - \Gamma^{\uparrow\uparrow\downarrow\downarrow}$. The physical content of the BSE is repeated scattering of particle-hole pairs. In the two channels the particle-hole pair has a definite total spin S and spin projection S_z. The density channel corresponds to the $S = 0$, $S_z = 0$ singlet channel, while Γ^m is the vertex in the $S = 1$, $S_z = 0$ triplet channel. In the magnetic channel, the collective excitations are magnons. The vertex $\Gamma^{\uparrow\downarrow\downarrow\uparrow}$ ($\Gamma^{\downarrow\uparrow\uparrow\downarrow}$) which corresponds to the $S_z = +1(-1)$ spin projection of the $S = 1$ channel must be equal to Γ^m in the paramagnetic state (longitudinal and transverse modes cannot be distinguished).

The BSE may be solved iteratively, starting from the approximation $\Gamma^{(0)} \approx \gamma$. Inserting this into the right-hand-side of Eq. 4.139 yields a new approximation $\Gamma^{(1)}$. Repeating this step successively generates a sum of all ladder diagrams with $1, \ldots, n+1$ irreducible rungs in the approximation $\Gamma^{(n)}$. In practice however, the BSE is solved by matrix inversion according to

$$[\Gamma^{\alpha}_{\omega\omega'\Omega}(q)]^{-1} = (\gamma^{\alpha}_{\omega\omega'\Omega})^{-1} + \frac{T}{N}\sum_{k}G_{\omega}(k)G_{\omega+\Omega}(k+q)\delta_{\omega\omega'}, \quad (4.140)$$

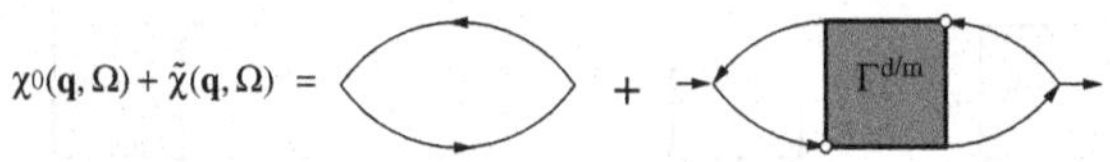

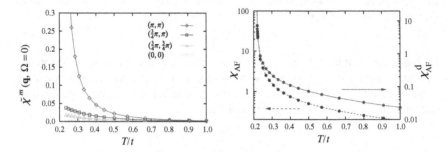

Fig. 4.29 Diagrammatic representation of the susceptibility, Eqs. 4.141, 4.142

Fig. 4.30 *Left*: Connected part of the dual magnetic susceptibility $\tilde{\chi}^m(q, \Omega = 0)$, Eq. 4.142, for different wave vectors at $U/t = 4$. The susceptibility diverges for the antiferromagnetic (AF) wave vector $q = (\pi, \pi)$, indicating the AF instability. *Right*: comparison of the antiferromagnetic dual and lattice fermion susceptibility for the same parameters. The two susceptibilities differ in absolute value, but diverge at the same temperature: Two-particle excitations and corresponding instabilities are the same for dual and lattice fermions

which corresponds to summing up the infinite series. The vertices are matrices in the fermionic Matsubara frequencies ω, ω'. They are diagonal with respect to Ω and q, since the center of mass energy and momentum of the particle-hole pair is conserved in scattering processes.

From the vertex, the spin and charge susceptibility is finally obtained as $\chi = \chi_0 + \tilde{\chi}$, where

$$\chi_0(q, \Omega) = -\frac{T}{N} \sum_{\omega} \sum_{k} G_{\omega}(k) G_{\omega+\Omega}(k + q) \tag{4.141}$$

and

$$\tilde{\chi}^{\alpha}(q, \Omega) = \frac{T^2}{N^2} \sum_{\omega\omega'} \sum_{kk'} G_{\omega}(k) G_{\omega+\Omega}(k + q) \Gamma^{\alpha}_{\omega\omega'\Omega} G_{\omega'}(k') G_{\omega'+\Omega}(k' + q). \tag{4.142}$$

If one is only interested in instabilities, which are signalled by the divergence of the corresponding susceptibility, it is sufficient to consider the dual quantities. The equivalence of two-particle excitations in terms of dual and lattice fermions ensures that the dual and lattice susceptibilities diverge at the same parameters (see Fig. 4.30). The lattice susceptibility is obtained using the exact relations between dual and lattice Green functions. The convolution of dual Green functions (4.141) is transformed using the relation (4.128) for the single-particle Green function. From (4.129) one immediately finds that $\tilde{\chi}$ is transformed by replacing the dual Green

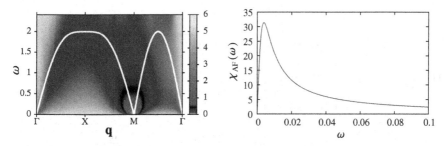

Fig. 4.31 *Left*: Dynamical susceptibility $\chi(\mathbf{q}, \omega)$ for $U/t = 4$ and $T/t = 0.19$, obtained from a dual fermion calculation and analytical continuation using Padé approximants. It shows the magnon spectrum in the paramagnetic state. The dispersion from spin wave theory with effective exchange coupling $J = 4t^2/U$ is shown for comparison. Values for $\chi > 6$ are excluded to improve the contrast. *Right*: Cross-section through the peak at the M-point. The displacement from zero is consistent with a small energy scale J/ξ, where ξ is the correlation length (in units of the lattice constant)

functions on the left (L) and right (R) of the vertex in Fig. 4.29 by the propagator-like functions

$$\mathcal{G}_\omega^L(\mathbf{k}) = (\Delta_\omega - h_\mathbf{k})^{-1} g_\omega^{-1} G_\omega^d(\mathbf{k}),$$
$$\mathcal{G}_\omega^R(\mathbf{k}) = G_\omega^d(\mathbf{k}) g_\omega^{-1} (\Delta_\omega - h_\mathbf{k})^{-1}. \tag{4.143}$$

In the context of DMFT, the susceptibility is obtained using relations similar to Eqs. 4.139, 4.141 and 4.142 [14]. The momentum dependence of the irreducible vertex is neglected in DMFT. It is further approximated by the irreducible vertex of the impurity model. Recall that the lattice Green function is exactly equal to the DMFT Green function when dual corrections to the self-energy are neglected and the dual Green function fulfills the self-consistency condition (4.130). It is an important property of the above equations that under the same conditions the lattice susceptibility calculated within the dual fermion approach is exactly equal to the DMFT susceptibility [83].

As a further illustration, we plot the dynamical susceptibility $\chi(\mathbf{q}, \omega)$ in Fig. 4.31. It clearly displays the magnon spectrum in the paramagnetic state. The dispersion from spin wave theory is shown for comparison. It is given by the expression [107] $\epsilon(\mathbf{k}) = 2zJS\sqrt{1 - \gamma(\mathbf{k})^2}$ where z is the coordination number, $S = 1/2$ is the spin of the fermions and $\gamma(\mathbf{k}) = \frac{1}{z}\sum_{NN} e^{i\mathbf{k}\mathbf{r}_{NN}} = (\cos k_x + \cos k_y)/2$ for the square lattice. The right panel of Fig. 4.31 shows a cross-section for the antiferromagnetic wave vector $\mathbf{q}_{AF} = (\pi, \pi)$ (M-point). The peak is broadened and slightly shifted from zero. Such a behavior is reminiscent of a 2D Heisenberg model at finite temperature, where long-range order with a correlation length $\xi \gg a$ takes place (a is the lattice constant) and a corresponding small energy scale of order Ja/ξ arises [91].

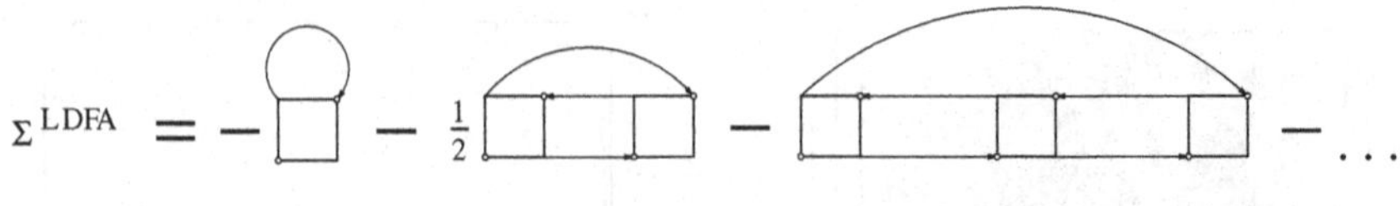

$$\Sigma^{\text{LDFA}} = -\bigcirc\hspace{-0.3em}\Box - \tfrac{1}{2}\,\overparen{\Box\Box} - \overparen{\Box\Box\Box} - \ldots$$

Fig. 4.32 Diagrammatic representation of the dual self-energy in the ladder approximation. The diagrams are shown with their corresponding signs and prefactors. All higher-order terms in the expansion of Σ^d have the same prefactor

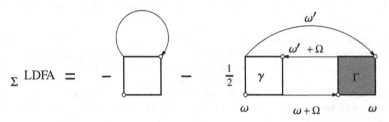

Fig. 4.33 Diagrammatic representation of the Schwinger-Dyson equation for the LDFA self-energy. The frequency labels are indicated

4.7.9 Ladder Approximation

The second-order correction to the dual self-energy turns out to be insufficient in the vicinity of a magnetic instability. In this case the paramagnon contribution to the self-energy should be properly taken into account. From the foregoing, it is clear that this can be accomplished by constructing a ladder approximation to the dual self-energy (Fig. 4.32). In general, the self-energy can be obtained by means of the Schwinger-Dyson equation (see Fig. 4.33), which relates it to the vertex function of the lattice. The vertex in turn is approximated by adding contributions from the different fluctuation channels as depicted in Fig. 4.34. This is similar to the construction of the DΓA self-energy [25] and can be motivated from the parquet equations. In the vicinity of the magnetic instability, it is sufficient to consider the contributions from the horizontal and vertical electron-hole channels. Inserting the approximation for the vertex $\Gamma = \Gamma^{\text{eh}} + \Gamma^{\text{v}} - \gamma$ into the Schwinger-Dyson equation generates the ladder approximation for the self-energy. The resulting self-energy explicitly reads

$$\Sigma_\sigma^{\text{LDFA}}(\omega, \mathbf{k}) =$$

$$-\frac{T}{N}\sum_{\omega'\mathbf{k}'}\sum_{\sigma'}\gamma_{\omega'\omega\Omega=0}^{\sigma\sigma'\sigma'\sigma'}G_{\omega'}(\mathbf{k}')$$

$$+\frac{1}{2}\frac{T}{N}\sum_{\Omega\mathbf{q}}\sum_{\omega'}\sum_{\sigma'}\gamma_{\omega'\omega\Omega}^{\sigma\sigma'\sigma'\sigma}G_{\omega+\Omega}(\mathbf{k}+\mathbf{q})\chi_{\omega'\Omega}^{0}(\mathbf{q})\left[2\Gamma_{\omega\omega'\Omega}^{\text{eh}\sigma\sigma'\sigma'}(\mathbf{q}) - \gamma_{\omega\omega'\Omega}^{\sigma\sigma'\sigma'}\right]$$

$$+\frac{1}{2}\frac{T}{N}\sum_{\Omega\mathbf{q}}\sum_{\omega'}\gamma_{\omega'\omega\Omega}^{\sigma\sigma\bar{\sigma}\bar{\sigma}}G_{\omega+\Omega}(\mathbf{k}+\mathbf{q})\chi_{\omega'\Omega}^{0}(\mathbf{q})\left[2\Gamma_{\omega\omega'\Omega}^{\text{eh}\sigma\bar{\sigma}\bar{\sigma}\sigma}(\mathbf{q}) - \gamma_{\omega\omega'\Omega}^{\sigma\bar{\sigma}\bar{\sigma}\sigma}\right].$$

$$(4.144)$$

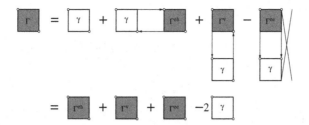

Fig. 4.34 Diagrammatic construction of an approximation to the full vertex in the ladder dual fermion approximation. The ladder sums are obtained as the solution of the Bethe–Salpeter equations in the tree different channels: (horizontal) electron-hole, vertical and electron-electron channel

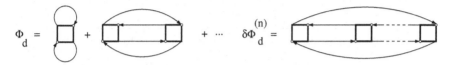

Fig. 4.35 The contributions to the dual Luttinger-Ward functional for the ladder approximation. The generic n-th order ring diagrams are included up to all orders n

where $\chi^0_{\omega\Omega}(q) := -(T/N) \sum_k G_\omega(k) G_{\omega+\Omega}(k+q)$. Note that because of the use of an antisymmetrized theory (the interaction γ is fully antisymmetric), the (horizontal) electron-hole and vertical channels are equivalent and can be transformed into each other, so that the self-energy is expressed in terms of the electron-hole vertex only. Since the latter already contains the full ladder sum, one may attempt to construct the corresponding self-energy by simply attaching a Green function to Γ^{eh}. This however leads to overcounting of the second-order contribution (4.123), which therefore is subtracted explicitly in (4.144). Such overcounting is also encountered in the case of Hugenholtz diagrams [63, 108].

Formally, the ladder approximation can be obtained by functional derivative of a suitable generating functional, which involves the generic ring diagrams depicted in Fig. 4.35. Hence the construction is similar to that of the fluctuation exchange approximation (FLEX) [109, 110]. Note however, that the ladder dual fermion approach (LDFA) goes far beyond the conventional FLEX. Most notably, as shown below, the LDFA is also applicable for strong coupling.

The relevance of the self-consistent ladder approximation (the dual self-energy is obtained by self-consistent renormalization) is best illustrated in the vicinity of the antiferromagnetic instability (AFI) in the 2D Hubbard model. To this end, we consider the eigenvalue problem derived from the BSE (4.139) (see below for more details). Solving the BSE is analogous to summing up a geometric series. A leading eigenvalue of $\lambda_{\text{max}} = 1$ implies the divergence of the ladder sum and the corresponding susceptibility. For the eigenvalue in the $q = (\pi, \pi)$ magnetic electron-hole channel this signals a transition to a state with long-range antiferromagnetic order. In Fig. 4.36 we show results for the leading eigenvalue in this channel within different approximations. At higher temperatures all approximations give similar

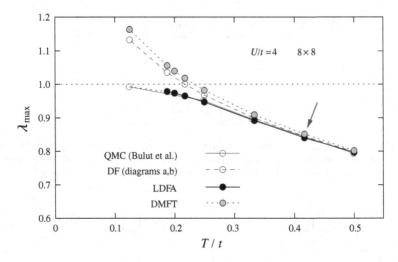

Fig. 4.36 Leading eigenvalue in the $q = (\pi, \pi)$ spin channel for different approximations. The QMC results have been taken from Ref. [111]. Close to the instability, it is essential to include the effect of exchange of spin fluctuations through the ladder approximation

results. As the temperature is lowered, first DMFT fails as documented through a finite Néel temperature of $T_{N}^{DMFT}/t = 0.233$. This is in contradiction to the lattice Quantum Monte Carlo (QMC) results, which do not predict an ordered state. It is an artifact of the mean-field approximation, which tends to stabilize the AF order. To be able to compare these results to those from a dual fermion calculation, the dual vertex and Green functions have to be transformed to the corresponding lattice quantities using the exact relations introduced earlier, before evaluating the eigenvalues. The thus obtained results from second-order dual perturbation theory include short-range spatial correlations beyond DMFT through the leading nonlocal diagram (b) of Fig. 4.17 and reduce the critical temperature down to $T_{N}^{DF}/t = 0.215$. Such small reduction is in accordance with the leading eigenvalue being close to unity, indicating a decelerated convergence and pointing to the importance of long-wavelength fluctuations in the vicinity of the AFI. These corrections are obviously included through the LDFA, which complies with QMC close to the AFI even on the two-particle level. This is quite remarkable because the results have been obtained perturbatively, starting from DMFT as a local approximation.

Figure 4.37 illustrates that including the paramagnon contribution to the self-energy changes the spectral properties of the system quite remarkably: While DMFT and the second-order dual perturbation theory display a quasiparticle peak in the local density of states for $U/t = 4$ and $T/t = 0.19$, the LDFA produces the antiferromagnetic pseudogap.

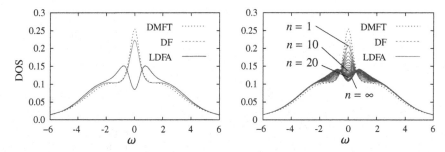

Fig. 4.37 *Left*: Local density of states (DOS) obtained within DMFT, a dual fermion calculation with diagrams (a) and (b) and the LDFA with selfconsistent renormalization of the self-energy for $T/t = 0.188$. While a quasiparticle peak is obtained in DMFT the LDFA DOS exhibits the antiferromagnetic pseudogap. *Right*: Local DOS for different approximations to the LDFA self-energy obtained by n iterations of the BSE and selfconsistent renormalization for $T/t = 0.2$. Solutions are shown for $n = 1, \ldots, 20$ in steps of 1 and from $n = 25, \ldots, 100$ in steps of 5. The self-energy includes all ladder diagrams up to order $n + 2$ in γ. DF is equivalent to the LDFA with $n = 0$. Diagrams at virtually every order contribute to the pseudogap

4.7.10 Convergence Properties

For a perturbative approach, the convergence properties are of paramount importance. For the present theory, the vertices appear as a small parameter in the expansion in the weak-coupling limit ($U \to 0$), because they vanish at least proportionally to U: $\gamma^{(4)} \sim U$, $\gamma^{(6)} \sim U^2$, ... On the other hand, for an expansion around the atomic limit ($\Delta \equiv 0$), the dual Green function is small near this limit: For h_k small, the bare dual Green function can be approximated as

$$G_\omega^{d0}(k) \approx g_\omega h_k g_\omega. \tag{4.145}$$

This enforces the convergence of the series in the opposite strong coupling limit. In contrast, IPT or FLEX, which operate with the bare interaction U, have to break down at intermediate to large U. In the general case, a fast convergence cannot be proven rigorously. Here we examine the convergence properties numerically in the vicinity of the antiferromagnetic instability (AFI) in the 2D Hubbard model. These can be characterized using the eigenvalue problem derived from the BSE (4.139)

$$-\frac{T}{N} \sum_{\omega' k'} \Gamma_{\omega\omega'\Omega}^{irr,m} G_{\omega'}(k') G_{\omega'+\Omega}(k'+q) \, \phi_{\omega'} = \lambda \phi_\omega. \qquad \boxed{\begin{array}{c} G_{\omega'+\Omega}(k'+q) \\ \Gamma_{\omega\omega'\Omega}^{irr,\,m} \end{array}}$$

$$\tag{4.146}$$

The matrix is the building block of the particle-hole ladder and may be thought of as the effective two-fermion interaction. For dual fermions, the irreducible vertex is given by the bare dual interaction $\Gamma_{\omega\omega'\Omega}^{irr,m} = \gamma_{\omega\omega'\Omega}^m = \gamma_{\omega\omega'\Omega}^{\uparrow\uparrow} - \gamma_{\omega\omega'\Omega}^{\uparrow\downarrow}$ in the magnetic channel and G stands for the full dual Green function. Here the focus is on the

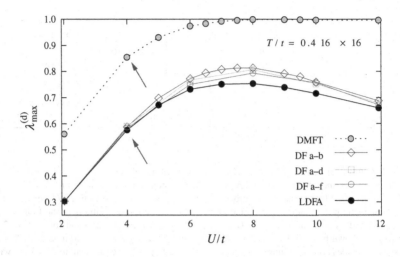

Fig. 4.38 Leading eigenvalue of the Bethe–Salpeter equation obtained within various approximations in the $q = (\pi, \pi)$ magnetic channel as a function of the interaction U. λ (λ^{d}) denotes lattice (dual) fermion eigenvalues. The diagrams included are indicated in the legend (labels are the same as in Fig. 4.17). The dual perturbation theory converges fast (i.e. the eigenvalues are small) in particular for weak and strong coupling. A straightforward diagrammatic expansion around DMFT breaks down for large U

leading eigenvalues in the vicinity of the AFI and hence $q = (\pi, \pi)$ and $\Omega = 0$. An eigenvalue of $\lambda_{\mathrm{max}} = 1$ implies a divergence of the ladder sum and hence a breakdown of the perturbation theory.

The results are displayed in Fig. 4.38. For weak coupling, the leading eigenvalue is small and implies a fast convergence of the diagrams in the electron-hole ladder. More significantly, the eigenvalues decrease and converge to the same intercept in the large U limit. This nicely illustrates that the dual perturbation theory smoothly interpolates between a standard perturbation expansion at small, and the cumulant expansion at large U, ensuring fast convergence in both regimes. From the figure it is clear that this also improves the convergence properties for intermediate coupling ($U \sim W$). Even here corrections from approximations involving higher-order diagrams remain small, including those from the LDFA. Diagrams involving the three-particle vertex give a negligible contribution.

For a straightforward diagrammatic expansion around DMFT, the building block of the particle-hole ladder is constructed from the irreducible impurity vertex $\gamma_{\omega\omega'\Omega}^{\mathrm{irr,m}}$ and DMFT Green functions. As seen in Fig. 4.38, the corresponding leading eigenvalue (and the effective interaction) is much larger than for dual fermions over the whole parameter range (e.g. at the red arrows). When transforming the leading eigenvalue back to lattice fermions (red arrow in Fig. 4.36), it is close to the DMFT value for these parameters (the data labeled DMFT at the red arrows is essentially the same in both plots). Hence convergence is enhanced for a perturbation theory in terms of dual fermions. Remarkably, for the intermediate to strong coupling region, a straightforward perturbation theory around DMFT breaks down (since the eigenvalue approaches one), while for a theory in terms of dual fermions, this is not the case. The fact that the leading eigenvalue for dual fermions is smaller is a generic feature. It is also observed away from half-filling and for the electron-electron channel (not shown). Note that the interaction in the dual fermion approach is given by the reducible (screened) vertex of the impurity. The frequency dependence accounts for the fact that the Coulomb interaction acts on short time scales in this approach. Strong local correlations are effectively separated (and treated non-perturbatively within the solution of the impurity model) from weaker spatial correlations, which are treated diagrammatically.

Acknowledgments We are greatly indebted to the input of our collaborators and colleagues Markus Aichhorn, Vladimir Anisimov, Matthias Balzer, Silke Biermann, Lewin Boehnke, Sergej Brener, Emanuel Gull, Václav Janiš, Christoph Jung, Martin Kecker, Gabriel Kotliar, Gang Li, Andrew Millis, Hartmut Monien, Alexander Poteryaev, Michael Potthoff, Leonid Pourovskii, Matouš Ringel and Philipp Werner.

References

1. Prange, R.E., Girvin, S.M.: The Quantum Hall Effect. Springer, New York (1997)
2. Stewart, G.R.: Heavy-fermion systems. Rev. Mod. Phys. **56**, 755 (1984)
3. Löhneysen, H.V., Rosch, A., Vojta, M., Wölfle, P.: Fermi-liquid instabilities at magnetic quantum phase transitions. Rev. Mod. Phys. **79**, 1015 (2007)
4. Hewson, A.C.: The Kondo Problem to Heavy Fermions. Cambridge University Press, Cambridge (1993)
5. Anderson, P.W.: The Theory of Superconductivity in High-T_c Cuprates. Princeton University Press, Princeton (1997)
6. Scalapino, D.J.: The case for $d_{x^2-y^2}$ pairing in the cuprate superconductors. Phys. Rep **250**, 329 (1995)
7. Biermann, S., Poteryaev, A., Lichtenstein, A.I., Georges, A.: Dynamical singlets and correlation-assisted Peierls transition in VO_2. Phys. Rev. Lett. **94**, 026404 (2005)
8. Kamihara, Y., Watanabe, T., Hirano, M., Hosono, H.: Iron-Based Layered Superconductor La $O_{1-x}F_x$ FeAs (x =0.05-0.12) with T_c = 26K. J. Am. Chem. Soc. **130**, 3296 (2008)
9. Imada, M., Fujimori, A., Tokura, Y.: Metal-insulator transitions. Rev. Mod. Phys. **70**, 1039 (1998)
10. Anisimov, V.I., Zaanen, J., Andersen, O.K.: Band theory and Mott insulators: Hubbard U instead of Stoner I. Phys. Rev. B **44**, 943 (1991)
11. Aryasetiawan, F., Gunnarsson, O.: The GW method. Rep. Prog. Phys. **61**, 237 (1998)
12. Kotliar, G., Savrasov, S.Y., Haule, K., Oudovenko, V.S., Parcollet, O., Marianetti, C.A.: Electronic structure calculations with dynamical mean-field theory. Rev. Mod. Phys. **78**, 865 (2006)
13. Kotliar, G., Vollhardt, D.: Strongly correlated materials: insights from dynamical mean-field theory. Phys. Today **57**, 53 (2004)
14. Georges, A., Kotliar, G., Krauth, W., Rozenberg, M.J.: Dynamical mean-field theory of strongly correlated fermion systems and the limit of infinite dimensions. Rev. Mod. Phys. **68**, 13 (1996)
15. Mott, N.F.: Metal-Insulator Transitions. Taylor and Francis, London (1974)

16. Anisimov, V.I., Poteryaev, A.I., Korotin, M.A., Anokhin, A.O., Kotliar, G.: First-principles calculations of the electronic structure and spectra of strongly correlated systems: dynamical mean-field theory. J. Phys.: Condens. Matter **9**, 7359 (1997)

17. Lichtenstein, A.I., Katsnelson, M.I.: Ab initio calculations of quasiparticle band structure in correlated systems: LDA++ approach. Phys. Rev. B **57**, 6884 (1998)

18. Lichtenstein, A.I., Katsnelson, M.I., Kotliar, G.: Finite-temperature magnetism of transition metals: an ab initio dynamical mean-field theory. Phys. Rev. Lett. **87**, 067205 (2001)

19. Lichtenstein, A.I., Katsnelson, M.I.: Antiferromagnetism and d-wave superconductivity in cuprates: a cluster dynamical mean-field theory. Phys. Rev. B **62**, R9283 (2000)

20. Kotliar, G., Savrasov, S.Y., Pálsson, G., Biroli, G.: Cellular dynamical mean field approach to strongly correlated systems. Phys. Rev. Lett. **87**, 186401 (2001)

21. Potthoff, M., Aichhorn, M., Dahnken, C.: Variational cluster approach to correlated electron systems in low dimensions. Phys. Rev. Lett. **91**, 206402 (2003)

22. Maier, T., Jarrell, M., Pruschke, T., Hettler, M.H.: Quantum cluster theories. Rev. Mod. Phys. **77**, 1027 (2005)

23. Irkhin, V.Y., Katanin, A.A., Katsnelson, M.I.: Robustness of the Van Hove scenario for high-T_c superconductors. Phys. Rev. Lett. **89**, 076401 (2002)

24. Slezak, C., Jarrell, M., Maier, T., Deisz, J.: Multi-scale extensions to quantum cluster methods for strongly correlated electron systems. J. Phys.: Condens. Matter **21**, 435604 (2009)

25. Toschi, A., Katanin, A.A., Held, K.: Dynamical vertex approximation: a step beyond dynamical mean-field theory. Phys. Rev. B **75**, 045118 (2007)

26. Kusunose, H.: Influence of spatial correlations in strongly correlated electron systems: extension to dynamical mean field approximation. J. Phys. Soc. Jpn **75**, 054713 (2006)

27. Georges, A., Kotliar, G.: Hubbard model in infinite dimensions. Phys. Rev. B **45**, 6479 (1992)

28. Metzner, W., Vollhardt, D.: Correlated lattice fermions in $d = \infty$ dimensions. Phys. Rev. Lett. **62**, 324 (1989)

29. Georges, A.: In: Avella A., and Mancini F. (eds.) Lectures on the Physics of Highly Correlated Electron Systems VIII, American Institute of Physics (2004) (cond-mat/0403123)

30. Bulla, R., Costi, T.A., Pruschke, T.: Numerical renormalization group method for quantum impurity systems. Rev. Mod. Phys. **80**, 395 (2008)

31. Kotliar, G.: Driving the electron over the edge. Science **302**, 67 (2003)

32. Lechermann, F., Georges, A., Poteryaev, A., Biermann, S., Posternak, M., Yamasaki, A., Andersen, O.K.: Dynamical mean-field theory using Wannier functions: a flexible route to electronic structure calculations of strongly correlated materials. Phys. Rev. B **74**, 125120 (2006)

33. Amadon, B., Lechermann, F., Georges, A., Jollet, F., Wehling, T.O., Lichtenstein, A.I.: Plane-wave based electronic structure calculations for correlated materials using dynamical mean-field theory and projected local orbitals. Phys. Rev. B **77**, 205112 (2008)

34. Miyake, T., Aryasetiawan, F.: Screened Coulomb interaction in the maximally localized Wannier basis. Phys. Rev. B **77**, 085122 (2008)

35. Anisimov, V.I., Aryasetiawan, F., Lichtenstein, A.I.: First-principles calculations of the electronic structure and spectra of strongly correlated systems: the LDA + U method. J. Phys.: Condens. Matter **9**, 767 (1997)

36. Pruschke, T., Bulla, R.: Hund's coupling and the metal-insulator transition in the two-band Hubbard model. Eur. Phys. J. B **44**, 217 (2005)

37. Pourovskii, L.V., Delaney, K.T., Vande Walle, C.G., Spaldin, N.A, Georges, A.: Role of atomic multiplets in the electronic structure of rare-earth semiconductors and semimetals. Phys. Rev. Lett. **102**, 096401 (2009)

38. Mo, S.-K., Denlinger, J.D., Kim, H.-D., Park, J.-H., Allen, J.W., Sekiyama, A., Yamasaki, A., Kadono, K., Suga, S., Saitoh, Y., Muro, T., Metcalf, P., Keller, G., Held, K., Eyert, V., Anisimov, V.I., Vollhardt, D.: Prominent quasiparticle peak in the photoemission spectrum of the metallic phase of V_2O_3. Phys. Rev. Lett. **90**, 186403 (2003)

39. Panaccione, G., Altarelli, M., Fondacaro, A., Georges, A., Huotari, S., Lacovig, P., Lichtenstein, A., Metcalf, P., Monaco, G., Offi, F., Paolasini, L., Poteryaev, A., Tjernberg, O., Sacchi, M.: Coherent peaks and minimal probing depth in photoemission spectroscopy of Mott-Hubbard systems. Phys. Rev. Lett. **97**, 116401 (2006)

40. Haule, K., Shim, J.H., Kotliar, G.: Correlated electronic structure of LaO_{1-x} F_xFeAs. Phys. Rev. Lett. **100**, 226402 (2008)

41. Haule, K., Kotliar, G.: Coherence–incoherence crossover in the normal state of iron oxypnictides and importance of Hund's rule coupling. New J. Phys. **11**, 025021 (2009)

42. Anisimov, V.I., Korotin, D.M., Korotin, M.A., Kozhevnikov, A.V., Kunes, J., Shorikov, A.O., Skornyakov, S.L., Streltsov, S.V.: Coulomb repulsion and correlation strength in LaFeAsO from density functional and dynamical mean-field theories. J. Phys.: Condens. Matter **21**, 075602 (2009)

43. Shorikov A.O., Korotin M.A., Streltsov S.V., Skornyakov S.L., Korotin D.M., Anisimov V.I.: Coulomb correlation effects in LaFeAsO: An LDA + DMFT(QMC) study, JETP **108**:121 (2009)

44. Anisimov, V.I., Korotin, D.M., Streltsov, S.V., Kozhevnikov, A.V., Kuneš, J., Shorikov, A.O., Korotin, M.A.: Coulomb parameter U and correlation strength in LaFeAsO. JETP Lett. **88**, 729 (2008)

45. Aichhorn, M., Pourovskii, L., Vildosola, V., Ferrero, M., Parcollet, O., Miyake, T., Georges, A., Biermann, S.: Dynamical mean-field theory within an augmented plane-wave framework: assessing electronic correlations in the iron pnictide LaFeAsO. Phys. Rev. B **80**, 085101 (2009)

46. Nakamura, K., Arita, R., Imada, M.: Ab initio Derivation of Low-Energy Model for Iron-Based Superconductors LaFeAsO and LaFePO. J. Phys. Soc. Jpn. **77**, 093711 (2008)

47. Miyake, T., Pourovskii, L., Vildosola, V., Biermann, S., Georges, A.: d- and f-orbital correlations in the REFeAsO compounds. J. Phys. Soc. Jpn. **77**, (Supp. c) 99 (2008)

48. Miyake, T., Nakamura, K., Arita, R., Imada, M.: Comparison of ab initio low-energy models for LaFePO, LaFeAsO, $BaFe_2As_2$, LiFeAs, FeSe, and FeTe, electron correlation and covalency. J. Phys. Soc. Jpn. **79**, 044705 (2010)

49. Craco, L., Laad, M.S., and Leoni, S.: α-FeSe as an orbital-selective incoherent metal: An LDA + DMFT study, arXiv:0910.3828, unpublished (2009).

50. Aichhorn, M., Biermann, S., Miyake, T., Georges, A., Imada, M.: Theoretical evidence for strong correlations and incoherent metallic state in FeSe. Phys. Rev. B **82**, 064504 (2010)

51. Werner, P., Gull, E., Troyer, M., Millis, A.J.: Spin freezing transition and non-Fermi-liquid self-energy in a three-orbital model. Phys. Rev. Lett. **101**, 166405 (2008)

52. Scalapino, D.J., Sugar, R.L.: Method for performing Monte Carlo calculations for systems with fermions. Phys. Rev. Lett. **46**, 519 (1981)

53. Blankenbecler, R., Scalapino, D.J., Sugar, R.L.: Monte Carlo calculations of coupled boson-fermion systems I. Phys. Rev. D **24**, 2278 (1981)

54. Hirsch, J.E.: Two-dimensional Hubbard model: numerical simulation study. Phys. Rev. B **31**, 4403 (1985)

55. Hirsch, J.E., Fye, R.M.: Monte Carlo method for magnetic impurities in metals. Phys. Rev. Lett. **56**, 2521 (1986)

56. Rubtsov, A.N., Savkin, V.V., Lichtenstein, A.I.: Continuous-time quantum Monte Carlo method for fermions. Phys. Rev. B **72**, 035122 (2005)

57. Rubtsov, A.N., Lichtenstein, A.I.: Continuous-time quantum Monte Carlo method for fermions: beyond auxiliary field framework. JETP Lett. **80**, 61 (2004)

58. Werner, P., Comanac, A., de'Medici, L., Troyer, M., Millis, A.J.: Continuous-time solver for quantum impurity models. Phys. Rev. Lett. **97**, 076405 (2006)

59. Werner, P., Millis, A.J.: Hybridization expansion impurity solver: general formulation and application to Kondo lattice and two-orbital models. Phys. Rev. B **74**, 155107 (2006)

60. Haule, K.: Quantum Monte Carlo impurity solver for cluster dynamical mean-field theory and electronic structure calculations with adjustable cluster base. Phys. Rev. B **75**, 155113 (2007)

61. Gull, E., Millis, A.J., Lichtenstein, A.I., Rubtsov, A.N., Troyer, M., and Werner, P.: Continuous-time Monte Carlo methods for quantum impurity models. Rev. Mod. Phys. **83**, 349 (2011).
62. Gull, E., Werner, P., Millis, A., Troyer, M.: Performance analysis of continuous-time solvers for quantum impurity models. Phys. Rev. B **76**, 235123 (2007)
63. Negele, J.W., Orland, H.: Quantum Many-Particle Systems. Westview Press, Boulder (1998)
64. Prokof'ev, N.V., Svistunov, B.V., Tupitsyn, I.S.: Exact quantum Monte Carlo process for the statistics of discrete systems. JETP Lett. **64**, 911 (1996)
65. Yoo, J., Chandrasekharan, S., Kaul, R.K., Ullmo, D., Baranger, H.U.: On the sign problem in the Hirsch & Fye algorithm for impurity problems. J. Phys. A: Math. Gen. **38**, 10307 (2005)
66. Gull E., Continuous-Time Quantum Monte Carlo Algorithms for Fermions. Ph.D. thesis, ETH Zurich (2008)
67. Läuchli, A.M., Werner, P.: Krylov implementation of the hybridization expansion impurity solver and application to 5-orbital models. Phys. Rev. B **80**, 235117 (2009)
68. Poteryaev, A.I., Lichtenstein, A.I., Kotliar, G.: Nonlocal Coulomb interactions and metal-insulator transition in Ti2O3: a cluster LDA+ DMFT approach. Phys. Rev. Lett. **93**, 086401 (2004)
69. Saha-Dasgupta, T., Lichtenstein, A., Hoinkis, M., Glawion, S., Sing, M., Claessen, R., Valenti, R.: Cluster dynamical mean-field calculations for TiOCl. New J. Phys. **9**, 380 (2007)
70. Fuhrmann, A., Okamoto, S., Monien, H., Millis, A.J.: Fictive-impurity approach to dynamical mean-field theory: a strong-coupling investigation. Phys. Rev. B **75**, 205118 (2007)
71. Okamoto, S., Millis, A.J., Monien, H., Fuhrmann, A.: Fictive impurity models: an alternative formulation of the cluster dynamical mean-field method. Phys. Rev. B **68**, 195121 (2003)
72. Biroli, G., Kotliar, G.: Cluster methods for strongly correlated electron systems. Phys. Rev. B **65**, 155112 (2002)
73. Potthoff, M.: Self-energy-functional approach to systems of correlated electrons. Eur. Phys. J. B **32**, 429 (2003)
74. Schiller, A., Ingersent, K.: Systematic $1/d$ corrections to the infinite-dimensional limit of correlated lattice electron models. Phys. Rev. Lett. **75**, 113 (1995)
75. Sadovskii, M.V., Nekrasov, I.A., Kuchinskii, E.Z., Pruschke, T., Anisimov, V.I.: Pseudogaps in strongly correlated metals: a generalized dynamical mean-field theory approach. Phys. Rev. B **72**, 155105 (2005)
76. Pairault, S., Sénéchal, D., Tremblay, A.-M.S.: Strong-coupling expansion for the Hubbard model. Phys. Rev. Lett. **80**, 5389 (1998)
77. Pairault, S., Sénéchal, D., Tremblay, A.-M.: Strong-coupling perturbation theory of the Hubbard model. Eur. Phys. J. B **16**, 85 (2000)
78. Sarker, S.K.: A new functional integral formalism for strongly correlated Fermi systems. J. Phys. C: Solid State Phys. **21**, L667 (1988)
79. Stanescu, T.D., Kotliar, G.: Strong coupling theory for interacting lattice models. Phys. Rev. B **70**, 205112 (2004)
80. Rubtsov, A.N.: Quality of the mean-field approximation: a low-order generalization yielding realistic critical indices for three-dimensional Ising-class systems. Phys. Rev. B **66**, 052107 (2002)
81. Rubtsov A.N., Small parameter for lattice models with strong interaction, arXiv:cond-mat/0601333, unpublished (2006)
82. Rubtsov, A.N., Katsnelson, M.I., Lichtenstein, A.I.: Dual fermion approach to nonlocal correlations in the Hubbard model. Phys. Rev. B **77**, 033101 (2008)
83. Hafermann, H.: Numerical Approaches to Spatial Correlations in Strongly Interacting Fermion Systems. Cuvillier Verlag, Göttingen (2010)
84. Schäfer, J., Schrupp, D., Rotenberg, E., Rossnagel, K., Koh, H., Blaha, P., Claessen, R.: Electronic quasiparticle renormalization on the spin wave energy scale. Phys. Rev. Lett. **92**, 097205 (2004)

85. Eschrig, M., Norman, M.R: Neutron Resonance: Modeling photoemission and tunneling data in the superconducting state of $Bi_2Sr_2CaCu_2O_{8+\delta}$. Phys. Rev. Lett. **85**, 3261 (2000)
86. Schachinger, E., Tu, J.J., Carbotte, J.P.: Angle-resolved photoemission spectroscopy and optical renormalizations: phonons or spin fluctuations. Phys. Rev. B **67**, 214508 (2003)
87. Claessen, R., Sing, M., Schwingenschlögl, U., Blaha, P., Dressel, M., Jacobsen, C.S.: Spectroscopic signatures of spin-charge separation in the quasi-one-dimensional organic conductor TTF-TCNQ. Phys. Rev. Lett. **88**, 096402 (2002)
88. Rubtsov, A.N., Katsnelson, M.I., Lichtenstein, A.I., Georges, A.: Dual fermion approach to the two-dimensional Hubbard model: antiferromagnetic fluctuations and Fermi arcs. Phys. Rev. B **79**, 045133 (2009)
89. Baym, G., Kadanoff, L.P.: Conservation laws and correlation functions. Phys. Rev. **124**, 287 (1961)
90. Abrikosov, A.A., Gor'kov, L.P., Dzyaloshinskii, I.E.: Methods of Quantum Field Theory in Statistical Physics. Pergamon Press, New York (1965)
91. Irkhin, V.Y., Katsnelson, M.I.: Current carriers in a quantum two-dimensional antiferromagnet. J. Phys.: Condens. Matter **3**, 6439 (1991)
92. Park, H., Haule, K., Kotliar, G.: Cluster dynamical mean field theory of the mott transition. Phys. Rev. Lett. **101**, 186403 (2008)
93. Macridin, A., Jarrell, M., Maier, T., Kent, P.R.C., D'Azevedo, E.: Pseudogap and antiferromagnetic correlations in the Hubbard Model. Phys. Rev. Lett. **97**, 036401 (2006)
94. Ferrero, M., Cornaglia, P.S., Leo, L.D., Parcollet, O., Kotliar, G., Georges, A.: Valence bond dynamical mean-field theory of doped Mott insulators with nodal/antinodal differentiation. Eur. phys. Lett. **85**, 57009 (2009)
95. Brener, S., Hafermann, H., Rubtsov, A.N., Katsnelson, M.I., Lichtenstein, A.I.: Dual fermion approach to susceptibility of correlated lattice fermions. Phys. Rev. B **77**, 195105 (2008)
96. Li, G., Lee, H., Monien, H.: Determination of the lattice susceptibility within the dual fermion method. Phys. Rev. B **78**, 195105 (2008)
97. Lee, H., Li, G., Monien, H.: Hubbard model on the triangular lattice using dynamical cluster approximation and dual fermion methods. Phys. Rev. B **78**, 205117 (2008)
98. Hafermann, H., Kecker, M., Brener, S., Rubtsov, A.N., Katsnelson, M.I., Lichtenstein, A.I.: Dual fermion approach to high-temperature superconductivity. J. Supercond. Nov. Magn. **22**, 45 (2009)
99. Hafermann, H., Brener, S., Rubtsov, A.N., Katsnelson, M.I., Lichtenstein, A.I.: Cluster dual fermion approach to nonlocal correlations. JETP Lett. **86**, 677 (2007)
100. Hafermann, H., Li, G., Rubtsov, A.N., Katsnelson, M.I., Lichtenstein, A.I., Monien, H.: Efficient perturbation theory for quantum lattice models. Phys. Rev. Lett. **102**, 206401 (2009)
101. Hafermann, H., Jung, C., Brener, S., Katsnelson, M.I., Rubtsov, A.N., Lichtenstein, A.I.: Superperturbation solver for quantum impurity models. Europhys. Lett. **85**, 27007 (2009)
102. Schollwöck, U.: The density-matrix renormalization group. Rev. Mod. Phys. **77**, 259 (2005)
103. Balzer, M., Hanke, W., Potthoff, M.: Mott transition in one dimension: benchmarking dynamical cluster approaches. Phys. Rev. B **77**, 045133 (2008)
104. Mishchenko, A.S., Prokof'ev, N.V., Sakamoto, A., Svistunov, B.V.: Diagrammatic quantum Monte Carlo study of the Fröhlich polaron. Phys. Rev. B **62**, 6317 (2000)
105. Migdal, A.B.: Theory of Finite Fermi Systems and Applications to Atomic Nuclei. Interscience Publishers, New York (1967)
106. Nozières, P.: Theory of Interacting Fermi Systems. Benjamin Day, New York (1964)
107. Auerbach, A. (eds): Interacting Electrons and Quantum Magnetism. Springer, New York (1998)
108. Hugenholtz, N.: Perturbation theory of large quantum systems. Physica **23**, 481 (1957)
109. Bickers, N.E., Scalapino, D.J., White, S.R.: Conserving approximations for strongly correlated electron systems: Bethe-Salpeter equation and dynamics for the two-dimensional Hubbard Model. Phys. Rev. Lett. **62**, 961 (1989)

110. Bickers, N.E., Scalapino, D.J.: Conserving approximations for strongly fluctuating electron systems. I. Formalism and calculational approach. Ann. Phys. **193**, 206 (1989)
111. Bulut, N., Scalapino, D.J., White, S.R.: Bethe-Salpeter eigenvalues and amplitudes for the half-filled two-dimensional Hubbard model. Phys. Rev. B **47**, 14599 (1993)

Chapter 5
Nonequilibrium Transport and Dephasing in Coulomb-Blockaded Quantum Dots

Alexander Altland and Reinhold Egger

Abstract We provide an introduction to the nonequilibrium physics encountered in quantum dots. A brief summary of the relevant Coulomb blockade physics and a concise account of the Keldysh functional integral method is followed by a derivation of the Keldysh Ambegaokar-Eckern-Schön action, which represents a prototypical model for charge transport through quantum dots. We show that the nonequilibrium current fluctuations cause a dephasing that can be probed via the tunneling density of states. We provide analytical and numerical estimates for the corresponding dephasing rates.

5.1 Introduction

Over the past few years, nonequilibrium phenomena have become a topic of ever-increasing interest in condensed matter physics. In nanoscale systems, such as quantum dots or molecules electrically contacted by electrodes, it is experimentally easy to reach a strongly out-of-equilibrium situation by simply applying a bias voltage [1]. At the same time, considerable theoretical progress has been achieved recently. In particular, exact fluctuation relations constraining the nonequilibrium dynamics of micro-reversible systems have been found [2], generalizing the standard fluctuation-dissipation theorem valid within the linear response regime. These relations impose symmetry relations for the full counting statistics of charge transport

A. Altland
Institut für Theoretische Physik,
Universität zu Köln, 50937 Köln, Germany
e-mail:alexal@thp.uni-koeln.de

R. Egger (✉)
Institut für Theoretische Physik,
Heinrich-Heine-Universität, 40225 Düsseldorf, Germany
e-mail:egger@thphy.uni-duesseldorf.de

D. C. Cabra et al. (eds.), *Modern Theories of Many-Particle Systems in Condensed Matter Physics*, Lecture Notes in Physics 843, DOI: 10.1007/978-3-642-10449-7_5,
© Springer-Verlag Berlin Heidelberg 2012

out of equilibrium, and thereby imply nontrivial connections among different cumulants. Furthermore, novel numerical methods have been developed very recently that allow to obtain numerically exact results for nonequilibrium problems characterized by strong correlations, see, e.g., Refs. [3–5] for different schemes. For special integrable models, one may even achieve exact solutions for the nonequilibrium case by using a scattering Bethe ansatz approach [6]. In these lecture notes, our goal is to provide a self-consistent and pedagogical introduction to (some of) the physics and theoretical methods in this rapidly developing and active field.

To establish the necessary background, we will first summarize some basic phenomena in quantum dots, such as Coulomb blockade and Kondo effect, in Sect. 5.2. More detailed textbooks [1] and reviews [7, 8] are available for further reading. The most general and powerful approach to nonequilibrium many-particle quantum physics is provided by the Keldysh formalism, and we provide a concise introduction to this approach in Sect. 5.3. A comprehensive and very detailed review has been given recently by Kamenev and Levchenko [9]. The Keldysh technique is usually formulated in the language of functional integrals, and we will also employ this language here. Recent work has also shown its usefulness for disordered interacting systems [10]. In a sense, the Keldysh formalism provides a "Theory of Everything" from which one can develop approximate simpler descriptions. For instance, in the classical limit one can establish contact to the Martin-Siggia-Rose formalism or to quasiclassical equations. For a recent textbook describing these connections, see Ref. [11].

The Keldysh formalism is then employed in Sect. 5.4 to develop a theory of nonequilibrium transport in large Coulomb-blockaded quantum dots. We focus mostly on the so-called weak Coulomb blockade regime, where the dot-to-lead contacts are of intermediate-to-high transparency [12]. This allows to quantitatively analyze the phenomenon of dephasing induced by nonequilibrium fluctuations, as we discuss in Sect. 5.5. Such dephasing processes have been discussed in detail for the nonequilibrium Kondo effect in quantum dots [13–15], in the context of the nonequilibrium Fermi edge singularity [16], for a spin-fermion model [17], and for Luttinger liquids [18] and quantum Hall edge states [19]. A convenient definition of the dephasing rate is based on the voltage-induced smearing of the zero-bias anomaly in the tunneling density of states [20]. Finally, we offer some concluding remarks in Sect. 5.6. We often employ units such that $\hbar = k_B = e = 1$.

5.2 Coulomb Blockade Phenomena in Quantum Dots

5.2.1 Basics

A closed quantum dot (i.e., without attached electrodes) corresponds to a confined region in space containing N unbound electrons (or holes). The dot is here assumed to be mesoscopic in size, with a typical linear dimension between several nanometers

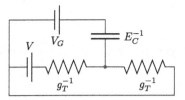

Fig. 5.1 Equivalent electric circuit diagram for a quantum dot—represented by the central node—connected to two leads by tunnel junctions—represented by Ohmic resistors—and capacitively coupled to a gate electrode

and several microns. In that case, quantum coherence can be maintained at sufficiently low temperatures. In semiconductor heterostructures, one either has lateral quantum dots which are tuned by employing top gates, or vertical dots in layered structures. Quantum dots can also be realized by contacting individual molecules or short nanotubes. The confinement is typically such that one has classically chaotic motion. Ignoring electron–electron interactions for the moment, a closed dot corresponds to a Schrödinger equation

$$\left(-\frac{1}{2m^*}\Delta + V_{\text{conf}}(\boldsymbol{r})\right)\psi_n(\boldsymbol{r}) = \epsilon_n\psi_n(\boldsymbol{r}). \tag{5.1}$$

The spectrum of the closed dot is then discrete, with average level spacing ΔE between subsequent ϵ_n in Eq. 5.1. At temperature $T < \Delta E$, these levels can be resolved. This limit finds an important application in quantum information science, where a spin qubit can be encoded in the spin state of an electron occupying this individual level. Note that for metallic grains, the Fermi wavelength is much smaller than the size of the grain and quantum effects are not so important. We here consider semiconducting or nanoscale molecular dots.

Consider now the equivalent circuit for a typical two-terminal setup, see Fig. 5.1, where the dot is contacted to left/right electron reservoirs (terminals). These are modelled as noninteracting fermions, with given temperature $T_{L/R}$ and chemical potential $\mu_{L/R}$. In addition, a third electrode (backgate) is capacitively coupled to the dot, and allows to shift energy levels on the dot via a backgate voltage V_G. Here we consider only $T_L = T_R = T$ (otherwise one can study thermal transport), and the applied bias voltage is $eV = \mu_L - \mu_R$. Since the discrete spectrum on the dot is influenced by external gates, one often speaks of (tunable) artificial atoms. We are interested in electrical transport through such artificial atoms, in particular when interaction effects are important. While in atomic physics, one has typically weak interactions and Hartree–Fock shell filling pictures are appropriate, in quantum dots one often has strong interaction effects because the electrostatic confinement potentials are typically rather shallow.

5.2.2 Energy Scales

Let us then address the important energy scales present in this problem.

5.2.2.1 Level Spacing ΔE

This scale describes the average spacing between subsequent single-particle energy levels ϵ_n. For typical linear dimension L of a d-dimensional dot, one can estimate $\Delta E \approx 1/\nu_d L^d$ with the average density of states (DoS) ν_d on the dot. In semiconductor dots, a typical value is $L \approx 100$–$1,000$ nm, leading to $\Delta E \approx 0.01$ –0.1 meV or temperature scales ≈ 0.1–1 K.

5.2.2.2 Thouless Energy E_T

This scale corresponds to the inverse propagation time of an electron through the (noninteracting) dot. It is useful to also define the dimensionless Thouless conductance $g_{Th} = E_T/\Delta E$. We will always consider clean or at most weakly disordered dots, such that $g_{Th} \gg 1$, i.e., the dot itself is presumed to be a good conductor. With Fermi velocity v_F, in a clean two-dimensional dot of linear size L, we have $E_T \approx v_F/L$ and $\Delta E \approx 1/m^* L^2$. Hence, with $k_F = m^* v_F$, we find $g_{Th} \approx k_F L \approx \sqrt{N}$, since the number of electrons in the dot is $N \approx (k_F L)^2$, as follows from simply counting the occupied momentum states.

5.2.2.3 Coulomb Charging Energy E_C

This important energy scale describes the electron–electron interaction cost to be paid for bringing an additional electron onto the dot. Writing $E_C = e^2/2C$ with the total capacitance C, the interaction for N electrons on the dot takes the form

$$V_{ee} = E_C(N - N_G)^2, \tag{5.2}$$

where $N_G = C_G V_G/e$ is externally controlled via the backgate voltage, and E_C is independent of N. The "constant-interaction model" (5.2) is the simplest possible model for interactions. When supplemented with a (subleading) exchange coupling term in the spin sector, one arrives at the so-called *universal Hamiltonian* [1]. As shown in Ref. [8], this treatment of interactions can be justified for $g_{Th} \gg 1$ (i.e., $N \gg 1$) and not too strong interactions, $E_C < E_T$. We will only discuss interaction effects within the framework of Eq. 5.2 in these notes. The typical size of E_C follows from the capacitance of a sphere, $C_{sphere} \propto L$ for radius L, i.e. $E_C \propto 1/L$. For linear size $L \approx 50$ nm of the dot, one finds $E_C \approx 1$ meV, corresponding to 10 K. In semiconductor dots, one usually has $E_C \gg \Delta E$, but in nanotube dots one can also achieve $E_C \approx \Delta E$. Finally, we note that E_C refers to the closed dot. For strong coupling to external leads (the "weak Coulomb blockade" limit), it is sometimes of advantage to introduce an effective (renormalized) charging energy scale $E_C^* < E_C$ [21, 22] which should not be confused with the scale employed here.

5.2.2.4 Hybridization Γ

Yet another energy scale comes from the coupling of the dot to the electrodes (reservoirs). This coupling effectively broadens the dot level energies by $\Gamma = \Gamma_L + \Gamma_R$, and the ϵ_n effectively acquire an imaginary part corresponding to escape processes to the leads. Here, the hybridization energies due to tunneling into/from the leads are

$$\Gamma_{L/R} = \pi \nu_{L/R} |t_{L/R}|^2, \tag{5.3}$$

where $\nu_{L/R}$ is the DoS in the L/R lead, and $t_{L/R}$ is the tunnel amplitude from the respective lead to a dot level. The contact conductance for the L/R tunnel junction is now defined as

$$G_{L/R} = \pi \nu_d \Gamma_{L/R} G_Q, \tag{5.4}$$

where $G_Q = 2e^2/h$ is the celebrated conductance quantum (the factor 2 describes the spin degeneracy) and ν_d the DoS of the dot. Note that $h/e^2 \simeq 25.8\,\mathrm{k\Omega}$.

5.2.3 Coulomb Blockade and Transport Regimes

The presence of Eq. 5.2 implies the possibility of quantized charge tunneling, commonly referred to as *Coulomb blockade*. For this to happen, one generally states that two conditions are necessary, namely temperature should be low compared to the charging scale, $T \ll E_C$, and the contacts to the left and right reservoir should have low transparency, $G_L, G_R \ll G_Q$. One can thus distinguish two important regimes: For small contact transparency, we have the regime of a *closed dot*, while large transparency implies an *open dot*. Coulomb blockade is most pronounced for the (almost) closed dot, but we shall see that remnants of Coulomb blockade (commonly referred to as "weak Coulomb blockade") are important even for open dots, see Sect. 5.5.

One quantity of primary experimental interest is the linear conductance through the dot in a setup like in Fig. 5.1,

$$G = \frac{I}{V}\bigg|_{V \to 0} = G(N_G, T). \tag{5.5}$$

Let us now discuss the behavior of the conductance in the regime

$$\Gamma \ll \Delta E \ll E_C, \tag{5.6}$$

resulting in four different temperature regimes. The first inequality in (5.6) implies an almost closed dot. For high temperatures, we have the classical Drude limit $T \gg E_C$, and the conductance $G = G_\infty$ is then expected to correspond to the series resistance of the contact resistances,

$$G_\infty^{-1} = G_L^{-1} + G_R^{-1}. \tag{5.7}$$

In that regime, we have Ohm's law, $I = G_\infty V$, with G_∞ being independent of N_G. Note that Eq. 5.6 implies $G_\infty \ll G_Q$.

For $T < E_C$, however, transport can be completely blocked because there is an energy barrier for adding or removing a single electron from the dot (Coulomb blockade). To understand that, consider the sequence of parabolas (5.2), $V_N(N_G)$. While N describes the number of electrons on the dot and as such must be integer, N_G is a polarization charge induced on the dot and hence can vary continuously. For given N_G that is not a half-integer, there is a unique N for which $V_N(N_G)$ is minimized. Adding or removing an electron, $N \to N \pm 1$, then comes with an energy cost of the order of E_C, and for $T \ll E_C$, this process is not possible anymore. The conductance through the dot is then expected to be very small. For half-integer N_G, however, we encounter charge degeneracy, and two values (N and $N+1$) minimize the energy. In that case, transport through the dot is possible. As a function of N_G, one thus gets periodic *Coulomb oscillations* of the linear conductance, with conductance peaks centered around the half-integers N_G^{res}. We now use the energy scale $\delta = 2E_C(N_G - N_G^{\text{res}})$ to describe the deviation from the peak center, and describe the lineshape $G(\delta)$ of a single peak. Note that even for $E_c \to 0$, this is a well-defined energy scale since $N_G - N_G^{\text{res}}$ is proportional to a capacitance (i.e., to $1/E_C$).

In fact, for $E_C = 0$ and $T \ll \Gamma$, the conductance peak corresponds to the Breit-Wigner resonant tunneling result, with a Lorentzian lineshape. In the symmetric case $\Gamma_L = \Gamma_R = \Gamma/2$, we have $G_{BW}(\delta) = G_Q \Gamma^2/(\Gamma^2 + \delta^2)$. This follows by computing the transmission probability for an electron traversing the double barrier, with a single relevant quantum level between the barriers. Note that for $\delta = 0$, one has $G = G_Q$ for any $\Gamma \neq 0$. This is the *unitary limit*, and the linear conductance through a single-level dot can never exceed this limit, even when including interactions. The width of the peak is here determined by Γ.

In actual quantum dots, however, E_C is important and modifies this picture drastically. A schematic view of the temperature dependence of the conductance for $\delta = 0$ ("peak") and $\delta \neq 0$ ("valley") under the condition (5.6) is shown in Fig. 5.2.

- For $T > E_C$, we have the classical limit and $G(\delta) = G_\infty$, see Eq. 5.7.
- For $\Delta E < T < E_C$, the *orthodox Coulomb blockade* regime is realized. Single-particle levels are thermally smeared, but charge quantization persists.
- For $\Gamma < T < \Delta E$, we enter the regime of *quantum Coulomb blockade*, where only a single dot level is involved in transport.
- Finally, for $T < \Gamma$, one may have the *Kondo effect* (see below) or enter a resonant tunneling situation.

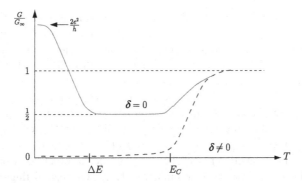

Fig. 5.2 Schematic temperature dependence of peak ($\delta = 0$) and the valley linear conductance ($\delta \neq 0$). For high $T > E_C$, both approach $G = G_\infty$, the classical Drude value (5.7). For $\Delta E < T < E_C$, we are in the orthodox Coulomb blockade regime, while for $\Gamma < T < \Delta E$, the quantum Coulomb blockade regime is realized

For $T > \Gamma$, one can use rate equations to derive the lineshape, with Fermi golden rule rates for tunneling into or out of the dot as the essential ingredients [1]. Here we only quote the relevant results. In the orthodox regime,

$$\frac{G_{OCB}(\delta)}{G_\infty} = \frac{\delta/T}{2\sinh(\delta/T)} \approx \frac{1}{2}\frac{1}{\cosh^2\left(\frac{\delta}{2.5T}\right)}, \tag{5.8}$$

where the second form holds to within 1% accuracy. In the quantum Coulomb blockade regime, one instead has

$$\frac{G_{QCB}(\delta)}{G_\infty} = \frac{\Delta E}{4T}\frac{1}{\cosh^2\left(\frac{\delta}{2T}\right)}, \tag{5.9}$$

which is very similar but with a prefactor $\propto 1/T$. For orthodox Coulomb blockade, the *peak conductance* is $G_\infty/2$ because of correlations: one first has to tunnel onto the dot and then out, or the other way around, depending on the charge state of the dot. In the high-temperature limit, however, both processes are allowed and uncorrelated, since they are not blocked by the charging energy. As a consequence, they occur twice as often. In the quantum Coulomb blockade regime, the peak conductance then increases when lowering T, but eventually saturates at the unitary limit $G = G_Q$ as $T \ll \Gamma$.

5.2.4 Kondo Effect in Quantum Dots

In the remainder of this section, we discuss the *valley conductance* for $T \ll E_C$, where N denotes the energetically preferred number of electrons on the dot. The above expressions would predict an exponentially small valley conductance $\propto e^{-cE_C/T}$

with some positive constant $c = c(\delta)$, because the occupation of a dot state costs an energy of order E_C. This expectation is, however, not quite correct. In particular, when the number N of electrons on the dot is odd, one can encounter the Kondo effect, and the actual conductance again approaches the unitary value G_Q at very low temperatures. To understand this correlation effect, we first need to discuss *cotunneling* processes. As we have just discussed, the valley conductance due to sequential tunneling processes—involving tunneling into the dot, a real occupation of it, and then tunneling out again—is exponentially small. However, one can also go from the left to the right reservoir by tunneling through the dot as a whole. Such "cotunneling" transition rates [1] come from second-order perturbation theory in $\Gamma_{L/R}$ and involve an energy denominator $\propto 1/E_C$. One can distinguish inelastic cotunneling processes, where electron-hole pairs on the dot can be excited, and elastic cotunneling, which involves only a single dot level. When including cotunneling processes in the rate equations, it turns out that the conductance is only algebraically suppressed in E_C/T [1, 8]

$$G \approx \frac{G_L G_R}{G_Q} \times \begin{cases} (T/E_C)^2, & T_{\text{el}} \ll T \ll E_C, \\ \Delta E/E_C, & T \ll T_{\text{el}}, \end{cases} \tag{5.10}$$

where T_{el} follows from comparing both contributions: $T_{\text{el}} = \sqrt{E_C \Delta E}$. Elastic cotunneling dominates the valley conductance for $T < T_{\text{el}}$, and even at $T = 0$, the conductance through the dot remains finite.

The Kondo effect can be understood as elastic cotunneling in all orders. Under the strong-interaction condition (5.6), it occurs at low temperatures provided that the dot level energy is such that the dot is always filled with an odd number N of electrons. The relevant topmost occupied level of the dot then hosts one electron, which can have spin $\sigma = \uparrow$ or $\sigma = \downarrow$. The interesting physics comes from this spin degree of freedom, while charge is gapped—empty and doubly occupied levels are energetically unfavorable. Hence, only a single quantum level on the dot plays a role, and one can reduce the description of the dot to the *Anderson model*. With $n_\sigma^2 = n_\sigma = d_\sigma^\dagger d_\sigma$, where d_σ is a dot fermion operator, we have up to an irrelevant constant

$$E_C(n_\uparrow + n_\downarrow - N_G)^2 = 2E_C n_\uparrow n_\downarrow + E_C[1 - 2N_G](n_\uparrow + n_\downarrow).$$

For a single level, the only interaction possible because of the Pauli principle is $U n_\uparrow n_\downarrow$, and thus (with $U = 2E_C$) we see that the constant-interaction model is perfectly adequate to describe the limit of a single level as well. (The more difficult part is a small dot with several levels.) Using the effective dot level $\epsilon_d = E_C(1 - 2N_G)$, which can be tuned by the gate voltage, the Anderson model reads

$$H = U n_\uparrow n_\downarrow + (n_\uparrow + n_\downarrow)\epsilon_d + \sum_{k,\sigma,\alpha=L/R} \left(\epsilon_{k\alpha} c_{k\sigma\alpha}^\dagger c_{k\sigma\alpha} - \left[t_{k,\text{ff}} c_{k\sigma\alpha}^\dagger d_\sigma + \text{h.c.} \right] \right). \tag{5.11}$$

It is customary and justified in many applications to employ the *wide-band approximation*, which consists of two steps. First, one assumes k-independent tunneling amplitudes $t_{L/R}$ connecting the lead electron states (corresponding to fermion operators $c_{k\sigma\alpha}$ and dispersion relation $\epsilon_{k\alpha}$) with the dot electron state. Second, the lead DoS $\nu_{L/R}$ is assumed to be independent of energy ϵ in the relevant energy regime. The hybridization $\Gamma_{L/R} = \pi \nu_{L/R} |t_{L/R}|^2$ is then also energy independent. For simplicity, let's now take identical and symmetrically coupled contacts, $\Gamma_L = \Gamma_R = \Gamma/2$. The dot is singly occupied when $-U < \epsilon_d < 0$ and $U/\Gamma \gg 1$. This is one condition for the occurence of the Kondo effect. In this regime, one can project out the empty and doubly-occupied dot states, and the dot's spin-1/2 operator $S = \frac{1}{2} \sum_{\sigma\sigma'} d_\sigma^\dagger \boldsymbol{\sigma}_{\sigma\sigma'} d_{\sigma'}$ becomes the relevant dynamical degree of freedom, similar to the magnetic impurity problem in a metal, i.e., the original Kondo problem. This procedure is the so-called Schrieffer-Wolff transformation [11], and the dot-lead coupling corresponds to an exchange coupling of S with the conduction electron spin density near the contacts, with $J \approx |t|^2/4U$. Below the Kondo temperature T_K, which for the Anderson model is given by

$$T_K = \sqrt{\frac{U\Gamma}{2}} \exp \frac{\pi \epsilon_d (\epsilon_d + U)}{2\Gamma U}, \qquad (5.12)$$

this Kondo exchange coupling implies entanglement of the dot and the conduction electron spins. The resulting strongly correlated many-body problem has been solved in the 1970s and 1980s by the numerical renormalization group (NRG) and by the Bethe ansatz for thermodynamic quantities; for details, see Refs. [1, 8] and references therein.

NRG calculations are also able to compute the full temperature dependence of the linear conductance, which in the Kondo regime turns out to be a universal function of T/T_K, i.e., the detailed parameter values (U, ϵ_d, Γ) only enter via the Kondo scale, and $G/G_Q = f(T/T_K)$. The scaling function $f(x)$ is sketched in Fig. 5.3, and has been experimentally observed as well. For $T \ll T_K$, analytical results from Nozières' Fermi liquid theory are available, $f(x \ll 1) = 1 - \frac{\pi^2}{3} x^2 + \mathcal{O}(x^4)$, and the conductance reaches the unitary limit at $T = 0$. In that case, a many-body resonance pinned to the Fermi level is formed, regardless of the precise parameter values, and electrons can now resonantly tunnel through this many-body level despite of the fact that we are located in a Coulomb blockade valley. The width of this resonance is not Γ but T_K. For $T > T_K$, on the other hand, one encounters $\ln(T/T_K)$ dependencies. To observe the Kondo effect, contacts should be of intermediate transparency. If the transparency is very small, also the Kondo temperature (5.12) is tiny. On the other hand, for good transparency, the single-occupancy condition $U \gg \Gamma$ will be hard to ensure, and the dot charge fluctuates so strongly that the Kondo effect is destroyed. Moreover, if N is even, the Kondo effect does not occur, and the valley conductance is given by the extremely small elastic cotunneling value (5.10). The resulting even-odd asymmetry was also experimentally observed.

Thanks to experimental progress, the Kondo effect in quantum dots has remained a topical issue. Under nonequilibrium conditions, it has not been completely settled how the voltage affects the Kondo resonance. Application of the voltage could split the resonance because of the attempt to simultaneously align the resonance

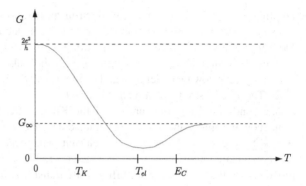

Fig. 5.3 Schematic temperature dependence of the valley linear conductance for the case of odd number of electrons N on the dot. For temperatures around T_K, logarithmic scaling due to the Kondo effect sets in. For $T \ll T_K$, the conductance approaches the unitary value $G_Q = 2e^2/h$

with both Fermi levels (of source and drain electrode). Alternatively, the voltage-induced dephasing broadens the Kondo resonance significantly, and could in fact imply just one broad Kondo peak centered around the mean Fermi level. In addition, more complicated Kondo effects occur when additional orbital quantum numbers are present. This happens, for instance, in ultraclean carbon nanotube dots, where electrons tunneling into or out of the nanotube may be able to preserve their valley ("K point") degeneracy. In such cases, non-Fermi liquid phases or exotic Fermi liquid Kondo effects (e.g. with $SU(4)$ symmetry) can take place. Finally, when coupled to superconductors with BCS gap Δ, the critical Josephson current is expected (and has been confirmed) to be a universal function of the ratio Δ/T_K. However, this function turns out to be difficult to compute even numerically, except in simple limiting cases.

5.3 Keldysh Approach

In this section, we will provide an introduction to the Keldysh formalism. After explaining the idea of the method, we discuss a single bosonic mode, and then mention the modifications for the fermionic case.

5.3.1 Basic Idea

First we remind ourselves that in quantum statistical mechanics, the time dependence of a quantum system is encoded in the dynamics of the density operator ρ. Expectation values of some Hermitian operator A are computed from

$$\langle A(t) \rangle = \text{Tr}[\rho(t)A], \quad \rho(t) = U(t)\rho U^\dagger(t), \tag{5.13}$$

where ρ is the initial density operator at time $t = 0$, and we used the unitary (forward) time-evolution operator $U(t)$; likewise, $U^\dagger(t)$ is the backward time-evolution operator. As discussed in standard time-dependent quantum mechanics,

Fig. 5.4 Keldysh contour $\mathcal{C}$
in time-discretized version

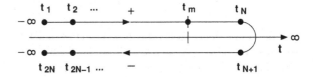

it can be expressed as a time-ordered exponential,[1]

$$U(t) = U(t, 0) = \mathcal{T} \exp\left(-i \int_0^t dt' H(t')\right),\qquad(5.14)$$

with the composition law for $t_3 > t_2 > t_1$,

$$U(t_3, t_2) U(t_2, t_1) = U(t_3, t_1).\qquad(5.15)$$

The forward-backward structure (5.13) now suggests to consider the dynamics along a closed-time loop as shown in Fig. 5.4. Specifically, we start from time $t_1 = 0$, propagate to time $t_f = t_N = t_{N+1}$ (where eventually $t_f \to \infty$), and then go back to the initial time $t_{2N} = t_1 = 0$. For $t_1 = t_{2N} = 0$, the system's density operator is supposed to be known, e.g., the equilibrium density operator for a "trivial" Hamiltonian $H(0) = H_0$ without "difficult" terms like interactions or tunneling terms connecting dot and leads. Such terms are then adiabatically switched on, so that after a transient time, $H(t) = H$ with the full Hamiltonian H. The system then evolves along the forward branch up to the final time t_f, and then along the backward branch. The "difficult" terms are switched off following the same protocol as $t \to 0$. Times $z = t^\alpha$ are then defined by the actual time and the branch $\alpha = \pm$ of the contour. We introduce a contour time-ordering ($z > z'$) along the contour. Now consider the time evolution operator along $\mathcal{C}$ between times z_1 and z_2 (where $z_2 > z_1$),

$$U(z_2, z_1) = \mathcal{T}_{\mathcal{C}} \exp\left(-i \int_{z_1}^{z_2} dz H(z)\right).$$

In particular, going along the full loop, we still have the composition law (5.15), and therefore

$$\mathcal{U} \equiv \mathcal{T}_{\mathcal{C}} \exp\left(-i \int_{\mathcal{C}} dz H(z)\right) = U(0^-, 0^+) = U(0^-, t_f) U(t_f, 0^+) = 1,\quad(5.16)$$

since $U(0^-, t_f) = U^{-1}(t_f, 0^+)$ by construction of the Hamiltonian $H(t)$. The Keldysh partition function is therefore automatically normalized to unity for any final time t_f: Exploiting cyclic invariance of the trace, we have

---

[1] The time-ordering operator is defined as $\mathcal{T}[A(t_1) B(t_2)] = \Theta(t_1 - t_2) A(t_1) B(t_2) + \Theta(t_2 - t_1) B(t_2) A(t_1)$.

$$\mathcal{Z} = \text{Tr}\rho(t_f) = \text{Tr}(U(t_f)\rho U^\dagger(t_f)) = \text{Tr}(\mathcal{U}\rho) = 1, \tag{5.17}$$

since the initial density operator ρ is normalized. This is a major advantage in the theory of disordered systems, and allows to circumvent the replica trick [10].

By including a source term into the Hamiltonian which breaks the symmetry between forward and backward contour, one can then see how useful this construction is. Let's add the piece $\eta\delta(z-t_m^+)A$ to the Hamiltonian, where the perturbation operator A acts only on the forward branch (see Fig. 5.4, with "measurement time" t_m) and the source parameter is $\eta \ll 1$. This perturbation breaks the symmetry between the two branches, and $\mathcal{Z} \neq 1$ is to be expected. The closed-loop time evolution operator $\mathcal{U}$ is then modified to first order in η as

$$\mathcal{U}(\eta) = \mathcal{U} - i\eta T_C \left(e^{-i \int_C dz H(z)} A_{t_m^+} \right) + \mathcal{O}(\eta^2),$$

and using the composition law (5.15), we can break up the exponential into a free evolution part from $z = 0^+ \rightarrow t_m^+$, the loop via t_f back to t_m^- (which gives just unity), and the evolution from $z = t_m^-$ to $z = 0^-$, which is the inverse of the first segment. Therefore, using the definition of the contour time ordering and $U(t_m) \equiv U(t_m^+, 0^+)$, we have

$$\mathcal{U}(\eta) = 1 - i\eta U^\dagger(t_m) A U(t_m) + \mathcal{O}(\eta^2).$$

The Keldysh generating function is then defined as

$$\mathcal{Z}(\eta) = \text{Tr}[\mathcal{U}(\eta)\rho] \tag{5.18}$$

and obviously satisfies the normalization condition $\mathcal{Z}(0) = 1$. We can extract observables by taking derivatives with respect to η and then putting $\eta = 0$. Exploiting the cyclic invariance of the trace, we get

$$i\partial_\eta \mathcal{Z}(\eta = 0) = \text{Tr}[A U(t_m)\rho U^\dagger(t_m)] = \text{Tr}[A\rho(t_m)] = \langle A(t_m)\rangle. \tag{5.19}$$

The closed-time construction then allows to conveniently compute the time-dependent expectation value (5.13).

5.3.2 Boson Mode

Let us consider as toy model a single boson mode, $H = \omega b^\dagger b$, to illustrate the construction of a Keldysh functional integral. Initially, the system is taken in a grand-canonical state ($\beta = 1/T$)

$$\rho = \frac{1}{Z_0} e^{-\beta(\omega-\mu)b^\dagger b}, \quad Z_0 = \text{Tr} e^{-\beta(\omega-\mu)b^\dagger b} = \frac{1}{1 - e^{-\beta(\omega-\mu)}}. \tag{5.20}$$

However, one can instead also take an arbitrary initial distribution, with straightforward modifications below. Consider now $\mathcal{Z} = \text{Tr}(\mathcal{U}\rho)$ and split up the time from 0 to t_f into $N-1$ steps of size $\delta_t = t_f/(N-1)$, with $t_1 = 0$ and $t_N = t_f$, see Fig. 5.4, and similarly on the backward branch with $t_{N+1} = t_N$ and $t_{2N} = t_1$. Eventually, we let $N \to \infty$, and afterwards send $t_f \to \infty$. At each discrete time t_j ($j = 1, \dots, 2N$), we then insert an over-complete coherent state resolution of the identity,

$$1_j = \int \frac{d(\text{Re}\phi_j)d(\text{Im}\phi_j)}{\pi} e^{-\phi_j^*\phi_j} |\phi_j\rangle\langle\phi_j|. \tag{5.21}$$

Coherent states [9, 11] are right eigenstates of the annihilation operator, $b|\phi\rangle = \phi|\phi\rangle$, parametrized by the complex number ϕ. Their explicit form is

$$|\phi\rangle = e^{\phi b^\dagger}|0\rangle, \quad \langle\phi| = \langle 0|e^{\phi^* b}.$$

Matrix elements of normal-ordered operators (such as the Hamiltonian) then take the form

$$\langle\phi|H(b^\dagger, b)|\phi'\rangle = H(\phi^*, \phi')\langle\phi|\phi'\rangle.$$

Below we need the auxiliary relation [9]

$$\langle\phi|e^{ab^\dagger b}|\phi'\rangle = e^{e^a \phi^*\phi'}, \tag{5.22}$$

which for $a = 0$ implies the overlap $\langle\phi|\phi'\rangle = e^{\phi^*\phi'}$. The integration over all $2N$ complex variables is denoted below by the shorthand notation $\int D(\phi^*, \phi)$.

Given the above relations, we can now construct the discretized Keldysh functional integral. Let us illustrate this for $N = 3$ in some detail. In $\text{Tr}(\mathcal{U}\rho)$, we encounter the string (read from right to left)

$$\langle\phi_6|e^{i\delta_t H}|\phi_5\rangle\langle\phi_5|e^{i\delta_t H}|\phi_4\rangle\langle\phi_4|1|\phi_3\rangle\langle\phi_3|e^{-i\delta_t H}|\phi_2\rangle\langle\phi_2|e^{-i\delta_t H}|\phi_1\rangle\langle\phi_1|\rho|\phi_6\rangle.$$

We can evaluate this now for $\delta_t \to 0$ as follows. With $H(b^\dagger, b)$ in normal-ordered form (which of course does not imply that $e^{\pm i\delta_t H}$ is also normal-ordered), we have

$$\langle\phi_{j+1}|e^{\pm i\delta_t H(b^\dagger, b)}|\phi_j\rangle \approx \langle\phi_{j+1}|[1 \pm i\delta_t H(b^\dagger, b)]|\phi_j\rangle$$
$$= \langle\phi_{j+1}|\phi_j\rangle\left[1 \pm i\delta_t H(\phi_{j+1}^*, \phi_j)\right]$$
$$\approx e^{\phi_{j+1}^*\phi_j \pm i\delta_t H(\phi_{j+1}^*, \phi_j)}.$$

Using Eq. 5.22, for $H = \omega b^\dagger b$, the matrix elements can be computed exactly, $e^{e^{\pm i\omega\delta_t}\phi_{j+1}^*\phi_j}$, which is in accordance with the $\delta_t \to 0$ result above. Moreover,

$$\frac{1}{Z_0}\langle\phi_1|e^{-\beta(\omega-\mu)b^\dagger b}|\phi_6\rangle = \frac{1}{Z_0}\exp\left(e^{-\beta(\omega-\mu)}\phi_1^*\phi_6\right).$$

Collecting everything, using a 2N-component state $\phi = (\phi_1, \ldots, \phi_{2N})^T$, the result can be written as

$$Z = \frac{1}{Z_0} \int D(\phi^*, \phi) e^{i\phi^* G^{-1} \phi} = \frac{1}{Z_0} \frac{1}{\det(-iG^{-1})}, \qquad (5.23)$$

where the standard Gaussian integration has been carried out in the second step. This introduces the $2N \times 2N$ inverse Green's function matrix (written out for $N = 3$)

$$-iG^{-1} = \begin{pmatrix} 1 & 0 & 0 & 0 & 0 & -e^{-\beta(\omega-\mu)} \\ -e^{-i\omega\delta_t} & 1 & 0 & 0 & 0 & 0 \\ 0 & -e^{-i\omega\delta_t} & 1 & 0 & 0 & 0 \\ 0 & 0 & -1 & 1 & 0 & 0 \\ 0 & 0 & 0 & -e^{i\omega\delta_t} & 1 & 0 \\ 0 & 0 & 0 & 0 & -e^{i\omega\delta_t} & 1 \end{pmatrix}. \qquad (5.24)$$

The diagonal entries result from the $e^{-|\phi|^2}$ weight factor in the coherent-state resolution of the identity. The upper right entry, $-e^{-\beta(\omega-\mu)}$ in the $(+-)$ block, reflects the initial distribution, while the -1 in the $(-+)$ block comes from the fusion of the upper and the lower branch at t_f, i.e., no time evolution happens between these two points. The diagonal blocks $(\pm\pm)$ have $-e^{\mp i\omega\delta_t}$ in the lower first off-diagonal. The determinant of the above matrix is for arbitrary N given by

$$\det(-iG^{-1}) = 1 - e^{-\beta(\omega-\mu)} = \frac{1}{Z_0},$$

in accordance with $Z = 1$. Note that the $(+-)$ and $(-+)$ blocks coupling the two branches each contain just one nonzero entry, which is crucial to obtain the correct determinant. In sloppy continuum notation, these may not show up explicitly but should always be kept in mind: they set the appropriate boundary conditions. From Eq. 5.23, we also see that G contains the correlation functions of the boson mode $\phi = (\phi_t^+, \phi_t^-)$,

$$G_{tt'}^{\alpha\alpha'} = -i\langle \phi_t^\alpha \phi_{t'}^{*\alpha'} \rangle. \qquad (5.25)$$

By inverting G^{-1}, which is a straightforward but not very illuminating exercise [9] in the discretized form, and then taking the continuum limit $N \to \infty$, we find

$$G_{tt'} = \begin{pmatrix} G_{tt'}^{++} & G_{tt'}^{+-} \\ G_{tt'}^{-+} & G_{tt'}^{--} \end{pmatrix} = -ie^{-i\omega(t-t')} \begin{pmatrix} \Theta(t-t') + n_b(\omega) & n_b(\omega) \\ 1 + n_b(\omega) & \Theta(t'-t) + n_b(\omega) \end{pmatrix}. \qquad (5.26)$$

The Bose function is $n_b(\epsilon) = 1/(e^{\beta(\epsilon-\mu)} - 1)$, and Θ is the Heaviside function.[2]

In Eq. 5.26, there is redundancy because of the relation $G^{++} + G^{--} = G^{+-} + G^{-+}$, which holds for any pair t, t'. We now perform a rotation in 2×2 Keldysh

---

[2] The value of $\Theta(0)$ follows from the discrete version (finite N). Consistent results follow, e.g., with $\Theta(0) = 1/2$.

space to remove this redundancy. This unitary transformation to (ϕ_c, ϕ_q) is defined as

$$\phi_c = (\phi^+ + \phi^-)/\sqrt{2}, \quad \phi_q = (\phi^+ - \phi^-)/\sqrt{2}, \qquad (5.27)$$

and the new field components are called "classical" (ϕ_c) and "quantum" (ϕ_q). This can be rationalized by noting that the quantum dynamics is governed by the product of two independent (forward vs backward) amplitudes. It is the difference between these that makes the dynamics quantum in character. The Green's function $G_{tt'}$ in rotated Keldysh space (continuum limit) follows after a short calculation:

$$G_{tt'} = -ie^{-i\omega(t-t')} \begin{pmatrix} 1 + 2n_b(\omega) & \Theta(t - t') \\ -\Theta(t' - t) & 0 \end{pmatrix} = \begin{pmatrix} G^K & G^R \\ G^A & 0 \end{pmatrix}_{tt'}, \qquad (5.28)$$

with the well-known retarded and advanced Green's functions,

$$G^{R/A}_{t-t'} = \mp i\Theta(\pm(t - t'))\langle[b(t), b^\dagger(t')]_-\rangle,$$

where $b(t) = e^{-i\omega t}b$ in Heisenberg representation. The connection to the unrotated components $G^{\alpha\alpha'}$ (with $\alpha, \alpha' = \pm$) is for bosonic Green's functions

$$G^{\alpha\alpha'} = \frac{1}{2}(G^K + \alpha G^A + \alpha' G^R). \qquad (5.29)$$

Notice that $(G^R)^\dagger = G^A$, while G^K is anti-Hermitian, $(G^K)^\dagger = -G^K$. This property and the structure in Eq. 5.28 are completely general. In particular, the qq matrix element is always zero. Note that only the Keldysh Green's function G^K contains information about the distribution function $n_b(\omega)$. In interacting models, the retarded and advanced Green's functions generally also contain information about the distribution function through self-energies. To lowest order in the interaction, however, this does not play a role.

We now switch to the energy representation of all Green's functions, with $t_f \to \infty^3$

$$G^{R/A/K}_\epsilon = \int_{-\infty}^{\infty} dt e^{i\epsilon t} G^{R/A/K}_{t,0}. \qquad (5.30)$$

When we evaluate these time integrals using Eq. 5.28, we have to ensure that correlations die out before they reach the end of contour. This is achieved by replacing $e^{-i\omega t} \to e^{-i\omega t - \kappa|t|}$ with positive infinitesimal κ such that $\kappa t_f \gg 1$. We then obtain from Eq. 5.28 the retarded/advanced Green's function in energy representation,

---

[3] We can extend the lower limit for the integral to $-\infty$, since we could have started with the interval $[-t_f/2, t_f/2]$.

$$G_\epsilon^{R/A} = \mp i \int dt \Theta(\pm t) e^{i(\epsilon-\omega)t - \kappa|t|} = \frac{1}{\epsilon \pm i\kappa - \omega}. \tag{5.31}$$

The Keldysh Green's function is with $G^R - G^A = -2\pi i \delta(\omega - \epsilon)$ given as

$$G_\epsilon^K = -2\pi i (1 + 2n_b(\omega)) \delta(\omega - \epsilon) = G_\epsilon^R F_b(\epsilon) - F_b(\epsilon) G_\epsilon^A, \tag{5.32}$$

where for the equilibrium case considered here,

$$F_b(\epsilon) = 1 + 2n_b(\epsilon) = \coth\left(\frac{\epsilon - \mu}{2T}\right). \tag{5.33}$$

In a nonequilibrium situation, F_b is not known a priori and the Keldysh Green's function is not simply determined by $G^{R/A}$ anymore. In any case, we have the Keldysh block structure

$$G_\epsilon^{-1} = \begin{pmatrix} 0 & (G^A)^{-1} \\ (G^R)^{-1} & (G^R)^{-1} F_b - F_b (G^A)^{-1} \end{pmatrix}, \tag{5.34}$$

which indeed is the inverse of G in Eq. 5.28. The Keldysh functional integral in energy representation then has the general structure of the action

$$\mathcal{Z} = Z_0^{-1} \int D(\phi^*, \phi) e^{iS}, \quad S = \int \frac{d\epsilon}{2\pi} (\phi_c^*, \phi_q^*)_\epsilon G_\epsilon^{-1} \begin{pmatrix} \phi_c \\ \phi_q \end{pmatrix}_\epsilon, \tag{5.35}$$

where the cc element is always zero. This form is not restricted to our toy model, where the matrix G_ϵ^{-1} reads

$$G_\epsilon^{-1} = \begin{pmatrix} 0 & \epsilon - \omega - i\kappa \\ \epsilon - \omega + i\kappa & 2i\kappa F_b(\epsilon) \end{pmatrix}.$$

Often the infinitesimal κ is "upgraded" to a finite self-energy due to interactions or the coupling to a heat bath.

5.3.3 Fermion Mode

For fermions, one can repeat the above steps in almost complete analogy. We consider a single fermion level, $H = \omega c^\dagger c$ with fermion operator c subject to the algebra $\{c, c^\dagger\} = 1$. Now $Z_0 = 1 + e^{-\beta(\omega - \mu)}$, and instead of boson coherent states we must use fermion coherent states $|\psi\rangle$ with anticommuting Grassmann variables ψ subject to the algebra $\{\psi, \psi'\} = \{\psi, c\} = 0$, see Ref. [11] for an introduction. Again $|\psi\rangle$ is an eigenstate of the annihilation operator, $c|\psi\rangle = \psi|\psi\rangle$, with the explicit form

$$|\psi\rangle = (1 - \psi c^\dagger)|0\rangle = e^{-\psi c^\dagger}|0\rangle.$$

Similarly, $\langle \psi | c^\dagger = \langle \psi | \bar{\psi}$, where $\bar{\psi}$ is another Grassmann number strictly independent of ψ (not the complex conjugate), and $\langle \psi | = \langle 0 | e^{-c\bar{\psi}} = \langle 0 | e^{\bar{\psi} c}$. The overlap is $\langle \psi | \psi' \rangle = 1 + \bar{\psi} \psi' = e^{\bar{\psi} \psi'}$ and the auxiliary relation (5.22) remains valid in the fermion case. Matrix elements of normal-ordered operators are again expressed as

$$\langle \psi | H(c^\dagger, c) | \psi' \rangle = H(\bar{\psi}, \psi') \langle \psi | \psi' \rangle,$$

and integrals are defined from $\int d\psi = 0$ and $\int d\psi\, \psi = 1$. Note that $\int d\bar{\psi} d\psi\, e^{-\bar{\psi}\psi} = 1$ without the factor π appearing in the boson integral. There are no convergence issues when integrating over Grassmann variables, and the multi-dimensional Gaussian integral now reads for complex invertible matrix A and Grassmann variables $\eta_i, \bar{\eta}_i$:

$$\int \left(\prod_{j=1} d\bar{\psi}_j d\psi_j \right) e^{-\sum_{i,j} \bar{\psi}_i A_{ij} \psi_j + \sum_j (\bar{\eta}_j \psi_j + \bar{\psi}_j \eta_j)} = \det(A) e^{\sum_{ij} \bar{\eta}_i A_{ij}^{-1} \eta_j}. \quad (5.36)$$

The fermion coherent-state resolution of the identity, inserted at time t_j in the construction of the Keldysh functional integral, reads

$$1_j = \int d\bar{\psi}_j d\psi_j e^{-\bar{\psi}_j \psi_j} | \psi_j \rangle \langle \psi_j |. \quad (5.37)$$

The Keldysh functional integral now involves an integration over Grassmann fields $\psi_t^\pm, \bar{\psi}_t^\pm$. The construction then proceeds completely analogous, and we only need to change a few signs, see Ref. [9]. The fermionic Green's function for $N \to \infty$ (unrotated basis) follows as in the Bose case, see Eq. 5.26, but $n_b \to -n_f$ with the Fermi function $n_f(\epsilon) = 1/(e^{\beta(\epsilon-\mu)} + 1)$. Still we have the relation $G^{++} + G^{--} = G^{+-} + G^{-+}$, and normalization $\mathcal{Z} = 1$ can be easily checked. It is a common convention to employ a slightly different non-unitary transformation for fermions, since there are no "classical" Grassmann fields and $\bar{\psi}$ is an independent field,

$$\psi_1 = \frac{1}{\sqrt{2}} (\psi^+ + \psi^-), \quad \bar{\psi}_1 = \frac{1}{\sqrt{2}} (\bar{\psi}^+ - \bar{\psi}^-), \quad (5.38)$$

$$\psi_2 = \frac{1}{\sqrt{2}} (\psi^+ - \psi^-), \quad \bar{\psi}_2 = \frac{1}{\sqrt{2}} (\bar{\psi}^+ + \bar{\psi}^-).$$

In the rotated basis, the Green's function is $\begin{pmatrix} G^R & G^K \\ 0 & G^A \end{pmatrix}$ with $G^{R/A}$ as before, but now $G^K = G^R F - F G^A$ involves the fermion distribution function

$$F_f(\epsilon) = 1 - 2n_f(\epsilon) = \tanh\left(\frac{\epsilon - \mu}{2T}\right). \quad (5.39)$$

The connection between unrotated and rotated fermionic Green's functions is similar yet different compared to the bosonic counterpart, see Eq. 5.29. We will always work with the rotated basis in the next section.

5.4 Ambegaokar-Eckern-Schön Action Within Keldysh Approach

In this section, as an application of the Keldysh functional integral approach, we study a quantum dot in the orthodox Coulomb blockade region, $\Delta E \ll T \ll E_C$. In particular, we shall analyze the role of the applied voltage bias $V = \mu_L - \mu_R$. A mesoscopic dot can easily be driven out of the linear-response regime, and a nonequilibrium theory is very useful. Since we want to keep quantum effects, the Keldysh functional integral is the most powerful machinery available and can in principle cover all limits of interest. Driving a current through the dot implies noise, i.e., current fluctuations. Noise in turn leads to dephasing and the (partial) destruction of quantum coherence. Such questions are currently investigated in many different contexts, and we will analyse them here for the probably simplest case of a Coulomb-blockaded dot ("single-electron transistor").

5.4.1 Model

For simplicity, consider spinless electrons described by the Hamiltonian

$$H = \sum_{\alpha=L/R} (H_\alpha + H_{\text{tun},\alpha}) + H_{\text{dot}}, \tag{5.40}$$

with lead Hamiltonian $H_\alpha = \sum_k \epsilon_{k\alpha} c_{k\alpha}^\dagger c_{k\alpha}$ (as for the Anderson model). The noninteracting (Fermi liquid) leads are characterized by the distribution functions

$$F_{L/R}(\epsilon) = 1 - 2n_f(\epsilon - \mu_{L/R}) = \tanh\left(\frac{\epsilon \mp V/2}{2T}\right), \tag{5.41}$$

with energies relative to the equilibrium Fermi energy, i.e. $\mu_{L/R} = \pm V/2$. The closed dot with $M \gg 1$ energy levels is described by

$$H_{\text{dot}} = \sum_{\mu,\nu=1}^{M} H_{\mu\nu} d_\mu^\dagger d_\nu + H_c, \tag{5.42}$$

with Hermitian $M \times M$ matrix $\hat{H}_d = (H_{\mu\nu})$ for the noninteracting part, e.g., a random matrix of appropriate symmetry if one treats a disordered dot. The charging energy (5.2) yields

$$H_c = E_C(N - N_G)^2, \quad N = \sum_\nu d_\nu^\dagger d_\nu. \tag{5.43}$$

Finally, the tunneling Hamiltonian now reads

$$H_{\text{tun},\alpha} = -\sum_{k,\nu} t_{k\alpha\nu} c_{k\alpha}^\dagger d_\nu + \text{h.c.}, \tag{5.44}$$

where we make the inessential simplification of choosing real tunneling amplitudes $t_{k\alpha\nu}$ between lead and dot states. Moreover, we will employ the wide-band approximation for both leads and dot, i.e. the $t_{k\alpha\nu}$ will be taken independent of k and ν later. We stress that the tunneling Hamiltonian can also describe the weak Coulomb blockade regime $G_{L,R} \gg G_Q$, since one has *many* weakly coupled channels ($M \gg 1$): each can be treated in lowest-order perturbation theory and still allow for junctions of good transparency. [The conductances $G_{L,R}$ are essentially given as the product of the (small) average tunneling probability, $|t_\nu|^2$, and the (large) channel number M.]

We now write down the full action for the Keldysh functional integral in the rotated basis. Remembering that $\kappa = 0^+$, using Grassmann fields $\Psi_{k,L/R}(t) = (\Psi_1, \Psi_2)^T$ and $\bar\Psi_{k,L/R}$ for the lead fermions, and $\psi_\mu(t) = (\psi_1, \psi_2)^T$ and $\bar\psi_\mu$ for the dot fermions, we can write

$$\mathcal{Z} \propto \int D(\bar\Psi_{k\alpha}, \Psi_{k\alpha}, \bar\psi_\nu, \psi_\nu) e^{i(S' + S_{\text{dot}})}$$

with

$$S' = \sum_{k\alpha} \int \frac{d\epsilon}{2\pi} \left(\bar\Psi_{k\alpha\epsilon} G_{k\alpha\epsilon}^{-1} \Psi_{k\alpha\epsilon} + t_\alpha \sum_\nu [\bar\Psi_{k\alpha\epsilon} \psi_{\nu\epsilon} + \bar\psi_{\nu\epsilon} \Psi_{k\alpha\epsilon}] \right),$$

$$G_{k\alpha\epsilon}^{-1} = \begin{pmatrix} \epsilon - \epsilon_{k\alpha} + i\kappa & 2i\kappa F_\alpha(\epsilon) \\ 0 & \epsilon - \epsilon_{k\alpha} - i\kappa \end{pmatrix}.$$

The $\bar\Psi \cdot \psi + \bar\psi \cdot \Psi$ structure of the second term follows from the Keldysh rotation (5.38) for fermions. The action S_{dot} of the isolated dot will be specified later.

We can now integrate out the lead fermions. To that end, we employ the Gaussian formula for Grassmann variables (5.36), where the matrix A_{ij} is diagonal and corresponds to $-iG_{k\alpha\epsilon}^{-1}$, while η_j corresponds to $it_\alpha \sum_\nu \psi_\nu$. Terms mixing the channel indices $\nu \neq \nu'$ in S' are assumed to be suppressed because of random phase factors between $t_{k\alpha\nu}$ and $t_{k\alpha\nu'}$. The result is then

$$S' = -\sum_{k\alpha\nu} t_\alpha^2 \int \frac{d\epsilon}{2\pi} \bar\psi_{\nu\epsilon} G_{k\alpha\epsilon} \psi_{\nu\epsilon}.$$

Now we invoke the wide-band approximation for the leads. With the constant lead DoS ν_α, we have $\sum_k \to \nu_\alpha \int d\epsilon_k$. Now there is a principal part contribution to the integral. The corresponding self-energy, however, only describes a shift of the energy scale. With the assumption of an energy-independent lead DoS, the wide-band approximation implies that this effect can be accomodated for by a simple shift of the dot energy levels, i.e., we may effectively disregard the principal part contribution ("pole approximation"). Using $\text{Im}(x \pm i\kappa)^{-1} = \mp\pi\delta(x)$, we thus obtain

$$\sum_k G_{k\alpha\epsilon} \simeq v_\alpha \int d\epsilon_k \begin{pmatrix} \frac{1}{\epsilon+i\kappa-\epsilon_k} & \left(\frac{1}{\epsilon+i\kappa-\epsilon_k} - \frac{1}{\epsilon-i\kappa-\epsilon_k}\right) F_\alpha(\epsilon) \\ 0 & \frac{1}{\epsilon-i\kappa-\epsilon_k} \end{pmatrix} \simeq -i\pi v_\alpha \Lambda_\alpha(\epsilon)$$

(5.45)

with the Keldysh matrices

$$\Lambda_{L/R}(\epsilon) = \begin{pmatrix} 1 & 2F_{L/R}(\epsilon) \\ 0 & -1 \end{pmatrix}.$$

(5.46)

They determine the self-energy due to the integration over all lead fermion degrees of freedom,

$$S' = i \sum_\alpha \Gamma_\alpha \int \frac{d\epsilon}{2\pi} \bar{\psi}_\epsilon \Lambda_\alpha(\epsilon) \psi_\epsilon,$$

(5.47)

where $\Gamma_\alpha = \pi v_\alpha t_\alpha^2$, see Eq. 5.3. We also introduce the Keldysh matrix $\Lambda_d(\epsilon)$ for the dot, where $F_{L/R}$ is replaced with the (a priori unknown) dot distribution function F_d.

For the action of the closed dot, $S_{dot} = S_0 + S_c$, which is composed of a noninteracting and the charging energy part, we stay in the time representation and perform a Hubbard-Stratonovich transformation to decouple the charging energy—S_c is quartic in the dot fermions—at the price of introducing the real-valued auxiliary fields $V_{c,q}(t)$,

$$S_0 = \int dt \sum_{\mu\nu} \bar{\psi}_\mu \left(G_d^{-1}\right)_{\mu\nu} \psi_\nu,$$

$$G_d^{-1} = \begin{pmatrix} i\partial_t + i\kappa - \hat{H}_d & 2i\kappa F_d \\ 0 & i\partial_t - i\kappa - \hat{H}_d \end{pmatrix},$$

$$S_c = \int dt \left(\frac{1}{2E_C} V_q V_c + N_G V_q - \sum_\nu \bar{\psi}_\nu \begin{pmatrix} V_c & V_q/2 \\ V_q/2 & V_c \end{pmatrix} \psi_\nu\right),$$

(5.48)

such that

$$\mathcal{Z} \propto \int D(V_q, V_c) \int D(\bar{\psi}_\nu, \psi_\nu) e^{i(S_0 + S_c + S')}.$$

One verifies in particular that the standard fermionic form of S_c, cf. Eq. (5.43), is recovered from Eq. 5.48 by integration over V_c, V_q. When switching back from rotated fields, $\psi_{1,2}$ and $\bar{\psi}_{1,2}$, to unrotated fields, $\psi^\pm$ and $\bar{\psi}^\pm$, note that $\bar{\psi}_1\psi_1 + \bar{\psi}_2\psi_2 = \bar{\psi}^+\psi^+ - \bar{\psi}^-\psi^-$. Similar relations follow for other combinations.

In the next step, it is convenient to perform a gauge transformation on the dot fermions, which is identical for all ψ_ν. With classical and quantum phase fields defined by $\partial_t \phi_{c,q} = V_{c,q}$, we transform according to

$$\psi(t) \rightarrow e^{-i\hat{\phi}(t)}\psi(t), \quad \bar{\psi} \rightarrow \bar{\psi}e^{+i\hat{\phi}}, \quad \hat{\phi} = \begin{pmatrix} \phi_c & \phi_q/2 \\ \phi_q/2 & \phi_c \end{pmatrix},$$

(5.49)

such that

$$e^{i\hat{\phi}} = e^{i\phi_c} \begin{pmatrix} \cos(\phi_q/2) & i\sin(\phi_q/2) \\ i\sin(\phi_q/2) & \cos(\phi_q/2) \end{pmatrix} \equiv \begin{pmatrix} c(t) & is(t) \\ is(t) & c(t) \end{pmatrix}. \tag{5.50}$$

Under this transformation, the ψ-dependent term in S_c, see Eq. 5.48, disappears since $\bar{\psi}e^{i\hat{\phi}}(i\partial_t)e^{-i\hat{\phi}}\psi = \bar{\psi}(i\partial_t + \partial_t\hat{\phi})\psi$. The transformation also generates from S_0 the extra piece $\delta S_0 = i\kappa \int dt dt' \bar{\psi}(t)e^{i\hat{\phi}(t)}[\Lambda_d(t-t'), e^{-i\hat{\phi}(t')}]_- \psi(t')$, which describes an interaction-induced renormalization of the dot distribution function. We consider weak interactions and neglect this effect here, cf. also the discussion below. The only change then appears in S', cf. Eq. 5.47, which now reads

$$S' = i\sum_\alpha \Gamma_\alpha \int dt dt' \bar{\psi}(t)e^{i\hat{\phi}(t)}\Lambda_\alpha(t-t')e^{-i\hat{\phi}(t')}\psi(t'). \tag{5.51}$$

What are the boundary conditions for the phase fields $\phi_{c,q}(t)$? Since the gauge transformation 5.49 should not alter the boundary conditions of the fermions at $t = 0$, the phase picked up when going around the closed Keldysh contour can only be a multiple of 2π. This implies [23] the condition $\phi^+(0) = \phi^-(0) + 2\pi W$ with the integer "winding number" W. This translates to an unconstrained classical field ϕ_c and boundary conditions for the quantum field ϕ_q: At $t = t_f$, we have $\phi_q(t_f) = 0$, while at $t = 0$, we impose $\phi_q(0) = 2\pi W$, where W is eventually summed over. The winding numbers are crucial in the strong Coulomb blockade regime.

At this stage, the essential degrees of freedom are encapsulated in the phase fields $\phi_{c,q}$, and the dot fermions can be integrated out to generate a determinant, i.e., a tracelog term in the action. (Recall that $\ln \det A = \text{tr}\ln A$.) We arrive at $\mathcal{Z} \propto \int D(\phi_{c,q})e^{iS[\hat{\phi}]}$ with the action

$$S[\hat{\phi}] = \int dt \left(\frac{1}{2E_C}\partial_t\phi_c\partial_t\phi_q + N_G\partial_t\phi_q\right) + S_T, \tag{5.52}$$

where (up to irrelevant constants)

$$S_T = -i\text{tr}\ln\left(1 + i\sum_\alpha \Gamma_\alpha G_d e^{i\hat{\phi}}\Lambda_\alpha e^{-i\hat{\phi}}\right).$$

The trace here extends over dot level space ($\{\nu\}$), energy (or time), and over Keldysh (2×2) space.

We next exploit that we have many dot levels, $M \gg 1$, and let each of these be very weakly coupled to the reservoirs. Then we can expand the tracelog,

$$S_T \simeq \sum_\alpha \Gamma_\alpha \text{tr}\left(\Lambda_\alpha e^{-i\hat{\phi}}G_d e^{i\hat{\phi}}\right).$$

In the orthodox Coulomb blockade regime, $T \gg \Delta E$, single-particle dot levels are thermally smeared out ("incoherent regime"), and the wide-band approximation can also be applied to the dot. (For the single-level Anderson model, one needs to proceed differently. One should then also avoid the above tunnel expansion.) The trace over the dot's Hilbert space (spanned by the $|\nu\rangle$ states) then yields, cf. Eq. 5.45, $\mathrm{tr}_\nu G_d(\epsilon) \simeq -i\pi \nu_d \Lambda_d(\epsilon)$. Using the contact tunneling conductances $G_{L/R} = \pi \nu_d \Gamma_\alpha G_Q$ as defined in Eq. 5.4, the *tunnel action* achieves the form

$$S_T = -i \sum_{\alpha=L/R} \frac{G_\alpha}{G_Q} \mathrm{tr}\left(\Lambda_\alpha e^{-i\hat\phi} \Lambda_d e^{i\hat\phi}\right). \tag{5.53}$$

Now F_d appears in the action but itself is an observable of the theory. This raises a complicated self-consistency problem, and we here "solve" it by choosing a trial distribution. Assuming that the coupling to the leads is more efficient in relaxing quasiparticles—which happens on timescales of order Γ^{-1} — than residual interactions beyond the constant interaction model on the dot, a reasonable *Ansatz* is to use the noninteracting result

$$F_d(\epsilon) = \frac{G_L F_L(\epsilon) + G_R F_R(\epsilon)}{G_L + G_R}. \tag{5.54}$$

The dot levels are then occupied not according to a standard one-step Fermi function but with a *double-step distribution*, $n_d(\epsilon) = [1 - F_d(\epsilon)]/2$. This is very accurate for the weak Coulomb blockade limit of main interest below, since then the good coupling to the leads ensures rapid equilibration to Eq. 5.54. However, in the limit of strong Coulomb blockade, a one-step Fermi function with effective temperature determined by the applied voltage and effective chemical potential may be more appropriate [1].

To simplify the discussion, we now assume identical contacts, $G_L = G_R = g_T G_Q$ with dimensionless transparency g_T. With $\Lambda_d = \frac{1}{2}(\Lambda_L + \Lambda_R)$, Eq. 5.53 yields

$$S_T = -\frac{i g_T}{2} \sum_{\alpha\alpha'} \mathrm{tr}\left(\Lambda_\alpha e^{-i\hat\phi} \Lambda_{\alpha'} e^{i\hat\phi}\right)$$

With the notation in Eq. 5.50, we get

$$S_T = \frac{g_T}{\pi} \int \frac{d\omega}{2\pi} (c^*, -is^*)_\omega \begin{pmatrix} 0 & -i\omega \\ i\omega & 2iK(\omega) \end{pmatrix} \begin{pmatrix} c \\ is \end{pmatrix}_\omega, \tag{5.55}$$

where the self-energy matrix appearing here has precisely the bosonic structure $\begin{pmatrix} 0 & \Sigma^A \\ \Sigma^R & \Sigma^K \end{pmatrix}$ like G^{-1} in Eq. 5.34. In particular, there is no cc matrix element. The kernel $K(\omega)$ contains the voltage effects,

$$K(\omega) = \frac{\omega}{2} F_b(\omega) + \frac{1}{4} \sum_{s=\pm} (\omega + sV) F_b(\omega + sV), \tag{5.56}$$

where $F_b(\omega) = \coth(\omega/2T)$. To derive Eq. 5.55 with (5.56), one writes the tunnel action in the form

$$S_T = -\frac{ig_T}{2} \sum_{\alpha\alpha'} \int \frac{d\omega}{2\pi} \frac{d\epsilon}{2\pi} \text{tr} \left[\Lambda(\epsilon - \omega - (\alpha - \alpha')V/2)(e^{-i\hat{\phi}})_\omega \Lambda(\epsilon)(e^{i\hat{\phi}})_\omega \right],$$

with $\Lambda(\epsilon) = \begin{pmatrix} 1 & 2F_f(\epsilon) \\ 0 & -1 \end{pmatrix}$, where the trace is over 2×2 Keldysh space only and $F_f(\epsilon) = \tanh(\epsilon/2T)$. In addition, we use the auxiliary relations

$$\int \frac{d\epsilon}{2\pi} (F_f(\epsilon)F_f(\epsilon - \omega) - 1) = -\frac{\omega}{\pi} F_b(\omega)$$

and

$$\int \frac{d\epsilon}{2\pi} (F_f(\epsilon) - F_f(\epsilon - \omega)) = \omega/\pi.$$

The Keldysh action (5.52) with S_T in Eq. 5.55 serves as starting point to treat the weak Coulomb blockade regime $g_T > 1$. It is often called *Ambegaokar-Eckern-Schön action* [1, 11], and has here been derived in its Keldysh variant. Apart from the present application, this prototypical action shows up in many other problems, e.g., in the description of Josephson junctions in the presence of quasiparticle dissipation, for quantum impurity problems in a Luttinger liquid, and for dissipative particle motion in a periodic potential.

5.5 Nonequilibrium Dephasing

As a prototypical observable, we will study the energy-dependent tunneling density of states (TDoS) of the dot for the "open" limit $G_{L/R} \gg G_Q$ of weak Coulomb blockade. Nevertheless, interaction effects remain important and may cause a pronounced *zero-bias anomaly*, manifest as a dip in the energy-dependent TDoS. This dip can even lead to a complete suppression of the TDoS at low energy scales. The voltage-induced dephasing then smears out this dip, and this effect in turn allows to read out the dephasing rate.

5.5.1 Tunneling Density of States

The TDoS is generally defined as

$$\nu(\epsilon) \equiv -\frac{1}{\pi} \text{Im} \, \text{tr} G^R(\epsilon). \tag{5.57}$$

This quantity probes the quantum-mechanical amplitude of quasiparticle (electron or hole) propagation on timescales $t \sim \epsilon^{-1}$, which will be affected by fluctuations, e.g., of thermal or nonequilibrium origin. It often exhibits the physics of the

Anderson orthogonality catastrophe: the state of the system with an additional electron right after the tunneling event is very different from the final asymptotic ground state. Because of such correlation effects, in order to reach the true ground state, the system then requires a large-scale readjustment to accomodate the incoming particle. At small energy scales, as a consequence the TDoS is suppressed and exhibits a dip, the zero-bias anomaly. Coulomb blockade is precisely a manifestation of such an anomaly: tunneling into the dot is suppressed under Coulomb blockade conditions, and the TDoS can be very small for low energies. This effect has been observed experimentally, and can be exploited even out of equilibrium to measure quasiparticle distribution functions [24, 25]. The equilibrium variant of the problem discussed below has been studied in Refs. [26, 27].

Exploiting the wide-band approximation, for the symmetric case $G_L = G_R = g_T G_Q$, the relevant Green's function appearing in Eq. 5.57 is $\mathrm{tr}_\nu G_d = -i\pi \nu_d \Lambda_d = -\frac{i\pi \nu_d}{2} \sum_\alpha \Lambda_\alpha$ in the noninteracting limit. Interactions are then encapsulated by phase factors $e^{\pm i\hat{\phi}(t)}$ dressing Λ_d after the gauge transformation (5.49). Extracting the retarded component of the dressed Λ_d requires some algebra but is not particularly illuminating; for details, see Ref. [12]. This gives for the TDoS

$$\frac{\nu(\epsilon)}{\nu_d} = 1 + \frac{1}{2} \sum_\pm \mathrm{Im} \int_{-\infty}^{\infty} d\tau\, e^{i(\epsilon \pm V/2)\tau}$$

$$\times F_\tau \left\langle e^{i[\phi_c(\bar{t}+\tau) - \phi_c(\bar{t})]} \sin\left(\frac{\phi_q(\bar{t}+\tau) + \phi_q(\bar{t})}{2}\right)\right\rangle, \qquad (5.58)$$

where F_τ is the Fourier transform of $F_f(\epsilon) = \tanh(\epsilon/2T)$; at zero temperature, $F_\tau = (i\pi\tau)^{-1}$. The time $\bar{t}$ is an arbitrary reference time that drops out of the final result, and the average in Eq. 5.58 refers to a functional integration over $\phi_{c,q}(t)$ with $e^{iS[\hat{\phi}]}$ as weight. In the noninteracting case ($E_C = 0$), the second term in Eq. 5.58 vanishes since then the phase fields themselves vanish, and $\nu(\epsilon) = \nu_d$ is structureless. Interactions then result in a zero-bias anomaly. We shall see below that this leads to two dips at $\epsilon = \pm V/2$.

The current flowing through the dot follows from $\nu(\epsilon)$ via the general expression [12, 28–30]

$$I(V) = G_\infty \int d\epsilon [n_f(\epsilon - V/2) - n_f(\epsilon + V/2)] \frac{\nu(\epsilon, V)}{\nu_d}. \qquad (5.59)$$

This intuitive relation can be derived via appropriate source fields from the Keldysh theory, but also follows from the general Meir–Wingreen formula [1]. We here do not go into the details of the derivation. In the noninteracting limit, this gives $I = G_\infty V$, corresponding to the series resistance of two contact resistors with $R = h/(g_T e^2)$, i.e. $G_\infty = 1/(2R)$. Knowledge of the TDoS $\nu(\epsilon)$ then also yields the temperature-dependent current–voltage characteristics. The energy-dependent TDoS contains, however, more information, and itself can be probed experimentally by scanning tunneling microscopy, or by adding a weakly coupled electrode and measuring the respective tunneling current.

5.5.2 Strong Tunneling: Open Dot

We now consider the strong tunneling case $g_T \gg 1$, where the phase fields fluctuate only weakly: the tunneling action dominates and tries to pin them. Therefore we may expand to quadratic order around the stationary value $\phi_c = \phi_q = 0$ (with $W = 0$ only). Anharmonic fluctuations turn out to be suppressed in g_T^{-1}, but they are singular at very low energies. For $\max(V, T) > g_T E_C e^{-g_T}$, however, anharmonic terms can be ignored and the quadratic approximation shown here is sufficient. In this case, the N_G term in Eq. 5.52 drops out, and charge quantization is thereby removed. Technically, for $W = 0$, we have $\phi_q(0, t_f) = 0$, and hence the only N_G-dependent term in the action (5.52) vanishes, $N_G \int_0^{t_f} dt \, \partial_t \phi_q = 0$. The Gaussian approximation for the action is then obtained in the form

$$S^{(2)}[\hat{\phi}] = \int dt \left(\frac{1}{2E_C} \partial_t \phi_c \partial_t \phi_q - \frac{g_T}{\pi} \phi_q \partial_t \phi_c \right) + \frac{ig_T}{2\pi} \int \frac{d\omega}{2\pi} \phi_{q,-\omega} K(\omega) \phi_{q,\omega},$$

(5.60)

with the kernel K in Eq. 5.56. One can derive Eq. 5.60 from Eq. 5.55 by expanding the action in ϕ_c, ϕ_q to second order.

In order to clarify the physics behind these equations, we can now establish a connection to Langevin theory. The quadratic piece in ϕ_q can be decoupled by a Hubbard-Stratonovich transformation introducing a real-valued noise field $\xi(t)$,

$$e^{-\frac{g_T}{2\pi} \int \frac{d\omega}{2\pi} K(\omega) \phi_{q,-\omega} \phi_{q,\omega}} = \int D\xi \, e^{-\int \frac{d\omega}{2\pi} \xi_{-\omega} \frac{\pi}{2g_T K(\omega)} \xi_\omega + i \int dt \phi_q \xi},$$

such that the action (5.60), after a partial integration, reads

$$S^{(2)}[\hat{\phi}, \xi] = \int dt \phi_q \left[\left(-\frac{1}{2E_C} \partial_t^2 - \frac{g_T}{\pi} \partial_t \right) \phi_c + \xi \right] + i \int \frac{d\omega}{2\pi} \xi_{-\omega} \frac{\pi}{2g_T K(\omega)} \xi_\omega.$$

(5.61)

Functional integration over ϕ_q nails the classical field to the solution of the semi-classical Langevin equation,

$$\left(\frac{1}{2E_C} \partial_t^2 + \frac{g_T}{\pi} \partial_t \right) \phi_c(t) = \xi(t),$$

(5.62)

with Gaussian noise $\xi(t)$ of zero mean, $\overline{\xi(t)} = 0$, and correlation function

$$\overline{\xi(t)\xi(t')} = \frac{g_T}{\pi} K(t - t') \equiv \frac{g_T}{\pi} \int \frac{d\omega}{2\pi} e^{-i\omega(t-t')} K(\omega).$$

(5.63)

Note that the voltage affects $K(t - t')$ and thereby causes the voltage-induced dephasing mechanism mentioned above.

Equation (5.62) is the Langevin equation for voltage fluctuations in the analogous classical resistor network in Fig. 5.1. The central node in Fig. 5.1 represents the quantum dot coupled by two resistors $R \equiv h/(e^2 g_T)$ to the bias voltage source.

The dot is also coupled to a capacitance $C \equiv e^2/(2E_C)$. With the RC time (since the dot is shunted by two parallel resistors, the relevant resistance is $R/2$)

$$\tau_{RC} = \frac{RC}{2} = \frac{\pi}{2g_T E_C}, \tag{5.64}$$

the Langevin equation (5.62) translates to an equation for the fluctuating voltage $U(t) = \partial_t \phi_c$ on the dot,

$$\left(\partial_t + \tau_{RC}^{-1}\right) U(t) = \eta(t), \quad \overline{\eta(t)\eta(t')} = \frac{1}{C\tau_{RC}} K(t - t'), \tag{5.65}$$

with the new noise field $\eta = \xi/C$. In the absence of fluctuations, $\eta = 0$, the voltage relaxes to its stationary value within the RC time (5.64). Now the role of the fluctuations is critical.

Let us first consider the equilibrium case, $V = 0$, and take the classical limit. This corresponds to sufficiently high temperatures such that $K(\omega) = \omega \coth(\omega/2T) \to 2T$. Then we have thermal white noise,

$$\overline{\eta(t)\eta(t')} = \frac{2T}{C\tau_{RC}} \delta(t - t').$$

The prefactor gives the noise strength and is in accordance with the classical fluctuation-dissipation theorem [11], relating the noise strength to the product of temperature T and the damping strength $1/\tau_{RC}$ appearing in Eq. 5.65. The steady-state distribution of voltage fluctuations resulting from Eq. 5.65 is then also Gaussian, $P(U) \propto \exp(-\frac{CU^2}{2T})$, corresponding to the Maxwell-Boltzmann distribution for an equilibrium system with capacitive energy $CU^2/2$. This thermal noise in the current through the dot is called *Johnson-Nyquist noise* [31].

The fluctuation-dissipation theorem only holds in equilibrium, and we next examine the opposite nonequilibrium limit, $T = 0$ but $V > 0$. The kernel $K(\omega)$ in Eq. 5.56 then has the limiting behavior

$$K(\omega) = \begin{cases} |\omega|, |\omega| \gg V, \\ V/2, |\omega| \ll V. \end{cases}$$

The low-frequency modes of the noise are therefore strongly influenced by the bias. For large V, we now obtain the asymptotic behavior of the noise correlator

$$\overline{\eta(t)\eta(t')} = \frac{1}{2} \frac{V}{C\tau_{RC}} \delta(t - t').$$

The current flowing through the quantum dot will then be noisy due to this correlator, and we have a situation dominated by *shot noise* [31].

This shot noise now causes voltage-induced dephasing, a direct consequence of the noise field $\eta(t)$. This dephasing shows up, for instance, in rounding off the zero-bias anomaly dips in the TDoS [20]. Using the above Langevin approach, or, equivalently, the Gaussian action (5.60), the TDoS (5.58) can be evaluated analytically. We focus on the $T = 0$ limit, where the result is

$$\frac{\nu(\epsilon)}{\nu_d} = 1 - \frac{\tau_{RC}}{2\pi C} \sum_{\pm} \int_0^{\infty} \frac{dt}{t} \cos[(\epsilon \pm V/2)t] \left(1 - e^{-t/\tau_{RC}}\right) e^{-S(t)}. \qquad (5.66)$$

The noise action due to the Gaussian η fluctuations is

$$S(t) = \frac{1}{C\tau_{RC}} \int \frac{d\omega}{2\pi} \frac{1 - \cos(\omega t)}{\omega^2(\omega^2 + \tau_{RC}^{-2})} K(\omega). \qquad (5.67)$$

In the equilibrium case, $V = 0$, the noise action can be estimated as

$$S(t; V = 0) \simeq \frac{\tau_{RC}}{2\pi C} \ln(1 + t^2/\tau_{RC}^2), \qquad (5.68)$$

and hence

$$\frac{\nu(\epsilon; V = 0)}{\nu_d} \simeq 1 - \frac{\tau_{RC}}{\pi C} \int_{\tau_{RC}}^{|\epsilon|^{-1}} \frac{dt}{t} (t/\tau_{RC})^{-\tau_{RC}/\pi C} = |\epsilon \tau_{RC}|^{1/g_T}. \qquad (5.69)$$

This Coulomb blockade power-law suppression of the TDoS is a prototypical example for a zero-bias anomaly. It results in a complete suppression of the TDoS due to interactions as $\epsilon \to 0$, despite of the fact that we are in the weak Coulomb blockade regime. Note that this effect originates from $S(t) \propto \ln t$ as $t \to \infty$, a direct consequence of the noisy fluctuating voltage.

How is this picture affected by the bias? First, the zero-bias anomaly is now split into a pair of dips at $\epsilon = \pm V/2$, reflecting the double-step distribution function for the dot. Second, the noise action $S(t)$ in Eq. 5.67 for $t \to \infty$ now goes like $S(t) \propto Vt$, indicating a much stronger noise level at low frequencies than the only logarithmic increase at $V = 0$. The consequence of this behavior is that the split anomaly dips are now smeared and less pronounced, a direct signature of shot noise due to the finite current flowing through the system, see Fig. 5.5. In particular, a full suppression of the TDoS is not possible anymore. The above results for the TDoS also allow to infer the *IV* characteristics for transport through the dot, see Eq. 5.59. For instance, to leading order in g_T^{-1}, one finds logarithmic corrections to the nonlinear conductance (at $T \ll V$) due to the Coulomb interaction [28–30],

$$\frac{1}{G_\infty} \frac{dI}{dV} = 1 - \frac{1}{4\pi g_T} \ln\left[1 + (V\tau_{RC})^{-2}\right]. \qquad (5.70)$$

This logarithmic suppression is a remnant of Coulomb blockade in the open dot.

One can now quantitatively define a voltage-induced dephasing rate $\Gamma(V)$ from the broadening of the zero-bias anomaly. Writing $\delta\epsilon = \epsilon - V/2$ with $|\delta\epsilon|\tau_{RC} \ll 1$,

Fig. 5.5 Energy dependence of the nonequilibrium TDoS for $g_T = 5$ and several bias voltages from the numerical integration of Eq. 5.66

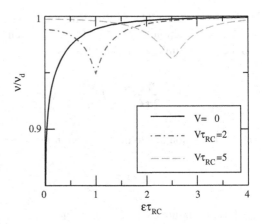

the dip is described by $\delta\nu(\delta\epsilon) = \nu_d - \nu(\epsilon)$. Parameterizing small deviations off the dip at $V/2$ in terms of the dephasing rate $[\delta\nu(0) - \delta\nu(\delta\epsilon)]/\delta\nu(0) \simeq \frac{1}{2}[\delta\epsilon/\Gamma(V)]^2$, Eq. 5.66 yields

$$\Gamma^2(V) = \frac{\int_0^\infty \frac{d\tau}{\tau} e^{-S(\tau)}(1 - e^{-\tau/\tau_{RC}})}{\int_0^\infty d\tau\, \tau e^{-S(\tau)}(1 - e^{-\tau/\tau_{RC}})}. \tag{5.71}$$

Analytical results for $\Gamma(V)$ can then be extracted from Eq. 5.71 in various limiting regimes [12]. In the fully shot-noise dominated limit mentioned above, one finds that the decoherence rate $\Gamma(V) \sim V/g_T$ is directly determined by the shot noise. For a more detailed discussion of the voltage-induced dephasing in this problem, see Ref. [12].

5.6 Conclusions

In these notes, we have discussed the Keldysh functional integral description of nonequilibrium transport through a Coulomb blockaded dot. Our principal aim was to provide an accessible and compact introduction to both the Keldysh method and to the physics of nonequilibrium dephasing in quantum dots. The formalism can easily be extended to study the full counting statistics [1] of such an interacting system.

Acknowledgments We acknowledge support by the SFB Transregio 12 by the Deutsche Forschungsgemeinschaft.

References

1. Nazarov, Y.V., Blanter, Y.M.: Quantum Transport: Introduction to Nanoscience. Cambridge University Press, Cambridge (2009)
2. Esposito, M., Harbola, U., Mukamel, S.: Nonequilibrium fluctuations, fluctuation theorems, and counting statistics in quantum systems. Rev. Mod. Phys. **81**, 1665 (2009)
3. Weiss, S., Eckel, J., Thorwart, M., Egger, R.: Iterative real-time path integral approach to nonequilibrium quantum transport. Phys. Rev. B **77**, 195316 (2008)
4. Anders, F.B.: Steady-state currents through nanodevices: a scattering-states numerical renormalization-group approach to open quantum systems. Phys. Rev. Lett. **101**, 066804 (2008)
5. Boulat, E., Saleur, H., Schmitteckert, P.: Twofold advance in the theoretical understanding of far-from-equilibrium properties of interacting nanostructures. Phys. Rev. Lett. **101**, 140601 (2008)
6. Metha, P., Andrei, N.: Nonequilibrium transport in quantum impurity models: the Bethe ansatz for open systems. Phys. Rev. Lett. **96**, 216802 (2006)
7. Alhassid, Y.: The statistical theory of quantum dots. Rev. Mod. Phys. **72**, 895 (2000)
8. Aleiner, I.L., Brouwer, P.W., Glazman, L.I.: Quantum effects in Coulomb blockade. Phys. Rep. **358**, 309 (2002)
9. Kamenev, A., Levchenko, A.: Keldysh technique and non-linear σ-model: basic principles and applications. Adv. Phys. **58**, 197 (2009)
10. Kamenev, A., Andreev, A.: Electron–electron interactions in disordered metals: Keldysh formalism. Phys. Rev. B **60**, 2218 (1999)
11. Altland, A., Simons, B.D.: Condensed Matter Field Theory, 2nd edn. Cambridge University Press, Cambridge (2010)
12. Altland, A., Egger, R.: Nonequilibrium dephasing in coulomb blockaded quantum dots. Phys. Rev. Lett. **102**, 026805 (2009)
13. Kaminski, A., Nazarov, Yu.V., Glazman, L.I.: Suppression of the Kondo effect in a quantum dot by external irradiation. Phys. Rev. Lett. **83**, 384 (1999)
14. Rosch, A., Paaske, J., Kroha, J., Wölfle, P.: Nonequilibrium transport through a Kondo dot in a magnetic field: perturbation theory and poor man's scaling. Phys. Rev. Lett. **90**, 076804 (2003)
15. Kehrein, S.: Scaling and decoherence in the nonequilibrium Kondo model. Phys. Rev. Lett. **95**, 056602 (2005)
16. Muzykantskii, B., d'Ambrumenil, N., Braunecker, B.: Fermi-edge singularity in a nonequilibrium system. Phys. Rev. Lett. **91**, 266602 (2003)
17. Mitra, A., Millis, A.J.: Coulomb gas on the Keldysh contour: Anderson–Yuval–Hamann representation of the nonequilibrium two-level system. Phys. Rev. B **76**, 085342 (2007)
18. Gutman, D.B., Gefen, Y., Mirlin, A.D.: Nonequilibrium Luttinger liquid: zero-bias anomaly and dephasing. Phys. Rev. Lett. **101**, 126802 (2008)
19. Neder, I., Marquardt, F.: Coherence oscillations in dephasing by non-Gaussian shot noise. New J. Phys. **9**, 112 (2007)
20. Gutman, D.B., Gefen, Y., Mirlin, A.D.: Nonequilibrium zero-bias anomaly in disordered metals. Phys. Rev. Lett. **100**, 086801 (2008)
21. König, J., Schoeller, H.: Strong tunneling in the single-electron box. Phys. Rev. Lett. **81**, 3511 (1998)
22. Göppert, G., Grabert, H., Prokof'ev, N.V., Svistunov, B.: Effect of tunneling conductance on the coulomb staircase. Phys. Rev. Lett. **81**, 2324 (1998)
23. Schön, G., Zaikin, A.D.: Quantum coherent effects, phase transitions, and the dissipative dynamics of ultra small tunnel junctions. Phys. Rep. **198**, 237 (1990)
24. Pothier, H., Guéron, S., Birge, N.O., Esteve, D., Devoret, M.H.: Energy distribution function of quasiparticles in mesoscopic wires. Phys. Rev. Lett. **79**, 3490 (1997)
25. Anthore, A., Pierre, F., Pothier, H., Esteve, D.: Magnetic-field-dependent quasiparticle energy relaxation in mesoscopic wires. Phys. Rev. Lett. **90**, 076806 (2003)
26. Kamenev, A., Gefen, Y.: Zero-bias anomaly in finite-size systems. Phys. Rev. B **54**, 5428 (1996)

27. Sedlmayr, N., Yurkevich, I.V., Lerner, I.V.: Tunnelling density of states at Coulomb-blockade peaks. Europhys. Lett. **76**, 109 (2006)
28. Nazarov, Yu.V.: Coulomb blockade without tunnel junctions. Phys. Rev. Lett. **82**, 1245 (1999)
29. Golubev, D.S., Zaikin, A.D.: Coulomb interaction and quantum transport through a coherent scatterer. Phys. Rev. Lett. **86**, 4887 (2001)
30. Golubev, D.S., Zaikin, A.D.: Electron transport through interacting quantum dots in the metallic regime. Phys. Rev. B **69**, 075318 (2004)
31. Blanter, Ya.M., Büttiker, M.: Shot noise in mesoscopic conductors. Phys. Rep. **336**, 1 (2000)

Chapter 6
Many-Body Physics from a Quantum Information Perspective

R. Augusiak, F. M. Cucchietti and M. Lewenstein

Abstract The quantum information approach to many-body physics has been very successful in giving new insights and novel numerical methods. In these lecture notes we take a vertical view of the subject, starting from general concepts and at each step delving into applications or consequences of a particular topic. We first review some general quantum information concepts like entanglement and entanglement measures, which leads us to entanglement area laws. We then continue with one of the most famous examples of area-law abiding states: matrix product states, and tensor product states in general. Of these, we choose one example (classical superposition states) to introduce recent developments on a novel quantum many-body approach: quantum kinetic Ising models. We conclude with a brief outlook of the field.

6.1 Introduction

There has been an explosion of interest in the interface between quantum information (QI) and many-body systems, in particular in the fields of condensed matter and ultracold atomic gases. Remarkable examples are Ref. [1], which proposed using ultracold atomic gases in optical lattices for QI (and stimulated interest in *distributed quantum information processing*), and Refs. [2–4], who discussed the first connections between entanglement and quantum phase transitions (QPT). Overall, the confluence of ideas has opened fundamentally deep questions about QPT's, as well as practical questions about how to use QI ideas in numerical simulations of many-body quantum systems. Here, we will (partially) review these two major themes. We will first introduce some basic notions and tools of quantum information

R. Augusiak · F. M. Cucchietti · M. Lewenstein (✉)
ICFO–Institut de Ciències Fotòniques,
Mediterranean Technology Park,
08860 Barcelona Castelldefels, Spain
e-mail:maciej.lewenstein@icfo.es

D. C. Cabra et al. (eds.), *Modern Theories of Many-Particle Systems in Condensed Matter Physics*, Lecture Notes in Physics 843, DOI: 10.1007/978-3-642-10449-7_6, © Springer-Verlag Berlin Heidelberg 2012

theory, focusing on entanglement and entanglement measures. We shall then discuss area laws, i.e. laws that characterize correlations and entanglement in physically relevant many-body states, and allow to make general statements about computational complexity of the corresponding Hamiltonians. Afterwards, we will explore the concept of matrix product states (MPS) and their generalizations (projected entangled pairs states, PEPS, and tensor networks states). These states provide not only a very useful ansatz for numerical applications, but also a powerful tool to understand the role of entanglement in the quantum many-body theory. We will review one particular example of a state with a straightforward MPS representation: the classical superposition state. The introduction of its parent Hamiltonian will lead us to the final subject of these lectures: quantum kinetic Ising models—an analytically solvable generalization of the popular classical many-body model described by a master equation.

6.2 Aspects of Quantum Information

Quantum theory contains elements that are radically different from our everyday ("classical") description of Nature: a most important example are the quantum correlations present in quantum formalism. Classically, complete knowledge of a system implies that the sum of the information of its subsystems makes up the total information for the whole system. In the quantum world, this is no longer true: there exist states of composite systems about which we have complete information, but we know nothing about its subsystems. We may even reach paradoxical conclusions if we apply a classical description to such "entangled" states—whose concept can be traced back to 1932 in manuscripts of Schrödinger.

What we have just realized during the last two decades is that these fundamentally nonclassical states (from hereon "entangled states") can provide us with more than just paradoxes: They may be *used* to perform tasks that cannot be achieved with classical states. As landmarks of this transformation in our view of such nonclassical states, we mention the spectacular discoveries of (entanglement-based) quantum cryptography [5], quantum dense coding [6], and quantum teleportation [7]. Even though our knowledge of entanglement is still far from complete, significant progress has been made in the recent years and very active research is currently underway (for a recent and very complete review see [8]).

In the next section, we will focus on bipartite composite systems. We will define formally what entangled states are, present some important criteria to discriminate entangled states from separable ones, and show how they can be classified according to their capability to perform some precisely defined tasks. However, before going into details, let us introduce the notation. In what follows we will be mostly concerned with bipartite scenarios, in which traditionally the main roles are played by two parties called Alice and Bob. Let $\mathcal{H}_A$ denote the Hilbert space of Alice's physical system, and $\mathcal{H}_B$ that of Bob's. Our considerations will be restricted to finite-dimensional Hilbert spaces, so we can set $\mathcal{H}_A = \mathbb{C}^{d_A}$ and $\mathcal{H}_B = \mathbb{C}^{d_B}$. Thus, the joint physical system of

Alice and Bob is described by the tensor product Hilbert space $\mathcal{H}_{AB} = \mathcal{H}_A \otimes \mathcal{H}_B = \mathbb{C}^{d_A} \otimes \mathbb{C}^{d_B}$. Finally, $\mathcal{B}(\mathcal{H})$ will denote the set of bounded linear operators from the Hilbert space $\mathcal{H}$ to $\mathcal{H}$.

6.2.1 Bipartite Pure States: Schmidt Decomposition

We start our study with pure states, for which the concepts are simpler. Pure states are either separable or entangled states according to the following definition:

Definition 1 Consider a pure state $|\psi_{AB}\rangle$ from $\mathcal{H}_A \otimes \mathcal{H}_B$. It is called separable if there exist pure states $|\psi_A\rangle \in \mathcal{H}_A$ and $|\psi_B\rangle \in \mathcal{H}_B$ such that $|\psi_{AB}\rangle = |\psi_A\rangle \otimes |\psi_B\rangle$. Otherwise we say that $|\psi_{AB}\rangle$ is entangled.

The most famous examples of entangled states in $\mathcal{H}_{AB}$ are the *maximally entangled states*, given by

$$|\psi_+^{(d)}\rangle = \frac{1}{\sqrt{d}} \sum_{i=0}^{d-1} |i\rangle_A \otimes |i\rangle_B \qquad (d = \min\{d_A, d_B\}), \tag{6.1}$$

where the vectors $\{|i\rangle_A\}$ and $\{|i\rangle_B\}$ form bases (in particular they can be the standard ones) in $\mathcal{H}_A$ and $\mathcal{H}_B$, respectively. In what follows, we also use $P_+^{(d)}$ to denote the projector onto $|\psi_+^{(d)}\rangle$. The reason why this state is called maximally entangled will become clear when we introduce entanglement measures.

In pure states, the *separability problem*—the task of judging if a given quantum state is separable—is easy to handle using the concept of Schmidt decomposition:

Theorem 1 *Let $|\psi_{AB}\rangle \in \mathcal{H}_{AB} = \mathbb{C}^{d_A} \otimes \mathbb{C}^{d_B}$ with $d_A \leq d_B$. Then $|\psi_{AB}\rangle$ can be written as a Schmidt decomposition*

$$|\psi_{AB}\rangle = \sum_{i=1}^{r} \lambda_i |e_i\rangle \otimes |f_i\rangle, \tag{6.2}$$

where $|e_i\rangle$ and $|f_i\rangle$ form a part of an orthonormal basis in $\mathcal{H}_A$ and $\mathcal{H}_B$, respectively, $\lambda_i > 0$, $\sum_{i=1}^{r} \lambda_i^2 = 1$, and $r \leq d_A$.

Proof A generic pure bipartite state $|\psi_{AB}\rangle$ can be written in the standard basis of $\mathcal{H}_A \otimes \mathcal{H}_B$ as $|\psi_{AB}\rangle = \sum_{i=0} \sum_{j=0} \alpha_{ij} |i\rangle \otimes |j\rangle$, where, in general, the coefficients α_{ij} form a $d_A \times d_B$ matrix Λ obeying $\operatorname{tr}(\Lambda^\dagger \Lambda) = 1$. Using singular-value decomposition, we can write $\Lambda = V D_\Lambda W^\dagger$, where V and W are unitary ($V^\dagger V = W^\dagger W = \mathbb{1}_A$) and D_Λ is diagonal matrix consisting of the eigenvalues λ_i of $|\Lambda| = \sqrt{\Lambda^\dagger \Lambda}$. Using this we rewrite $|\psi_{AB}\rangle$ as

$$|\psi_{AB}\rangle = \sum_{i=0}^{d_A-1} \sum_{j=0}^{d_B-1} \sum_{k=1}^{r} V_{ik} \lambda_k U_{jk}^* |i\rangle |j\rangle, \tag{6.3}$$

where $r \leq d_A \leq d_B$ denotes the rank of Λ. By reshuffling terms, and defining $|e_k\rangle = \sum_{i=0}^{d_A-1} V_{ik}|i\rangle$ and $|f_k\rangle = \sum_{j=0}^{d_B-1} U_{jk}^*|j\rangle$ we get the desired form Eq. 6.2. To complete the proof, we notice that due to the unitarity of V and W, vectors $|e_i\rangle$ and $|f_i\rangle$ satisfy $\langle e_i|e_j\rangle = \langle f_i|f_j\rangle = \delta_{ij}$, and constitute bases of $\mathcal{H}_A$ and $\mathcal{H}_B$, respectively. In fact, $\{\lambda_i^2, |e_i\rangle\}$ and $\{\lambda_i^2, |f_i\rangle\}$ are eigensystems of the first and second subsystem of $|\psi_{AB}\rangle$. Moreover, since $\mathrm{tr}(\Lambda^\dagger\Lambda) = 1$ it holds that $\sum_i \lambda_i^2 = 1$. $\square$

The numbers $\lambda_i > 0$ $(i = 1, \ldots, r)$ are called *the Schmidt coefficients*, and r *the Schmidt rank* of $|\psi_{AB}\rangle$. One can also notice that $\{\lambda_i^2, |e_i\rangle\}$ and $\{\lambda_i^2, |f_i\rangle\}$ are eigensystems of the first and second subsystem of $|\psi_{AB}\rangle$, and that the Schmidt rank r denotes the rank of both subsystems. Then, comparison with definition 1 shows that bipartite separable states are those with Schmidt rank one. Thus, to check if a given pure state is separable, it suffices to check the rank r of one of its subsystems. If $r = 1$ (the corresponding subsystem is in a pure state) then $|\psi_{AB}\rangle$ is separable; otherwise it is entangled. Notice that the maximally entangled state (6.1) is already written in the form (6.2), with $r = d$ and all the Schmidt coefficients equal to $1/\sqrt{d}$.

6.2.2 Bipartite Mixed States: Separable and Entangled States

The easy-to-handle separability problem in pure states complicates considerably in the case of mixed states. In order to understand the distinction between separable and entangled mixed states—first formalized by Werner in 1989 [9]—let us consider the following state preparation procedure. Suppose that Alice and Bob are in distant locations and can produce and manipulate any physical system in their laboratories. Moreover, they can communicate using a classical channel (for instance a phone line). However, they do not have access to quantum communication channels, i.e. they are not allowed to exchange quantum states. These two capabilities, i.e. local operations (LO) and classical communication (CC), are frequently referred to as LOCC.

Suppose now that in each round of the preparation scheme, Alice generates with probability p_i a random integer i $(i = 1, \ldots, K)$, which she sends to Bob. Depending on this number, in each round Alice prepares a pure state $|e_i\rangle$, and Bob a state $|f_i\rangle$. After many rounds, the result of this preparation scheme is of the form

$$\varrho_{AB} = \sum_{i=1}^K p_i |e_i\rangle\langle e_i| \otimes |f_i\rangle\langle f_i|, \tag{6.4}$$

which is the most general one that can be prepared by Alice and Bob by means of LOCC. In this way we arrive at the formal definition of separability in the general case of mixed states.

Definition 2 We say that a mixed state ϱ_{AB} acting on $\mathcal{H}_{AB}$ is separable if and only if it can be represented as a convex combination of the product of projectors on local states as in Eq. 6.4. Otherwise, the mixed state is said to be entangled.

The number of pure separable states K necessary to decompose any separable state according to Eq. 6.4 is limited by the Caratheodory theorem as $K \leq (d_A d_B)^2$ (see Refs. [10, 8]). No better bound is known in general, however, for two-qubit and qubit-qubit systems it was shown that $K \leq 4$ [11] and $K \leq 6$ [12], respectively.

By definition, entangled states cannot be prepared locally by two parties even after communicating over a classical channel. To prepare entangled states, the physical systems must be brought together to interact.[1] Mathematically, a nonlocal unitary operator[2] must *necessarily* act on the physical system described by $\mathcal{H}_A \otimes \mathcal{H}_B$ to produce an entangled state from an initial separable state.

The question whether a given bipartite state is separable or not turns out to be quite complicated. Although the general answer to the separability problem still eludes us, there has been significant progress in recent years, and we will review some such directions in the following paragraphs.

6.2.3 Entanglement Criteria

An operational necessary and sufficient criterion for detecting entanglement still does not exist. However, over the years the whole variety of criteria allowing for detection of entanglement has been worked out. Below we review some of the most important ones, while for others the reader is referred to Ref. [14]. Note that, even if we do not have necessary and sufficient separability criteria, there are numerical checks of separability: semidefinite programming was used to show that separability can be tested in a finite number of steps, although this number can become too large for big systems [15, 16]. In general—without a restriction on dimensions—the separability problem belongs to the NP-hard class of computational complexity [17].

6.2.4 Partial Transposition

Let us start with an easy–to–apply necessary criterion based on the transposition map recognized by Choi [18] and then independently formulated directly in the separability context by Peres [19].

Let ϱ_{AB} be a state on the product Hilbert space $\mathcal{H}_{AB}$, and $T : \mathcal{B}(\mathbb{C}^d) \to \mathcal{B}(\mathbb{C}^d)$ a transposition map with respect to the some real basis $\{|i\rangle\}$ in $\mathbb{C}^d$, defined through $T(X) \equiv X^T = \sum_{i,j} x_{ij} |j\rangle\langle i|$ for any $X = \sum_{i,j} x_{ij} |i\rangle\langle j|$ from $\mathcal{B}(\mathbb{C}^d)$. Let us

---

[1]   Due to entanglement swapping [13], one must suitably enlarge the notion of preparation of entangled states. So, an entangled state between two particles can be prepared if and only if either the two particles (call them A and B) themselves come together to interact at a time in the past, or two *other* particles (call them C and D) do the same, with C having interacted beforehand with A and D with B.

[2]   A unitary operator on $\mathcal{H}_A \otimes \mathcal{H}_B$ is said to be "nonlocal" if it is not of the form $U_A \otimes U_B$, where U_A is a unitary operator acting on $\mathcal{H}_A$ and U_B acts on $\mathcal{H}_B$.

now consider an extended map $T \otimes I_B$ called hereafter *partial transposition*, where I_B is the identity map acting on the second subsystem. When applied to ϱ_{AB}, the map $T \otimes I_B$ transposes the first subsystem leaving the second one untouched. More formally, writing ϱ_{AB} as

$$\varrho_{AB} = \sum_{i,j=0}^{d_A-1} \sum_{\mu,\nu=0}^{d_B-1} \varrho_{ij}^{\mu\nu} |i\rangle\langle j| \otimes |\mu\rangle\langle\nu|, \tag{6.5}$$

where $\{|i\rangle\}$ and $\{|\mu\rangle\}$ are real bases in Alice and Bob Hilbert spaces, respectively, we have

$$(T \otimes I_B)(\varrho_{AB}) \equiv \varrho_{AB}^{T_A} = \sum_{i,j=0}^{d_A-1} \sum_{\mu,\nu=0}^{d_B-1} \varrho_{ij}^{\mu\nu} |j\rangle\langle i| \otimes |\mu\rangle\langle\nu|. \tag{6.6}$$

Similarly, one may define partial transposition with respect to the Bob's subsystem (denoted by $\varrho_{AB}^{T_B}$). Although the partial transposition of ϱ_{AB} depends upon the choice of the basis in which ϱ_{AB} is written, its eigenvalues are basis independent. The applicability of the transposition map in the separability problem can be formalized by the following statement [19].

Theorem 2 *If a state ρ_{AB} is separable, then $\rho_{AB}^{T_A} \geq 0$ and $\rho_{AB}^{T_B} \geq 0$.*

Proof Since ϱ_{AB} is separable, according to definition 2 it has the form (6.4). Then, performing the partial transposition with respect to the first subsystem, we have

$$\rho_{AB}^{T_A} = \sum_{i=1}^{K} p_i \left(|e_i\rangle\langle e_i|\right)^{T_A} \otimes |f_i\rangle\langle f_i| = \sum_{i=1}^{K} p_i |e_i^*\rangle\langle e_i^*| \otimes |f_i\rangle\langle f_i|. \tag{6.7}$$

In the second step we used that $A^\dagger = (A^*)^T$ for all A. The above shows that $\rho_{AB}^{T_A}$ is a proper (and also separable) density matrix implying that $\rho_{AB}^{T_A} \geq 0$. The same reasoning leads to the conclusion that $\rho_{AB}^{T_B} \geq 0$, finishing the proof. □

Due to the identity $\varrho_{AB}^{T_B} = (\varrho_{AB}^{T_A})^T$, and the fact that global transposition does not change eigenvalues, partial transpositions with respect to the A and B subsystems are equivalent from the point of view of the separability problem.

In conclusion, we have a simple criterion (*partial transposition criterion*) for detecting entanglement. More precisely, if the spectrum of one of the partial transpositions of ϱ_{AB} contains at least one negative eigenvalue then ϱ_{AB} is entangled. As an example, let us apply the criterion to pure entangled states. If $|\psi_{AB}\rangle$ is entangled, it can be written as (6.2) with $r > 1$. Then, the eigenvalues of $|\psi_{AB}\rangle\langle\psi_{AB}|^{T_A}$ will be λ_i^2 ($i = 1, \ldots, r$) and $\pm\lambda_i\lambda_j$ ($i \neq ji, j = 1, \ldots, r$). So, an entangled $|\psi_{AB}\rangle$ of Schmidt rank $r > 1$ has partial transposition with $r(r-1)/2$ negative eigenvalues violating the criterion stated in theorem 2.

The partial transposition criterion allows to detect in a straightforward manner all entangled states that have non–positive partial transposition (hereafter called

NPT states). However, even if this is a large class of states, it turns out that—as pointed out in Refs. [10, 20]—there exist entangled states with positive partial transposition (called *PPT states*) (cf. Fig. 6.2). Moreover, the set of PPT entangled states does not have measure zero [21]. It is, therefore, important to have further independent criteria that identify entangled PPT states. Remarkably, PPT entangled states are the only known examples of *bound entangled states*, i.e., states from which one cannot distill entanglement by means of LOCC, even if the parties have an access to an unlimited number of copies of the state [8, 20]. The conjecture that there exist NPT "bound entangled" states is one of the most challenging open problems in quantum information theory [22, 23]. Note also that both separable as well as PPT states form convex sets.

Theorem 2 is a necessary condition of separability in any arbitrary dimension. However, for some special cases, the partial transposition criterion is both a necessary and sufficient condition for separability [24]:

Theorem 3 *A state ϱ_{AB} acting on $\mathbb{C}^2 \otimes \mathbb{C}^2$ or $\mathbb{C}^2 \otimes \mathbb{C}^3$ is separable if and only if $\varrho_{AB}^{T_A} \geq 0$.*

We will prove this theorem later. Also, we will see that Theorem 2 is true for a whole class of maps (of which the transposition map is only a particular example), which also provide a *sufficient* criterion for separability. Before this, let us discuss the dual characterization of separability *via* entanglement witnesses.

6.2.5 Entanglement Witnesses from the Hahn–Banach Theorem

Central to the concept of entanglement witnesses is the corollary from the Hahn–Banach theorem (or Hahn–Banach separation theorem), which we will present here limited to our needs and without proof (which the reader can find e.g. in Ref. [25]).

Theorem 4 *Let S be a convex compact set in a finite–dimensional Banach space. Let ρ be a point in this space, however, outside of the set S ($\rho \notin S$). Then there exists a hyperplane[3] that separates ρ from S.*

The statement of the theorem is illustrated in Fig. 6.1. In order to apply it to our problem let S denote now the set of all separable states acting on $\mathcal{H}_A \otimes \mathcal{H}_B$. This is a convex compact subset of the Banach space of all the linear operators $\mathcal{B}(\mathcal{H}_A \otimes \mathcal{H}_B)$. The theorem implies that for any entangled state ϱ_{AB} there exists a hyperplane separating it from S.

Let us introduce a coordinate system located within the hyperplane (along with an orthogonal vector W chosen so that it points towards S). Then, every state ϱ_{AB} can be characterized by its "distance" from the plane, here represented by the

---

[3]   A hyperplane is a linear subspace with dimension smaller by one than the dimension of the space itself.

Fig. 6.1 Schematic picture
of the Hahn-Banach
theorem. The (unique) unit
vector orthonormal to the
hyperplane can be used to
define *right* and *left* with
respect to the hyperplane by
using the sign of the scalar
product

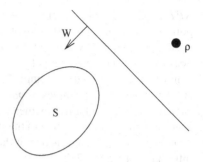

Hilbert–Schmidt scalar product.[4] According to our choice of the coordinate system (see Fig. 6.1), for any such hyperplane W every separable state has a positive "distance", while there are some entangled states with a negative "distance". More formally, theorem (4) implies the following seminal result [24].

Theorem 5 *Let ϱ_{AB} be some entangled state acting on $\mathcal{H}_{AB}$. Then there exists a Hermitian operator $W \in \mathcal{B}(\mathcal{H}_A \otimes \mathcal{H}_B)$ such that $\mathrm{tr}(\varrho_{AB}W) < 0$ and $\mathrm{tr}(\sigma_{AB}W) \geq 0$ for all separable $\sigma_{AB} \in \mathcal{B}(\mathcal{H}_A \otimes \mathcal{H}_B)$.*

It is then clear that all the operators W representing such separating hyperplanes deserve special attention as they are natural candidates for entanglement detectors. That is, given some Hermitian W, if $\mathrm{tr}(W\varrho_{AB}) < 0$ and simultaneously $\mathrm{tr}(W\sigma_{AB}) \geq 0$ for all separable σ_{AB}, we know that ϱ_{AB} is entangled. One is then tempted to introduce the following definition [26].

Definition 3 We call the Hermitian operator W an entanglement witness if $\mathrm{tr}(W\sigma_{AB}) \geq 0$ for all separable σ_{AB} and there exists an entangled state ϱ_{AB} such that $\mathrm{tr}(W\varrho_{AB}) < 0$.

Example 1 Let us discuss how to construct entanglement witnesses for all NPT states. If ϱ_{AB} is NPT then its partial transposition has at least one negative eigenvalue. Let $|\psi_i\rangle$ denote the eigenstates of $\varrho_{AB}^{T_B}$ corresponding to its negative eigenvalues $\lambda_i < 0$. Then the Hermitian operator $W_i = |\psi_i\rangle\langle\psi_i|^{T_B}$ has negative mean value on ϱ_{AB}, i.e., $\mathrm{tr}(\varrho_{AB}|\psi_i\rangle\langle\psi_i|^{T_B}) = \mathrm{tr}(\varrho_{AB}^{T_B}|\psi_i\rangle\langle\psi_i|) = \lambda_i < 0$. Simultaneously, using the identity $\mathrm{tr}(AB^T) = \mathrm{tr}(A^TB)$ obeyed by any pair of matrices A and B, it is straightforward to verify that $\mathrm{tr}(W_i\sigma_{AB}) \geq 0$ for all i and separable σ_{AB}. One notices also that any convex combination of W_i and in particular $\varrho_{AB}^{T_B}$ itself are also entanglement witnesses.

Let us comment shortly on the properties of entanglement witnesses. First, it is clear that they have negative eigenvalues, as otherwise their mean value on all entangled states would be positive. Second, since entanglement witnesses are Hermitian, they can be treated as physical observables—which means that separability criteria

---

[4] Let $\mathcal{H}$ be some Hilbert space. Then the set $\mathcal{B}(\mathcal{H})$ of linear bounded operators acting on $\mathcal{H}$ is also a Hilbert space with the Hilbert-Schmidt scalar product $\langle A|B\rangle = \mathrm{tr}(A^\dagger B)$ $(A, B \in \mathcal{B}(\mathcal{H}))$.

based on entanglement witnesses are interesting from the experimental point of view. Third, even if conceptually easy, entanglement witnesses depend on states in the sense that there exist entangled states that are only detected by different witnesses. Thus, in principle, the knowledge of all entanglement witnesses is necessary to detect all entangled states.

6.2.6 Positive Maps and the Entanglement Problem

Transposition is not the only map that can be used to deal with the separability problem. It is rather clear that the statement of theorem 2 remains true if, instead of the transposition map, one uses any map that when applied to a positive operator gives again a positive operator (a *positive map*). Remarkably, as shown in Ref. [24], positive maps give not only necessary but also sufficient conditions for separability and entanglement detection. Moreover, *via* the Jamiołkowski–Choi isomorphism, theorem 5 can be restated in terms of positive maps. To see this in more detail we need to review a bit of terminology.

We say that a map $\Lambda : \mathcal{B}(\mathcal{H}_A) \rightarrow \mathcal{B}(\mathcal{H}_B)$ is linear if $\Lambda(\alpha X + \beta Y) = \alpha \Lambda(X) + \beta \Lambda(Y)$ for any pair of operators X, Y acting on $\mathcal{H}_A$ and complex numbers α, β. We also say that Λ is Hermiticity–preserving (trace–preserving) if $\Lambda(X^\dagger) = [\Lambda(X)]^\dagger$ ($\mathrm{tr}[\Lambda(X)] = \mathrm{tr}(X)$) for any Hermitian $X \in \mathcal{B}(\mathcal{H}_A)$.

Definition 4 A linear map $\Lambda : \mathcal{B}(\mathcal{H}_A) \rightarrow \mathcal{B}(\mathcal{H}_B)$ is called positive if for all positive $X \in \mathcal{B}(\mathcal{H}_A)$ the operator $\Lambda(X) \in \mathcal{B}(\mathcal{H}_B)$ is positive.

As every Hermitian operator can be written as a difference of two positive operators, any positive map is also Hermiticity–preserving. On the other hand, a positive map does not have to be necessarily trace–preserving.

It follows immediately from the above definition that positive maps applied to density matrices give (usually unnormalized) density matrices. One could then expect that positive maps are sufficient to describe all quantum operations (as for instance measurements). This, however, is not enough, as it may happen that the considered system is only part of a larger one and we must require that any quantum operation on our system leaves the global system in a valid physical state. This requirement leads us to the notion of completely positive maps:

Definition 5 Let $\Lambda : \mathcal{B}(\mathcal{H}_A) \rightarrow \mathcal{B}(\mathcal{H}_B)$ be a positive map and let $I_d : M_d(\mathbb{C}) \rightarrow M_d(\mathbb{C})$ denote an identity map. Then, we say that Λ is completely positive if for all d the extended map $I_d \otimes \Lambda$ is positive.

Let us illustrate the above definitions with some examples.

Example 2 (Hamiltonian evolution of a quantum state) Let $\mathcal{H}_A = \mathcal{H}_B = \mathcal{H}$ and let $\Lambda_U : \mathcal{B}(\mathcal{H}) \rightarrow \mathcal{B}(\mathcal{H})$ be defined as $\Lambda_U(X) = UXU^\dagger$ for any $X \in \mathcal{B}(\mathcal{H})$, with U being some unitary operation acting on $\mathcal{H}$. Since unitary operations do not change eigenvalues when applied to X, it is clear that Λ_U is positive for any such U.

Furthermore, Λ_U is completely positive: an application of the extended map $I_d \otimes \Lambda_U$ to $X \in \mathcal{B}(\mathcal{H} \otimes \mathcal{H})$ gives $(I_d \otimes \Lambda_U)(X) = (\mathbb{1}_d \otimes U)X(\mathbb{1}_d \otimes U)^\dagger$, where $\mathbb{1}_d$ denotes identity acting on $\mathcal{H}$. Therefore, the extended unitary $\tilde{U} = \mathbb{1}_d \otimes U$ is also unitary. Thus, if $X \geq 0$, then $\tilde{U} X \tilde{U}^\dagger \geq 0$. The commonly known example of Λ_U is the unitary evolution of a quantum state $\varrho(t) = U(t)\varrho(0)[U(t)]^\dagger = \Lambda_{U(t)}(\varrho(0))$.

Example 3 (Transposition map) The second example of a linear map is the already considered transposition map T. It is easy to check that T is Hermiticity and trace–preserving. However, the previously discussed example of partially transposed pure entangled states shows that it cannot be completely positive.

To complete the characterization of positive and completely positive maps let us just mention the *Choi–Kraus–Stinespring representation*. Recall first that any linear Hermiticity–preserving (and so positive) map $\Lambda : \mathcal{B}(\mathbb{C}^d) \rightarrow \mathcal{B}(\mathbb{C}^d)$ can be represented as [27]:

$$\Lambda(X) = \sum_{i=1}^{k} \eta_i V_i X V_i^\dagger, \tag{6.8}$$

where $k \leq d^2$, $\eta_i \in \mathbb{R}$, and $V_i : \mathbb{C}^d \rightarrow \mathbb{C}^d$ are orthogonal in the Hilbert–Schmidt scalar product $\mathrm{tr}(V_i^\dagger V_j) = \delta_{ij}$. In this representation, completely positive maps are those (and only those) that have $\eta_i \geq 0$ for all i. As a result, by replacing $W_i = \sqrt{\eta_i} V_i$ (which preserves the orthogonality of W_i), we arrive at the aforementioned form for completely positive maps [28–30].

Theorem 6 *A linear map* $\Lambda : \mathcal{B}(\mathbb{C}^d) \rightarrow \mathcal{B}(\mathbb{C}^d)$ *is completely positive iff admits the Choi–Kraus–Stinespring form*

$$\Lambda(X) = \sum_{i=1}^{k} V_i X V_i^\dagger, \tag{6.9}$$

where $k \leq d^2$ *and* $V_i : \mathbb{C}^d \rightarrow \mathbb{C}^d$, *called usually Kraus operators, are orthogonal in the Hilbert–Schmidt scalar product.*

Finally, let us recall the so-called Choi–Jamiołkowski isomorphism [28, 31]: every linear operator X acting on $\mathbb{C}^d \otimes \mathbb{C}^D$ can be represented as $X = (I \otimes \Lambda)(P_+^{(d)})$ with some linear map $\Lambda : \mathcal{B}(\mathbb{C}^d) \rightarrow \mathcal{B}(\mathbb{C}^D)$. With this isomorphism, entanglement witnesses correspond to positive maps. Notice also that the dual form of this isomorphism reads $\Lambda(X) = \mathrm{tr}_B[W(\mathbb{1}_A \otimes X^T)]$.

Equipped with new definitions and theorems, we can now continue with the relationship between positive maps and the separability problem. It should be clear by now that theorem 2 is just a special case of a more general necessary condition for separability: if ϱ_{AB} acting on $\mathcal{H}_A \otimes \mathcal{H}_B$ is separable, then $(I \otimes \Lambda)(\varrho_{AB})$ is positive for any positive map Λ. In a seminal paper in 1996 [24], the Horodeckis showed that positive maps also give a *sufficient* condition for separability. More precisely, they proved the following [24]:

Theorem 7 *A state $\rho_{AB} \in \mathcal{B}(\mathbb{C}^{d_A} \otimes \mathbb{C}^{d_B})$ is separable if and only if the condition*

$$(I \otimes \Lambda)(\rho_{AB}) \geq 0 \qquad (6.10)$$

holds for all positive maps $\Lambda : \mathcal{B}(\mathbb{C}^{d_B}) \to \mathcal{B}(\mathbb{C}^{d_A})$.

Proof The "only if" part goes along exactly the same lines as proof of theorem 2, where instead of the transposition map we put Λ. On the other hand, the "if" part is much more involved. Assuming that ϱ_{AB} is entangled, we show that there exists a positive map $\Lambda : \mathcal{B}(\mathbb{C}^{d_B}) \to \mathcal{B}(\mathbb{C}^{d_A})$ such that $(I \otimes \Lambda)(\varrho_{AB}) \not\geq 0$. For this we can use theorem 5, which says that for any entangled ϱ_{AB} there always exists entanglement witness W detecting it, i.e., $\mathrm{tr}(W\varrho_{AB}) < 0$. Denoting by $L : \mathcal{B}(\mathbb{C}^{d_A}) \to \mathcal{B}(\mathbb{C}^{d_B})$ a positive map corresponding to the witness W via the Choi–Jamiołkowski isomorphism, i.e., $W = (I \otimes L)(P_+^{(d_A)})$, we can rewrite this condition as

$$\mathrm{tr}[(I \otimes L)(P_+^{(d_A)})\varrho_{AB}] < 0. \qquad (6.11)$$

As L is positive it can be represented as in Eq. 6.8, and hence the above may be rewritten as $\mathrm{Tr}[P_+^{(d_A)}(I \otimes L^\dagger)(\varrho_{AB})]$ with $L^\dagger : \mathcal{B}(\mathbb{C}^{d_B}) \to \mathcal{B}(\mathbb{C}^{d_A})$ called the dual map of L. One immediately checks that dual maps of positive maps are positive. This actually finishes the proof since we showed that there exists a positive map $\Lambda = L^\dagger$ such that $(I \otimes \Lambda)(\varrho_{AB}) \not\geq 0$. □

In conclusion, we have two equivalent characterizations of separability in bipartite systems, in terms of either entanglement witnesses or positive maps. However, on the level of a particular entanglement witness and the corresponding map, both characterizations are no longer equivalent. This is because usually maps are stronger in detection than entanglement witnesses (see Ref. [32]). A good example comes from the two qubit case. On one hand, theorem 3 tell us that the transposition map detects all the two-qubit entangled states. On the other hand, it is clear that the corresponding witness, the so-called swap operator (see Ref. [9]) $V = P_+^{(2)\Gamma}$ does not detect all entangled states—as for instance $\mathrm{tr}(P_+^{(2)} V) \geq 0$.

Let us also notice that an analogous theorem was proven in Ref. [32], which gave a characterization of the set of the fully separable multipartite states

$$\varrho_{A_1...A_N} = \sum_i p_i \varrho_{A_1}^{(i)} \otimes \cdots \otimes \varrho_{A_N}^{(i)} \qquad (6.12)$$

in terms of multipartite entanglement witnesses. Here, however, instead of positive maps one deals with maps which are positive on products of positive operators.

6.2.7 Positive Maps and Entanglement Witnesses: Further Characterization and Examples

We discuss here the relationship between positive maps (or the equivalent entanglement witnesses) and the separability problem.

Definition 6 Let $\Lambda : \mathcal{B}(\mathcal{H}_A) \to \mathcal{B}(\mathcal{H}_B)$ be a positive map. We call it decomposable if it admits the form[5] $\Lambda = \Lambda_1^{CP} + \Lambda_2^{CP} \circ T$, where Λ_i^{CP} ($i = 1, 2$) are some completely positive maps. Otherwise Λ is called indecomposable.

It follows from this definition that decomposable maps are useless for detection of PPT entangled states. To see this explicitly, assume that ϱ_{AB} is PPT entangled. Then it holds that $(I \otimes \Lambda)(\varrho_{AB}) = (I \otimes \Lambda_1^{CP})(\varrho_{AB}) + (I \otimes \Lambda_2^{CP})(\varrho_{AB}^{T_B}) = (I \otimes \Lambda_1^{CP})(\varrho_{AB}) + (I \otimes \Lambda_2^{CP})(\widetilde{\varrho}_{AB})$, where $\widetilde{\varrho}_{AB} = \varrho_{AB}^{T_B}$ is some quantum state. Since Λ_i^{CP} are completely positive, both terms are positive and thus $(I \otimes \Lambda)(\varrho_{AB}) \geq 0$ for any decomposable Λ and PPT entangled ϱ_{AB}.

The simplest example of a decomposable map is the transposition map, with both Λ_i^{CP} ($i = 1, 2$) being just the identity map. It is then clear that, from the point of view of entanglement detection, the transposition map is also the most powerful example of a decomposable map. Furthermore, as shown by Woronowicz [33], all positive maps from $\mathcal{B}(\mathbb{C}^2)$ and $\mathcal{B}(\mathbb{C}^3)$ to $\mathcal{B}(\mathbb{C}^2)$ are decomposable. Therefore, the partial transposition criterion is necessary and sufficient in two-qubit and qubit-qutrit systems as stated in theorem 3.

Using the Jamiołkowski–Choi isomorphism we can check the form of entanglement witnesses corresponding to the decomposable positive maps. One immediately sees that they can be written as $W = P + Q^{T_B}$, with P and Q being some positive operators. Following the nomenclature of positive maps, such witnesses are called decomposable.

It is then clear that PPT entangled states can only be detected by indecomposable maps, or, equivalently indecomposable entanglement witnesses (cf. Fig. 6.2). Still, however, there is no criterion that allows to judge unambiguously if a given PPT state is entangled.

To support the above discussion, we give particular examples of positive maps and corresponding entanglement witnesses.

Example 4 Let $\Lambda_r : \mathcal{B}(\mathbb{C}^d) \to \mathcal{B}(\mathbb{C}^d)$ be the so-called *reduction map* map defined through $\Lambda_r(X) = \text{tr}(X)\mathbb{1}_d - X$ for any $X \in \mathcal{B}(\mathbb{C}^d)$. It was introduced in Ref. [34] and considered first in the entanglement context in Refs. [35, 36]. One immediately finds that Λ_r is positive, but not completely positive, as it detects entanglement of $P_+^{(d)}$. Moreover, $\Lambda_r = \Lambda^{CP} \circ T$, where Λ^{CP} is a completely positive map with Kraus operators (cf. theorem 6) given by $V_{ij} = |i\rangle\langle j| - |j\rangle\langle i|$ ($i < j, i, j = 0, \ldots, d-1$), meaning that the reduction map is decomposable.

---

[5] By $\Lambda_1 \circ \Lambda_2$ we denote the composition of two maps Λ_i ($i = 1, 2$), i.e., a map that acts on a given operator X as $\Lambda_1 \circ \Lambda_2(X) = \Lambda_1(\Lambda_2(X))$.

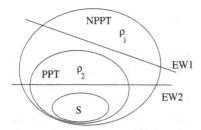

Fig. 6.2 Schematic view of the Hilbert-space with two states ρ_1 and ρ_2 and two witnesses *EW1* and *EW2*. *EW1* is a decomposable EW, and it detects only NPT states like ρ_1. *EW2* is an indecomposable EW, and it detects also some PPT states like ρ_2. Note that none of the witnesses detect *all* entangled states

Example 5 Let $\Lambda_{ext}^{U} : \mathcal{B}(\mathbb{C}^d) \to \mathcal{B}(\mathbb{C}^d)$ be the so-called extended reduction map [37, 38] defined by $\Lambda_{ext}^{U}(X) = \mathrm{tr}(X)\mathbb{1}_d - X - U X^T U^\dagger$, where U obeys $U^T = -U$ and $U^\dagger U \leq \mathbb{1}_d$. It is obviously positive but not completely positive. However, unlike the reduction map, this one is indecomposable as examples of PPT entangled states detected by Λ_{ext}^{U} can be found [37, 38].

Let us summarize our considerations with the following two theorems. First, using the definitions of decomposable and indecomposable entanglement witnesses, we can restate the consequences of the Hahn–Banach theorem in several ways [18, 24, 39–41]:

Theorem 8 *The following statements hold.*

1. *A state ρ_{AB} is entangled iff there exists an entanglement witness W such that* $\mathrm{tr}(W\rho_{AB}) < 0$.
2. *A state ρ_{AB} is PPT entangled iff there exists an indecomposable entanglement witness W such that* $\mathrm{tr}(W\rho_{AB}) < 0$.
3. *A state σ_{AB} is separable iff* $\mathrm{tr}(W\sigma_{AB}) \geq 0$ *for all entanglement witnesses.*

Notice that the Jamiołkowski–Choi isomorphism between positive maps and entanglement witnesses allows to rewrite immediately the above theorem in terms of positive maps. From a theoretical point of view, the theorem is quite powerful. However, it does not give any insight on how to construct for a given state ρ, the appropriate witness operator.

Second, the relations between maps and witnesses can be collected as follows [24, 31, 39–41].

Theorem 9 *Let W be a Hermitian operator and Λ_W map defined as $\Lambda_W(X) = \mathrm{tr}_B[W(\mathbb{1}_A \otimes X^T)]$. Then the following statements hold.*

1. $W \geq 0$ *iff Λ_W is a completely positive map.*
2. W *is an entanglement witness iff Λ_W is a positive map.*
3. W *is a decomposable entanglement witness iff Λ_W is decomposable map.*

6.2.8 Entanglement Measures

The criteria discussed above allow to check if a given state ϱ_{AB} is entangled. However, in general they do not tell us directly *how much* ϱ_{AB} is entangled. In what follows we discuss several methods to quantify entanglement of bipartite states. This quantification is necessary, at least partly because entanglement is viewed as a resource in quantum information theory. There are several complementary ways to quantify entanglement (see Refs. [8, 42–51] and references therein). We will present here three possible ways to do so.

Let us just say few words about the definition of entanglement measures.[6] The main ingredient in this definition is the monotonicity under LOCC operations. More precisely, if Λ denotes some LOCC operation, and E our candidate for the entanglement measure, E has to satisfy

$$E(\Lambda(\varrho)) \le E(\varrho) \tag{6.13}$$

or

$$\sum_i p_i E(\varrho_i) \le E(\varrho), \tag{6.14}$$

where ϱ_i are states resulting from the LOCC operation Λ appearing with probabilities p_i (as in the case of e.g. projective measurements). Both requirements follow from the very intuitive condition saying that entanglement should not increase under local operations and classical communication. It follows also that if E is convex, then the condition (6.14) implies (6.13), but not vice versa—therefore (6.14) gives a stronger condition for the monotonicity. For instance, the three examples of measures presented below satisfy this condition. Finally, notice that from the monotonicity under LOCC operations one also concludes that E is invariant under unitary operations, and gives a constant value on separable states (see e.g. Ref. [8]).

6.2.8.1 Entanglement of Formation

Consider a bipartite pure state $|\psi_{AB}\rangle \in \mathbb{C}^{d_A} \otimes \mathbb{C}^{d_B}$ shared between Alice and Bob. As shown by Bennett et al. [52], given $nE(|\psi_{AB}\rangle)$ copies of the maximally entangled state, Alice and Bob can by LOCC transform them into n copies of $|\psi_{AB}\rangle$, if n is large. Here

$$E(|\psi_{AB}\rangle) = S(\varrho_A) = S(\varrho_B) \tag{6.15}$$

with ϱ_A and ϱ_B being the local density matrices of $|\psi_{AB}\rangle$ and $S(\rho)$ stands for the von Neumann entropy of ρ given by $S(\rho) = -\mathrm{tr}\rho \log_2 \rho$. It clearly follows from theorem

---

[6] For a more detailed axiomatic description, and other properties of entanglement measures, the reader is encouraged to consult, e.g., Refs. [8, 50, 51].

1 that E is zero iff $|\psi_{AB}\rangle$ is separable, while its maximal value $\log_2 \min\{d_A, d_B\}$ is attained for the maximally entangled states (6.1).

For the two-qubit maximally entangled state $|\psi_+^{(2)}\rangle$, the function E gives one: an amount of entanglement also called *ebit*. With this terminology, one can say that $|\psi_{AB}\rangle$ has $E(|\psi_{AB}\rangle)$ ebits. Since $E(\psi_{AB})$ is the number of singlets required to prepare a copy of the state $|\psi_{AB}\rangle$, it is called *entanglement of formation* of $|\psi_{AB}\rangle$. We are therefore using the amount of entanglement of the singlet state as our unit of entanglement.

Following Ref. [42], let us now extend the definition of entanglement of formation to all bipartite states. By definition, any mixed state is a convex combination of pure states, i.e., $\varrho = \sum_i p_i |\psi_i\rangle\langle\psi_i|$, where probabilities p_i and pure states (not necessarily orthogonal) $|\psi_i\rangle$ constitute what is called an ensemble. A particular example of such an ensemble is the eigendecomposition of ϱ. Thus, it could be tempting to define the entanglement of formation of ϱ as an averaged cost of producing pure states from the ensemble, i.e., $\sum_i p_i E(|\psi_i\rangle)$. One knows, however, that there exist an infinite number of ensembles realizing any given ϱ. A natural solution is then to minimize the above function over all such ensembles—with which we arrive at the definition of entanglement of formation for mixed states [42]:

$$E(\varrho_{AB}) = \min_{\{p_i, |\psi_i\rangle\}} \sum_i p_i E(|\psi_i\rangle), \tag{6.16}$$

with the minimum taken over all ensembles $\{p_i, |\psi_i\rangle\}$ such that $\sum_i p_i |\psi_i\rangle\langle\psi_i| = \varrho_{AB}$.

In general, the above minimization makes the calculation of entanglement of formation extremely difficult. Nevertheless, it was determined for two-qubits [53, 54], or states having some symmetries, as the so-called isotropic [55] and Werner [56] states. In the first case it amounts to

$$E_F(\varrho_{AB}) = H\left(\frac{1 + \sqrt{1 - C^2(\varrho_{AB})}}{2}\right), \tag{6.17}$$

where $H(x) = -x \log_2 x - (1 - x) \log_2(1 - x)$ is the binary entropy function. The function C is given by

$$C(\varrho_{AB}) = \max\{0, \lambda_1 - \lambda_2 - \lambda_3 - \lambda_4\}. \tag{6.18}$$

with $\lambda_1, \ldots, \lambda_4$ the eigenvalues of the Hermitian matrix $[(\varrho_{AB})^{1/2} \tilde{\varrho}_{AB} (\varrho_{AB})^{1/2}]^{1/2}$ in decreasing order, and $\tilde{\varrho}_{AB} = \sigma_y \otimes \sigma_y \varrho_{AB}^* \sigma_y \otimes \sigma_y$. Note that the complex conjugation over ϱ is taken in the σ_z eigenbasis, and σ_y denotes the well-known Pauli matrix.[7] The function C, called *concurrence*, can also be used to quantify entanglement of more general quantum states. Although Eq. 6.18 gives the explicit form of concurrence only for two-qubit states, it can also be defined for arbitrary bipartite states—as we shall discuss in the following section.

---

[7] In the standard basis σ_y is given by $\sigma_y = -i|0\rangle\langle 1| + i|1\rangle\langle 0|$.

6.2.8.2 Concurrence

For any $|\psi_{AB}\rangle \in \mathbb{C}^{d_A} \otimes \mathbb{C}^{d_B}$ we define concurrence as $C(|\psi_{AB}\rangle) = \sqrt{2(1 - \mathrm{tr}\varrho_r^2)}$ where ϱ_r is one of the subsystems of $|\psi_{AB}\rangle$ (note that the value of C does not depend on the choice of subsystems) [57]. In the case $d_A = d_B = d$, one sees that its value for pure states ranges from 0 for separable states to $\sqrt{2(1 - 1/d)}$ for the maximally entangled state.

The extension to mixed states goes in exactly the same way as in the case of entanglement of formation,

$$C(\varrho_{AB}) = \min_{\{p_i, |\psi_i\rangle\}} \sum_i p_i C(|\psi_i\rangle), \tag{6.19}$$

where again the minimization is taken over all the ensembles that realize ϱ_{AB}. For the same reason, as in the case of EOF, concurrence is calculated analytically only in few instances like two-qubit states [53, 54] and isotropic states [58].

Seemingly, the only difference between E and C lies in the function taken to define both measures for pure states. However, the way concurrence is defined enables one to determine it experimentally for pure states [59, 60], provided that two copies of the state are available simultaneously.

6.2.8.3 Negativity and Logarithmic Negativity

Based on the previous examples of entanglement measures, one may get the impression that all of them are difficult to determine. Even if this is true in general, there are entanglement measures that can be calculated for arbitrary states. The examples we present here are *negativity* and *logarithmic negativity*. The first one is defined as [21, 61]:

$$N(\varrho_{AB}) = \frac{1}{2} \left(\left\| \varrho_{AB}^\Gamma \right\| - 1 \right). \tag{6.20}$$

The calculation of N even for mixed states reduces to determination of eigenvalues of $\varrho_{AB}^{T_B}$, and amounts to the sum of the absolute values of negative eigenvalues of $\varrho_{AB}^{T_B}$. This measure has a disadvantage: partial transposition does not detect PPT entangled states; therefore N is zero not only for separable states but also for all PPT states.

The logarithmic negativity is defined as [61]:

$$E_N(\varrho_{AB}) = \log_2 \left\| \varrho_{AB}^\Gamma \right\| = \log_2[2N(\varrho_{AB}) + 1]. \tag{6.21}$$

It was shown in Ref. [62] that it satisfies condition (6.14). Moreover, logarithmic negativity is additive, i.e., $E(\varrho_{AB} \otimes \sigma_{AB}) = E(\varrho_{AB}) + E(\sigma_{AB})$ for any pair of density matrices ϱ_{AB} and σ_{AB}, which is a desirable feature. However, this comes at a cost: E_N is not convex [62]. Furthermore, for the same reason as negativity it cannot be used to quantity entanglement of PPT entangled states. Finally, let us notice that these measures range from zero for separable states, to $(d-1)/2$ for negativity and $\log_2 d$ for logarithmic negativity.

6.3 Area Laws

Area laws play a very important role in many areas of physics, since generically relevant states of physical systems described by local Hamiltonians (both quantum and classical) fulfill them. This goes back to the seminal work on the free Klein–Gordon field [63, 64], where it was suggested that the area law of geometric entropy might be related to the physics of black holes, and in particular the Bekenstein-Hawking entropy that is proportional to the area of the black hole surface [65, 66, 67]. The related *holographic principle* [68] says that information about a region of space can be represented by a theory which lives on a boundary of that region. In recent years there has been a wealth of studies of area laws, and there are excellent reviews [69] and special issues [70] about the subject. As pointed out by the authors of Ref. [69], the interest in area laws is particularly motivated by the four following issues:

- The holographic principle and the entropy of black holes,
- Quantum correlations in many-body systems,
- Computational complexity of quantum many-body systems,
- Topological entanglement entropy as an indicator of topological order in certain many-body systems

6.3.1 Mean Entanglement of Bipartite States

Before we turn to the area laws for physically relevant states let us first consider a *generic* pure state in the Hilbert space in $\mathbb{C}^m \otimes \mathbb{C}^n$ ($m \leq n$). Such a generic state (normalized, i.e. unit vector) has the form

$$|\Psi\rangle = \sum_{i=1}^{m} \sum_{j=1}^{n} \alpha_{ij}|i\rangle|j\rangle, \tag{6.22}$$

where the complex numbers α_{ij} may be regarded as random variables distributed uniformly on a hypersphere, i.e. distributed according to the probability density

$$P(\alpha) \propto \delta\left(\sum_{i=1}^{m} \sum_{j=1}^{n} |\alpha_{ij}|^2 - 1\right), \tag{6.23}$$

with the only constraint being the normalization. As we shall see, such a generic state fulfills on average a "volume" rather than an area law. To this aim we introduce a somewhat more rigorous description, and we prove that on average, the entropy of one of subsystems of bipartite pure states in $\mathbb{C}^m \otimes \mathbb{C}^n$ ($m \leq n$) is almost maximal for sufficiently large n. In other words, typical pure states in $\mathbb{C}^m \otimes \mathbb{C}^n$ are almost maximally entangled. This "typical behavior" of pure states happens to be completely atypical for ground states of local Hamiltonians with an energy gap between ground and first excited eigenstates.

Rigorously speaking, the average with respect to the distribution (6.23) should be taken with respect to the unitarily invariant measure on the projective space $\mathbb{C}P^{mn-1}$. It is a unique measure generated by the Haar measure on the unitary group by applying the unitary group on an arbitrarily chosen pure state. One can show then that the eigenvalues of the first subsystem of a randomly generated pure state $|\psi_{AB}\rangle$ are distributed according to the following probability distribution [71–73] (see also Ref. [74]):

$$P_{m,n}(\lambda_1, \ldots, \lambda_m) = C_{m,n}\delta\left(\sum_i \lambda_i - 1\right)\prod_i \lambda_i^{n-m}\prod_{i<j}(\lambda_i - \lambda_j)^2, \qquad (6.24)$$

where the delta function is responsible for the normalization, and the normalization constant reads (see e.g. Ref. [74])

$$C_{m,n} = \frac{\Gamma(mn)}{\prod_{i=0}^{m-1}\Gamma(n-i)\Gamma(m-i+1)} \qquad (6.25)$$

with Γ being the Euler gamma function.[8]

Theorem 10 *Let $|\psi_{AB}\rangle$ be a bipartite pure state from $\mathbb{C}^m \otimes \mathbb{C}^n$ ($m \leq n$) drawn at random according to the Haar measure on the unitary group and $\varrho_A = \mathrm{tr}_B|\psi_{AB}\rangle\langle\psi_{AB}|$ be its subsystem acting on $\mathbb{C}^m$. Then,*

$$\langle S(\varrho_A)\rangle \approx \log m - \frac{m}{2n}. \qquad (6.27)$$

Proof Let us give here just an intuitive proof without detailed mathematical discussion (which can be found e.g. in Refs. [71–77]). Our aim is to estimate the following quantity

$$\langle S(\varrho_A)\rangle = -\int\left(\sum_{i=1}^m \lambda_i \log \lambda_i\right)P(\lambda_1, \ldots, \lambda_m)\mathrm{d}\lambda_1 \ldots \mathrm{d}\lambda_1, \qquad (6.28)$$

---

[8]   In general the gamma function is defined through

$$\Gamma(z) = \int_0^\infty t^{z-1}e^{-t}\mathrm{d}t \qquad (z \in \mathbb{C}). \qquad (6.26)$$

For z being positive integers $z=n$ the gamma function is related to the factorial function via $\Gamma(n) = (n-1)!$

where the probability distribution $P(\lambda_1, \ldots, \lambda_m)$ is given by Eq. 6.24. We can always write the eigenvalues λ_i as $\lambda_i = 1/m + \delta_i$, where $\delta_i \in \mathbb{R}$ and $\sum_i \delta_i = 0$. This allows us to expand the logarithm into the Taylor series in the neighborhood of $1/m$ as

$$\log\left(\frac{1}{m} + \delta_i\right) = -\log m + \sum_{k=1}^{\infty} \frac{(-1)^{k-1}}{k}(m\delta_i)^k, \tag{6.29}$$

which after application to Eq. 6.28 gives the following expression for the mean entropy

$$\langle S(\varrho_A)\rangle = \log m - \frac{m}{1 \cdot 2}\left\langle \sum_i \delta_i^2\right\rangle + \frac{m^2}{2 \cdot 3}\left\langle \sum_i \delta_i^3\right\rangle - \frac{m^3}{3 \cdot 4}\left\langle \sum_i \delta_i^4\right\rangle - \cdots . \tag{6.30}$$

Let us now notice that $\mathrm{tr}\varrho_A^2 = \sum_i \lambda_i^2 = \sum_i (\delta_i + 1/m)^2 = \sum_i \delta_i^2 + 1/m$, and therefore $\sum_i \delta_i^2 = \mathrm{tr}\varrho_A^2 - 1/m$. This, after substitution in the above expression, together with the fact that for sufficiently large n we can omit terms with higher powers of δ_i (cf. [71]), leads us to

$$\langle S(\varrho_A)\rangle \approx \log m - \frac{m}{2}\left\langle \mathrm{tr}\varrho_A^2 - \frac{1}{m}\right\rangle. \tag{6.31}$$

One knows that $\mathrm{tr}\varrho_A^2$ denotes the purity of ϱ_A. Its average was calculated by Lubkin [71] and reads

$$\left\langle \mathrm{tr}\varrho_A^2\right\rangle = \frac{m+n}{mn+1}. \tag{6.32}$$

Substitution in Eq. 6.31 leads to the desired results, completing the proof.   □

Two remarks should be made before discussing the area laws. First, it should be pointed out that it is possible to get analytically the exact value of $\langle S\rangle$. There is a series of papers [75–77] presenting different approaches leading to

$$\langle S(\varrho)\rangle = \Psi(mn+1) - \Psi(n+1) - \frac{m-1}{2n} \tag{6.33}$$

with Ψ being the bigamma function.[9] Using now the fact that $\Psi(z+1) = \Psi(z)+1/z$, and the asymptotic properties of bigamma function, $\Psi(z) \approx \log z$, we get (6.27).

---

[9] The bigamma function is defined as $\Psi(z) = \Gamma'(z)/\Gamma(z)$ and for natural $z = n$ it takes the form

$$\Psi(n) = -\gamma + \sum_{k=1}^{n-1} \frac{1}{n} \tag{6.34}$$

with γ being the Euler constant, of which exact value is not necessary for our consideration as it vanishes in Eq. 6.33.

Fig. 6.3 Schematic
representation of a lattice
system L, an arbitrary region
R (denoted in light *grey*
background), and its
boundary ∂R (denoted in
dark *grey* background)

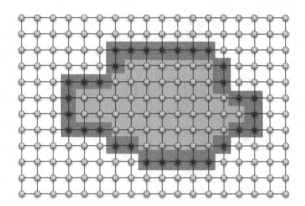

Second, notice that the exact result of Lubkin (6.32) can be estimated by relaxing
the normalization constraint in the distribution (6.23), and replacing it by a product
of independent Gaussian distributions, $P(\alpha) = \prod_{i,j}(nm/\pi)\exp[-nm|\alpha_{ij}|^2]$, with
$\langle \alpha_{ij} \rangle = 0$, and $\langle |\alpha_{ij}|^2 \rangle = 1/nm$. The latter distribution, according to the central
limit theorem, tends for $nm \to \infty$ to a Gaussian distribution for $\sum_{i=1}^{m}\sum_{j=1}^{n}|\alpha_{ij}|^2$
centered at 1, with width $\simeq 1/\sqrt{nm}$. One obtains then straightforwardly $\langle \mathrm{tr}\varrho_A \rangle = 1$,
and after a little more tedious calculation $\langle \mathrm{tr}\varrho_A^2 \rangle = (n+m)/nm$, which agrees
asymptotically with the Lubkin result for $nm \gg 1$.

6.3.2 Area Laws in a Nutshell

In what follows we shall be concerned with lattices L in D spatial dimensions,
$L \subseteq \mathbb{Z}^D$. At each site we have a d-dimensional physical quantum system (one
can, however, consider also classical lattices, with a d-dimensional classical spin
at each site with the configuration space $\mathbb{Z}_d = \{0, \ldots, d-1\}$) at each site.[10]
The distance between two sites x and y of the lattice is defined as

$$\mathcal{D}(x, y) = \max_{1 \leq i \leq D}|x_i - y_i|. \qquad (6.35)$$

Accordingly, we define the distance between two disjoint regions X and Y of L as
the minimal distance between all pairs of sites $\{x, y\}$, where $x \in X$ and $y \in Y$; i.e.,
$\mathcal{D}(X, Y) = \min_{x \in X}\min_{y \in Y}\mathcal{D}(x, y)$. If R is some region of L, we define its boundary
∂R as the set of sites belonging to R whose distance to $L \setminus R$ (the complement of R)
is one. Formally, $\partial R = \{x \in R | \mathcal{D}(x, L \setminus R) = 1\}$. Finally, by $|R|$ we denote number
of sites (or volume) in the region R (see Fig. 6.3).

Now, we can add some physics to our lattice by assuming that interactions between
the sites of L are governed by some hamiltonian H. We can divide the lattice L into

---

[10] For results concerning other kind of systems one can consult Ref. [69].

two parts, the region R and its complement $L \setminus R$. Roughly speaking, we aim to understand how the entropy of the subsystem R scales with its size. In particular, we are interested in the entropy of the state ϱ_R reduced from a ground state or a thermal state of the Hamiltonian H. We say that the entropy satisfies an area law if it scales at most as the boundary area,[11] i.e.,

$$S(\varrho_R) = O(|\partial R|). \tag{6.36}$$

6.3.2.1 One-Dimensional Systems

Let us start with the simplest case of one-dimensional lattices, $L = \{1, \ldots, N\}$. Let R be a subset of L consisting of n contiguous spins starting from the first site, i.e., $R = \{1, \ldots, n\}$ with $n < N$. In this case the boundary ∂R of the region R contains one spin for open boundary conditions, and two for periodic ones. Therefore, in this case the area law is extremely simple:

$$S(\varrho_R) = O(1). \tag{6.37}$$

The case of $D = 1$ seems to be quite well understood. In general, all local gapped systems (away from criticality) satisfy the above law, and there might be a logarithmic divergence of entanglement entropy when the system is critical. To be more precise, let us recall the theorem of Hastings leading to the first of the above statements, followed by examples of critical systems showing a logarithmic divergence of the entropy with the size of R.

Consider the nearest-neighbor interaction Hamiltonian

$$H = \sum_{i \in L} H_{i,i+1}, \tag{6.38}$$

where each $H_{i,i+1}$ has a nontrivial support only on the sites i and $i+1$. We assume also that the operator norm of all the terms in Eq. 6.38 are upper bounded by some positive constant J, i.e., $\|H_{i,i+1}\| \leq J$ for all i (i.e., we assume that the interaction strength between ith site and its nearest-neighbor is not greater that some constant). Under these assumptions, Hastings proved the following [78]:

Theorem 11 *Let L be a one-dimensional lattice with N d-dimensional sites, and let H be a local Hamiltonian as in Eq. 6.38. Assuming that H has a unique ground state separated from the first excited states by the energy gap $\Delta E > 0$, the entropy of any region R satisfies*

$$S(\varrho_R) \leq 6c_0 \xi 2^{6\xi \log d} \log \xi \log d \tag{6.39}$$

---

[11] Let us shortly recall that the notation $f(x) = O(g(x))$ means that there exist a positive constant c and $x_0 > 0$ such that for any $x \geq x_0$ it holds that $f(x) \leq cg(x)$.

with c_0 denoting some constant of the order of unity and $\xi = \min\{2v/\Delta E, \xi_C\}$. Here, v denotes the sound velocity and is of the order of J, while ξ_C is a length scale of order unity.

Let us remark that both constants appearing in the above theorem come from the Lieb–Robinson bound [79] (see also Ref. [80] for a recent simple proof of this bound).

This theorem tells us that when the one-dimensional system with the local inter-action defined by Eq. 6.38 is away from the criticality ($\Delta E > 0$), the entropy of R is bounded by some constant independent of $|R|$—even if this bound does not have to be tight. Of course, we can naturally ask if there exist gapped systems with long-range interaction violating (6.37). This was answered in the affirmative in Ref. [81, 82], which gave examples of one-dimensional models with long–range interactions, nonzero energy gap, and scaling of entropy diverging logaritmically with n.

The second question one could pose is about the behavior of the entropy when the gap ΔE goes to zero and the system becomes critical. Numerous analytical and numerical results show that usually one observes a logarithmic divergence of $S(\varrho_R)$ with the size of the region R. Here we recall only the results obtained for the so-called XY model in a transverse magnetic field (for the remaining ones we refer the reader to recent reviews [69, 83], and to the special issue of J. Phys. A devoted to this subject [70]).

The Hamiltonian for the XY model reads

$$H_{XY} = -\frac{1}{2}\sum_{i \in L}\left(\frac{1+\gamma}{2}\sigma_i^x\sigma_{i+1}^x + \frac{1-\gamma}{2}\sigma_i^y\sigma_{i+1}^y\right) - \frac{h}{2}\sum_{i \in L}\sigma_i^z, \qquad (6.40)$$

with $0 \le \gamma \le 1$ the anisotropy parameter, and h the magnetic field. In the case of vanishing anisotropy ($\gamma = 0$), we have the isotropic XY model called shortly XX model, while for $\gamma = 1$ one recovers the well–known Ising Hamiltonian in a transverse field. The Hamiltonian H_{XY} is critical when either $\gamma = 0$ and $|h| \le 1$ (the critical XX model) or for $|h| = 1$.

It was shown in a series of papers [84–87] that for the critical XY model (that is when $\gamma \ne 0$ and $|h| = 1$) the entropy of the region $R = \{1, \ldots, n\}$ scales as

$$S(\varrho_R) = \frac{1}{6}\log_2 n + O(1), \qquad (6.41)$$

while for the critical XX model, the constant multiplying the logarithms becomes one–third. Then, in the case of the critical Ising model ($\gamma = 1$), it can be shown that the entropy scales at least logaritmically,[12] i.e., $S(\varrho_R) = \Omega(\log_2 n)$ [69, 88].

Concluding, let us mention that there is an extensive literature on the logarithmic scaling of the block entropy using conformal field theory methods (see Ref. [89] for a

---

[12] The notation $f(x) = \Omega(g(x))$ means that there exist $c > 0$ and $x_0 > 0$ such that $f(x) \ge cg(x)$ for all $x \ge x_0$.

very good overview of these results). Quite generally, the block entropy at criticality scales as

$$S(\varrho_R) = \frac{c}{3} \log_2 \left(\frac{|R|}{a} \right) + O(1), \qquad (6.42)$$

or, more in general for the Rényi entropy[13]

$$S_\alpha(\varrho_R) = (c/6)\,(1 + 1/\alpha) \log_2(|R|/a) + O(1), \qquad (6.44)$$

where c is called the *central charge* of the underlying conformal field theory, and a is the cutoff parameter (the lattice constant for lattice systems).

6.3.2.2 Higher-Dimensional Systems

The situation is much more complex in higher spatial dimensions $(D > 1)$. The boundary ∂R of the general area law, Eq. 6.36, is no longer a simple one or two–element set and can have a rather complicated structure. Even if there are no general rules discovered so far, it is rather believed that (6.36) holds for ground states of local gapped Hamiltonians. This intuition is supported by results showing that for quadratic quasifree fermionic and bosonic lattices the area law (6.36) holds [69]. Furthermore, for critical fermions the entropy of a cubic region $R = \{1, \ldots, n\}^D$ is bounded as $\gamma_1 n^{D-1} \log_2 n \leq S(\varrho_R) \leq \gamma_2 n^{D-1}(\log_2 n)^2$ with γ_i $(i = 1, 2)$ denoting some constants [90–92]. Let us notice that the proof of this relies on the fact that logarithmic negativity (see Eq. 6.21) upper bounds the von Neumann entropy, i.e., for any pure bipartite state $|\psi_{AB}\rangle$, the inequality $S(\varrho_{A(B)}) \leq E_N(|\psi_{AB}\rangle)$ holds. This in turn is a consequence of monotonicity of the Rényi entropy S_α with respect to the order α, i.e., $S_\alpha \leq S_{\alpha'}$ for $\alpha \geq \alpha'$. This is one of the instances where insights from quantum information help to deal with problems in many–body physics.

Interestingly, very recently Masanes [80] showed that the ground state (and also low–energy eigenstates) entropy of a region R (even a disjoint one) always scales at most as the size of the boundary of R with some correction proportional to $(\log |R|)^D$—as long as the Hamiltonian H is of the local form

$$H = \sum_{i \in L} H_i, \qquad (6.45)$$

where each H_i has nontrivial support only on the nearest-neighbors of the ith site, and satisfies as previously $\|H_i\| \leq J$ for some $J > 0$. Thus, the behavior of entropy

---

[13]   Recall that the quantum Rényi entropy is defined as

$$S_\alpha = \frac{1}{1 - \alpha} \log_2 \left[\mathrm{Tr} \left(\varrho^\alpha \right) \right] \qquad (6.43)$$

where $\alpha \in [0, \infty]$. For $\alpha = 0$ one has $S_0(\varrho) = \log_2 \mathrm{rank}(\varrho)$ and $S_\infty = -\log_2 \lambda_{max}$ with λ_{max} being the maximal eigenvalue of ϱ.

which is considered to be a violation of the area law (6.36) can in fact be treated as an area law itself. This is because in this case[14] $[|\partial R|(\log|R|)^k]/|R| \to 0$ for $|R| \to \infty$ with some $k > 0$, meaning that still this behavior of entropy is very different from the typical behavior following from theorem 10. That is, putting $m = d^{|R|}$ and $n = d^{|L \setminus R|}$ with $|L| \gg |R|$ one has that $S(\varrho_R)/|R|$ is arbitrarily close to $\log d$ for large $|R|$.

Let R_1 and R_2 be two disjoint regions of the lattice such that $|R_1| \le |R_2|$, and let l denote the distance between these regions. Let us call Γ a function that bounds from above the correlations between two operators X and Y ($\|X\|, \|Y\| \le 1$) acting respectively on R_1 and R_2, i.e., $C(X, Y) = |\langle XY \rangle - \langle X \rangle \langle Y \rangle| \le \Gamma(l, |R_1|)$. The first assumption leading to the results of Ref. [80] is that if the mean values in C are taken in the ground state of H, Γ is given by

$$\Gamma(l, |R_1|) = c_1(l - \xi \log|R_1|)^{-\mu} \tag{6.46}$$

with some constants c_1, ξ, and $\mu > D$. Notice that this function decays polynomially in L, meaning that this first assumption is weaker than the property of exponential decay observed in Ref. [93] for gapped Hamiltonians.

Let now H_R denote a part of the global Hamiltonian H which acts only on sites in some region R. It has its own eigenvalues and eigenstates, denoted by e_n and $|\psi_n\rangle$ respectively, with e_0 denoting the lowest eigenvalue. The second assumption made in Ref. [80] is that there exist constants c_2, τ, γ, and η such that for any region R and energy $e = 2J3^D|\partial R| + e_0 + 40v$ (notice that 3^D is the number of the first neighbours in a cubic lattice), the number of eigenenergies of H_R lower than e is upper bounded as

$$\Omega_R(e) \le c_2(\tau|R|)^{\gamma(e-e_0)+\eta|\partial R|}. \tag{6.47}$$

Now, we are in position to formulate the main result of Ref. [80].

Theorem 12 *Let R be some arbitrary (even disjoint) region of L. Then, provided the assumptions (6.46) and (6.47) hold, the entropy of the reduced density matrix ϱ_R of the ground state of H satisfies*

$$S(\varrho_R) \le C|\partial R|(10\xi \log|R|)^D + O(|\partial R|(\log|R|)^{D-1}), \tag{6.48}$$

where C collects the constants D, ξ, γ, J, η, and d. If R is a cubic region, the above statement simplifies, giving $S(\varrho_R) \le \widetilde{C}|\partial R| \log|R| + O(|\partial R|)$ with $\widetilde{C}$ being some constant.

Leaving out the first assumption, however, at the cost of extending the second assumption to all energies e (not only the ones bounded by $2J3^D|\partial R| + e_0 + 40v$), leads to the following simple area law:

---

[14] It should be noticed that one can have much stronger condition for such scaling of entropy. To see this explicitly, say that R is a cubic region $R = \{1, \ldots, n\}^D$ meaning that $|\partial R| = n^{D-1}$ and $|R| = n^D$. Then since $\lim_{n \to \infty}[(\log n)/n^\epsilon] = 0$ for any (even arbitrarily small) $\epsilon > 0$, one easily checks that $S(\varrho_R)/|\partial R|^{1+\epsilon} \to 0$ for $|\partial R| \to \infty$.

Theorem 13 *Let R be an arbitrary region of the lattice L. Assuming that the above number of eigenvalues $\Omega_R(e)$ satisfies condition (6.47) for all e, then*

$$S(\varrho_R) \leq C|\partial R| \log |R| + O(|\partial R|). \tag{6.49}$$

Proof Let $|\psi_i\rangle$ and e_i denote the eigenvectors and ordered eigenvalues ($e_0 \leq e_1 \leq \cdots \leq e_n \leq \cdots$) of H_R. Then, it is clear that the ground state $|\Psi_0\rangle$ of H can be written as $|\Psi_0\rangle = \sum_{i,j} \alpha_{ij} |\psi_i\rangle |\varphi_j\rangle$, where the vectors $|\varphi_j\rangle$ constitute some basis in the Hilbert space corresponding to the region $L \setminus R$. One may always denote $\sqrt{\mu_i} |\widetilde{\varphi}_i\rangle = \sum_j \alpha_{ij} |\varphi_j\rangle$, and then

$$|\Psi_0\rangle = \sum_i \sqrt{\mu_i} |\psi_i\rangle |\widetilde{\varphi}_i\rangle, \tag{6.50}$$

where $\mu_i = 1/\langle \widetilde{\varphi}_i | \widetilde{\varphi}_i \rangle = 1/\sum_j |\alpha_{ij}|^2 \geq 0$ and they add up to unity. The vectors $|\widetilde{\varphi}_i\rangle$ in general do not have to be orthogonal, therefore Eq. 6.50 should not the confused with the Schmidt decomposition of $|\Psi_0\rangle$. Nevertheless, one may show that tracing out the $L \setminus R$ subsystem the entropy of the density matrix acting on R is upper bounded as (see Ref. [52])

$$S(\varrho_R) \leq -\sum_i \mu_i \log \mu_i. \tag{6.51}$$

We now aim to maximize the right-hand side of the above equation under the following conditions imposed on μ_i : First, the locality of our Hamiltonian means that $\langle H_R \rangle \leq e_0 + J3^D |\partial R|$, implying that the probabilities μ_i obey

$$\sum_i \mu_i \widetilde{e}_i \leq J3^D |\partial R|, \tag{6.52}$$

with $\widetilde{e}_i = e_i - e_0$. Second, the modified version of the second assumption allows to infer that for any eigenvalues e_i the inequality $i \leq c_2(\tau |R|)^{\gamma \widetilde{e}_i + \eta |\partial R|}$ holds. Substitution of the above in Eq. 6.52 gives

$$\sum_i \mu_i \log i \leq C|\partial R| \log |R| + O(|\partial R|) \tag{6.53}$$

where C contains the constants η, γ, J, and D. Eventually, following the standard convex optimization method (see e.g. Ref. [94]) with two constraints (normalization and the inequality (6.53)) one gets (6.49). $\qquad \square$

6.3.2.3 Are Laws for Mutual Information: Classical and Quantum Gibbs States

So far, we considered area laws only for ground states of local Hamiltonians. In addition, it would be very interesting to ask similar questions for nonzero temperatures. Here, however, one cannot rely on the entropy of a subsystem, as in the case of mixed states it looses its meaning. A very good quantity measuring the total amount of correlation in bipartite quantum systems is the *quantum mutual information* [95] defined as

$$I(A : B) = S(\varrho_A) + S(\varrho_B) - S(\varrho_{AB}), \tag{6.54}$$

where ϱ_{AB} is some bipartite state with its subsystems $\varrho_{A(B)}$. It should be noticed that for pure states the mutual information reduces to twice the amount of entanglement of the state.

Recently, it was proven that thermal states $\varrho_\beta = e^{-\beta H}/\mathrm{tr}[e^{-\beta H}]$ with local Hamiltonians H obey an area law for mutual information. Interestingly, a similar conclusion was drawn for classical lattices, in which at each site we have a classical spin with the configuration space $\mathbb{Z}_d$, and instead of density matrices one deals with probability distributions. In the following we review these two results, starting from the classical case.

To quantify correlations in classical systems, we use the classical mutual information, defined as in Eq. 6.54 with the von Neumann entropy substituted by the Shannon entropy $H(X) = -\sum_x p(x) \log_2 p(x)$, where p stands for a probability distribution characterizing random variable X. More precisely, let A and $B = S \setminus A$ denote two subsystems of some classical physical system S. Then, let $p(x_A)$ and $p(x_B)$ be the marginals of the joint probability distribution $p(x_{AB})$ describing S (x_a denotes the possible configurations of subsystems $a = A$, B, AB). The correlations between A and B are given by

$$I(A : B) = H(A) + H(B) - H(AB). \tag{6.55}$$

We are now ready to formulate and prove the following theorem [96].

Theorem 14 *Let L be a lattice with d–dimensional classical spins at each site. Let p be a Gibbs probability distribution coming from finite–range interactions on L. Then, dividing L into regions A and B, one has*

$$I(A : B) \leq |\partial A| \log d. \tag{6.56}$$

Proof First, notice that the Gibbs distributions coming from finite–range interactions have the property that if a region C separates A from B in the sense that no interaction is between A and B then $p(x_A|x_C, x_B) = p(x_A|x_C)$, which we rewrite as

$$p(x_A, x_B, x_C) = \frac{p(x_A, x_C)p(x_B, x_C)}{p(x_C)}. \tag{6.57}$$

Now, let A and B be two regions of L, and let $\partial A \subset A$ and $\partial B \subset B$ be boundaries of A and B, respectively, collecting all sites interacting with their exteriors. Finally, let $\overline{A} = A \setminus \partial A$ and $\overline{B} = B \setminus \partial B$. Since ∂A separates A from ∂B (there is no interaction between A and ∂B), we can use Eq. 6.57 to obtain

$$H(AB) = H(\bar{A} \partial A B) = - \sum_{x_{\bar{A}}, x_{\partial A}, x_B} p(x_{\bar{A}}, x_{\partial A}, x_B) \log_2 p(x_{\bar{A}}, x_{\partial A}, x_B)$$

$$= - \sum_{x_{\bar{A}}, x_{\partial A}} p(x_{\bar{A}}, x_{\partial A}) \log_2 p(x_{\bar{A}}, x_{\partial A})$$

$$- \sum_{x_{\partial A}, x_B} p(x_{\partial A}, x_B) \log_2 p(x_{\partial A}, x_B)$$

$$+ \sum_{x_{\partial A}} p(x_{\partial A}) \log_2 p(x_{\partial A})$$

$$= H(A) + H(\partial A B) - H(\partial A). \tag{6.58}$$

Since ∂B separates ∂A from B, the same reasoning may be applied to the second term of the right-hand side of the above, obtaining $H(\partial AB) = H(\partial A \partial B) + H(B) - H(\partial B)$. This, together with Eq. 6.58, gives

$$H(AB) = H(A) + H(B) + H(\partial A \partial B) - H(\partial A) - H(\partial B), \tag{6.59}$$

which in turn after application to Eq. 6.55 allows us to write

$$I(A : B) = I(\partial A : \partial B). \tag{6.60}$$

It means that whenever the probability distribution p has the above Markov property, correlations between A and B are the same as between their boundaries.

Now, we know that the mutual information can be expressed through the conditional Shannon entropy[15] as $I(X : Y) = H(X) - H(X|Y)$. Since $H(X|Y)$ is always nonnegative, we have the following inequality

$$I(\partial A : \partial B) \leq H(\partial A) \log d. \tag{6.61}$$

To get Eq. 6.56 it suffices to notice that $H(A)$ is upper bounded by the Shannon entropy of independently and identically distributed probability $p(x_A) = 1/d^{|A|}$, which means that $H(A) \leq |A| \log d$. $\square$

Let us now show that a similar conclusion can be drawn in the case of quantum thermal states [96], where the Markov property does not hold in general.

Theorem 15 *Let L be a lattice consisting of d-dimensional quantum systems divided into parts A and B ($L = A \cup B$). Thermal states ($T > 0$) of local Hamiltonians H obey the following area law*

$$I(A : B) \leq \beta \mathrm{tr}[H_\partial(\varrho_A \otimes \varrho_B - \varrho_{AB})]. \tag{6.62}$$

Proof The thermal state $\varrho_\beta = e^{-\beta H}/\mathrm{tr}(e^{-\beta H})$ minimizes the free energy $F(\varrho) = \mathrm{tr}(H\varrho) - (1/\beta)S(\varrho)$, and therefore $F(\varrho_\beta) \leq F(\varrho_\beta^A \otimes \varrho_\beta^B)$ with ϱ_β^A and ϱ_β^B subsystems of ϱ_β. This allows us to estimate the entropy of the thermal state as

---

[15] The conditional Shannon entropy is defined as $H(A|B) = H(A, B) - H(B)$.

$$S(\varrho_\beta) = \beta \left[\mathrm{tr}(H\varrho_\beta) - F(\varrho_\beta) \right]$$
$$\geq \beta \left[\mathrm{tr}(H\varrho_\beta) - F(\varrho_\beta^A \otimes \varrho_\beta^B) \right]$$
$$= \beta \left[\mathrm{tr}(H\varrho_\beta) - \mathrm{tr}(H\varrho_\beta^A \otimes \varrho_\beta^B) \right] + S(\varrho_\beta^A \otimes \varrho_\beta^B)$$
$$= \beta \left[\mathrm{tr}(H\varrho_\beta) - \mathrm{tr}(H\varrho_\beta^A \otimes \varrho_\beta^B) \right] + S(\varrho_\beta^A) + S(\varrho_\beta^B), \qquad (6.63)$$

where the last equality follows from additivity of the von Neumann entropy $S(\rho \otimes \sigma) = S(\rho) + S(\sigma)$. Putting Eq. 6.63 into the formula for mutual information we get

$$I(A : B) \leq \beta \left[\mathrm{tr}(H\varrho_\beta^A \otimes \varrho_\beta^B) - \mathrm{tr}(H\varrho_\beta) \right]. \qquad (6.64)$$

Let us now write the Hamiltonian as $H = H_A + H_\partial + H_B$, where H_A and H_B denote all the interaction terms within the regions A and B, respectively, while H_∂ stands for interaction terms connecting these two regions. Then one immediately notices that $\mathrm{tr}[H_{A(B)}(\varrho_\beta^A \otimes \varrho_\beta^B - \varrho_\beta)] = 0$ and only the H_∂ part of the Hamiltonian H contributes to the right-hand side of Eq. 6.64. This finishes the proof.   □

Let us notice that the right–hand side of Eq. 6.62 depends only on the boundary, and therefore it gives a scaling of mutual information similar to the classical case (6.61). Moreover, for the nearest-neighbor interaction, Eq. 6.62 simplifies to $I(A : B) \leq 2\beta \|h\| |\partial A|$ with $\|h\|$ denoting the largest eigenvalue of all terms of H crossing the boundary.

6.4 The Tensor Network Product World

Quantum many-body systems are, in general, difficult to describe: specifying an arbitrary state of a system with N-two level subsystems requires 2^N complex numbers. For a classical computer, this presents not only storage problems, but also computational ones, since simple operations like calculating the expectation value of an observable would require an exponential number of operations. However, we know that completely separable states can be described with about N parameters—indeed, they correspond to classical states. Therefore, what makes a quantum state difficult to describe are quantum correlations, or entanglement. We saw already that even if in general the entropy of a subsystem of an arbitrary state is proportional to the volume, there are some special states which obey an entropic area law. Intuitively, and given the close relation between entropy and information, we could expect that states that follow an area law can be described (at least approximately) with much less information than a general state. We also know that such low entanglement states are few, albeit interesting—we only need an efficient and practical way to describe and parameterize them.

6.4.1 The Tensor Network Representation of Quantum States

Consider a general state of a system with N d-level particles,

$$|\psi\rangle = \sum_{i_1,i_2,\dots,i_N=1}^{d} c_{i_1 i_2 \dots i_N} |i_1, i_2, \dots, i_N\rangle. \tag{6.65}$$

When the state has no entanglement, then $c_{i_1 i_2 \dots i_N} = c_{i_1}^{(1)} c_{i_2}^{(2)} \dots c_{i_N}^{(N)}$ where all c's are scalars. The locality of the information (the set of coefficients c for each site is independent of the others) is key to the efficiency with which separable states can be represented. How can we keep this locality while adding complexity to the state, possibly in the form of correlations but only to nearest-neighbors? As we shall see, we can do this by using a tensor at each site of our lattice, with one index of the tensor for every physical neighbor of the site, and another index for the physical states of the particle. For example, in a one-dimensional chain we would assign a matrix for each state of each particle, and the full quantum state would write as

$$|\psi\rangle = \sum_{i_1,i_2,\dots,i_N=1}^{d} \mathrm{tr}\left[A_{i_1}^{[1]} A_{i_2}^{[2]} \dots A_{i_N}^{[N]} \right] |i_1, i_2, \dots i_N\rangle, \tag{6.66}$$

where $A_{i_k}^{[k]}$ stands for a matrix with dimensions $D_k \times D_{k+1}$. A useful way of understanding the motivations for this representation is to think of a valence bond picture [97]. Imagine that we replace every particle at the lattice by a pair (or more in higher dimensions) of particles of dimensions D that are in a maximally entangled state with their corresponding partners in a neighboring site (see Fig. 6.4). Then, by applying a map from this virtual particles into the real ones,

$$\mathcal{A} = \sum_{i=1}^{d} \sum_{\alpha,\beta=1}^{D} A_{\alpha,\beta}^{[i]} |i\rangle\langle\alpha, \beta|, \tag{6.67}$$

we obtain a state that is expressed as Eq. 6.66. One can show that any state $|\psi\rangle \in \mathbb{C}^{dN}$ can be written in this way with $D = \max_m D_m \le d^{N/2}$. Furthermore, a matrix product state can always be found such that [98]

- $\sum_i A_i^{\dagger[k]} A_i^{[k]} = 1_{D_k}$, for $1 \le k \le N$,
- $\sum_i A_i^{\dagger[k]} \Lambda^{[k-1]} A_i^{[k]} = \Lambda^{[k]}$, for $1 \le k \le N$, and
- For open boundary conditions $\Lambda^{[0]} = \Lambda^{[N]} = 1$, and $\Lambda^{[k]}$ is a $D_{k+1} \times D_{k+1}$ positive diagonal matrix, full rank, with $\mathrm{tr}\Lambda^{[k]} = 1$.

In fact, $\Lambda^{[k]}$ is a matrix whose diagonal components λ_n^k, $n = 1, \dots, D_k$, are the non-zero eigenvalues of the reduced density matrix obtained by tracing out the particles from $k+1$ to N, i.e., the Schmidt coefficients of a bipartition of the system at site k. A MPS with these properties is said to be in its canonical form [99].

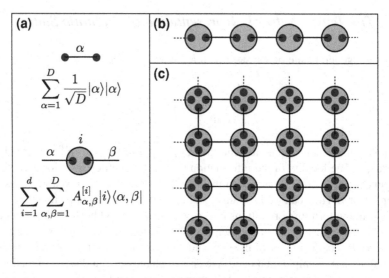

Fig. 6.4 Schematic representation of tensor networks. In panel **a** we show the meaning of the elements in the representation, namely the solid line joining two virtual particles in different sites means the maximally entangled state between them, and the *grey circle* represents the map from virtual particles in the same site to the physical index. In panel **b** we see a one-dimensional tensor network or MPS, while in **c** we show how the scheme can be extended intuitively to higher dimensions—in the two-dimensional example shown here, a PEPS that contains four virtual particles per physical site

Therefore, Eq. 6.66 is a representation of all possible states—still cumbersome. It becomes an efficient representation when the virtual bond dimension D is small, in which case it is typically said that the state has a matrix product state (MPS) representation. In higher dimensions we talk about projected entangled pair states (PEPS) [100]. When entanglement is small (but finite), most of the Schmidt coefficients are either zero or decay rapidly to zero [84]. Then, if $|\psi\rangle$ contains little entanglement, we can obtain a very good approximation to it by truncating the matrices A to a rank D much smaller than the maximum allowed by the above theorem, $d^{N/2}$. In fact, we can demonstrate the following

Lemma 1 Reference [99] *There exists a MPS $|\psi_D\rangle$ with bond dimension D such that $\||\psi\rangle - |\psi_D\rangle\|^2 < 2 \sum_{\alpha=1}^{N-1} \epsilon_\alpha(D)$, where $\epsilon_\alpha(D) = \sum_{i=D+1}^{d^{\min(\alpha, N-\alpha)}} \lambda_i^{[k]}$.*

Proof Let us assume that the MPS is in its canonical form with $D = 2^{N/2}$. Defining a projector into the virtual bond dimension $P = \sum_{k=1}^{D} |k\rangle\langle k|$, and a TPCM map $\$_m(X) = \sum_i A_i^{[m]\dagger} X A_i^{[m]}$, we can write the overlap

$$\langle\psi|\psi_D\rangle = \mathrm{Tr}\left[\$_2(\dots\$_{N-2}(\$_{N-1}(\Lambda^{[N-1]}P)P)P)\dots)P\right]. \tag{6.68}$$

By defining $Y^{[k]} = \$_k(Y^{[k+1]}P)$, with $Y^{[N-1]} = \Lambda^{[N-1]}P$, and using that $\mathrm{Tr}|\$(X)| \leq \mathrm{Tr}|X|$, we can see that

$$\mathrm{tr}\left|\Lambda^{[k]} - Y^{[k]}\right| = \mathrm{tr}\left|\$_k(\Lambda^{[k+1]} - Y^{[k+1]}P)\right|$$
$$\leq \mathrm{tr}\left|\Lambda^{[k+1]} - Y^{[k+1]}\right| + \mathrm{tr}\left|\Lambda^{[k+1]}(1 - P)\right|, \tag{6.69}$$

where the last term is equal to $\sum_{\alpha=D+1}^{2^{N/2}} \lambda_\alpha^{[k]}$. Finally, applying this last inequality recursively from $N - 1$ to 2, and using that $\langle\psi_D|\psi_D\rangle \leq 1$, we can obtain the desired bound on $\langle\psi|\psi_D\rangle$. $\qquad\square$

Lemma (1) is most powerful in the context of numerical simulations of quantum states: it gives a controllable handle on the precision of the approximation by MPS. In practical terms, for the representation to be efficient the Schmidt coefficients λ need to decay faster than polynomially. However, we can be more precise and give bounds on the error of the approximation in terms of entropies [101]:

Lemma 2 *Let $S_\alpha(\rho) = \log(\mathrm{tr}\rho^\alpha)/(1 - \alpha)$ be the Rényi entropy of a reduced density matrix ρ, with $0 < \alpha < 1$. Denote $\epsilon(D) = \sum_{i=D+1}^\infty \lambda_i$, with λ_i being the eigenvalues of ρ in nonincreasing order. Then,*

$$\log(\epsilon(D)) \leq \frac{1 - \alpha}{\alpha}\left(S_\alpha(\rho) - \log\frac{D}{1 - \alpha}\right). \tag{6.70}$$

The question now is when can we find systems with relevant states that can be written efficiently as a MPS; i.e. how broad is the simulability of quantum states by MPS. For example, one case of interest where we could expect the method to fail is near quantum critical points where correlations (and entanglement) are singular and might diverge. However, at least in 1D systems, we can state the following [99]:

Lemma 3 *In one dimension there exists a scalable, efficient MPS representation of ground states even at criticality.*

Proof In one dimension, the worst case growth of entropy of a subsystem of size L, exactly at criticality, is given by

$$S_\alpha(\rho_L) \simeq \frac{c + \tilde{c}}{12}\left(1 + \frac{1}{\alpha}\right)\log L. \tag{6.71}$$

Let us take the length L to be half the chain, $N = 2L$. By means of the previous discussion, we can find a MPS $|\psi_D\rangle$ such that its distance with the ground state $|\psi_{GS}\rangle$ is bounded as $\||\psi_{GS}\rangle - |\psi_D\rangle\|^2 \leq \epsilon_0/L$, with ϵ_0 constant. Now, let D_L be the minimal virtual bond dimension needed for this precision, i.e. $\||\psi_{GS}\rangle - |\psi_D\rangle\|^2 \leq 2 \times 2L\epsilon_{max}(D)$. We demand that

$$\epsilon_{max}(D) \leq \frac{\epsilon_0}{4L^2}$$
$$\leq \exp\left[\frac{1 - \alpha}{\alpha}\left(\frac{c + \tilde{c}}{12}\frac{1 + \alpha}{\alpha}\log L - \log\frac{D_L}{1 - \alpha}\right)\right], \tag{6.72}$$

from which we can extract

Table 6.1 Relation between scaling of block Rényi entropies and approximability by MPS [101]. In the "undetermined" region denoted with question marks, nothing can be said about approximability just from looking at the scaling

$S_\alpha \sim$	const	$\log L$	$L^\kappa (\kappa < 1)$	L
$S_\alpha < 1$	OK	OK	?	?
$S \equiv S_1$	?	?	?	NO
$S_\alpha > 1$	?	?	NO	NO

$$ D_L \leq \text{const} \left(\frac{L^2}{\epsilon_0}\right)^{\frac{\alpha}{1-\alpha}} L^{\frac{c+\tilde{c}}{12} \frac{1+\alpha}{\alpha}} \propto \text{poly}(L). \tag{6.73} $$

□

Establishing that there exists an efficient representation of the ground state is not enough: we must also know if it is possible to find it efficiently too. In one dimensional gapped systems, the gap Δ typically scales polynomially, which means that DMRG and MPS methods should converge reasonably fast. One can, however, formalize the regime of efficiency of MPS as a function of how the different Rényi entropies scale with subsystem size [101]. In Table 6.1 we summarize the currently known regimes where the MPS approach is an appropriate one or not.

6.4.2 Examples

Here we present a few models and states that are fine examples of the power of MPS representations[99].

Example 6 A well known model with a finite excitation gap and exponentially decaying spin correlation functions was introduced by Affleck, Kennedy, Lieb, and Tasaki [102, 103]—the so called AKLT model. The model Hamiltonian is

$$ H = \sum_i S_i \cdot S_{i+1} + \frac{1}{3} (S_i \cdot S_{i+1})^2 . \tag{6.74} $$

For $S=1$, the local Hilbert space of each spin has three states, thus $d=3$. The ground state of this Hamiltonian can be written compactly using a translationally invariant MPS with bond dimension $D=2$, specifically

$$ A_{-1} = \sigma_x, A_0 = \sqrt{2}\sigma^+, A_1 = -\sqrt{2}\sigma^-. \tag{6.75} $$

Example 7 A paradigmatic example of a frustrated one dimensional spin chain is the Majumdar–Ghosh [104] model with nearest and next nearest-neighbor interactions:

$$ H = \sum_i 2\sigma_i \cdot \sigma_{i+1} + \sigma_i \cdot \sigma_{i+2}, \tag{6.76} $$

The model is equivalent to the or $J_1 - J_2$ Heisenberg model with $J_1/J_2 = 2$. The ground state of this model is composed of singlets between nearest-neighbor spins. However, since the state must be translationally invariant, we must include a superposition of singlets between even-odd spins, and "shifted" singlets between odd–even spins. The state can be written compactly in MPS form using $D=3$,

$$A_0 = \begin{pmatrix} 0 & 1 & 0 \\ 0 & 0 & -1 \\ 0 & 0 & 0 \end{pmatrix}, \ A_1 = \begin{pmatrix} 0 & 0 & 0 \\ 1 & 0 & 0 \\ 0 & 1 & 0 \end{pmatrix} \tag{6.77}$$

Example 8 A relevant state for quantum information theory is the Greenberger–Horne–Zeilinger (GHZ) state, which for N spin 1/2 particles can be written as

$$|GHZ\rangle = \frac{|0\rangle^{\otimes N} + |1\rangle^{\otimes N}}{\sqrt{2}}. \tag{6.78}$$

GHZ states are considered important because for many entanglement measures they are maximally entangled, however by measuring or tracing out any qubit a classical state is obtained (although with correlations). GHZ states can be written using $D=2$ MPS, specifically $A_{0,1} = 1 \pm \sigma_z$. Also the "antiferromagnetic" GHZ state is simple, $A_{0,1} = \sigma^{\pm}$.

Example 9 Cluster states are relevant for one-way quantum computing. They are the ground state of

$$H = \sum \sigma_{i-1}^z \sigma_i^x \sigma_{i+1}^z, \tag{6.79}$$

and can be represented using a $D=2$ MPS,

$$A_0 = \begin{pmatrix} 0 & 0 \\ 1 & 1 \end{pmatrix}, \ A_1 = \begin{pmatrix} 1 & -1 \\ 0 & 0 \end{pmatrix} \tag{6.80}$$

Example 10 (Classical superposition MPS) Imagine we have a classical Hamiltonian

$$H = \sum_{(i,j)} h(\sigma_i, \sigma_j), \tag{6.81}$$

where $\sigma_i = 1, \ldots, d$, and $h(\sigma_i, \sigma_j)$ are local interactions. The partition function of such a model at a given inverse temperature β is

$$Z = \sum_{\{\sigma\}} \exp\left[-\beta H(\sigma)\right], \tag{6.82}$$

where the sum is over all possible configurations of the vector σ. Let us now define a quantum state $|\psi_\beta\rangle$ whose amplitude for a given state of the computational basis corresponds to the term in the partition function for that state, i.e.

$$|\psi_\beta\rangle = \frac{1}{\sqrt{Z}} \sum_{\{\sigma\}} \exp\left[-\frac{\beta}{2} H(\sigma)\right] |\sigma_1 \ldots \sigma_N\rangle$$

$$= \frac{1}{\sqrt{Z}} \sum_{\{\sigma\}} \prod_{(i,j)} \exp\left[-\frac{\beta}{2} h(\sigma_i, \sigma_j)\right] |\sigma_1 \ldots \sigma_N\rangle. \tag{6.83}$$

We shall now define a map P—in the same manner as in valence bond states—that goes from $\mathbb{C}^{d^2}$ to $\mathbb{C}^2$ such that

$$P|s, k\rangle = |s\rangle\langle\varphi_s|k\rangle, \tag{6.84}$$

where we have defined

$$\sum_{\alpha=1}^{d} \langle\varphi_s|\alpha\rangle\langle\varphi_{\tilde{s}}|\alpha\rangle = \exp\left[-\frac{\beta}{2} h(s, \tilde{s})\right]. \tag{6.85}$$

To visualize what happens when we insert these back into the classical superposition state $|\psi_\beta\rangle$, let us concentrate for a moment on a one-dimensional system:

$$|\psi_\beta\rangle = \frac{1}{\sqrt{Z}} \sum_{\sigma_1,\ldots,\sigma_N} \exp\left[-\frac{\beta}{2} h(\sigma_1, \sigma_2)\right] \ldots \exp\left[-\frac{\beta}{2} h(\sigma_{N-1}, \sigma_N)\right] |\sigma_1 \ldots \sigma_N\rangle,$$

$$|\psi_\beta\rangle = \frac{1}{\sqrt{Z}} \sum_{\sigma_1,\ldots,\sigma_N} \left(\sum_{\alpha_1=1}^{d} \langle\varphi_{\sigma_1}|\alpha_1\rangle\langle\varphi_{\sigma_2}|\alpha_1\rangle \sum_{\alpha_2=1}^{d} \langle\varphi_{\sigma_2}|\alpha_2\rangle\langle\varphi_{\sigma_3}|\alpha_2\rangle \right.$$

$$\left. \times \ldots \times \sum_{\alpha_N=1}^{d} \langle\varphi_{\sigma_N}|\alpha_N\rangle\langle\varphi_{\sigma_1}|\alpha_N\rangle \right) |\sigma_1 \ldots \sigma_N\rangle, \tag{6.86}$$

$$|\psi_\beta\rangle = \frac{1}{\sqrt{Z}} \sum_{\sigma_1,\ldots,\sigma_N} \sum_{\alpha_1,\ldots,\alpha_N=1}^{d} \left[\langle\varphi_{\sigma_1}|\alpha_N\rangle\langle\varphi_{\sigma_1}|\alpha_1\rangle\right] \left[\langle\varphi_{\sigma_2}|\alpha_1\rangle\langle\varphi_{\sigma_2}|\alpha_2\rangle\right]$$

$$\times \ldots \times \left[\langle\varphi_{\sigma_N}|\alpha_{N-1}\rangle\langle\varphi_{\sigma_N}|\alpha_N\rangle\right] |\sigma_1 \ldots \sigma_N\rangle, \tag{6.87}$$

and we can replace $A^{(i)}_{s_i,\alpha,\beta} = \langle\varphi_{s_i}|\alpha\rangle\langle\varphi_{s_i}|\beta\rangle$, thus expressing the classical thermal superposition state as a MPS.

These states have some important properties:

1. They obey strict area laws,
2. They allow to calculate classical and quantum correlations, and
3. They are ground states of local Hamiltonians.

Property (1) should be obvious by now, since we have explicitly shown the MPS form of the state. We can show easily property (2) for Ising models. A classical correlation function f must be evaluated with the partition function, $\langle f(\sigma)\rangle =$

$\sum_\sigma f(\sigma)e^{-\beta H(\sigma)}/Z$, but this is just the expectation value of an operator made of changing the argument of f into σ_z operators, and evaluated with $|\psi_\beta\rangle$. Since it is the expectation value of a MPS, it is efficient to compute. Finally, we will demonstrate property (3) at length in the next section, because it will lead us into the final topic of this lectures: Quantum kinetic models.

6.4.3 Classical Kinetic Models

Our goal in this section is to show the local Hamiltonians whose ground state is the classical superposition state defined in the previous section. As we shall see, these Hamiltonians will arise from the master equation of a classical system that is interesting in its own right, so we will first spend some time on it.

Let us consider a system made out of N classical spins interacting through a Hamiltonian H. If σ_i denotes the state of ith spin, we will label the configurations of the system by $\sigma = (\sigma_1, \ldots, \sigma_N)$, and the probability of finding at time t the system in state σ (given that it was in state σ_0 at time t_0) by $P(\sigma, t) = P(\sigma, t|\sigma_0, t_0)$. In what follows we focus on this probability distribution, whose dynamics is described by a master equation:

$$\dot{P}(\sigma, t) = \sum_{\sigma'} W(\sigma, \sigma')P(\sigma', t) - \sum_{\sigma'} W(\sigma', \sigma)P(\sigma, t), \qquad (6.88)$$

where $W(\sigma, \sigma')$ is the transition probability from state σ' to state σ. This equation defines the class of kinetic models, and it clearly describes a Markov process— the instantaneous change of $P(\sigma, t)$ does not depend on its history.

We will only consider systems that obey a detailed balance condition, i.e.

$$W(\sigma, \sigma')e^{-\beta H(\sigma')} = W(\sigma', \sigma)e^{-\beta H(\sigma)}. \qquad (6.89)$$

With this condition, the stationary state of the master equation (the one that fulfills $\dot{P}_{st}(\sigma, t) = 0$) is simply $P_{st}(\sigma) = e^{-\beta H(\sigma)}/Z$, with Z being the partition function. This state in particular will map into the classical superposition state defined above, but we still have not found its parent Hamiltonian. For this, we will rewrite Eq. 6.88 in the form of a matrix Schrödinger equation (albeit with imaginary time) from which we can identify a Hamiltonian.

Let us apply the transformation $\psi(\sigma, t) = e^{\beta H(\sigma)/2}P(\sigma, t)$, which leads to

$$\dot{\psi}(\sigma, t) = \sum_{\sigma'} e^{\beta H(\sigma)/2}W(\sigma, \sigma')e^{-\beta H(\sigma')/2}\psi(\sigma', t) - W(\sigma', \sigma)\psi(\sigma, t)$$

$$= -\sum_{\sigma'} H_\beta(\sigma, \sigma')\psi(\sigma', t) \qquad (6.90)$$

with

$$H_\beta(\sigma, \sigma') = \sum_{\sigma''} W(\sigma'', \sigma')\delta_{\sigma\sigma'} - e^{\beta H(\sigma)/2}W(\sigma, \sigma')e^{-\beta H(\sigma')/2}. \qquad (6.91)$$

Notice that the detailed balance condition guarantees that the matrix H_β is Hermitian, so we can interpret it as a Hamiltonian. Furthermore, because of the conservation of probability, H_β can only have non-negative eigenvalues, which means that the state ψ_{st} associated to the stationary state P_{st} with eigenvalue zero must be a ground state: the classical superposition MPS that we were looking for.

Remarkably, we have not said anything yet about H. A famous example of such kinetic model is a single spin-flip model considered by Glauber [105], for which H is the Ising Hamiltonian[16]

$$H(\sigma) \equiv H_{\text{Ising}}(\sigma) = -J \sum_{\langle i,j \rangle} \sigma_i^z \sigma_j^z \qquad (J > 0). \qquad (6.92)$$

Denoting by P_i the flip operator of the ith spin, i.e., $P_i\sigma_i = -\sigma_i$, the general master Eq. 6.88 reduces in this case to

$$\dot{P}(\sigma, t) = \sum_i [W(\sigma, P_i\sigma)P(P_i\sigma, t) - W(P_i\sigma, \sigma)P(\sigma, t)] \qquad (6.93)$$

with $W(\sigma, P_i\sigma)$ now called spin rates. It was shown in [105] that the most general form of spin rates with symmetric interaction with both nearest-neighbors, and satisfying the detailed balance condition (6.89), is given by

$$w(P_i\sigma, \sigma) = \Gamma(1 + \delta\sigma_{i-1}\sigma_{i+1})[1 - (1/2)\gamma\sigma_i(\sigma_{i-1} + \sigma_{i+1})] \qquad (6.94)$$

with $\Gamma > 0$, $-1 \le \delta \le 1$, and $0 \le \gamma \le 1$. The $\delta = 0$ case was thoroughly investigated by Glauber [105], who showed that all the relevant quantities can be derived analytically—including the dynamical exponent that turned out to be $z=2$. The more general case of nonzero δ was treated in a series of papers [106–108], that showed for instance that the choice $\delta = \gamma/(2-\gamma)$ leads to an interesting dynamical exponent $z \ne 2$.

If we rewrite the single spin-flip master equation in the form of the Schrödinger equation, we obtain an associated quantum Hamiltonian

$$H_\beta(\delta, \gamma) = -\Gamma \sum_i \left[\left(A(\delta, \gamma) - B(\delta, \gamma)\sigma_{i-1}^z\sigma_{i+1}^z \right)\sigma_i^x \right.$$
$$\left. -(1 + \delta\sigma_{i-1}^z\sigma_{i+1}^z)\left(1 - (1/2)\gamma\sigma_i^z\left(\sigma_{i-1}^z + \sigma_{i+1}^z\right) \right) \right], \qquad (6.95)$$

where

---

[16] This is the reason why the Glauber model is also known as the kinetic Ising model (KIM).

$$A(\delta, \gamma) = \frac{(1 + \delta)\gamma^2}{2(1 - \sqrt{1 - \gamma^2})} - \delta, \qquad B(\delta, \gamma) = 1 - A(\delta, \gamma) \qquad (6.96)$$

and σ^z and σ^x are the standard Pauli matrices. For $\delta = 0$ this Hamiltonian was diagonalized in Ref. [109], and independently in Ref. [110].

The Hamiltonian $H_\beta(\delta, \gamma)$, and also the other ones that can be derived in this way, are typically gapped except at a critical temperature β_c where the gap vanishes with the critical exponent z that characterizes the model. In one dimension $\beta_c = \infty$, but for larger dimensions this model has a finite critical temperature.

We have seen thus far how the master equation of a classical spin model (that obeys the detailed balance condition) can be associated to a quantum Hamiltonian with some interesting critical properties—for example, its ground state obeys a strict area law and can be written efficiently as a MPS. Nevertheless, the underlying model is still classical. In the next section, we will see one way in which we can generalize the initial model to be quantum, while retaining the same structure that leads to associated Hamiltonians that obey area laws.

6.5 Quantum Kinetic Ising Models

Here we discuss ways to generalize the kinetic Eq. 6.88 to a quantum master equation, but in such a way that its diagonal part reproduces the corresponding kinetic model. A similar approach was taken in Ref. [111], where a quantum master equation that reproduced a kinetic Ising model was proposed (see also Ref. [112]). However, no attempts aiming at fully solving such QMEs are known so far. Our purpose is to give quantum generalizations of the classical kinetic models that can be solved analytically.

Recently, we presented such a generalization [113] for the single spin-flip model, Eq. 6.93, with the spin rates of Eq. 6.94. In Ref. [113] we were able to decouple the master equation for the density matrix of a quantum system into 2^N master equations with the same structure as the ones studied above. Here, we will only show the associated Hamiltonians (and their spectra) obtained in these models. However, we will demonstrate how to approach the problem but in a different model that allows transitions that flip two consecutive spins.

6.5.1 A Two Spin Flip Model

First, let us particularize the classical kinetic Eq. 6.88 to the case where the flip operator acts on pairs of consecutive spins of the chain, i.e.

$$\frac{\partial P(\sigma, t)}{\partial t} = \sum_i \left[w_i(F_{i,i+1}\sigma \to \sigma) P(F_{i,i+1}\sigma, t) - w_i(\sigma \to F_{i,i+1}\sigma) P(\sigma, t) \right],$$

(6.97)

where $F_{i,i+1}$ denote spin flips at positions i and $i+1$, while the spin rates are given by $w_i(F_{i,i+1}\sigma, \sigma) = \Gamma[1 - (1/2)\gamma(\sigma_{i-1}\sigma_i + \sigma_{i+1}\sigma_{i+2})]$ with $0 < \Gamma < \infty$ and $\gamma = \tanh 2\beta J$. This model was investigated in Ref. [114], where the associated Hamiltonian was found and diagonalized using the Jordan-Wigner transformation [115] followed by Fourier and Bogoliubov-Valatin [116, 117] transformations. In particular, Hilhorst et al. were able to show that, despite the complexity of the transformations, one can easily compute expectation values such as magnetization, energy density, or correlations, and that they have a relatively simple exponential behavior [114].

Here we will define through a master equation a quantum model that resembles the kinetic model above. For this, we will replace classical probabilities with the quantum density matrices, and classical operators with quantum ones (e.g. σ^x is the qubit flip operator). Consider the following master equation

$$\partial_t \varrho(t) = \sum_i \left[\sigma_i^x \sigma_{i+1}^x \sqrt{w_i(\sigma^z)} \varrho(t) \sqrt{w_i(\sigma^z)} \sigma_i^x \sigma_{i+1}^x - \frac{1}{2} \{ w_i(\sigma^z), \varrho(t) \} \right], \quad (6.98)$$

where $\{ \cdot, \cdot \}$ denotes the anticommutator and $w_i(\sigma^z)$ are quantum mechanical generalizations of the spin rates (6.94), now written in terms of the σ^z operators,

$$w_i(\sigma^z) = \Gamma \left[1 - \frac{1}{2}\gamma(\sigma_{i-1}^z \sigma_i^z + \sigma_{i+1}^z \sigma_{i+2}^z) \right]. \quad (6.99)$$

Although it looks complicated, the quantum kinetic model above can still be solved with techniques similar to the classical case [113]: the key ingredient is to find a large number of constants of motion that allow to split the master equation into a set of ordinary Schrödinger equations. To see this, we must represent the density matrix $\varrho(t)$ as a vector in an expanded Hilbert space. This follows from a simple isomorphism between linear operators from $M_d(\mathbb{C})$ and vectors from $\mathbb{C}^{d^2}$. In other words, writing our density matrix in the computational basis in $(\mathbb{C}^2)^{\otimes N}$ as $\varrho(t) = \sum_{\sigma, \tilde{\sigma}} [\varrho(t)]_{\sigma, \tilde{\sigma}} |\sigma\rangle\langle\tilde{\sigma}|$, we can treat it as a vector $|\varrho(t)\rangle = \sum_{\sigma, \tilde{\sigma}} [\varrho(t)]_{\sigma, \tilde{\sigma}} |\sigma\rangle|\tilde{\sigma}\rangle$ from $(\mathbb{C}^2)^{\otimes N} \otimes (\mathbb{C}^2)^{\otimes N}$. Even if formally we are enlarging the number of spins from N to $2N$, the advantage is that now we deal with "pure states" instead of density matrices which allows us to find many conserved quantities. This, in turn, shows that the effective Hilbert space used is much smaller than the initial one.

To be consistent, operators that appear to the right of $\varrho(t)$ must be replaced with "tilded" operators that act on the right subsystem of the expanded space, while operators on the left of the density matrix ("untilded") act on the left subsystem (for instance $\sigma_i^x \tilde{\sigma}_i^x |s\rangle|\tilde{s}\rangle = \sigma_i |s\rangle \tilde{\sigma}_i^x |\tilde{s}\rangle$). This notation allows us to rewrite the master equation (6.98) as the following matrix equation

$$|\dot{\varrho}(t)\rangle = \sum_i \left[\sigma_i^x \sigma_{i+1}^x \tilde{\sigma}_i^x \tilde{\sigma}_{i+1}^x \sqrt{w_i(\sigma^z) w_i(\tilde{\sigma}^z)} - \frac{1}{2} [w_i(\sigma^z) + w_i(\tilde{\sigma}^z)] \right] |\varrho(t)\rangle.$$

(6.100)

As was the case for the initial classical master equation, the matrix appearing on the right-hand side of Eq. 6.100 is not Hermitian. In order to bring it to Hermitian form we can use the detailed balance condition, which suggests the transformation

$$|\varrho(t)\rangle = \exp\left[-(\beta/4)[\mathcal{H}(\sigma) + \mathcal{H}(\widetilde{\sigma})]\right]|\psi(t)\rangle, \tag{6.101}$$

with $\mathcal{H}$ denoting the quantum generalization of the Ising Hamiltonian $\mathcal{H} = -J \sum_i \sigma_i^z \sigma_{i+1}^z$. With this transformation, and denoting

$$v_i(\sigma^z) = w_i(\sigma^z)\exp[(\beta J)\sigma_i^z(\sigma_{i-1}^z + \sigma_{i+1}^z)], \tag{6.102}$$

Eq. 6.100 can be written as

$$|\dot{\psi}(t)\rangle = \sum_i \left[\sigma_i^x \sigma_{i+1}^x \widetilde{\sigma}_i^x \widetilde{\sigma}_{i+1}^x \sqrt{v_i(\sigma^z)v_i(\widetilde{\sigma}^z)} - \frac{1}{2}[v_i(\sigma^z) + v_i(\widetilde{\sigma}^z)]\right]|\psi(t)\rangle \tag{6.103}$$

which we can see as a Schrödinger equation $|\dot{\psi}(t)\rangle = -H|\psi(t)\rangle$ with Hermitian H.

We have reached the point where all these changes of notation payoff: indeed, the form of H makes it clear that it commutes with $\sigma_i^z \sigma_{i+1}^z \widetilde{\sigma}_i^z \widetilde{\sigma}_{i+1}^z$ ($i = 1, \ldots, N$). Therefore, we can introduce new variables $\tau_i = \sigma_i^z \sigma_{i+1}^z \widetilde{\sigma}_i^z \widetilde{\sigma}_{i+1}^z$ ($i = 1, \ldots, N$) which are constants of motion and reduce the number of degrees of freedom. In particular, tilded variables can be expressed by σ and the new variables τ as $\widetilde{\sigma}_i^z \widetilde{\sigma}_{i+1}^z = \tau_i \sigma_i^z \sigma_{i+1}^z$ for any i. In other words, we have replaced σ and $\widetilde{\sigma}$ by τ and σ, of which τ is conserved. To each configuration of τ's we associate a natural number from 0 to $2^N - 1$, which corresponds to a particular correlation between the σ and $\widetilde{\sigma}$ variables. For example, $\tau = 0$ corresponds to all τ-spins up ($\tau_i = 1$ for $i = 1, \ldots, N$), while $\tau = 2^N - 1$ means that $\tau_i = -1$ for $i = 1, \ldots, N$. With this notation, each value of τ is associated to a Hamiltonian H_τ that acts only in the space of N spins and is of the form

$$H_\tau = -\sum_i \left[\sigma_i^x[v_i(\sigma^z)]^{\frac{1}{2}}[v_i(\tau\sigma^z)]^{\frac{1}{2}} - \frac{1}{2}[w_i(\sigma^z) + w_i(\tau\sigma^z)]\right], \tag{6.104}$$

where $\tau\sigma^z$ denotes $\tau_i\sigma_i^z$ ($i = 1, \ldots, N$). Because these Hamiltonians are independent from each other, we have converted the problem of solving the general master equation (6.98) to the problem of diagonalizing 2^N Hamiltonians, each of dimension $2^N \times 2^N$. Now, we have that

$$|\psi(t)\rangle = \bigotimes_{\tau=0}^{2^N-1} |\psi_\tau(t)\rangle, \qquad H = \bigotimes_{\tau=0}^{2^N-1} H_\tau. \tag{6.105}$$

After simple algebra one sees that the explicit form of H_τ is

$$H_\tau = -\sum_i \left[\left(A_i(\varphi) - B_i(\varphi)\sigma_{i-1}^z \sigma_i^z \sigma_{i+1}^z \sigma_{i+2}^z\right)\sigma_i^x \sigma_{i+1}^x\right.$$
$$\left. - \left[1 - \frac{1}{2}\gamma\left(f(\tau_{i-1})\sigma_{i-1}^z \sigma_i^z + f(\tau_{i+1})\sigma_{i+1}^z \sigma_{i+2}^z\right)\right]\right], \tag{6.106}$$

where

$$A_i(\varphi) = \begin{cases} \cos^2\varphi, & \tau_{i-1}\tau_{i+1} = 1 \\ \sqrt{\cos 2\varphi}, & \tau_{i-1}\tau_{i+1} = -1 \end{cases}, \quad B_i(\varphi) = \begin{cases} \sin^2\varphi, & \tau_{i-1}\tau_{i+1} = 1 \\ 0, & \tau_{i-1}\tau_{i+1} = -1, \end{cases} \tag{6.107}$$

with

$$\cos\varphi = \frac{\cosh\beta J}{(\cosh^2\beta J + \sinh^2\beta J)^{1/2}}, \quad \sin\varphi = \frac{\sinh\beta J}{(\cosh^2\beta J + \sinh^2\beta J)^{1/2}}, \tag{6.108}$$

and $f(x) = (1/2)(1 + x)$. Here the angle ranges from zero (which corresponds to infinite temperature) to $\pi/4$ (which corresponds to $T=0$) and in this notation $\gamma = \sin 2\varphi$. Let us notice that for $\tau = 0$ Eqs. 6.106 and 6.107, as it should be, reproduce the Hamiltonian derived in [114]. This, however, contrary to the single spin-flip case, is not the case for $\tau = 2^N - 1$, where one of the terms in the square brackets vanishes and the Hamiltonian reduces to

$$H_{2^N-1} = -\sum_i \left[\left(A_i(\varphi) - B_i(\varphi)\sigma_{i-1}^z \sigma_i^z \sigma_{i+1}^z \sigma_{i+2}^z \right) \sigma_i^x \sigma_{i+1}^x - 1 \right] \tag{6.109}$$

Let us discuss now some of the properties of H_τ. Below we show that for all τ they are always positive operators. We also find all the cases with respect to φ and τ for which the Hamiltonians can have zero-energy ground states.

Lemma 4 *The Hamiltonians H_τ are positive for any $\tau = 0, \ldots, 2^N - 1$.*

Proof Let us denote by $H_\tau^{(i)}$ the ith term appearing in the sum in Eq. 6.106. The idea is to show that all $H_\tau^{(i)}$ are positive, and the positivity of H_τ follows immediately. Of course, the form of $H_\tau^{(i)}$ changes depending on τ-spins at positions i-1 and i+1. Therefore, we distinguish several cases with respect to different possible configurations of these spins. For $\tau_{i-1} = \tau_{i+1}$, one easily infers from Eqs. 6.106 and 6.107 that

$$H_\tau^{(i)} = 1 - \frac{1}{2}\gamma \left[f(\tau_{i-1})\sigma_{i-1}^z \sigma_i^z + f(\tau_{i+1})\sigma_{i+1}^z \sigma_{i+2}^z \right]$$
$$- \left(\cos^2\varphi - \sin^2\varphi \sigma_{i-1}^z \sigma_i^z \sigma_{i+1}^z \sigma_{i+2}^z \right) \sigma_i^x \sigma_{i+1}^x. \tag{6.110}$$

In the case when both spins τ_{i-1} and τ_{i+1} are down, the function f is zero and both terms in square brackets vanish and the above operator becomes $1 - (\cos^2\varphi - \sin^2\sigma_{i-1}^z \sigma_i^z \sigma_{i+1}^z \sigma_{i+2}^z)\sigma_i^x \sigma_{i+1}^x$. It is clear then that its minimal eigenvalue is zero. In the case when $\tau_{i-1} = \tau_{i+1} = 1$ these terms do not vanish, however, still this is effectively a 16×16 matrix which can be shown to be positive computationally: using the software Mathematica we can easily see that the minimal eigenvalue is zero.

For $\tau_{i-1} = -\tau_{i+1}$, one of the values $f(\tau_{i-1})$ or $f(\tau_{i+1})$ is zero. Assuming that $f(\tau_{i-1}) = 0$ (the case of $f(\tau_{i+1}) = 0$ leads to the same eigenvalues), one has

$$H_\tau^{(i)} = 1 - \frac{1}{2} \sin 2\varphi \sigma_{i+1}^z \sigma_{i+2}^z - \sqrt{\cos 2\varphi} \sigma_i^x \sigma_{i+1}^x. \tag{6.111}$$

When constrained to three consecutive spins (i-1, i, and i+1) this $H_\tau^{(i)}$ is just a 8 by 8 matrix (on the remaining spins it acts as the identity matrix) and its eigenvalues can be obtained using Mathematica. One then checks that its minimal eigenvalue is $1 - (1/2)\sqrt{4 \cos 2\varphi + (\sin 2\varphi)^2}$ with $\varphi \in [0, \pi/4]$. Simple analysis shows that this is a nonnegative function of φ and gives zero only when $\varphi = 0$. In conclusion, $H_\tau^{(i)} \geq 0$ for all τ s and $\varphi \in [0, \pi/4]$ and therefore our Hamiltonians H_τ are positive. $\square$

Based on the above analysis, let us now distinguish all the cases with respect to τ and φ when $H_\tau(\varphi)$ can have zero-energy eigenstates. It clearly follows from the proof of lemma 1 that if $\tau \neq 0$ or $\tau \neq 2^N - 1$ there exists i such that $\tau_{i-1} \neq \tau_{i+1}$ and then the corresponding $H_\tau(\varphi)$ can have zero eigenvalues only when $\varphi = 0$. Let us now discuss this case. It follows from Eqs. 6.106 and 6.107 that for $\varphi = 0$, which corresponds to infinite temperature, the dependence on τ vanishes and one obtains

$$H_\tau(0) \equiv \overline{H} = \sum_i \left(1 - \sigma_i^x \sigma_{i+1}^x\right), \tag{6.112}$$

which has a doubly degenerate ferromagnetic ground state.

For $\tau = 0$ one gets the Hamiltonian obtained in [114], that is

$$H_0(\varphi) = - \sum_i \left[\left(\cos^2\varphi - \sin^2\varphi \sigma_{i-1}^z \sigma_i^z \sigma_{i+1}^z \sigma_{i+2}^z\right) \sigma_i^x \sigma_{i+1}^x \right. \\ \left. - \left(1 - (1/2)\sin 2\varphi (\sigma_{i-1}^z \sigma_i^z + \sigma_{i+1}^z \sigma_{i+2}^z)\right)\right]. \tag{6.113}$$

The ground state of this Hamiltonian is doubly degenerate for all values of φ, except for $\varphi = \pi/4$ (zero temperature) where also the first excited state becomes degenerate with the ground state [114]. For many values of τ this statement holds, except that the ground state has a positive energy—implying that the off-diagonal elements of the QME decay in time. In other cases we find that the ground state is unique for all values of φ, even $\pi/4$. Typical spectra for some values of τ in finite systems are shown in Fig. 6.5.

6.5.2 The Single Flip Model

For comparison only, we reproduce here the associated Hamiltonians that are obtained when single flip processes are allowed in the quantum master equation [113]. Again, a set of conserved quantities allows us to break the QME into 2^N Schrödinger equations labeled by a parameter τ

$$|\dot\psi_\tau(t)\rangle = -H_\tau|\psi_\tau(t)\rangle \quad (\tau = 0, \dots, 2^N - 1), \tag{6.114}$$

Fig. 6.5 Low energy states of the Hamiltonians (6.106) associated to the two flip quantum master equation for a system with $N=16$ spins as a function of φ. The panels are **a** $\tau = 2^8 - 1$ (half τ-spins up and half down), **b** $\tau = 2^8$ (only one τ-spin up, the others down), and **c** $\tau = 2^9 + 2^8$ (two neighboring τ-spins up, the others down). Only in case **c** the ground state is fully degenerate for all values of φ, in the other two the first excited state energy is very close but not equal to the ground state. In case **b** the ground state is not degenerate at $\varphi = \pi/4$, while in the other two cases it is

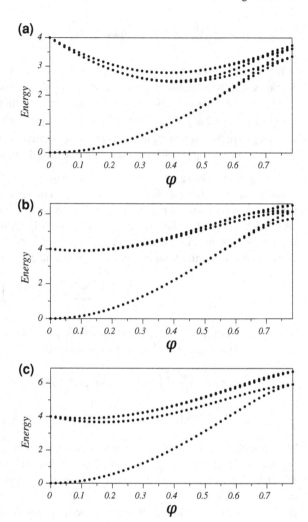

where the Hamiltonians H_τ are given by

$$H_\tau \equiv H_\tau(\delta, \gamma) = -\Gamma \sum_i \left[\left(\widetilde{A}_i(\delta, \gamma) - \widetilde{B}_i(\delta, \gamma) \sigma_{i-1}^z \sigma_{i+1}^z \right) \sigma_i^x \right.$$

$$- 1 + \frac{\gamma}{2}(1+\delta)\sigma_i^z \left(f(\tau_{i-1}\tau_i)\sigma_{i-1}^z + \right.$$

$$\left. f(\tau_i\tau_{i+1})\sigma_{i+1}^z \right) - \delta f(\tau_{i-1}\tau_{i+1})\sigma_{i-1}^z \sigma_{i+1}^z \right], \quad (6.115)$$

where

$$\widetilde{A}_i(\gamma, \delta) = \begin{cases} \dfrac{(1+\delta)\gamma^2}{2(1-\sqrt{1-\gamma^2})} - \delta, & \tau_{i-1} = \tau_{i+1}, \\[2mm] \sqrt{1-\delta^2}\sqrt[4]{1-\gamma^2}, & \tau_{i-1} = -\tau_{i+1} \end{cases} \quad (6.116)$$

and

$$\tilde{B}_i(\gamma, \delta) = \begin{cases} 1 - \dfrac{(1+\delta)\gamma^2}{2(1 - \sqrt{1-\gamma^2})}, & \tau_{i-1} = \tau_{i+1}, \\ 0, & \tau_{i-1} = -\tau_{i+1}. \end{cases} \qquad (6.117)$$

Here, each τ means a configuration of the conserved quantities that is different than the one shown above for the two flip model – however, we still use its binary representation so that τ is a shorthand notation for N variables $(\tau_1, \ldots, \tau_N)$, each taking values ± 1. The Eq. in (6.114) for $\tau = 0$ corresponds to the diagonal elements of $\varrho(t)$, while for the remaining $\tau \neq 0$, they describe the off-diagonal elements of the density matrix.

Let us shortly comment on the above model. First, it is easy to notice that for $\tau = 0$ or $\tau = 2^N - 1$, from Eqs. 6.115, 6.116, and 6.117 one recovers the Hamiltonian (6.95). Since, as shown in Ref. [109] the Hamiltonian (6.95) has a ground state with zero energy, it means that there exist off-diagonal elements surviving the evolution. On the other hand for $\tau \neq 0, 2^N - 1$ one gets (6.95), however, with some impurities. After substitution of bond variables (see e.g. [110]) one can map H_τ to disordered Heisenberg chains meaning that for some particular values of the involved parameters the model can be solved analytically. On the other hand, one may always treat this model numerically through matrix product states.

We show in Fig. 6.6 the spectra for some of these Hamiltonians, which is to be contrasted with the spectra from the two spin flip models, Fig. 6.5. In the single flip model the ground state is always unique except at zero temperature, where for all of the associated Hamiltonians one observes criticality.

6.6 Discussion and Outlook

In these lectures we have seen how quantum information theory can bring about a fresh perspective into many-body physics. However, the field is much bigger than what we have reviewed. Let us just mention here a few relevant topics that we have not covered, but that have received plenty of attention from the community, and that certainly have contributed to sizable advances in our understanding of many body physics.

One interesting application of entanglement is to critical phenomena. We briefly saw how block entanglement entropy scales differently at a gapless critical point. However, many entanglement measures display some kind of special behavior around quantum criticality—which was first observed [4] in the concurrence of nearest-neighbor spins of an Ising chain (see Ref. [118] for a recent review of activity in this field). Quantum criticality, in fact, is a very active subject in the condensed matter community, and has been studied using other quantum information approaches like the ground state fidelity [119] and the Loschmidt echo [120], whose usefulness in practice has been demonstrated experimentally [121, 122].

Fig. 6.6 Low energy states of the single flip Hamiltonians studied in Ref. [113] as a function of the temperature parameter $\gamma = \tanh 2\beta J$ for the same parameters as in Fig. 6.5. The three panels correspond to the same τ-spin configurations, even though the variables τ are defined differently. Notice that in this case the spectra becomes degenerate always at $\gamma = 1$, and that the ground state is always unique

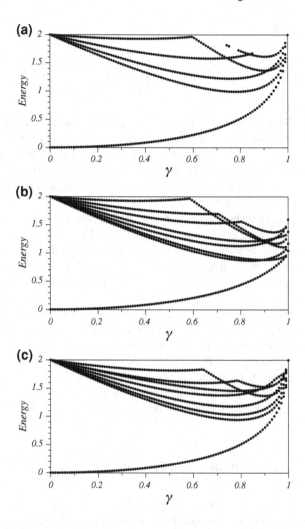

Another problem that is gaining interest is that of topological order, which we mentioned briefly as one of the motivations for studying area laws. One interesting recent development is the study of entanglement spectra [123, 124], defined through the Schmidt decomposition in such a way that each Schmidt coefficient λ_α of a bipartition is interpreted as a dimensionless energy $\xi_\alpha = -\log \lambda_\alpha$. This approach allows to generalize the von Neumann block entropy by introducing a virtual temperature, and study the structure of entanglement with more detail. In particular, it appears that gapped systems with topological order always have a gapless entanglement spectrum [123]. As a characterization of entanglement, the whole spectrum promises to be better than just entropy—simply because a set of numbers contains more information than a single one.

Although we concentrated mostly on the theoretical aspect of matrix and tensor product states, the field is also strongly geared to the practical application of simulation of many-body systems in classical computers. On the theory side, the tensor product approach has given successful advances in the theory of computational complexity applied to quantum mechanics [101], and in the recent theory of entanglement renormalization [125].

On the computational side, MPS algorithms have expanded the effective DMRG methods, and tremendous progress is being done in the simulation of strongly correlated particle systems. Bosonic particles can be represented straightforwardly[126] by mapping the d internal levels of the spins to the occupation number at each lattice site—thus truncating the Hilbert space to the subspace with at most d-1 particles in each site. Fermionic models, however, require some extra care when contracting the indices in the network so that fermionic commutation relations are respected [127–132]. In any case, tensor networks are still computationally efficient with strongly correlated electron systems, which puts these algorithms at an advantage over quantum Montecarlo type techniques—who suffer from the so called "sign problem" in this type of systems [133]. Therefore, tensor network techniques are important to support the large experimental efforts towards implementing quantum simulations of fermionic models (mainly with trapped ions [134] and ultracold atomic systems [135]).

The quantum kinetic Ising models discussed in the last section hold plenty of potential for the near future in at least two fronts. First, they represent a whole new class of many-body systems amenable to analytical solution, and can therefore bring new insight into our understanding of complex quantum many-body dynamics, as well as some classical reaction-diffusion problems [136]. Second, they have a close relationship and could be useful to the recent ideas on "environment design" [137–139]: crafting and/or manipulating the environment of a system so that it is driven to an interesting quantum many-body state, usually with a dynamics given by a quantum master equation. Because quantum kinetic models can be well understood and controlled, they might provide the foundation on top of which more elaborated systems are designed.

Acknowledgments We are grateful to Ll. Masanes for helpful discussion. We acknowledge the support of Spanish MEC/MINCIN projects TOQATA (FIS2008-00784) and QOIT (Consolider Ingenio 2010), ESF/MEC project FERMIX (FIS2007-29996-E), EU Integrated Project SCALA, EU STREP project NAMEQUAM, ERC Advanced Grant QUAGATUA, Caixa Manresa, AQUTE, and Alexander von Humboldt Foundation Senior Research Prize.

References

1. Jaksch, D., Briegel, H.-J., Cirac, J.I., Gardiner, C.W., Zoller, P.: Entanglement of atoms via cold controlled collisions. Phys. Rev. Lett. **82**, 1975 (1999)
2. Osborne, T.J., Nielsen, M.A.: Entanglement, quantum phase transitions, and density matrix renormalization. Quantum Inf. Proc. **1**, 45 (2002)

3. Osborne, T.J., Nielsen, M.A.: Entanglement in a simple quantum phase transition. Phys. Rev. A **66**, 032110 (2002)
4. Osterloh, A., Amico, L., Falci, G., Fazio, R.: Scaling of entanglement close to a quantum phase transition. Nature **416**, 608 (2002)
5. Ekert, A.K.: Quantum cryptography based on Bell's theorem. Phys. Rev. Lett. **67**, 661 (1991)
6. Bennett, C.H., Wiesner, S.J.: Communication via one- and two-particle operators on Einstein-Podolsky-Rosen states. Phys. Rev. Lett. **69**, 2881 (1992)
7. Bennett, C.H., Brassard, G., Crépeau, C., Jozsa, R., Peres, A., Wootters, W.K.: Teleporting an unknown quantum state via dual classical and Einstein-Podolsky-Rosen channels. Phys. Rev. Lett. **70**, 1895 (1992)
8. Horodecki, R., Horodecki, M., Horodecki, P., Horodecki, K.: Quantum entanglement. Rev. Mod. Phys. **81**, 865 (2009)
9. Werner, R.F.: Quantum states with Einstein–Podolsky–Rosen correlations admitting a hidden-variable model. Phys. Rev. A **40**, 4277 (1989)
10. Horodecki, P.: Separability criterion and inseparable mixed states with positive partial transposition. Phys. Lett. A **232**, 333 (1997)
11. Sanpera, A., Terrach, R., Vidal, G.: Local description of quantum inseparability. Phys. Rev. A **58**, 826 (1998)
12. Samsonowicz, J., Kuś, M., Lewenstein, M.: Phys. Rev. A **76**, 022314 (2007)
13. Żukowski, M., Zeilinger, A., Horne, M.A., Ekert, A.K.: "Event-ready-detectors" Bell experiment via entanglement swapping. Phys. Rev. Lett. **71**, 4287 (1993)
14. Gühne, O., Tóth, G.: Entanglement detection. Phys. Rep. **474**, 1 (2009)
15. Doherty, A.C., Parrilo, P.A., Spedalieri, F.M.: Distinguishing separable and entangled states. Phys. Rev. Lett. **88**, 187904 (2002)
16. Hulpke, F., Bruß, D.: A two-way algorithm for the entanglement problem. J. Phys. A: Math. Gen. **38**, 5573 (2005)
17. Gurvits, L.: Classical complexity and quantum entanglement. STOC **69**, 448 (2003)
18. Choi, M.-D.: Positive linear maps. Proc. Symp. Pure Math. **38**, 583 (1982)
19. Peres, A.: Separability criterion for density matrices. Phys. Rev. Lett **77**, 1413 (1996)
20. Horodecki, M., Horodecki, P., Horodecki, R.: Mixed–state entanglement and distillation: is there a "bound" entanglement in nature?. Phys. Rev. Lett. **80**, 5239 (1998)
21. Życzkowski, K., Horodecki, P., Sanpera, A., Lewenstein, M.: Volume of the set of separable states. Phys. Rev. A **58**, 883 (1998)
22. DiVincenzo, D.P., Shor, P.W., Smolin, J.A., Terhal, B.M., Thapliyal, A.V.: Evidence for bound entangled states with negative partial transpose. Phys. Rev. A **61**, 062312 (2000)
23. Dür, W., Cirac, J.I., Lewenstein, M., Bruß, D.: Distillability and partial transposition in bipartite systems. Phys. Rev. A **61**, 062313 (2000)
24. Horodecki, M., Horodecki, P., Horodecki, R.: Separability of mixed states: necessary and sufficient conditions. Phys. Lett. A **223**, 1 (1996)
25. Bishop, E., Bridges, D.: Constructive Analysis. Springer, Berlin (1985)
26. Terhal, B.M.: Bell inequalities and the separability criterion. Phys. Lett. A **271**, 319 (2000)
27. Gorini, V., Kossakowski, A., Sudarshan, E.C.G.: Completely positive dynamical semigroups of N–level systems. J. Math. Phys. **17**, 821 (1976)
28. Choi, M.-D.: Completely positive linear maps on complex matrices. Linear Alg. Appl. **10**, 285 (1975)
29. Kraus, K.: States, Effects and Operations: Fundamental Notions of Quantum Theory. Springer, Berlin (1983)
30. Stinespring, W.F.: Positive functions on C*-algebras. Proc. Am. Math. Soc. **6**, 211 (1955)
31. Jamiołkowski, A.: Linear transformations which preserve trace and positive semidefiniteness of operators. Rep. Math. Phys. **3**, 275 (1972)
32. Horodecki, M., Horodecki, P., Horodecki, R.: Separability of n-particle mixed states: necessary and sufficient conditions in terms of linear maps. Phys. Lett. A **283**, 1 (2001)

33. Woronowicz, S.L.: Positive maps of low dimensional matrix algebras. Rep. Math. Phys. **10**, 165 (1976)
34. Tanahashi, K., Tomiyama, J.: Indecomposable positive maps in matrix algebras. Can. Math. Bull **31**, 308 (1988)
35. Horodecki, M., Horodecki, P.: Reduction criterion of separability and limits for a class of distillation protocols. Phys. Rev. A **59**, 4206 (2000)
36. Cerf, N.J., Adami, C., Gingrich, R.M.: Reduction criterion for separability. Phys. Rev. A **60**, 898 (1999)
37. Breuer, H.-P.: Optimal entanglement criterion for mixed states. Phys. Rev. Lett. **97**, 080501 (2006)
38. Hall, W.: A new criterion for indecomposability of positive maps. J. Phys. A **39**, 14119 (2006)
39. Woronowicz, S.: Nonextendible positive maps. Comm. Math. Phys. **51**, 243 (1976)
40. Lewenstein, M., Kraus, B., Cirac, J.I., Horodecki, P.: Optimization of entanglement witnesses. Phys. Rev. A **62**, 052310 (2000)
41. Terhal, B.M.: A family of indecomposable positive linear maps based on entangled quantum states. Lin. Alg. Appl. **323**, 61 (2001)
42. Bennett, C.H., DiVincenzo, D.P., Smolin, J.A., Wootters, W.K.: Mixed-state entanglement and quantum error correction. Phys. Rev. A **54**, 3824 (1996)
43. Vedral, V., Plenio, M.B., Rippin, M.A., Knight, P.L.: Quantifying entanglement. Phys. Rev. Lett **78**, 2275 (1997)
44. DiVincenzo, D.P., Fuchs, C.A., Mabuchi, H., Smolin, J.A., Thapliyal, A., Uhlmann, A.: In Proceedings of the first NASA International Conference on Quantum Computing and Quantum Communication. Springer (1998)
45. Laustsen, T., Verstraete, F., van Enk, S.J.: Local vs. joint measurements for the entanglement of assistance. Quantum Inf. Comput. **3**, 64 (2003)
46. Nielsen, M.A.: Conditions for a class of entanglement transformations. Phys. Rev. Lett. **83**, 436 (1999)
47. Vidal, G.: Entanglement monotones. J. Mod. Opt. **47**, 355 (2000)
48. Jonathan, D., Plenio, M.B.: Minimal conditions for local pure-state entanglement manipulation. Phys. Rev. Lett. **83**, 1455 (1999)
49. Horodecki, M., Sen(De), A., Sen, U.: Dual entanglement measures based on no local cloning and no local deleting. Phys. Rev. A **70**, 052326 (2004)
50. Horodecki, M.: Distillation and bound entanglement. Quantum Inf. Comput. **1**, 3 (2001)
51. Plenio, M.B., Virmani, S.: An introduction to entanglement measures. Quant. Inf. Comp. **7**, 1 (2007)
52. Bennett, C.H., Bernstein, H.J., Popescu, S., Schumacher, B.: Concentrating partial entanglement by local operations. Phys. Rev. A **53**, 2046 (1996)
53. Hill, S., Wootters, W.K.: Entanglement of a pair of quantum bits. Phys. Rev. Lett. **78**, 5022 (1997)
54. Wootters, W.K.: Entanglement of formation of an arbitrary state of two qubits. Phys. Rev. Lett. **80**, 2245 (1998)
55. Terhal, B.M., Vollbrecht, K.G.H.: Entanglement of formation for isotropic states. Phys. Rev. Lett. **85**, 2625 (2000)
56. Vollbrecht, K.G.H., Werner, R.F.: Entanglement measures under symmetry. Phys. Rev. A **64**, 062307 (2001)
57. Rungta, P., Bužek, V.V., Caves, C.M., Hillery, M., Milburn, G.J.: Universal state inversion and concurrence in arbitrary dimensions. Phys. Rev. A **64**, 042315 (2001)
58. Rungta, P., Caves, C.M.: Concurrence-based entanglement measures for isotropic states. Phys. Rev. A **67**, 012307 (2003)
59. Aolita, L., Mintert, F.: Measuring multipartite concurrence with a single factorizable observable. Phys. Rev. Lett. **97**, 050501 (2006)
60. Walborn, S.P., Ribero, P.H.S., Davidovich, L., Mintert, F., Buchleitner, A.: Experimental determination of entanglement with a single measurement. Nature **440**, 1022 (2006)

61. Vidal, G., Werner, R.F.: Computable measure of entanglement. Phys. Rev. A **65**, 032314 (2002)
62. Plenio, M.B.: Logarithmic negativity: a full entanglement monotone that is not convex. Phys. Rev. Lett. **95**, 090503 (2005)
63. Bombelli, L., Koul, R.K., Lee, J., Sorkin, R.D.: Quantum source of entropy for black holes. Phys. Rev. D **34**, 373 (1986)
64. Srednicki, M.: Entropy and area. Phys. Rev. Lett. **71**, 666 (1993)
65. Bekenstein, J.D.: Black holes and entropy. Phys. Rev. D **7**, 2333 (1973)
66. Bekenstein, J.D.: Black holes and information theory. Contemp. Phys. **45**, 31 (2004)
67. Hawking, S.W.: Black hole explosions?. Nature **248**, 30 (1974)
68. Bousso, R.: The holographic principle. Rev. Mod. Phys. **74**, 825 (2002)
69. Eisert, J., Cramer, M., Plenio, M.B.: Area laws for the entanglement entropy – a review. Rev. Mod. Phys. **82**, 277 (2010)
70. Calabrese, P., Cardy, J., Doyon, B.: Special issue: entanglement entropy in extended quantum systems. J. Phys. A **42**, 500301 (2009)
71. Lubkin, E.: Entropy of an n–system from its correlation with a k–reservoir. J. Math. Phys. **19**, 1028 (1978)
72. Lloyd, S., Pagels, H.: Complexity as thermodynamic depth. Ann. Phys. **188**, 186 (1988)
73. Page, D.N.: Average entropy of a subsystem. Phys. Rev. Lett. **71**, 1291 (1993)
74. Bengtsson, I., Życzkowski, K.: Geometry of Quantum States. Cambridge University Press, Cambridge, MA (2006)
75. Foong, S.K., Kanno, S.: Proof of a Page's conjecture on the average entropy of a subsystem. Phys. Rev. Lett. **72**, 1148 (1994)
76. Sen, S.: Average entropy of a quantum subsystem. Phys. Rev. Lett. **77**, 1 (1996)
77. Sanchez-Ruíz, J.: Simple Proof of Page's conjecture on the average entropy of a subsystem. Phys. Rev. E **52**, 5653 (1995)
78. Hastings, M.B.: An area law for one-dimensional quantum system. J. Stat. Mech. Theory Exp. **2007**, 08024 (2007)
79. Eisert, E.H., Robinson, D.W.: The finite group velocity of quantum spin systems. Comm. Math. Phys. **28**, 251 (1972)
80. Masanes, L.: Area law for the entropy of low-energy states. Phys. Rev. A **80**, 052104 (2009)
81. Dür, W., Hartmann, L., Hein, M., Lewenstein, M., Briegel, H.-J.: Entanglement in spin chains and lattices with long-range Ising-type interactions. Phys. Rev. Lett. **94**, 097203 (2005)
82. Eisert, J., Osborne, T.: General entanglement scaling laws from time evolution. Phys. Rev. Lett. **97**, 150404 (2006)
83. Latorre, J.I., Riera, A.: A short review on entanglement in quantum spin systems. J. Phys. A **42**, 504002 (2009)
84. Vidal, G., Latorre, J.I., Rico, E., Kitaev, A.: Entanglement in quantum critical phenomena. Phys. Rev. Lett. **90**, 227902 (2003)
85. Jin, B.-Q., Korepin, V.E.: Quantum spin chain, Toeplitz deteminants and the Ficher–Hartwig conjecture. J. Stat. Phys. **116**, 79 (2004)
86. Its, A.R., Jin, B.-Q., Korepin, V.E.: Entanglement in the XY spin chain. J. Phys. A: Math. Gen. **38**, 2975 (2005)
87. Keating, J.P., Mezzadri, F.: Entanglement in quantum spin chains, symmetry classes of random matrices, and conformal field theory. Phys. Rev. Lett. **94**, 050501 (2005)
88. Eisert, J., Cramer, M.: Single-copy entanglement in critical quantum spin chains. Phys. Rev. A **72**, 042112 (2005)
89. Calabrese, P., Cardy, J.: Entanglement entropy and conformal field theory. J. Phys. A **42**, 504005 (2009)
90. Wolf, M.M.: Violation of the entropic area law for fermions. Phys. Rev. Lett. **96**, 010404 (2006)
91. Gioev, D., Klich, I.: Entanglement entropy of fermions in any dimension and the Widom conjecture. Phys. Rev. Lett. **96**, 100503 (2006)

92. Farkas, S., Zimboras, Z.: The von Neumann entropy asymptotics in multidimensional fermionic systems. J. Math. Phys. **48**, 102110 (2007)
93. Hastings, M.B.: Locality in quantum and Markov dynamics on lattices and networks. Phys. Rev. Lett. **93**, 140402 (2004)
94. Boyd, S., Vanderberghe, L.: Convex Optimization. Cambridge University Press, Cambridge, MA (2004)
95. Groisman, B., Popescu, S., Winter, A.: Quantum, classical, and total amount of correlations in a quantum state. Phys. Rev. A **72**, 032317 (2005)
96. Wolf, M.M., Verstraete, F., Hastings, M.B., Cirac, J.I.: Area laws in quantum systems: mutual information and correlations. Phys. Rev. Lett. **100**, 070502 (2008)
97. Verstraete, F., Popp, M., Cirac, J.I.: Entanglement versus correlations in spin systems. Phys. Rev. Lett. **92**, 027901 (2004)
98. Vidal, G.: Efficient classical simulation of slightly entangled quantum computations. Phys. Rev. Lett. **91**, 147902 (2003)
99. Perez-García, D., Verstraete, F., Wolf, M.M., Cirac, J.I.: Matrix product state representation. Quantum Inf. Comput. **7**, 401 (2007)
100. Verstraete, F., Cirac J.I.: Renormalization algorithms for Quantum-Many Body Systems in two and higher dimensions. cond-mat/0407066 (2004)
101. Schuch, N., Wolf, M.M., Verstraete, F., Cirac, J.I.: Computational complexity of projected entangled pair states. Phys. Rev. Lett. **98**, 140506 (2007)
102. Affleck, I., Kennedy, T., Lieb, E.H., Tasaki, H.: Rigorous results on valence-bond ground states in antiferromagnets. Phys. Rev. Lett. **59**, 799 (1987)
103. Affleck, I., Kennedy, T., Lieb, E.H., Tasaki, H.: Valence bond ground states in isotropic quantum antiferromagnets. Commun. Math. Phys. **115**, 477 (1988)
104. Majumdar, C.K., Ghosh, D.K.: On next-nearest-neighbor interaction in linear chain. I. J. Math. Phys. **10**, 1388 (1969)
105. Glauber, R.J.: Time-dependent statistics of the Ising model. J. Math. Phys **4**, 294 (1963)
106. Deker, U., Haake, F.: Renormalization group transformation for the master equation of a kinetic Ising chain. Z. Phys. B **35**, 281 (1979)
107. Kimball, J.C.: The kinetic Ising model: exact susceptibilities of two simple examples. J. Stat. Phys. **21**, 289 (1979)
108. Haake, F., Thol, K.: Universality classes for one dimensional kinetic Ising models. Z. Phys. B **40**, 219 (1980)
109. Felderhof, B.U.: Spin relaxation of the Ising chain. Rep. Math. Phys. **1**, 215 (1971)
110. Siggia, E.D.: Pseudospin formulation of kinetic Ising models. Phys. Rev. B **16**, 2319 (1977)
111. Heims, S.P.: Master equation for Ising model. Phys. Rev. **138**, A587 (1965)
112. Kawasaki, K. In: Domb, C., Green, M.S. (eds.) Phase Transition and Critical Phenomena, vol. 2, pp. 443–501. Academic Press, London (1972)
113. Augusiak, R., Cucchietti, F.M., Haake, F., Lewenstein, M.: Quantum kinetic Ising models. New J. Phys. **12**, 025021 (2010)
114. Hilhorst, H.J., Suzuki, M., Felderhof, B.U.: Kinetics of the stochastic Ising chain in a two–flip model. Physica **60**, 199 (1972)
115. Jordan, P., Wigner, E.: Über das Paulische Aequivalenzverbot. Z. Phys. **47**, 631 (1928)
116. Bogoliubov, N.N.: On a new method in the theory of superconductivity. Nuovo Cimento **7**, 794 (1958)
117. Valatin, J.G.: Comments on the theory of superconductivity. Nuovo Cimento **7**, 843 (1958)
118. Amico, L., Fazio, R., Osterloh, A., Vedral, V.: Entanglement in many-body systems. Rev. Mod. Phys. **80**, 517 (2008)
119. Zanardi, P., Paunković, N.: Ground state overlap and quantum phase transitions. Phys. Rev. E **74**, 031123 (2006)
120. Quan, H.T., Song, Z., Liu, X.F., Zanardi, P., Sun, C.P.: Decay of Loschmidt echo enhanced by quantum criticality. Phys. Rev. Lett. **96**, 140604 (2006)

121. Zhang, C., Tewari, S., Lutchyn, R., Sarma, S.D.: $px+ipy$ Superfluid from s-wave interactions of fermionic cold atoms. Phys. Rev. Lett. **101**, 160401 (2008)

122. Zhang, J., Cucchietti, F.M., Chandrashekar, C.M., Laforest, M., Ryan, C.A., Ditty, M., Hubbard, A., Gamble, J.K., Laflamme, R.: Direct observation of quantum criticality in Ising spin chains. Phys. Rev. A **79**, 012305 (2009)

123. Li, H., Haldane, F.D.M.: Entanglement spectrum as a generalization of entanglement entropy: identification of topological order in non-Abelian fractional quantum Hall effect states. Phys. Rev. Lett. **101**, 010504 (2008)

124. Calabrese, P., Lefevre, A.: Entanglement spectrum in one-dimensional systems. Phys. Rev. A **78**, 032329 (2008)

125. Vidal, G.: Entanglement renormalization. Phys. Rev. Lett. **99**, 220405 (2007)

126. Clark, S.R., Jaksch, D.: Dynamics of the superfluid to Mott-insulator transition in one dimension. Phys. Rev. A **70**, 043612 (2004)

127. Kraus, C.V., Schuch, N., Verstraete, F., Cirac, J.I.: Fermionic projected entangled pair states. Phys. Rev. A **81**, 052338 (2010)

128. Corboz, P., Vidal, G.: Fermionic multiscale entanglement renormalization ansatz. Phys. Rev. B **80**, 165129 (2009)

129. Corboz, P., Evenbly, G., Verstraete, F., Vidal, G.: Simulation of interacting fermions with entanglement renormalization. Phys. Rev. A **81**, 010303 (2010)

130. Barthel, T., Pineda, C., Eisert, J.: Contraction of fermionic operator circuits and the simulation of strongly correlated fermions. Phys. Rev. A **80**, 042333 (2009)

131. Corboz, P., Orús, R., Bauer, B., Vidal, G.: Simulation of strongly correlated fermions in two spatial dimensions with fermionic projected entangled-pair states. Phys. Rev. B **81**, 165104 (2010)

132. Pineda, C., Barthel, T., Eisert, J.: Unitary circuits for strongly correlated fermions. Phys. Rev. A **81**, 050303 (2010)

133. Troyer, M., Wiese, U.-J.: Computational complexity and fundamental limitations to fermionic quantum Monte Carlo simulations. Phys. Rev. Lett. **94**, 170201 (2005)

134. Kim, K., Chang, M.-S., Korenblit, S., Islam, R., Edwards, E.E., Freericks, J.K., Lin, G.-D., Duan, L.-M., Monroe, C.: Quantum simulation of frustrated Ising spins with trapped ions. Nature **465**, 590 (2010)

135. Jördens, R., Tarruell, L., Greif, D., Uehlinger, T., Strohmaier, N., Moritz, H., Esslinger, T., DeLeo, L., Kollath, C., Georges, A., Scarola, V., Pollet, L., Burovski, E., Kozik, E., Troyer, M.: Quantitative determination of temperature in the approach to magnetic order of ultracold fermions in an optical lattice. Phys. Rev. Lett. **104**, 180401 (2010)

136. Temme, K., Wolf, M.M., Verstraete, F.: Stochastic exclusion processes versus coherent transport. e-print arXiv:0912.0858 (2009)

137. Verstraete, F., Wolf, M.M., Cirac, J.I.: Quantum computation and quantum-state engineering driven by dissipation. Nat. Phys. **5**, 633 (2009)

138. Kraus, B., Büchler, H.P., Diehl, S., Kantian, A., Micheli, A., Zoller, P.: Preparation of entangled states by quantum Markov processes. Phys. Rev. A **78**, 042307 (2008)

139. Diehl, S., Micheli, A., Kantian, A., Kraus, B., Büchler, H.P., Zoller, P.: Quantum states and phases in driven open quantum systems with cold atoms. Nat. Phys. **4**, 878 (2008)

Chapter 7
Statistical Mechanics of Classical and Quantum Computational Complexity

C. R. Laumann, R. Moessner, A. Scardicchio and S. L. Sondhi

Abstract The quest for quantum computers is motivated by their potential for solving problems that defy existing, classical, computers. The theory of computational complexity, one of the crown jewels of computer science, provides a rigorous framework for classifying the hardness of problems according to the computational resources, most notably time, needed to solve them. Its extension to quantum computers allows the relative power of quantum computers to be analyzed. This framework identifies families of problems which are likely hard for classical computers ("NP-complete") and those which are likely hard for quantum computers ("QMA-complete") by indirect methods. That is, they identify problems of comparable worst-case difficulty without directly determining the individual hardness of any given instance. Statistical mechanical methods can be used to complement this classification by directly extracting information about particular families of instances— typically those that involve optimization—by studying random ensembles of them. These pose unusual and interesting (quantum) statistical mechanical questions and the results shed light on the difficulty of problems for large classes of algorithms as well as providing a window on the contrast between typical and worst case complexity. In these lecture notes we present an introduction to this set of ideas with older work on classical satisfiability and recent work on quantum satisfiability

C. R. Laumann · S. L. Sondhi
Department of Physics,
Princeton University, Princeton, NJ 08544, USA

R. Moessner (✉)
Max-Planck-Institut für Physik komplexer Systeme,
01187 Dresden, Germany
e-mail: moessner@pks.mpg.de

A. Scardicchio
Abdus Salam International Centre for Theoretical Physics,
Strada Costiera 11, 34014 Trieste, Italy

D. C. Cabra et al. (eds.), *Modern Theories of Many-Particle Systems in Condensed Matter Physics*, Lecture Notes in Physics 843, DOI: 10.1007/978-3-642-10449-7_7,
© Springer-Verlag Berlin Heidelberg 2012

as primary examples. We also touch on the connection of computational hardness with the physical notion of glassiness.

7.1 Introduction

A large and exciting effort is underway to build quantum computers. While the roots of this effort lie in the deep insights of pioneers such as Feynman and Deutsch, what triggered the growth was the discovery by Shor that a quantum computer could solve the integer factoring problem efficiently—a feat currently beyond the reach of classical computers. In addition to the desire to create useful devices intrinsically more powerful than existing classical computers, the challenge of creating large quantum systems subject to precise control has emerged as the central challenge of quantum physics in the last decade.

The first and primary objective—that of enhanced computational power—has in turn spurred the founding of quantum computer science and the development of a rigorous theory of the (potential) power of quantum computers: quantum complexity theory. This theory builds on the elegant ideas of classical complexity theory to classify computational problems according to the resources needed to solve them as they become large. The distinction between polynomial scaling of resources, most notably time, and super-polynomial scaling (e.g. exponential) generates a robust distinction between easy and hard problems.

While this distinction is easily made in principle, in practice complexity theory often proceeds by the powerful technique of assigning guilt by association: more precisely, that of classifying problems by mapping between them. This allows the isolation of sets of problems that encapsulate the difficulty of an entire class: for example, the so-called satisfiability problem (SAT) captures the difficulty of all problems whose solution can be easily checked by a classical computation; the quantum satisfiability problem (QSAT) does the same for quantum computers, as we will explain later in these notes. The solution of these problems would thus enable the solution of the vast set of of all checkable problems. This implication is a powerful argument that both SAT and QSAT must be hard.

This kind of indirect reasoning is very different from the way physicists normally approach problems: one of the purposes of these notes is to help physics readers appreciate the power of the computer science approach. However, the direct approach of examining actual problem instances and attempting to come up with algorithms for them is, of course, also important and this is where physicists are able to bring their own methods to bear. Specifically, physicists have applied themselves to the task of trying to understand problems such as the two satisfiability problems. These can be expressed as optimization problems and thus look much like the Hamiltonian problems the field is used to. Even more specifically, they have studied ensembles of these problems with a variety of natural probability measures in order to reveal features of the hard instances.

As is familiar from the statistical mechanical theory of disordered systems such as spin glasses, studying a random ensemble brings useful technical simplifications that allow the structure of a typical instance to be elucidated with less trouble than for a specific capriciously picked instance. This has enabled the identification of phase transitions as parameters defining the ensembles are changed—exactly the kind of challenge to warm a statistical physicist's heart. A further major product of such work, thus far largely in the classical realm, has been the identification of obstacles to the solution of such typical instances by large classes of algorithms and the construction of novel algorithms that avoid these pitfalls. We note that this focus on typical instances also usefully complements the standard results of complexity theory which are necessarily controlled by the worst cases—instances that would take the longest to solve but which may be very unusual. This is then an independent motivation for studying such ensembles.

Certainly, the flow of ideas and technology from statistical mechanics to complexity theory has proven useful. In return, it is useful to reflect on what complexity theory has to say about physical systems. Here the central idea—whose precise version is the Church–Turing hypothesis—is that a physical process is also a computation in that it produces an output from an input. Thus if a complexity theoretic analysis indicates that a problem is hard, any physical process that encodes its solution must take a long time. More precisely, the existence of hard optimization problems implies the existence of a class of Hamiltonians whose ground states can only be reached in a time that scales exponentially in the volume of the system *irrespective* of the processes used.

This sounds a lot like what physicists mean by glassiness. We remind the reader that physical systems typically exhibit symmetric, high temperature or large quantum fluctuation phases, with a characteristic equilibration time that is independent of the size of the system. At critical points, or in phases with broken continuous symmetries, algebraic dependences are the norm. But glassy systems exhibit much slower relaxation and thus present a challenge to experimental study in addition to theoretical understanding. Indeed, there is no settled understanding of laboratory glassiness. Consequently, the complexity theoretic arguments that imply the existence of glassy Hamiltonians, both in the classical and quantum cases, ought to be interesting to physicists. That said, we hasten to add that the connection is not so simple for two reasons. First, complexity theoretic results do not always hold if we restrict the degrees of freedom to live in Euclidean space—say on regular lattices—and require spatial locality. Thus hard Hamiltonians in complexity theory can look unphysical to physicists. Nonetheless, there are many interesting low dimensional (even translation invariant) problems which are hard in the complexity theoretic sense [1, 2]. Second, the physical processes intrinsic to a given system can sometimes be slow for reasons of locality or due to energetic constraints which are ignored when one is considering the full set of algorithms that can solve a given optimization problem. Still, we feel this is a direction in which Computer Science has something to say to Physics and we refer readers to a much more ambitious manifesto along this axis by Aaronson [3] for stimulation.

In the bulk of these notes we provide an introduction to this complex of ideas, which we hope will enable readers to delve further into the literature to explore the themes that we have briefly outlined above. In the first part, we provide a tutorial on the basics of complexity theory including sketches of the proofs of the celebrated proofs of NP and QMA completeness for the satisfiability and quantum satisfiability problems. In the second part, we show how statistical methods can be applied to these problems and what information has been gleaned. Unsurprisingly, the quantum part of this relies on recent work and is less developed than the classical results. In the concluding section we list some open questions stemming from the quantum results to date.

7.2 Complexity Theory for Physicists

Complexity theory classifies how "hard" it is to compute the solution to a problem as a function of the *input size N* of the problem instance. As already mentioned above, algorithms are considered *efficient* if the amount of time they take to complete scales at most polynomially with the size of the input and *inefficient* otherwise. The classification of algorithms by asymptotic efficiency up to polynomial transformations is the key to the robustness of complexity theoretic results, which includes the independence from an underlying model of computation.

In this line of reasoning, if $P \neq NP$, as is the current consensus, there are natural classes of problems which cannot be solved in polynomial time by any computational process, including any physical process which can be simulated by computer. Complexity theory provides its own guide to focusing our attention on a certain set of NP problems, those termed NP-complete, which capture the full hardness of the class NP. In particular, the problem of Boolean satisfiability of 3-bit clauses, 3-SAT, is NP-complete and therefore can encode the full hardness of the class NP.

The advent of the quantum computer modifies the above reasoning only slightly. It appears that quantum computers are somewhat more powerful than their classical counterparts, so that we must introduce new quantum complexity classes to characterize them. Nonetheless, analogous statements hold within this new framework: quantum polynomial (BQP) is larger than classical polynomial (P) but not powerful enough to contain classical verifiable (NP), nor quantum verifiable (QMA). If we wish to study the particularly hard quantum problems, we may turn to the study of QMA-complete problems such as LOCAL HAMILTONIAN and the closely related QSAT.

In this section, we provide a concise review of the key concepts for the above argument culminating in a discussion of worst-case hardness and the Cook–Levin theorem, showing the existence of NP-complete problems, and the quantum analogues due to Kitaev and Bravyi. This story motivates and complements the statistical study of 'typical' instances of 3-SAT, 3-QSAT and other classical and quantum hard optimization problems, which are discussed in the following sections.

7.2.1 Problems, Instances, Computers and Algorithms

The success of complexity theory as a classification scheme for the "hardness" of problems is in part due to the careful definitions employed. Here we sketch the most important aspects of these concepts and leave the rigorous formalism for the textbooks, of which we particularly recommend Arora and Barak to the interested reader [4]. We have taken a particular path through the forest of variations on the ideas below and do not pretend to completeness.

Throughout these notes we focus on so-called *decision problems*, that is Yes/No questions such as "Does the Hamiltonian H have a state with energy less than E?" rather than more general questions like "What is the ground state energy of H?" This restriction is less dramatic than it seems—many general questions may be answered by answering a (reasonably short) sequence of related Yes/No questions, much like playing the game twenty questions—and it significantly simplifies the conceptual framework we need to introduce. Moreover, many of the essential complexity theoretic results arise already within the context of decision problems.

A decision problem, then, is a question that one may ask of a class of *instances*. For example, the DIVIDES problem asks "Does a divide b?" for integers a and b. Leaving the variables a and b unspecified, we clearly cannot yet answer this question. An *instance* of DIVIDES might be "Does 5 divide 15?" A moment's thought now reveals that a definitive answer exists: Yes. We refer to instances of a problem as Yes-instances (No-instances) if the answer is Yes (No). We follow computer science convention by giving problems fully capitalized names.[1]

What does it mean to solve a problem? We would not feel we had solved the problem if we could only answer a few specific instances. On the other hand, we certainly could not expect to have a book containing the (infinite) table of answers to all possible instances for easy reference. Thus, we want a general *algorithm* which, given an arbitrary instance, provides us with a step-by-step recipe by which we can compute the answer to the instance. Some physical object must carry out the algorithmic instructions and it is this object that we call a *computer*—whether it is a laptop running C code, ions resonating in an ion trap or a sibling doing long division with pencil and paper. Thus, a solution to a decision problem is an algorithm which can decide arbitrary instances of the problem when run on an appropriate computer. Such an algorithm for a decision problem is often called a *decision procedure*.

Clearly it is less work to answer the DIVIDES problem for small numbers than for large. "Does 5 divide 15?" takes essentially no thought at all while "Does 1437 divide 53261346150?" would take a few moments to check. We therefore define the *input size* (or just *size*) of an instance as the number of symbols we need to specify the instance. In the DIVIDES problem, we could take the size as the number of symbols needed to specify the pair (a, b). The size N of $(5, 15)$ would be 6 while that of $(1437, 53261346150)$ is 18.

Computer scientists measure the *efficiency* of an algorithm by considering the asymptotic scaling of its resource consumption with the input size of the problem.

---

[1]   We trust this will not give physics readers PROBLEMS.

More precisely, consider the finite but large collection of all possible problem instances whose size is no greater than N. For each of these instances, the algorithm will take some particular amount of time. For the finite collection at size N, there will be a worst-case instance which takes more time T than any of the others at that size. Complexity theory generally focusses on the scaling of this worst-case time T as a function of input size N as $N \to \infty$. Clearly, the slower the growth of T with N, the more efficient the algorithm for large inputs. Indeed, algorithmic procedures are considered efficient so long as T grows at most polynomially with N, for any particular polynomial we like. Thus both linear and quadratic growth are efficient, even though linear growth clearly leeds to faster computations, at least for sufficiently large N. Anything slower, such as $T = O(e^N)$, is inefficient.

For example, the most famous decision procedure for DIVIDES—long division— takes of order $T = O(\log b \times \log a) \leq O(N^2)$ arithmetic steps to perform the division and check the remainder is 0. That T grows with a and b logarithmically corresponds nicely to our intuition that bigger numbers are harder to divide, but not too much harder. It is instructive to consider a different, inefficient, algorithm for the same problem. Suppose we had not yet learned how to divide but knew how to multiply. We might try the following decision procedure: try to multiply a with every number c from 1 up to b and check if $ac = b$. This trial-and-error approach would take $T = O(\log a \times \log b \times b) \approx O(e^{cN})$ to try out all the possibilities up to b. Even for relatively small instances, this approach would quickly become prohibitively time consuming—simply enumerating all of the numbers of up to 30 digits at one per nanosecond would take longer than the age of the universe!

Finally, we lift our classification of algorithm efficiency to a classification of problem hardness: A problem is *tractable* if there exists an efficient algorithm for solving it and it is *intractable* otherwise. By this definition, DIVIDES is tractable (long division solves it efficiently), despite the existence of alternative slower algorithms. As we will discuss further in the following sections, we can rarely *prove* that no efficient algorithm exists for a given problem, but complexity theory nonetheless offers strong arguments that certain large classes of problems are intractable in this sense.

Computers are clearly central to the determination of the difficulty of problems— we classify problems according to the efficiency of the computational algorithms that exist for treating them. In addition to the time taken, we can measure the resource requirements in various implementation dependent ways—memory consumed, number of gates required, laser pulse bandwidth, quantity of liquid Helium evaporated. One might expect that the kind of computer that we use would greatly influence any complexity classification. At the very least, your laptop will be faster than your brother at dividing 1,000 digit numbers. The beauty of the definition of efficiency by polynomial scaling is that many of these implementation dependent details drop out and we really can focus on the time efficiency as an overall measure of difficulty.[2]

---

[2] In practice the amount of memory or the number of cores in a workstation regularly limits its ability to do computations. Since in finite time, even a parallel computer can only do a finite amount of work or address a finite amount of memory, a polynomial bound on T also provides a

The robustness of these definitions follows from one of the great ideas of complexity theory: up to polynomial overheads, any reasonable classical computer may be simulated by any other. This is known as the strong Church–Turing hypothesis and, as its name suggests, is only a conjecture. Nonetheless, it has been examined and confirmed for many particular models of classical computation[3] and is widely believed to hold more generally. This is the reason for defining efficiency up to polynomial scaling: since any computer can simulate the operation of any other up to polynomial overheads, all computers can solve the same problems efficiently. In Sect. 7.2.3 below, we consider the most important classical *complexity classes* that arise from these coarse but robust definitions of efficiency.

The careful reader will have noticed that we restricted our statement of the Church–Turing hypothesis to *classical* computers. It is widely believed that classical computers *cannot* efficiently simulate quantum systems. Certainly, directly simulating Schrödinger's equation on a polynomially sized classical computer is problematic since the Hilbert space is exponentially large. On the other hand, if we had a quantum computer with which to do our simulation, the state space of our computer would also be a Hilbert space and we could imagine representing and evolving complex states of the system by complex states and evolutions of the quantum computer. This reasoning leads to the strong *quantum* Church–Turing hypothesis: that any reasonable quantum computer may be efficiently simulated by any other. With this hypothesis in hand, we may proceed to develop a robust classification of *quantum* complexity classes, as in Sect. 7.2.4.

7.2.2 Polynomial Reductions and Worst-Case Behavior

Reduction is the most important tool in complexity theory. A decision problem A reduces to another problem B if there is a polynomial time algorithm which can transform instances of A into instances of B such that Yes-instances (No-instances) of A map to Yes-instances (No-instances) of B. In this case, B is at least as hard as A: any algorithm which could efficiently decide B would be able to efficiently decide A as well. Just use the transformation to convert the given instance of A into an instance of B and then apply the efficient algorithm for B.

Reductions formalize the interrelationships between problems and allow us to show that new problems are actually part of known classes. Obviously, if we can reduce a problem A to a problem B that we know how to solve efficiently, we have just shown how to solve A efficiently as well. Conversely, if we have a problem C which we believe is *intractable*—that is, not solvable by an efficient algorithm—and

---

(Footnote 2 continued)
polynomial bound on the space requirements. Likewise, finite parallelization only provides constant time improvements. More refined classifications can be made by restricting resource consumption more tightly but we will not consider them here.

[3] For example, Turing machines, Boolean circuit models and your laptop.

we can reduce it to another problem D, that suggests D should also be intractable. Using this logic we can try to show that all kinds of interesting problems ought to be intractable if we can find one to start with.

7.2.3 Classical: P and NP

The most important complexity class is known as P—this is the class of decision problems which a classical computer can decide efficiently. More precisely, a decision problem is in P if there exists an algorithm that runs in polynomial time as a function of the input size of the instance and outputs Yes or No depending on whether the instance is a Yes-instance or No-instance of the problem. From a logical point of view, to show that a given problem is in P we need to provide an efficient procedure to decide arbitrary instances. Colloquially, P is the class of problems that are easy to solve.

We have already discussed one example, the DIVIDES problem, for which long division constitutes a polynomial time decision procedure. Another example is given by the energy evaluation problem: "Does a specific configuration σ of a classical Ising Hamiltonian $H = \sum J_{ij}\sigma_i\sigma_j$ have energy less than E?" Here the instance is specified by a configuration made of N bits, a Hamiltonian function with N^2 coupling terms and a threshold energy E (where all real numbers are specified with some fixed precision). Since we can evaluate the energy $H(\sigma)$ using of order N^2 multiplications and additions and then compare it to E, this problem is also in P.

The second most important complexity class is NP: this is the class of decision problems for which there exists a scheme by which Yes-instances may be efficiently verified by a classical computation. We may think of this definition as a game between a prover and a verifier in which the prover attempts, by hook or by crook, to convince the verifier that a given instance is a Yes-instance. The prover provides the verifier with a proof of this claim which the verifier can efficiently check and either Accept or Reject. NP places no restrictions on the power of the prover—only that Yes-instances must have Acceptable proofs, No-instances must not have Acceptable proofs and that the verifier can decide the Acceptability of the proof efficiently. We note that there is an intrinsic asymmetry in the definition of NP: we do not need to be able to verify that a No-instance is a No-instance.

For example, the problem "Does the ground state of the Hamiltonian $H = \sum J_{ij}\sigma_i\sigma_j$ have energy less than E?" has such an efficient verification scheme. If the prover wishes to show that a given H has such low energy states, he can prove it to the verifier by providing some configuration σ which he claims has energy less than E. The skeptical verifier may efficiently evaluate $H(\sigma)$ using the energy evaluation algorithm outlined above and if indeed $H(\sigma) < E$, the skeptic would Accept the proof. If H did not have such low energy states, then no matter what the prover tried, he would be unable to convince the verifier to Accept.

At first brush, NP seems a rather odd class—why should we be so interested in problems whose Yes-instances may be efficiently checked? Of course, any problem we can decide efficiently (in P) can be checked efficiently (because we can simply decide it!). What of problems outside of NP? These do not admit efficient verification schemes and thus certainly cannot have efficient decision procedures. Moreover, even if, by some supernatural act of intuition (not unusual in theoretical physics), we guess the correct answer to such a problem, we would not be able to convince anybody else that we were correct. There would be no efficiently verifiable proof! Thus, NP is the class of problems that we could ever hope to be convinced about.

Since 1971, the outstanding question in complexity theory (worth a million dollars since the new millennium), has been "Is P = NP?" This would be an astonishing result: it would state that all of the difficulty and creativity required to come up with the solution to a tough problem could be automated by a general purpose algorithm running on a computer. The determination of the truth of theorems would reduce to the simple matter of asking your laptop to think about it. Since most scientists believe that there are hard problems, beyond the capability of general purpose algorithms, the consensus holds that P ≠ NP.[4]

7.2.4 Quantum: BQP and QMA

The most important quantum complexity class is BQP—this is the class of decision problems which a quantum computer can decide efficiently with bounded error (the B in the acronym). Since general quantum algorithms have intrinsically stochastic measurement outcomes, we have no choice but to allow for some rate of false-positive and false-negative outcomes. As long as these rates are bounded appropriately (say by 1/3), a few independent repetitions of the quantum computation will exponentially suppress the probability of determining the incorrect result. Thus, BQP is the quantum analogue of P and plays a similar role in the classification of decision problems. Since a quantum computer can simulate any classical computation, P is contained in BQP.[5]

The most important example of a BQP problem that is not known to be in P is integer factoring. As a decision problem, this asks "Given N and M, does the integer N have a factor p with $1 < p \leq M$?" In the 1990s, Peter Shor famously proved that factoring is in BQP by developing a quantum factoring algorithm. There is no proof that factoring is classically hard (outside of P)—nonetheless, many of the cryptography schemes on which society relies for secure communication over the internet are only secure if it is. Shor's algorithm renders all of these schemes useless if a large scale quantum computer is ever built.

The quantum analogue of NP is the class QMA, Quantum Merlin–Arthur, which is the class of decision problems whose Yes-instances can be efficiently checked by a

---

[4] This may of course be the bias of the scientists who don't want to be replaced by omniscient laptops.

[5] For the expert, we note that a closer analogue of BQP is BPP, the class of decision problems which can be efficiently decided by a randomized classical algorithm with bounded error. In an attempt to minimize the onslaught of three letter acronyms, we have left this complication out.

quantum computer given a quantum state as a proof (or witness). The colorful name comes from the description of this class in terms of a game: Merlin, all-powerful but less than trustworthy, wishes to prove to Arthur, a fallible but well-intentioned individual who happens to have access to a quantum computer, that a particular instance of a problem is a Yes-instance. Merlin, using whatever supernatural powers he enjoys, provides Arthur with a quantum state designed to convince Arthur of this claim. Arthur then uses his quantum computer to decide, with some bounded error rate (say 1/3), whether to accept or reject the proof.

There are three primary differences between NP and QMA: (1) the verifier is a quantum computer, (2) the proof is a quantum state, and, (3) the verification is allowed a bounded error rate. The first two differences provide the class with its additional quantum power; that the verifier is allowed a bounded error rate is necessary due to quantum stochasticity, but not believed to be the source of its additional power. We note that the particular error bound is again somewhat arbitrary—Arthur can exponentially improve the accuracy of a noisy verification circuit by requesting Merlin provide him multiple copies of the proof state and running his verifier multiple times [5]. Thus, even a verifier which falsely accepts No-instances with probability up to $1/2 - 1/\text{poly}(N)$ while accepting valid proofs with probability 1/2 only slightly larger can be turned into an efficient bounded error QMA verifier through repetition.

An example of a QMA problem is given by the k-LOCAL HAMILTONIAN problem:

Input: A quantum Hamiltonian $H = \sum_m A_m$ composed of M bounded operators, each acting on k qubits of an N qubit Hilbert space. Also, two energies $a < b$, separated by at worst a polynomially small gap $b - a > 1/\text{poly(N)}$.
Output: Does H have an energy level below a?
Promise: Either H has an energy level below a or all states have energies above b.

Here we have introduced the notion of a 'promise' in a decision problem. Promises are a new feature in our discussion: they impose a restriction on the instances that a questioner is allowed to present to a decision procedure. The restriction arises because the algorithms and verification procedures we use to treat promise problems need not be correct when presented with instances that do not satisfy the promise— an efficient solver for LOCAL HAMILTONIAN could in fact fail on Hamiltonians with ground state energies in the *promise gap* between a and b and we would still consider LOCAL HAMILTONIAN solved.

Heuristically, it is clear why we need the promise gap for LOCAL HAMILTONIAN to be QMA: suppose we had a quantum verifier which took a quantum state $|\psi\rangle$ and tried to measure its energy $\epsilon = \langle\psi|H|\psi\rangle$ through a procedure taking time T. Time-energy uncertainty suggests that we should not be able to resolve ϵ to better than 1/T. Thus, if T is to be at most polynomially large in N, the verifier would not be able to determine whether an ϵ exponentially close to a is above or below the threshold.

The actual construction of a verification circuit for the LOCAL HAMILTONIAN problem is somewhat more subtle than simply 'measuring' the energy of a given state. As we will provide a very closely related construction for the QSAT problem below, we do not include the verifier for LOCAL HAMILTONIAN in these notes and instead refer the interested reader to Ref. [5].

The quantum analogue of the classical claim that P $\neq$ NP is that BQP $\neq$ QMA—a conjecture that is strongly believed for many of the same reasons as in the classical case.

7.2.5 NP-Completeness: Cook–Levin

In the early 1970s, Cook and Levin independently realized that there are NP problems whose solution captures the difficulty of the entire class NP. These are the so-called *NP-complete* problems. What does this mean?

A problem is NP-complete if it is both (a) in NP (efficiently verifiable) and (b) any problem in NP can be reduced to it efficiently. Thus, if we had an algorithm to solve an NP-complete problem efficiently, we could solve any problem whatsoever in NP efficiently. This would prove P = NP with all of the unexpected consequences this entails. Assuming on the contrary that P $\neq$ NP, any problem which is NP-complete must be intractable.

Let us sketch a proof of the Cook–Levin theorem showing the existence of NP-complete problems. In particular, we will show that classical 3-satisfiability, 3-SAT, is NP-complete. 3-SAT is the decision problem which asks whether a given Boolean expression composed of the conjunction of clauses, each involving at most 3 binary variables, has a satisfying assignment. Re-expressed as an optimization problem, 3-SAT asks, "Does the energy function

$$H = \sum_m E_m(\sigma_{m_1}, \sigma_{m_2}, \sigma_{m_3}), \tag{7.1}$$

acting on N binary variables σ_i in which each local energy term E_m takes values 0 or 1, have a zero energy (satisfying) ground state?"[6]

7.2.5.1 3-SAT is in NP

First, it is clear that 3-SAT is itself efficiently verifiable and therefore in NP. If a prover wishes to prove that a particular instance H is satisfiable, she could provide a verifier a zero energy configuration. The verifier would take this configuration and evaluate its energy (using arithmetic in a polynomial number of steps) and thus be able to check the validity of the claim. If H is satisfiable, such a configuration exists.

---

[6] The interactions E_m in 3-SAT are usually defined to penalize exactly one of the $2^3 = 8$ possible configurations of its input variables—but allow each of the $\binom{N}{3}$ possible 3-body interactions to appear multiple times in the sum. Thus, our definition is equivalent up to absorbing these terms together, which modifies the excited state spectrum but not the counting of zero energy satisfying states.

Fig. 7.1 Circuit representing an NP verifier. The circuit depends on the particular instance and must be efficiently constructible by a polynomial time circuit drawing algorithm

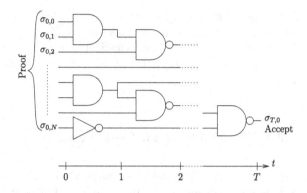

On the other hand, if H is not satisfiable, the prover would not be able to convince the verifier otherwise because all states would have energy greater than zero.

7.2.5.2 3-SAT is NP-Complete

The tricky part is to show that 3-SAT is as hard as the entire class NP. We need to show that *any* possible NP problem can be reduced to a 3-SAT problem by a polynomial transformation. What is the only thing that all NP problems have in common? By definition, they all have polynomial size verification procedures which take as input a proposed proof that an instance is a Yes-instance and output either Accept or Reject based on whether the proof is valid. This verification procedure is what we will use to provide the reduction to 3-SAT.

Let us think of the verification procedure for a particular instance A of some NP problem as a polynomially sized Boolean circuit as in Fig. 7.1. The input wires encode the proposed proof that A is a Yes-instance and the output wire tells us whether to Accept or Reject the proof. The gates in the figure are simply the usual Boolean logic gates such as NAND and NOR, which take two input bits and provide one output bit. Any Boolean circuit may be written using such binary operations with arbitrary fan-out, so we assume that we can massage the verification circuit into the form shown. Now we will construct an instance of 3-SAT encoding the operation of this circuit. That is, if the instance is satisfiable, then there exists a proof that the verifier accepts showing that the original NP problem is a Yes-instance and conversely, if the instance is not satisfiable, then no such proof exists and the original NP problem is a No-instance.

The 3-SAT instance is very simple to construct if we simply change our point of view on the picture in Fig. 7.1. Instead of viewing it as a Boolean circuit operating from left to right, let us view it as the interaction graph for a collection of $O(N \times T)$ binary bond variables—one for each of the wires in the circuit: the input bits of the proof, the output bit and each of the intermediate variables. Each gate then specifies a 3-body interaction E_m for the adjacent variables which we define to take the value 0

Fig. 7.2 Interpretation of Boolean AND gate as three-body interaction

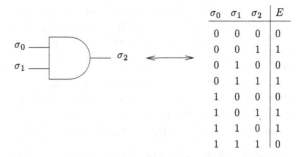

σ_0	σ_1	σ_2	E
0	0	0	0
0	0	1	1
0	1	0	0
0	1	1	1
1	0	0	0
1	0	1	1
1	1	0	1
1	1	1	0

for configurations in which the variables are consistent with the operation of the gate and 1 otherwise. See Fig. 7.2. We now add a final 1-body term on the output bit of the verification circuit which penalizes the Reject output. We now have an Ising-like model with polynomially many 3-body interactions and non-negative energy.

That's it. If the 3-SAT instance described by Fig. 7.1 has a zero energy ground state, then there is a configuration of the wire variables such that the circuit operates correctly and produces an Accept output. In this state, the input wires represent a valid proof showing that the original instance was a Yes-instance. On the other hand, if the 3-SAT instance is not satisfiable, no state exists such that the circuit operates correctly and the output produced is always REJECT. Thus we have shown that all problems in NP can be efficiently reduced to 3-SAT.

Now that we have one problem, 3-SAT, which is NP-complete, it is straightforward to show the existence of many other NP-complete problems: we need only find reductions from 3-SAT to those problems. Indeed, a veritable menagerie of NP-complete problems exists (see e.g. [6]) including such famous examples as the traveling salesman problem and graph coloring. A more physics oriented example is that of determining the ground state energy of the $\pm J$ Ising model in 3 or more dimensions [1].

7.2.6 QMA-Completeness: Kitaev

The complexity class QMA provides the quantum analogue to NP and, just like NP, it contains complete problems which capture the difficulty of the entire class. Kitaev first introduced the QMA-complete problem 5-LOCAL HAMILTONIAN in the early '00s and proved its completeness using a beautiful idea due to Feynman: that of the history state, a superposition over computational histories. The quantum Cook–Levin proofs are somewhat more complicated than the classical case and we will only sketch them here (see [7] for more details). For simplicity and to connect with the statistical study undertaken in the later sections, we restrict our attention to the slightly simpler problem of k-QSAT, which is QMA_1-complete for $k \geq 4$. QMA_1 is the variant of

QMA in which the verification error is one-sided: Yes-instances may be verified with no errors while invalid proofs still occasionally get incorrectly accepted.

First, let us define k-QSAT a bit more carefully:

Input: A quantum Hamiltonian $H = \sum_m \Pi_m$ composed of M projectors, each acting on at most k qubits of an N qubit Hilbert space.
Promise: Either H has a zero energy state or all states have energy above a promise gap energy $\Delta > 1/\text{poly}(N)$.
Question: Does H have a zero energy ground state?

Now, we sketch the proof that QSAT is QMA_1-complete.

7.2.6.1 QSAT is QMA_1

To show that QSAT is QMA_1, we need to find an efficient quantum verification scheme such that (a) there exist proofs for Yes-instances which our verifier always accepts and (b) any proposed proof for a No-instance will be rejected with probability at least $\epsilon = 1/\text{poly}(N)$. This rather weak requirement on the bare false-Acceptance rate can be bootstrapped into an arbitrarily accurate verification scheme by repetition, as sketched in Sect. 7.2.4 above.

Given an instance $H = \sum_m \Pi_m$, the obvious proof is for Merlin to provide a state $|\Psi\rangle$ which he alleges is a zero energy state. Arthur's verification procedure will be to check this claim. The verifier works by measuring each of the Π in some pre-specified order on the state. That this can be done efficiently follows from the fact that Π acts on no more than k qubits and therefore its measurement can be encoded in an N-independent number of quantum gates. Clearly, if $|\Psi\rangle$ is a zero-energy state, it is a zero-energy eigenstate of each of the Π and therefore all of these measurements will produce 0 and the verifier accepts. This checks condition (a) above and we say our verification scheme is complete.[7]

On the other hand, if H is a No-instance, it has a ground state energy above the promise gap Δ and $|\Psi\rangle$ necessarily has overlap with the positive eigenspaces of at least some of the Π. It is a short computation to show that the probability that all of the measurements return 0 will then be bounded above by $1 - \Delta/N^k \sim 1 - 1/\text{poly}(N)$. Thus, No-instances will be rejected with probability at least $\epsilon = 1/\text{poly}(N)$ and our verification scheme is sound.

7.2.6.2 QSAT is QMA_1-Complete

Just as in the classical Cook-Levin proof, we need to show that *any* QMA_1 problem can be reduced to solving an instance of QSAT. We again exploit the only thing

---

[7] The feature that one can do these measurements by local operations and that they provide probability 1 verification of ground states is a special feature of the QSAT Hamiltonian which allows it somewhat to evade the heuristic expectations of time-energy uncertainty.

Fig. 7.3 QMA verification
circuit. The circuit depends
on the instance and must be
constructible by an efficient
algorithm given the instance.
We have drawn the circuit so
that there is a single local
gate per time step, so
$T = \text{poly}(N)$

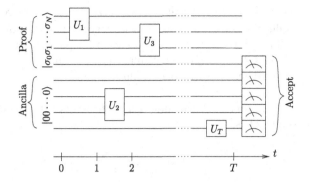

that all QMA$_1$ problems have in common: their quantum verification algorithm. We
will take the quantum circuit representing this verifier and construct from it a QSAT
Hamiltonian whose ground state energy is directly related to the maximal acceptance
probability of the verification circuit.

Let A be an arbitrary instance of a QMA$_1$ problem L. Then A has a polynomial
sized quantum verification circuit as in Fig. 7.3. This circuit takes as input a quantum
state encoding a proof that A is a Yes-instance of L, along with some ancilla work
qubits in a fiducial 0 state, then performs a sequence of T unitary one and two qubit
gates and finally measures the output state of some subset of the qubits. If A is a
Yes-instance, then there exists a valid input state such that all of the output bits will
yield 0 with probability 1. Otherwise, at least one of the output bits will read 1 with
probability polynomially bounded away from 0 for any input state $|\sigma_0 \cdots \sigma_N\rangle$.[8]

We now construct a single particle hopping Hamiltonian whose ground state
encodes the operation of this quantum circuit. We introduce a *clock* particle hopping
on a chain of length T and label its position states by $|t\rangle$. We endow this particle with
an enormous 'spin': a full N qubit Hilbert space (dimension 2^N). As the particle
hops, 'spin-orbit' coupling induces rotations on the N qubit space which correspond
to the unitary gates in the verification circuit. To wit:

$$H_p = \frac{1}{2} \sum_{t=0}^{T-1} \left(-|t+1\rangle\langle t| \otimes U_{t+1} - |t\rangle\langle t+1 \otimes U_{t+1}^\dagger + |t\rangle\langle t| + |t+1\rangle\langle t+1| \right)$$

The terms of this Hamiltonian have been normalized and shifted such that each is
a projector with energies 0 and 1, but otherwise it is just a 1-D hopping problem with
Neumann boundary conditions.[9] Indeed, there is a simple basis transformation in

---

[8] The observant reader will notice the addition of ancillae qubits. These are necessary when
computation is done by reversible gates, as in unitary circuit computation. We leave it as an exercise
to figure out why the absence of ancillae would make the verification circuit unsound.

[9] In fact, Neumann boundary conditions (which stipulate the value of the derivative of the solution
to a differential equation) apply to the problem obtained in the time continuum limit of the discrete
hopping Hamiltonian under consideration.

which the spin-orbit coupling disappears entirely. This consists of rotating the basis
of the position t spin space by a sequence of unitary transformations $U_1^\dagger U_2^\dagger \cdots U_t^\dagger$.

In this representation, we see that the 2^N spin components decouple and the system
is really 2^N copies of the Neumann chain. Thus, the spectrum is $(1 - \cos k)/2$, a
cosine dispersion with bandwidth 1, ground state energy 0 at wave vector $k = 0$ and
allowed $k = n\pi/(T + 1)$.

The propagation Hamiltonian has (zero energy) ground states of the form (in the
original basis):

$$|\psi\rangle = \frac{1}{\sqrt{T+1}} \sum_t |t\rangle \otimes U_t U_{t-1} \cdots U_1 |\xi\rangle \tag{7.2}$$

where $|\xi\rangle$ is an arbitrary 'input state' for the N qubit space. This state $|\psi\rangle$ is called
the *history state* of the computation performed by the verification circuit given input
$|\xi\rangle$. It is a sum over the state of the quantum computation at each step in the circuit
evolution. Any correct computation corresponds to a zero energy history state—
incorrect computations will have a non-zero overlap with higher energy hopping
states.

Now, we have a Hamiltonian that encodes the correct operation of the verification
circuit. We simply need to add terms that will penalize computations which do not
correspond to appropriate input states and accepting output states. These terms affect
the computational state at times $t=0$ and T, so they are boundary fields from the point
of view of the hopping problem. In general, they should split the 2^N degeneracy of
the pure hopping in H_p, and since they are positive operators, lift the ground state
energy.

The initialization term is simply a sum over the ancilla qubits of projectors penal-
izing $|\xi\rangle$ with incorrectly zeroed ancillae:

$$H_i = \sum_{j \in \text{Ancilla}} |0\rangle\langle 0|_t \otimes |1\rangle\langle 1|_j \tag{7.3}$$

Similarly, the output term penalizes output states which overlap $|1\rangle$ on the measured
output bits:

$$H_o = \sum_{j \in \text{Accept}} |T\rangle\langle T|_t \otimes |1\rangle\langle 1|_j \tag{7.4}$$

Now we consider the full Hamiltonian

$$H = H_i + H_p + H_o \tag{7.5}$$

If $|\psi\rangle$ is a zero energy state of H, then it is a zero energy state of each of the three
pieces. Hence, it will be a history state in which the input state $|\xi\rangle$ has appropriately
zeroed ancillae and the output state has no overlap with $|1\rangle$ on the measured qubits—
thus, $|\xi\rangle$ is a proof that the verifier accepts with probability 1 and the original instance

A is a Yes-instance. Conversely, if such a proof state $|\xi\rangle$ exists then the history state built from it will have zero energy.

It is somewhat more work to show the soundness of the construction: A is a No-instance if and only if the Hamiltonian H has ground state energy bounded polynomially away from 0 [5]. The intuition is straightforward—the strength of the boundary fields in the basis transformed hopping problem for a given spin state corresponds to the acceptance probability of the associated input state. Since these repulsive fields lift Neumann conditions, they raise the ground state energy quadratically in $1/T$—they effectively force the ground state wavefunction to bend on the scale of T. For a No-instance, since no spin sector is both valid and accepting, all states must gain this inverse quadratic energy.[10]

To be a bit more precise, we assume for contradiction that we have a state $|\psi\rangle$ with energy exponentially small in N (hence smaller than any polynomial in N or T):

$$\langle\psi|H|\psi\rangle = \langle\psi|H_i|\psi\rangle + \langle\psi|H_p|\psi\rangle + \langle\psi|H_o|\psi\rangle \le O(e^{-N}) \quad (7.6)$$

Since each term is positive, each is bounded by the exponential. The hopping Hamiltonian H_p has a gap of order $1/T^2$ for chains of length T, thus if we decompose $|\psi\rangle$ into a zero energy piece (a history state) and an orthogonal complement,

$$|\psi\rangle = \frac{\sqrt{1-\alpha^2}}{\sqrt{T+1}} \sum_t |t\rangle \otimes U_t \cdots U_1|\xi\rangle + \alpha|\text{Exc}\rangle \quad (7.7)$$

we must have exponentially small overlap onto the complement:

$$O(e^{-N}) > \langle\psi|H_p|\psi\rangle = \alpha^2\langle\text{Exc}|H_p|\text{Exc}\rangle > \alpha^2 O(1/T^2) \quad (7.8)$$

In other words,

$$|\psi\rangle = \frac{1}{\sqrt{T+1}} \sum_t |t\rangle U_t \cdots U_1|\xi\rangle + O(e^{-N}) \quad (7.9)$$

The input term H_i has energy 0 for valid input states $|\xi^V\rangle$ and energy at least 1 for invalid states $|\xi^I\rangle$. Thus, decomposing $|\xi\rangle = \sqrt{1-\beta^2}|\xi^V\rangle + \beta|\xi^I\rangle$, we find (abusing notation and dropping explicit reference to the $|t = 0\rangle$ sector on which H_i acts):

$$O(e^{-N}) > \langle\psi|H_i|\psi\rangle = \frac{\beta^2}{T+1}\langle\xi^I|H_i|\xi^I\rangle + O(e^{-N}) \ge \frac{\beta^2}{T+1} + O(e^{-N}) \quad (7.10)$$

In other words,

---

[10] This is overly simplified: in the absence of the output term H_o, the gauge transformed problem can be thought of as 2^N decoupled hopping chains, some fraction of which have boundary fields at $t=0$. The output term is not simply a field on these chains—it couples them and in principle allows hopping between them as a star of chains. The upshot is that the repulsive (diagonal) piece outweighs the off-diagonal mixing.

$$|\psi\rangle = \frac{1}{\sqrt{T+1}} \sum_t |t\rangle U_t \cdots U_1 |\xi^V\rangle + O(e^{-N}) \qquad (7.11)$$

Finally, considering the output term we find:

$$\langle \psi | H_o | \psi \rangle = \frac{1}{T+1} \langle \xi^V | U_1^\dagger \cdots U_t^\dagger H_o U_t \cdots U_1 | \xi^V \rangle + O(e^{-N})$$

$$= \frac{p_r}{T+1} + O(e^{-N}) \qquad (7.12)$$

where p_r is the probability that the original QMA_1 verifier rejects the proposed proof $|\xi^V\rangle$ for the No-instance A. Since this rejection probability is bounded below by a constant, the state $|\Psi\rangle$ cannot possibly have exponentially small energy.

We have reduced the arbitrary QMA_1 instance A to asking about the zero energy states of a hopping Hamiltonian H constructed out of projectors. This is almost what we want. We have a Hamiltonian constructed out of a sum of projectors but they each act on three qubits tensored with a (large) particle hopping space rather than on a small collection of qubits.

The final step in the reduction to k-QSAT is to represent the single particle Hilbert space in terms of a single excitation space for a chain of clock qubits in such a way that we guarantee the single particle sector is described by H above and that it remains the low energy sector. We refer the interested reader to the literature for more details on these clock constructions. Each of the projectors of H becomes a joint projector on one or two of the computational (spin) qubits and some number of the clock qubits (two in [7]). The final 4-QSAT Hamiltonian will then be given by a sum of projectors involving at most 4 qubits

$$H = H_i + H_p + H_o + H_c \qquad (7.13)$$

where H_c acts on the clock qubits to penalize states which have more than one clock particle. This concludes our brief overview of complexity theory. We next turn to a review of results obtained by applying ideas from (quantum) statistical mechanics to random ensembles of classical and quantum k-SAT.

7.3 Physics for Complexity Theory

There are two main ways for physicists to contribute to complexity theory. One is to bring to bear their methods to answer some of the questions posed by complexity theorists. Another is to introduce concepts from physics to ask new types of questions, thereby providing an alternative angle, permitting a broader view and new insights. This section is devoted to the illustration of this point, using the k-SAT problem introduced above as a case in point. In particular, we discuss both classical k-SAT and its quantum generalisation k-QSAT [7, 8].

7.3.1 Typical Versus Worst-Case Complexity

As explained above k-SAT is NP complete for $k \geq 3$. Thus, for any given algorithm, we expect that there are instances which will take an exponentially long time to solve. However, we ought not be too discouraged—some instances of k-SAT may be parametrically easier to solve than others, and these may be the ones of interest in a given context. To make this more precise, it is useful to introduce the concept of typical, as opposed to worst-case, complexity.

In order to define typicality, one can consider an ensemble in which each problem instance is associated with a probability of actually occurring. Typical quantities are then given by stochastic statements, e.g. about a median time required for solving problem instances, which may differ substantially from the corresponding average, or indeed the worst-case, quantities when the latter have a sufficiently small weight in the ensemble. Precisely what quantities to calculate depends on the aspects of interest. For instance, a median run-time is not much affected by a small fraction of exponentially long runs, while these may dominate the expectation value of the run-time.

It is worth emphasizing again that the polynomial reductions discussed in Sect. 7.2 provide a characterization of the *worst case* difficulty of solving problems. The reductions and algorithms in this context must work for *all* instances of a problem. Reductions however may transform typical instances of A into rather 'atypical' instances of B. Whether a useful framework of reductions can be defined that preserve typicality is an open question (see Chap. 22 of Ref. [4]), but the study of typical instances of particular hard problems has itself been a fruitful activity, as we will discuss in the following.

7.3.2 Classical Statistical Mechanics of k-SAT

We now give an account of an analysis of such an ensemble for classical k-SAT. For completeness, let us begin with reviewing the original definition of classical k-SAT, expanding on the brief definition provided in Sect. 7.2.5. Indeed, the original computer science definition of satisfiability looks somewhat different from the Hamiltonian problem we introduced. Consider a set of N Boolean variables $\{x_i \mid i = 1 \ldots N\}$, i.e. each variable x_i can take two values, true or false (in which cases the negation $\bar{x}_i$ is false or true, respectively). Classical k-SAT asks the question, "Does the Boolean expression:

$$C = \bigwedge_{m=1}^{M} C_m \qquad (7.14)$$

evaluate to true for some assignment of the x_i ?" Here, each clause is composed of a disjunction of k literals, e.g. for $k = 3$:

$$C_m = x_{m_1} \vee \bar{x}_{m_2} \vee x_{m_3} \qquad (7.15)$$

where each variable occurs either affirmed (x_{m_1}) or negated ($\bar{x}_{m_2}$). Hence, there are 2^k possible clauses for a given k-tuplet $\{x_{m_j} \mid j = 1 \ldots k\}$.[11]

This definition is equivalent to the definition in terms of the k-body interacting spin Hamiltonian of Eq. 7.1. To obtain a spin Hamiltonian from the collection of clauses, Eq. 7.14, we convert the Boolean variables x_i into Ising spins $\sigma_i = \pm 1$, with $\sigma_i = +1(-1)$ representing x_i being true (false). A clause then becomes a k-spin interaction designed such that the satisfying assignments evaluate to energy 0, and the forbidden assignment to energy 1. For instance, the clause given in Eq. 7.15 turns into:

$$H_m = 2^{-3} \left(1 - \sigma_{i_{m_1}}\right) \left(1 + \sigma_{i_{m_2}}\right) \left(1 - \sigma_{i_{m_3}}\right). \qquad (7.16)$$

(It is this formulation of classical k-SAT that will lend itself naturally to a quantum generalisation, which we describe below.)

The k-SAT ensemble is now random in two ways:

(R1) each k-tuple occurs in H with probability $p = \alpha N \Big/ \binom{N}{k}$

(R2) each k-tuple occurring in H is randomly assigned one of the 2^k possible clauses.

Here, we have introduced a parameter α for the number of clauses, $M = \alpha N$, which is proportional to the number of variables[12] because there are $\binom{N}{k}$ possible k-tuples. The 'interactions' can be pictorially represented by an interaction graph, Fig. 7.4. This is a bipartite graph, one sublattice of which has N nodes, denoted by circles representing the x_i, and the M nodes of the other sublattice denoted by triangles represent the clauses C_m. Each triangle is connected to the k variables participating in the clause it represents, whereas each variable is connected to all clauses it participates in, which implies an average coordination of αk, with a Poissonian distribution. The random graph thus constructed contains all the information on a given problem instance if we label each triangle with which of the 2^k possible clauses it represents. This graph will be used for random quantum k-SAT as well, where the Boolean variables and clauses will be replaced by appropriate quantum generalisations.

---

[11]   The symbols $\wedge$ and $\vee$ are the Boolean operators 'and' and 'or'.

[12]   Actually, only the expectation value of M equals αN. The Poissonian distribution for M of course has vanishing relative fluctuations ($\langle M^2 \rangle - \langle M \rangle^2)/\langle M \rangle^2$ as $N \to \infty$.

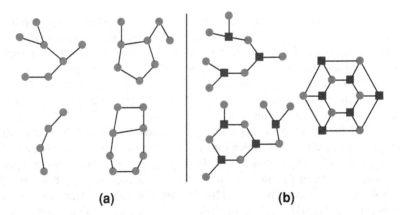

Fig. 7.4 Examples of random interaction graphs for **a** 2-SAT and **b** 3-SAT, respectively. The *circles* represent qubits. **a** The clusters, clockwise from *bottom left*, are chain, tree, clusters with one and two closed loops ("figure eight"). The short closed loops, as well as the planarity of the graphs, are not representative of the large-N limit. **b** Each *square* represents a clause connected to 3 nodes. Clockwise from *top left* are a tree, a graph with nontrivial core and a graph with simple loops but no core

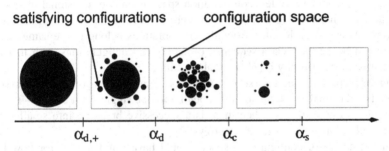

Fig. 7.5 Schematic phase diagram for random classical k-SAT ($k \geq 4$) [9]. Actually, configuration space is very high-dimensional (an N-dimensional hypercube), and the cartoons are only suggestive of the actual structure of the space of solutions, for the real complexity of which our everyday intuition from low dimensions may be quite inadequate

7.3.3 Schematic Phase Diagram of Classical Random k-SAT

In Fig. 7.5 , we show a schematic phase diagram for random k-SAT. The first question one might ask is: is there a well-defined phase transition, at some value $\alpha = \alpha_s(k)$, such that instances for $\alpha < \alpha_s$ are satisfiable, and those for $\alpha > \alpha_s$ are not? It has been shown that there exists such a transition for the random ensemble. This does not mean that *all* instances with $\alpha < \alpha_s$ are satisfiable: given an UNSAT instance with N sites and αN clauses, one could simply add N disconnected sites to get a new UNSAT instance with $\alpha' = \alpha/2$. What is true instead is that the probability of having such an UNSAT graph with $\alpha' < \alpha_s$ is exponentially small in N, so that for $N \to \infty$, such graphs do not arise with a probability approaching 1.

It is easy to provide a very rough estimate for where this happens, by adapting an idea of Pauling's which, amusingly, was devised for estimating the configurational entropy of the protons in water ice. We consider the clauses as constraints, each of which 'decimates' the number of allowed configurations by a factor $(1-2^{-k})$: only 1 out of the 2^k possible configurations of variables of any given clause is ruled out. For $M = \alpha N$ such constraints, one is left with $2^N (1 - 2^{-k})^{\alpha N}$ solutions. In the thermodynamic limit, this number vanishes for $\alpha > \alpha_{wb} = -1/\log_2(1 - 2^{-k}) \sim 2^k \log 2$. In the k-SAT literature, this is known as the 'first-moment bound', for which there is a straightforward rigorous derivation. To find rigorous *upper bounds* one should instead employ different techniques, with inequalities coming from the analysis of the *second moment* of the number of solutions (after appropriately restricting the ensemble to reduce the fluctuations) [10]. It is interesting here to note that for large k the upper bounds and lower bounds converge, becoming a prediction for the actual location of the threshold.

The SAT-UNSAT transition is not the only transition of this problem, though. As indicated in Fig. 7.5, statistical mechanical methods imported from the study of spin glasses have been used to establish finer structure in the SAT phase. This plot shows a set of cartoons of configuration space, indicating the location of satisfying assignments. For N variables, configuration space is an N-dimensional hypercube and this plot indicates, in a two-dimensional 'projection', how 'close' satisfying assignments are to each other. Roughly, two solutions belong to the same cluster if they can be reached via a sequence of satisfying configurations such that two consecutive ones differ by $O(N^\beta)$ variables with $\beta < 1$ [9].

Figure 7.5 thus documents a set of transitions in the clustering of satisfying assignments. For the smallest α, all solutions belong to one single giant cluster—the full hypercube for $\alpha = 0$—and then there is a successive break-up into smaller, and increasingly numerous clusters as α grows [9].

This structure of configuration space should have ramifications for how hard it is to solve the corresponding instances: small-scale clusters indicate a rugged energy landscape, with numerous local minima providing opportunities for search algorithms to get stuck. Indeed, all known algorithms slow down near α_s. That said, many simple approaches to random 3-SAT problems actually do quite well even in the clustered phases and the detailed relationship between clustering in configuration space and algorithmic difficulty is subtle, somewhat detail dependent and an ongoing research topic. It is particularly worth noting that even the simplest random greedy algorithms typically work across most of these transitions, at least for k-SAT. Indeed, while the identification of distinct phases has grown to give a phase diagram replete with fine structure, the portion of the phase diagram containing 'hard' instances— those for which deciding satisfiability takes exponentially long typically— has by now been pushed back to a tiny sliver at α_s of width $\delta \alpha < \alpha_s/100$. In the meantime, however, the action has started to shift to other problem ensembles which at the time of writing have proven more robustly difficult.

The derivation of this phase diagram was obtained using methods imported from the study of spin glasses, in particular the *cavity method* [9, 11]. The insights thus gained have lead to the development of an impressive arsenal of techniques for not only determining whether or not a k-SAT problem instance is soluble, but also for actually finding solutions in the form of satisfying assignments [12–15]. In the following section, we provide a brief introduction to cavity analysis.

7.3.4 Cavity Analysis

The cavity method is a cluster of techniques and heuristics for solving statistical models on sparse, tree-like interaction graphs G. In this approach, one determines the behavior of the model on G by first analyzing the restriction of the model to so-called *cavity graphs*. A cavity graph $G\backslash\{i\}$ is formed by carving site i out of G:

The neighbors of i, ∂i, which now sit at the boundary of the cavity in $G\backslash\{i\}$, are called *cavity spins*. The central assumption of the cavity method is that cavity spins are statistically independent in the absence of site i because they sit in disconnected components of $G\backslash\{i\}$. This assumptions massively simplifies the evaluation of observables in such models, ultimately leading to efficient procedures for finding ground states, evaluating correlation functions and determining thermodynamic free energies, phase diagrams, and clustering phenomena in models with quenched disorder.

The ensemble of interaction graphs G that arise from the rule (R1) have loops and thus do not fall into disconnected pieces when a cavity site i is removed. Nonetheless, for large N, any finite neighborhood of a randomly chosen point in G is a tree with high probability. That is to say, G does not contain short closed loops and we call it *locally tree-like*. For such G, we can hope that the cavity assumption will hold at least to a good approximation.

That neighborhoods in G are trees can be seen as follows: the subgraph consisting of site i and its neighbors has on average $n_1 = (1 + \alpha k)$ out of the N sites. The αk neighbors will in turn have αk further neighbors, so that the subgraph containing those as well has approximately $n_2 = 1 + (\alpha k) + (\alpha k)^2$ sites. Up to the γ th nearest neighbors, the resulting subgraph grows exponentially, $n_\gamma \sim (\alpha k)^\gamma$. Obviously, $n_\gamma \leq N$, so that closed loops must appear at length $\gamma_c = (\ln N)/\ln(\alpha k)$. For $\gamma < \gamma_c$, $n_\gamma \ll N$ due to the exponential growth of n_γ, so that the randomly chosen interaction partners are overwhelmingly likely to be drawn from the sites not yet included in the neighborhood. Thus, the length of loops on G deverges with N, although excruciatingly (logarithmically) slowly.

Let us make these considerations more precise. Consider carving a cavity into a large regular random graph G with N spins and M edges representing two body

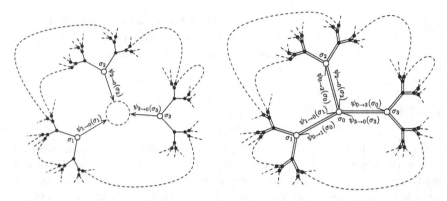

Fig. 7.6 (*Left*) Schematic of $q = 3$ regular random graph with a cavity carved out at spin σ_0. (*Right*) The belief propagation equations for a graph G involve $2M$ cavity distributions, or *beliefs*, $\psi_{i \to j}$, one for each of the two directions of a link in the graph

interactions, as in Fig. 7.6. We will not explicitly consider the straightforward generalization to k-body interactions with $k > 2$; it needlessly complicates the notation. The statistical connection at temperature $1/\beta$[13] between the removed spin σ_0 and the rest of the graph is entirely mediated by the *joint* cavity distribution

$$\psi_{G\setminus\{0\}}(\sigma_1, \sigma_2, \sigma_3) = \frac{1}{Z_{G\setminus\{0\}}} \prod_{j \notin \{0,1,2,3\}} \sum_{\sigma_j} e^{-\beta H_{G\setminus\{0\}}} \qquad (7.17)$$

That is, the thermal distribution for σ_0 in the original model is given by:

$$\psi_0(\sigma_0) = \frac{1}{Z_0} \sum_{\sigma_1,\sigma_2,\sigma_3} e^{-\beta(H_{01}+H_{02}+H_{03})} \psi_{G\setminus\{0\}}(\sigma_1, \sigma_2, \sigma_3) \qquad (7.18)$$

If, on carving out the cavity, the neighboring spins become independent then the joint cavity distribution factors:

$$\psi_{G\setminus\{0\}}(\sigma_1, \sigma_2, \sigma_3) = \psi_{1 \to 0}(\sigma_1)\psi_{2 \to 0}(\sigma_2)\psi_{3 \to 0}(\sigma_3) \qquad (7.19)$$

On trees, the independence is exact because the cavity spins sit in disconnected clusters; on locally tree-like graphs, the cavity spins are only connected through long (divergent in system size) paths and thus we might expect Eq. 7.19 to hold approximately.

Thus the objects of interest are the $2M$ cavity distributions $\psi_{i \to j}$, see Fig. 7.6. These are also known as messages or beliefs: $\psi_{i \to j}(\sigma_i)$ is a message passed from site i to site j which indicates site i's beliefs about what it should do in the absence of site j, and thus also its beliefs about what site j should do to optimize the free

---

[13] We may eventually take the temperature to 0 ($\beta \to \infty$) in order to find actual ground state solutions of the optimization problem.

energy. Crucially, when the cavity distributions are independent, they also satisfy the iteration relation:

$$\psi_{i \to j}(\sigma_i) = \frac{1}{Z_{i \to j}} \prod_{k \in \partial i \backslash \{j\}} \sum_{\sigma_k} e^{-\beta H_{ik}(\sigma_i, \sigma_k)} \psi_{k \to j}(\sigma_k) \qquad (7.20)$$

These are the Belief Propagation (BP) equations and they are really just the Bethe–Peierls self-consistency equations in more formal, generalized, notation. Indeed, if we parameterize the functions $\psi_{i \to j}$ by cavity fields $h_{i \to j}$

$$\psi_{i \to j}(\sigma_i) = \frac{1}{2 \cosh(\beta h_{i \to j})} e^{-\beta h_{i \to j} \sigma_i}, \qquad (7.21)$$

and specialize to an Ising Hamiltonian $H = \sum_{\langle ij \rangle} J_{ij} \sigma_i \sigma_j$, then the BP equation becomes:

$$h_{i \to j} = \frac{1}{\beta} \sum_{k \in \partial i \backslash \{j\}} \tanh^{-1} [\tanh(\beta J_{ki}) \tanh(\beta h_{k \to i})] \qquad (7.22)$$

which should be familiar from Ising mean field theory.

There are linearly many BP equations, each of which is a simple relation involving a finite summing out procedure on the right to define a cavity distribution on the left. We may now consider two approaches to solving them: (a) take a thermodynamic limit and find the statistics of their solutions; and (b) iteratively solve them for finite N using a computer. The former approach leads to the so-called cavity equations and the estimates regarding the thermodynamic phase diagram of Fig. 7.5. The latter leads to the belief propagation algorithm for solving particular instances of optimization problems. Indeed, once the solution of the BP equations is known for a particular instance, one can obtain a solution of the SAT formula by using a decimation heuristic which fixes the variables with positive (resp. negative) total of incoming messages to 1 (resp. 0) and re-running BP on the remaining formula if necessary [12].

There are as many BP equations as unknown cavity distributions and thus we generically expect to find discrete solutions. However, there can be more than one solution. For instance in the low temperature phase of the Ising ferromagnet, there are three: the (unstable) paramagnet and the two (symmetry related) magnetized solutions. For spin glass models with quenched disorder, there may be exponentially many solutions, each corresponding to a macroscopically distinct magnetization pattern in the system. When this occurs, the belief propagation algorithm may fail to converge. In this case, one needs to take into account the presence of multiple solutions of the BP equations, which can be done statistically using an algorithm known as Survey Propagation and in the thermodynamic limit using the 'replica symmetry breaking' cavity equations, which arise as a hierarchy of distributional equations. These equations describe the statistics of solutions of the BP equations. Although these analyses are important for a correct understanding of many types of glassy optimization problems on tree-like graphs [16], we will not consider these technical generalizations further. We note that the jargon of 'replica symmetry breaking' arises in a completely different approach to solving mean field glasses based on the so-called replica trick and Parisi ansatz. The terms have no intrinsic meaning in the context of the cavity approach.

For quantum stoquastic (or Frobenius–Perron) Hamiltonians—those for which a basis is known for which all off-diagonal matrix elements are negative, which guarantees that there exists a ground state wavefunction for which all components can be chosen to have a positive amplitude—BP has been generalized in some recent works [17–19] and has since been taken up in the study of a number of quantum models on trees [20–23]. Spin-glasses with a transverse field are a case in point. This might be relevant for the study of the performance of some quantum adiabatic algorithms for solving classical instances of k-SAT, which we turn to next.

7.3.5 Adiabatic Quantum Algorithm for Classical k-SAT

Before we move on to the problem of quantum satisfiability k-QSAT, let us first ask whether we could use a quantum algorithm to solve classical k-SAT efficiently. One possible strategy for this is based on a protocol known as adiabatic quantum computing [24], which in its full generality is equivalent to computation based on circuits. Here we will discuss a particularly simple member of this class of algorithms.

Consider a time-dependent quantum Hamiltonian, with real time t parametried by $s(t)$ (with $0 \le s \le 1$):

$$H(s) = (1 - s)H_\Gamma + s H_0, \tag{7.23}$$

where H_Γ is a transverse-field term and H_0 is obtained from Eq. 7.1

$$H_\Gamma = -\Gamma \sum \sigma^{X_i}$$
$$H_0 = \sum_m E_m \left(\{\sigma_i^z\} \right), \tag{7.24}$$

by replacing the σ_i's by Pauli operators σ_i^z.

The ground state space of $H(s)$ at time $s = 1$ is spanned by (one of) lowest-energy classical configurations (ground states) of the k-SAT problem; it is these which we are after, but which can be hard to find. By contrast, at time $s = 0$, the quantum ground state has all spins polarized in the x-direction. This is both easy to describe and to prepare. If we start the system in its ground state at $s = 0$, and change $H(s)$ sufficiently slowly, the state of the system will evolve adiabatically, and reach the desired state at time $s = 1$.

However, we do not want to change $H(s)$ arbitrarily slowly, as this would be no gain over an exponentially long classical run-time.

What is it that limits the sweep-rate? A non-zero sweep rate can induce transitions to excited states. Careful derivation of the adiabatic theorem reveals that the probability of nonadiabatic transitions tends to zero so long as the sweep rate is slower than the minimal adiabatic gap Δ squared.[14] That is, the run time T must be greater than or of order $O(1/\Delta^2)$ in order to ensure adiabaticity.

Heuristically, we need to be concerned about avoided level crossings in the course of the evolution and in particular, the avoided crossing at the location of the minimal gap Δ.[15] As two levels approach closely, we get an effective two-level problem:

$$H_2 = \begin{pmatrix} \alpha(t - t_0) & \Delta/2 \\ \Delta/2 & -\alpha(t - t_0) \end{pmatrix}. \tag{7.25}$$

At the closest approach, at $t = t_0$, the ground state is separated from the excited state by a gap Δ. For $|\alpha(t - t_0)| \gg \Delta$, variation of t has very little effect on the adiabatic eigenstates and the Schrödinger evolution remains adiabatic even for fast sweeping. It is only the time spent in the interaction region $|\alpha(t - t_0)| < \Delta$ where the adiabatic states rotate significantly and nonadiabatic transitions may arise. Thus, the interaction time is $t_I \sim \Delta/\alpha$ and the dimensionless figure of merit for adiabatic behavior should be $\Delta \cdot t_I \sim \Delta^2/\alpha$. In particular, for low sweep rates $\alpha \ll \Delta^2$ or long run times $T \geq O(1/\Delta^2)$, we expect to have purely adiabatic evolution. We note that the nonadiabatic transition probability P for this two-level model was calculated some eighty years ago by Landau [26] and Zener[27] whose exact result:

$$P = 1 - e^{-\pi \Delta^2/4\hbar\alpha} \tag{7.26}$$

quantifies the physical intuition in this case.

The quantum adiabatic algorithm has been studied extensively since its introduction ten years ago on a number of hard random constraint satisfaction problems closely related to 3-SAT [24, 28]. The critical question is simple: how does the typical minimal gap encountered during the procedure scale with increasing instance size N? Analytic work on simple (classically easy) problem ensembles found several polynomial time quantum adiabatic algorithms. Moreover, early numerical studies for very small instances of harder problems held promise that the gap would scale only polynomially [24, 29]. Unfortunately, subsequent numerical studies on larger systems indicate that the gap eventually becomes exponentially small due to a first order transition between a quantum paramagnet and a glassy state [30]. Even worse, recent perturbative work argues that many-body localization leads to a plethora of exponentially narrow avoided crossings throughout the glassy phase [31, 32] and thus the algorithm discussed here does not produce an efficient solution of random 3-SAT or related random constraint satisfaction problems. We note that this is consistent with other evidence that quantum computers will not enable the efficient solution of NP-complete problems.

---

[14] In fact, there is some controversy in the rigorous literature about whether the asymptotic sweep rate must be slower than $1/\Delta^2$ or $1/\Delta^{2+\delta}$ for some arbitrarily small constant δ. See [25].

[15] In the absence of any symmetries and fine-tuning, all level crossings are avoided as a function of a single adiabatic parameter s.

7.4 Statistical Mechanics of Random k-QSAT

Let us now turn to the study of random instances of quantum satisfiability k-QSAT. As discussed in Sect. 7.2.6, k-QSAT is QMA$_1$-complete and thus should be generally intractable. As in the classical case, one might hope to gain some insight into the nature of the difficulty of the quantum problem by studying a random ensemble of its instances. Moreover, the richness of phenomena exhibited by the classical random satisfiability problem—and the many important spin-off techniques that have been developed in their study—encourages us to seek analogous behaviors hiding in the quantum system.

7.4.1 Random k-QSAT Ensemble

Let us recap the definition of k-QSAT from Sect. 7.2.6:

Input: A quantum Hamiltonian $H = \sum_m \Pi_m$ composed of M projectors, each acting on at most k qubits of an N qubit Hilbert space.

Promise: Either H has a zero energy state or all states have energy above a promise gap $\Delta > 1/\mathrm{poly}(N)$.

Question: Does H have a zero energy ground state?

Quantum satisfiability is a natural generalization of the classical satisfiability problem: bits become qubits and k-body clauses become k-body projectors. In key contrast to the classical case, where the binary variables and clauses take on discrete values, their quantum generalizations are continuous: the states of a qubit live in Hilbert space, which allows for linear combinations of $|0\rangle$ and $|1\rangle$.

Thinking of a Boolean clause as forbidding one out of 2^k configurations leads to its quantum generalization as a projector $\Pi_\phi^I \equiv |\phi\rangle\langle\phi|$, which penalizes any overlap of a state $|\psi\rangle$ of the k qubits in set I with a state $|\phi\rangle$ in their 2^k dimensional Hilbert space. Indeed, if we restrict the Π_m to project onto computational basis states, k-QSAT reduces back to k-SAT—all energy terms can be written as discrete 0 or 1 functions of the basis state labels and the promise gap is automatically satisfied since all energies are integers.

As in the classical problem, we make two random choices in order to specify an instance:

(R1) each k-tuple occurs in H with probability $p = \alpha N \big/ \binom{N}{k}$

(R2) each k-tuple occurring in H is assigned a projector $\Pi_m = |\phi\rangle\langle\phi|$, uniformly chosen from the space of projectors of rank r. For these notes, we will mostly consider the case $r = 1$, although higher rank ensembles can be studied [33].

The first rule is identical to that of the classical random ensemble and thus the geometry of the interaction graphs (Fig. 7.4) is the same—locally tree-like with long loops for sufficiently high clause density α.

The second rule, however, rather dramatically changes the nature of our random ensemble—the measure on instances is now continuous rather than discrete. This turns out to be a major simplification for much of the analysis: *Generic* choices of projectors reduce quantum satisfiability to a graph, rather than Hamiltonian, property. This "geometrization" property allows us to make strong statements about the quantum satisfiability of Hamiltonians associated with both random and non-random graphs and even non-generic choices of projector, both analytically and numerically. For the remainder of these notes, we use the term *generic* to refer to the continuous choice of projectors and *random* to refer to the choice of the graph. See Sect. 7.4.3 below for a more detailed discussion of geometrization.

7.4.2 Phase Diagram

The first step in understanding the random ensemble is to compute the statistics of this decision problem as a function of α. Specifically we would like to know if there are phase transitions in the satisfying manifold as α is varied: these include both the basic SAT-UNSAT transition as well as any transitions reflecting changes in the structure of the satisfying state manifold. Additionally, we would like to check that the statistics in the large N limit are dominated by instances that automatically satisfy the promise gap.

The current state-of-the-art QSAT phase diagram is shown in Fig. 7.7. Let us walk through a few of the features indicated. First, we have separated the $k=2$ case from the higher connectivity cases because it is significantly simpler. For $k=2$, we can solve the satisfiability phase diagram rigorously and even estimate energy exponents for the (non-zero) ground state energy above the satisfiability transition. Our ability to do so is consistent with the fact that 2-QSAT is in P—instances of 2-QSAT can be efficiently decided by *classical* computers! In particular, it can be shown that the zero energy subspace can be spanned, if it is nontrivial, by *product* states. Since these are much simpler to specify classically (a product state needs only $2N$ complex numbers instead of 2^N), it is perhaps not surprising that we can decide whether or not such states exist that satisfy a given instance H. In any event, the only significant feature of the phase diagram is that for $\alpha < \alpha_s = 1/2$, we have a PRODSAT phase—that is a phase which is satisfiable by unentangled product states—and for $\alpha > \alpha_s$, we have an UNSAT phase with finite ground state energy density. The transition coincides with a geometric transition: $\alpha_s = 1/2$ corresponds to the emergence of a giant component in the underlying interaction graph.

For $k \geq 3$, the phase diagram is somewhat more interesting. Again, at low $\alpha < \alpha_{ps}(k) \sim 1$ there exists a PRODSAT regime in which satisfying product states are guaranteed to exist. Above α_{ps}, there are no satisfying product states, but the system remains SAT—thus, there is an "entanglement" transition in the ground state space as a function of α. Finally, above some $\alpha_c \sim 2^k$, there is an UNSAT phase in which it can be shown that there are no zero energy satisfying states. We note that the emergence of a giant component in the underlying interaction graph happens at

Fig. 7.7 Phase diagram of k-QSAT

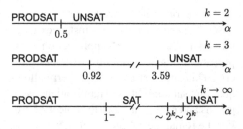

$\alpha_{gc} = \frac{1}{k(k-1)} \ll \alpha_{ps} \ll \alpha_c$—the various relevant transitions are well-separated at large k. A variety of different techniques go in to showing the existence of these transitions and phases—we sketch a few of these arguments in Sect. 7.4.4.

7.4.3 Geometrization Theorem

One of the most useful tools for studying random quantum satisfiability is geometrization. That is, the satisfiability of a generic instance of QSAT is a purely geometric property of the underlying interaction graph. This point of view extends to many properties of generic instances of QSAT—such as whether they are product satisfiable or not. Results about generic QSAT thus follow from identifying the right geometric properties in the interaction graph ensemble. We discuss a few of these examples in the sections to follow. Here, we provide an elementary proof of geometrization and a few immediate corollaries.

Geometrization Theorem 1 *Given an instance H of random k-QSAT with interaction graph G, the degeneracy of zero energy states $R(H) = \dim(\ker(H))$ takes a particular value R_G for almost all choices of clause projectors. R_G is minimal with respect to the choice of projectors.*

Proof For a fixed interaction graph G with M clauses, $H = H_\phi = \sum_{i=1}^{M} \Pi_i = \sum_{i=1}^{M} |\phi_i\rangle\langle\phi_i|$ is a matrix valued function of the $2^k M$ components of the set of M vectors $|\phi_i\rangle$. In particular, its entries are polynomials in those components. Choose $|\phi\rangle$ such that H has maximal rank D. Then there exists an $D \times D$ submatrix of H such that $\det(H|_{D \times D})$ is nonzero. But this submatrix determinant is a polynomial in the components of $|\phi\rangle$ and therefore is only zero on a submanifold of the $|\phi\rangle$ of codimension at least 1. Hence, generically H has rank D and the degeneracy $R_G = \dim(\ker(H)) = 2^N - D$. □

The theorem holds for general rank r problems as well by a simple modification of the argument to allow extra ϕ' s to be associated to each edge.

A nice corollary of this result is an upper bound on the size of the SAT phase at any k. Consider any assignment of classical clauses on a given interaction graph: this is a special instance of k-QSAT where the projectors are all diagonal in the computational basis. As this is a non-generic choice of projectors, the dimension of

Fig. 7.8 Example of a $k = 3$ interaction graph with $M < N$, *circles* indicate qubits and *squares* indicate clause projectors that act on adjacent qubits (*left*); a *dimer shaded*) covering that covers all clauses (*right*)

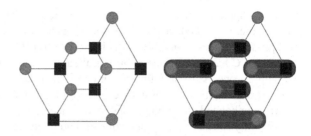

its satisfying manifold is an upper bound on the dimension for generic choices. We conclude then that the classical UNSAT threshold is an upper bound on the quantum threshold. Indeed, if we can identify the most frustrated assignment of classical clauses on a given interaction graph, i.e. the assignment that minimizes the number of satisfying assignments, we could derive an even tighter bound.

Corollary 1 *The generic zero state degeneracy is bounded above by the number of satisfying assignments of the most constrained classical k-SAT problem on the same graph.*

It is easy to construct example instances in which the quantum problem has fewer ground states than the most frustrated classical problem on the same interaction graph. Thus, the bound of corollary 1 is not tight.

7.4.4 A Few Details of Phases and Transitions

There are three flavors of arguments that have been used to pin down the phase diagram of $k \geq$ 3-QSAT: construction of satisfying product states [8, 34]; combinatorial upper bounds on the zero state degeneracy [8, 35]; and, a non-constructive invocation of the quantum Lovasz local lemma to establish the entangled SAT phase[36]. All three ultimately rely on establishing a correspondence between some geometric feature of the interaction graph G (or its subgraphs) and the properties of zero energy states for generic instances through geometrization. We sketch each of the three kinds of results below and refer the motivated reader to the relevant literature for details Fig. 7.8.

7.4.4.1 Product States

Perhaps the most direct approach to establishing a SAT phase in the phase diagram is to attempt explicitly to construct satisfying product states [8]. This is sufficient completely to determine the $k=2$ SAT phase, but only proves the existence of the PRODSAT phase of the $k \geq 3$ phase diagram.

In this approach, one constructs product states by a "transfer matrix"-like procedure: if there is a product state on some interaction graph G, then we can extend it to a product state on a graph G' which is G plus one additional clause C, so long as the clause C has at least one qubit not already in G. Thus, if a given graph G can be built up one clause at a time in an order such that there is always a previously unconstrained qubit brought in by each additional clause, G is PRODSAT.

Moving beyond our explicit construction, a complete characterization of product satisfiability can be found by analyzing the equations which a satisfying product state must obey [34]. This is a system of M algebraic equations in N complex unknowns and naive constraint counting suggests that $M \leq N$ should have solutions while $M > N$ should not. Since the system is sparse, however, a somewhat more detailed analysis is required to show:

Theorem 1 G is PRODSAT for generic choices of projectors Π_m if and only if its interaction graph has a dimer covering of its clauses.

Here a "a dimer covering of its clauses" is a pairing between qubits and clauses such that every clause appears paired with exactly one qubit and no qubit or clause appears more than once. The proof relies on 'product state perturbation theory', or in other words, the smoothness of the complex manifolds defining the projector space and the product state space.

If we apply the dimer covering characterization to the random interaction graph ensemble for G, we find the α_{ps} indicated in the phase diagram of Fig. 7.7. In particular, for $\alpha < \alpha_{ps}$, such dimer coverings exist w.p. 1 in the thermodynamic limit while for $\alpha > \alpha_{ps}$ they do not. The location of the geometric transition for the existence of dimer coverings in known in the literature [37]. Thus, for $\alpha > \alpha_{ps}$ there are no satisfying product states, although there may still be satisfying entangled states.

We note that the dimer covering characterization of product states provides an explicit mapping between dimer coverings and generic product states. In the case $M = N$, this mapping is one-to-one and provides a handle on counting the number of product states and, with more work, the ability to estimate their linear dependence. There are many avenues to explore here.

7.4.4.2 Bounding the Degeneracy

The existence of an UNSAT regime for large α follows immediately from the geometrization theorem and the existence of an UNSAT regime for classical SAT. That is, for $\alpha > \alpha_s^{\text{Classical}}$, typical graphs G are classically UNSAT and therefore, since the generic dimension R_G of the zero energy state space is minimal, they are also generically quantum UNSAT. In other words, the quantum SAT-UNSAT transition $\alpha_s \leq \alpha_s^{\text{Classical}}$.

This estimate of the SAT-UNSAT transition is not tight: the quantum UNSAT phase begins at a lower α than the classical UNSAT phase. This can be seen using another approach to bounding the ground state degeneracy. In this approach, one builds up a given graph G out of small clusters, each of which decimates the satisfying eigenspace by some known fraction. Indeed, if we consider two interaction graphs K and H on N qubits with that respective generic zero energy dimensions R_K and R_H, it is straightforward to show [35]

$$R_{K \cup H} \le R_K \frac{R_H}{2^N} \qquad (7.27)$$

for generic choices of projectors on K and H. As an example, let us build a graph G with M clauses, one clause at a time. Each individual clause H has $R_H = (1 - 1/2^k)2^N$ because it penalizes 1 out of the 2^k states in the local k-qubit space and leaves the other 2^{N-k} alone. Thus, adding each additional clause H_m decimates the satisfying subspace by at least a factor $(1 - 1/2^k)$ and:

$$R_G \le 2^N (1 - 1/2^k)^M \qquad (7.28)$$

Plugging in $M = \alpha N$ and taking N to infinity, we find that for $\alpha > -1/\log_2(1 - 1/2^k)$, R_G must go to zero. This simply reproduces the Pauling bound mentioned for classical k-SAT in Sect. 7.3.2.

However, one can do better by taking somewhat larger clusters H, calculating R_H exactly for these small clusters and then working out how many such clusters appear in the random graph G. This leads to much tighter bounds on α_s from above and in particular, as shown in Ref. [35], $\alpha_s(k) < \alpha_s^{\text{Classical}}(k)$ for all connectivities k.

7.4.4.3 Quantum Lovasz Local Lemma

The final technique that has been used to fill in the phase diagram of Fig. 7.7 is the development of a quantum version of the Lovasz local lemma [36]. This lemma provides a nonconstructive proof that satisfying states must exist for k-QSAT instances built out of interaction graphs with sufficiently low connectivity—that is, graphs in which the degree of every qubit is bounded by $2^k/(ek)$. The QSAT ensemble which we study in fact has average degree αk but the degree distribution has an unbounded tail. By cutting the graph into low and high connectivity subgraphs, using the product state characterization on the high connectivity part and the Lovasz lemma on the low connectivity part, and carefully glueing these results back together, it is then possible to show that satisfying states exist for $\alpha < 2^k/(12ek^2)$.

For sufficiently large k, this result proves that $\alpha_s \ge O(2^k/k^2) \gg \alpha_{ps}$, establishing the entangled SAT regime indicated in Fig. 7.7.

We now sketch the idea behind the classical and quantum Lovasz local lemmas.

Suppose we have some classical probability space and a collection of M events B_m, each with a probability of occurring $P(B_m) \le p < 1$. We think of these as low probability 'bad' events, such as "the m'th clause of a k-SAT instance is not satisfied by σ" given a uniformly chosen configuration σ. In this particular case, $P(B_m) = p = 1/2^k$ for all clauses m. If there is a positive probability that no bad event comes to pass, then there is clearly an overall assignment of σ which satisfies all of the clauses. Thus, we would like to show this probability is positive.

If the events B_m are independent, this is clearly possible:

$$Pr\left(\bigwedge_m \neg B_m\right) = \prod_{m=1}^{M} (1 - Pr(B_m)) \geq (1 - p)^M > 0 \qquad (7.29)$$

In the k-SAT example, clauses are independent if they do not share any bits—thus this argument provides us the rather obvious result that k-SAT instances composed of only completely disconnected clauses are satisfiable. On the other hand, if the events B_m are dependent, it is clear that we can make

$$Pr\left(\bigwedge_m \neg B_m\right) = 0. \qquad (7.30)$$

For instance, simply take a 3-SAT instance with 3 qubits and 8 clauses, each of which penalizes a different configuration. These clauses still have individually low probability ($p = 1/2^k$) but at least one of them is violated by any configuration.

The classical Lovasz local lemma [38, 39] provides an elementary method for relaxing the independence requirement a little bit. In particular, if each event B_m depends on no more than d other events $B_{m'}$ where

$$p \, e \, d \leq 1 \qquad (7.31)$$

(Euler's constant $e \approx 2.7182\ldots$ being the basis of the natural logarithm) then the local lemma tells us that there is indeed a positive probability that no bad event happens:

$$Pr\left(\bigwedge_m \neg B_m\right) > 0 \qquad (7.32)$$

This means that for connected k-SAT instances of sufficiently low degree, Lovasz proves the existence of satisfying configurations.

In the quantum generalization of the Lovasz lemma, probability is replaced by the relative dimension of satisfying subspaces. That is, for a QSAT projector Π of rank 1, the "probability of the clause being satisfied" is

$$\frac{\text{Dim}(SAT)}{\text{Dim}(\mathcal{H})} = \frac{2^k - 1}{2^k} = 1 - \frac{1}{2^k}. \qquad (7.33)$$

With the right definitions in hand, the generalization is also elementary and the result looks nearly identical to the classical case. However, now a positive probability that all projectors are satisfied tells us that there exists a (potentially quite entangled) quantum state of an N-qubit Hilbert space which satisfies the instance of k-QSAT.

The Lovasz local lemma is nonconstructive because it works by bounding (from below) the satisfying space degeneracy as the graph G is built up, so long as each additional clause does not overlap too many other clauses. In some sense this is dual to the arguments used to prove the UNSAT phase exists by bounding this degeneracy from above, but the technical details are somewhat more subtle since they require a more careful consideration of the interaction between additional clauses and the existing constraints.

In the last few years, computer scientists have developed a *constructive* version of the classical Lovasz local lemma. That is, there are now proofs that certain probabilistic algorithms will actually efficiently construct the Lovasz satisfying states [39]. Recent work suggests that a quantum generalization of this constructive approach may also be possible [40].

7.4.5 Satisfying the Promise

Ideally, we would like to study an ensemble of k-QSAT instances which always satisfy the promise. Such an ensemble would only contain Yes-instances with strictly zero energy and No-instances with energy bounded away from zero energy by a polynomially small promise gap. Such an ensemble is hard to construct as one does not know *a priori* which instances have zero energy or not, let alone whether their energy might be exponentially small. The best we can hope to do is choose a random ensemble in which the promise is satisfied statistically—perhaps with probability 1 in the thermodynamic limit.

Physical arguments suggest that the k-QSAT ensemble that we study here satisfies the promise in this statistical sense and for $k = 2$ it can be proven. On the SAT side of the phase diagram, all of the arguments that have been constructed to date show the existence of strict zero-energy states in the thermodynamic limit. These arguments all rely on geometrization: the existence of generic zero energy states is a graph property and such properties are either present or not in the thermodynamic limit of the random graph at a given α. Hence the zero energy phase as determined by such arguments is a strictly zero energy phase.

As statistical physicists, we expect that the UNSAT phase of k-QSAT has extensive ground state energy with relatively vanishing fluctuations for any k. If this is true, the promise that $E \geq O(N^{-a})$ fails to be satisfied only with exponentially small probability by Chebyshev's inequality. More generally, so long as the average ground state energy is bounded below by a polynomially small scale $E \geq O(N^{-b})$ with relatively vanishing fluctuations, the promise will be violated with only exponentially small probability for $a > b$.

For $k = 2$, it can be shown rigorously that the expected ground state energy for $\alpha > \alpha_s = 1/2$ is bounded below by a nearly extensive quantity (i.e. $E \geq O(N^{1-\epsilon})$ for any $\epsilon > 0$). Also, we know that the SAT phase extends to $\alpha = 1/2$ because we can show that satisfying zero energy product states exist up to this clause density. Thus, the ensemble satisfies the promise with high probability in both phases. At the critical point, things are not quite so clear, but one might expect fluctuations around $E=0$ at the scale $O(\sqrt{N})$ so that if the promise gap is chosen to be $O(N^{-1})$, the weight of the ensemble below the gap scale goes to zero.

7.4.6 Open Questions

In closing let us take stock of where we are at in the analysis of QSAT with the set of results on SAT as our template. First, the phase diagram clearly needs more work starting with more precise estimates for the SAT-UNSAT boundary. Within the SAT phase we have identified one phase transition where the satisfying states go from being products to being entangled and the key question is whether there are any others and whether they involve a clustering of quantum states in some meaningful fashion. Second, we have not said anything about the performance of algorithms for QSAT or about the relationship between phase structure and algorithm performance. Apart from some preliminary work on the adiabatic algorithm for 2-SAT [41], this direction is wide open for exploration.

7.5 Conclusion

In this review, we have tried to provide a reasonably self-contained introduction to the statistical mechanics of classical and quantum computational complexity, starting at the venerable subject of classical complexity theory, and ending at an active current research frontier at the intersection in quantum computing, quantum complexity theory and quantum statistical mechanics. We hope that this review will not only encourage some of its readers to contribute to these fields of study, but that it will also have provided them with some of the background necessary for getting started.

Acknowledgments We very gratefully acknowledge collaborations with Andreas Läuchli, in particular on the work reported in Ref. [34]. Chris Laumann was partially supported by a travel award of ICAM-I2CAM under NSF grant DMR-0844115.

References

1. Barahona, F.: On the computational complexity of Ising spin glass models. J. Phys. A: Math. Gen. **15**, 3241 (1982)
2. Aharonov, D., Gottesman, D., Kempe, J.: The power of quantum systems on a line. commun. Math. Phys. **287**, 41 (2009)
3. Aaronson, S.: Guest column: NP-complete problems and physical reality. SIGACT News. **36**, 30–52 (2005)
4. Arora, S., Barak, B.: Complexity Theory: A Modern Approach. Cambridge University Press, Cambridge, MA (2009)
5. Aharonov, D., and Naveh, T.: Quantum NP—A Survey, arXiv:quant-ph/0210077v1

6. Garey, M.R., Johnson, D.S.: Computers and Intractability: A Guide to the Theory of NP-Completeness. Series of Books in the Mathematical Sciences. W. H. Freeman & Co Ltd, San Francisco, CA (1979)
7. Bravyi, S.: Efficient algorithm for a quantum analogue of 2-SAT, arXiv:quant-ph/0602108v1
8. Laumann, C.R., Moessner, R., Scardicchio, A., Sondhi, S.L.: Phase transitions and random quantum satisfiability. Quant. Inf. Comp. **10**, 0001 (2010)
9. Krzakala, F., Montanari, A., Ricci-Tersenghi, F., Semerjian, G., Zdeborova, L.: Gibbs states and the set of solutions of random constraint satisfaction problems. Proc. Nat. Acad. Sci. USA **104**, 10318 (2007)
10. Dubois, O.: Upper bounds on the satisfiability threshold. Theor. Comput. Sci. **265**, 187 (2001)
11. Hartmann, A.K., Weigt, M.: Phase Transitions in Combinatorial Optimization Problems: Basics, Algorithms and Statistical Mechanics. Wiley, Weinheim (2005)
12. Braunstein, A., Mezard, M., Zecchina, R.: Survey propagation: an algorithm for satisfiability. Rand. Struct. Alg. **27**, 201–226 (2005)
13. Mézard, M.: Physics/computer science: passing messages between disciplines. Science **301**, 1685 (2003)
14. Mézard, M., Zecchina, R.: Random K-satisfiability problem: from an analytic solution to an efficient algorithm. Phys. Rev. E **66**, 056126 (2002)
15. Mézard, M., Parisi, G., Zecchina, R.: Analytic and algorithmic solution of random satisfiability problems. Science **297**, 812 (2002)
16. Mézard, M., Montanari, A.: Information, Physics and Computation. Oxford University Press Inc, New York (2009)
17. Laumann, C.R., Scardicchio, A., Sondhi, S.L.: Cavity method for quantum spin glasses on the Bethe lattice. Phys. Rev. B **78**, 134424 (2008)
18. Hastings, M.B.: Quantum belief propagation: an algorithm for thermal quantum systems. Phys. Rev. B **76**(20), 201102–201104 (2007)
19. Leifer, M., Poulin, D.: Quantum graphical models and belief propagation. Ann. Phys. **323**, 1899 (2008)
20. Semerjian, G., Tarzia, M., Zamponi, F.: Exact solution of the Bose-Hubbard model on the Bethe lattice. Phys. Rev. B **80**, 014524 (2009)
21. Krzakala, F., Rosso, A., Semerjian, G., Zamponi, G.: Path-integral representation for quantum spin models: application to the quantum cavity method and Monte Carlo simulations. Phys. Rev. B **78**, 134428 (2008)
22. Carleo, G., Tarzia, M., Zamponi, F.: Bose-Einstein condensation in quantum glasses. Phys. Rev. Lett. **103**, 215302 (2009)
23. Laumann, C.R., Parameswaran, S.A., Sondhi, S.L., Zamponi, F.: AKLT models with quantum spin glass ground states. Phys. Rev. B **81**, 174204 (2010)
24. Farhi, E., Goldstone, J., Gutmann, S., Lapan, J., Lundgren, A., Preda, D.: A quantum adiabatic evolution algorithm applied to random instances of an NP-complete problem. Science **292**, 472 (2001)
25. Aharonov, D., van Dam, W., Kempe, J., Landau, Z., Lloyd, S., Regev, O.: Adiabatic quantum computation is equivalent to standard quantum computation. Siam. Rev. **50**, 755 (2008)
26. Landau, L.: Zur theorie der Energieubertragung II. Phys. Sov. Union **2**, 46 (1932)
27. Zener, C.: Non-adiabatic crossing of energy levels. Proc. Roy. Soc. London: Ser. A **137**, 696 (1932)
28. Smelyanskiy, V., Knysh, S., Morris, R.: Quantum adiabatic optimization and combinatorial landscapes. Phys. Rev. E **70**, 036702 (2004)
29. Young, A.P., Knysh, S., Smelyanskiy, V.N.: Size dependence of the minimum excitation gap in the quantum adiabatic algorithm. Phys. Rev. Lett. **101**, 170503 (2008)
30. Young, A.P., Knysh, S., Smelyanskiy, V.N.: First-order phase transition in the quantum adiabatic algorithm. Phys. Rev. Lett. **104**, 020502 (2010)
31. Altshuler, B., Krovi, H., and Roland, J.: Adiabatic quantum optimization fails for random instances of NP-complete problems, arXiv:0908.2782v2

32. Altshuler, B., Krovi, H., Roland, J.: Anderson localization makes adiabatic quantum optimization fail. PNAS **107**, 12446 (2010)
33. Movassagh, R., Farhi, E., Goldstone, J., Nagaj, D., Osborne, T.J., Shor, P.W.: Unfrustrated qudit chains and their ground states. Phys. Rev. A **82**, 012318 (2010)
34. Laumann, C.R., Läuchli, A.M., Moessner, R., Scardicchio, A., Sondhi, S.L.: Product, generic, and random generic quantum satisfiability. Phys. Rev. A **81**, 062345 (2010)
35. Bravyi, S., Moore, C., Russell A.: Bounds on the quantum satisfibility threshold, arXiv:0907.1297v2
36. Ambainis, A., Kempe, J., Sattath, O.: in 42nd Annual ACM Symposium on Theory of Computing (2009)
37. Mézard, M., Ricci-Tersenghi, F., Zecchina, R.: Two solutions to diluted p-spin models and XORSAT problems. J. Stat. Phys. **111**, 505 (2003)
38. Erdös, P., Lovász, L.: Problems and results on 3-chromatic hypergraphs and some related questions. Infinite finite sets **2**, 609 (1975)
39. Moser, R.A., Tardos, G.: A constructive proof of the general Lovász local lemma. J. ACM. **57**, 1 (2010)
40. Arad, I., Cubitt, T., Kempe, J., Sattath, O., Schwarz, M., Verstraete F.: Private communication (2010)
41. Govenius, J.: Junior Paper: Running Time Scaling of a 2-QSAT Adiabatic Evolution Algorithm. Princeton Junior Paper, Princeton (2008)

Chapter 8
Non-Perturbative Methods in (1+1) Dimensional Quantum Field Theory

Giuseppe Mussardo

Abstract In recent years there has been an enormous progress in low-dimensional quantum field theory. The most important results concern the conformal properties of the critical points of the Renormalization Group and the scaling region nearby. In this respect a crucial role is played by integrable deformations of Conformal Field Theories, which can be solved using bootstrap methods coming from *S*-matrix theory. In these lectures I present the Form-Factor Approach to the computation of correlation functions. Non-perturbative methods of both Conformal and Integrable Field Theories find remarkable applications in low-dimensional quantum systems.

8.1 Introduction

One of the fundamental problems of statistical mechanics and its quantum field theory formulation is the characterization of the order parameters and the computation of their correlation functions. Beside the intrinsic interest of this problem, the correlation functions are the key quantities in the determination of the universal ratios of the Renormalization Group and therefore they can have a direct experimental confirmation [1]. I will briefly review below the properties of the free energy nearby a critical point, the definition of the universal ratios and their relation with the correlation functions.

It should be pointed out that the computation of correlation functions is quite often a difficult task, usually achieved with partial success through perturbative methods. An exact determination of the operator content and the correlation functions of a two-dimensional theory can be obtained only when the model is at its critical point. In this case, in fact, one has a classification of the order parameters in terms of

G. Mussardo (✉)
SISSA, Via Beirut 1,
34100 Trieste, Italy
e-mail:mussardo@sissa.it

D. C. Cabra et al. (eds.), *Modern Theories of Many-Particle Systems in Condensed Matter Physics*, Lecture Notes in Physics 843, DOI: 10.1007/978-3-642-10449-7_8, © Springer-Verlag Berlin Heidelberg 2012

the irreducible representation of the Virasoro algebra of Conformal Field Theory and, moreover, one can get an exact expression of the correlators solving the linear differential equations that they satisfy [2, 3].

Unfortunately, the elegant theoretical scheme of the critical points cannot be generalized once we move away from criticality. In this case, the problem has to be faced with different techniques. Significant progress can be made when we deal with integrable theories: these theories are characterized by an elastic S-matrix and the exact spectrum of the massive excitations [4]. The central quantities are in this case the matrix elements of the various operators on the asymptotic states of the theory, called the *Form Factors* [5, 6]. The precise definition of these quantities is given below. The general properties related to the unitarity and crossing symmetry lead to a set of functional equations for the Form Factors that can be explicitly solved in many interesting cases. Once the matrix elements of the operators are known, their correlation functions can be recovered in terms of spectral representation series. It is worth mentioning that these series present remarkable convergence properties (see, for instance [7–10]) whose explanation was given in [11].

Hence, anticipating the main results of these lecture, the success of the Form Factor method relies on two points: (a) the possibility to determine exactly the matrix elements of the order parameters on the asymptotic states of the theory, identified by the scattering theory; (b) the fast convergence properties of the spectral series. These two steps lead to the determination of the correlation functions away from criticality with a precision that cannot be obtained by other methods. Before entering into the details of this approach, let us first discuss the behavior of the free energy near the fixed points and the associated universal ratios.

8.2 Functional Form of the Free Energy

The linearized form of the Renormalization Group equations permits to easily derive the scaling form of the free energy in the vicinity of the fixed point and the relationships between the critical exponents. Consider a statistical system with n relevant coupling constants λ_i and conjugated fields $\phi_i(x)$. In the field theory formulation, in the vicinity of the fixed point the action is given by

$$S = S^* + \sum_i^n \lambda_i \int d^d x \phi_i(x). \tag{8.1}$$

In the Ising model, for instance, there are two relevant variables, given by the magnetic field $h \equiv \lambda_1$ and by the displacement of the temperature from the critical value $T - T_c \equiv \lambda_2$: the conjugate fields are $\phi_1(x)$, that corresponds to the continuum limit of the spin variable s_i, and $\phi_2(x)$, associated to the continuum limit of the energy density, given on the lattice by $\sum_j s_i s_{i+\hat{e}_j}$.

Since the variables λ_j in the action (8.1) have dimensions $[\lambda_j] = a^{y_j}$, the theory has a finite correlation length. Selecting one of the couplings, say λ_i, in the thermodynamic limit the correlation length can be expressed as

$$\xi(\{\lambda_j\}) = a(K_i\lambda_i)^{-\frac{1}{y_i}} L_i\left(\frac{K_1\lambda_j}{(K_i\lambda_i)^{\eta_{1i}}}, \cdots, \frac{K_j\lambda_j}{(K_i\lambda_i)^{\eta_{ji}}}, \cdots\right), \tag{8.2}$$

where $K_i \simeq 1/\lambda_i^{(0)}$ are some non-universal metric terms that depend on the unity by which we measure the coupling constants, L_i are universal homogeneous functions of the $(n-1)$ ratios $\frac{K_j\lambda_j}{(K_i\lambda_i)^{\eta_{ji}}}$, with $j \neq i$ and finally

$$\eta_{ji} = \frac{y_j}{y_i}, \tag{8.3}$$

are the so-called *crossover exponents*. There are many (but equivalent) ways of expressing this scaling law of the correlation length, according to which coupling constant we choose as prefactor. Each way selects its own scaling function L of the above ratio of the couplings. When $\lambda_k \to 0$ ($k \neq i$) with $\lambda_i \neq 0$, Eq. 8.2 can be written as

$$\xi_i = a\xi_i^0\lambda_i^{-\frac{1}{y_i}}, \quad \xi_i^0 \sim K_i^{-\frac{1}{y_i}}. \tag{8.4}$$

Consider now the free energy of the system, $f[\lambda_i]$, defined by

$$Z[\{\lambda_i\}] = \int \mathcal{D}\phi_i e^{-[\mathcal{S}^* + \sum_{i=1}^{n} \lambda_i \int \phi_i(x)d^dx]} \equiv e^{-Nf(\lambda_i)}. \tag{8.5}$$

Making a Renormalization Group transformation we have

$$e^{-Nf(\{\lambda\})} = e^{-Np(\{\lambda\}) - N'f(\{\lambda'\})},$$

where $p(\{\lambda\})$ is an additive constant related to the degrees of freedom over which we have integrated. Since the new number of sites is $N' = b^{-d}N$, we have the functional equation

$$f(\{\lambda\}) = p(\{\lambda\}) + b^{-d}f(\{\lambda'\}). \tag{8.6}$$

The function $p(\{\lambda\})$ is an analytic function of the coupling constants, since it involves a sum over a finite number of spins. If we are interested in studying the singular behavior of the free energy, we can safely discard this term and arrive to a functional equation that involves only the singular part of f

$$f_s(\{\lambda\}) = b^{-d}f_s(\{\lambda'\}). \tag{8.7}$$

Substituting in it the expression of the new coupling constants given by the Renormalization Group transformations, we have

$$f_s(\{\lambda_k\}) = b^{-d}f_s(\{b^{y_k}\lambda_i\}). \tag{8.8}$$

Iterating this equation, the irrelevant variables go to zero (this is a manifestation of the universality of the critical behavior) and the free energy, as function of the relevant variables alone, satisfies

$$f_s(\{\lambda_j\}) = b^{-nd} f_s(\{b^{ny_j}\lambda_i)\}. \tag{8.9}$$

As for the correlation length, there are many ways to express the general solution of this equation. Selecting once more one of the couplings, say λ_i, we have

$$f(\{\lambda_i\}) = f_i[\{\lambda_j\}] \equiv (K_i\lambda_i)^{-\frac{d}{y_i}} F_i\left(\frac{K_1\lambda_j}{(K_i\lambda_i)^{\eta_{1i}}}, \cdots, \frac{K_j\lambda_j}{(K_i\lambda_i)^{\eta_{ji}}}, \cdots\right)^{-\frac{d}{y_i}}. \tag{8.10}$$

The functions F_i are universal homogeneous functions of the $(n-1)$ ratios $\frac{K_j g_j}{(K_i g_i)^{\eta_{ji}}}$. As we will see below, there are some obvious advantages in considering different expressions for these scaling functions, obtained by changing the selected variable λ_i. In fact, in several physical applications, there is only one coupling constant kept different from zero till the end, and the best choice of expressing the free energy depends on this situation. As we are going to show, even in the absence of an explicit expression of the F_i's (that can be explicitly found only by solving exactly the model by other methods), the functional dependence of the free energy is sufficient to obtain useful information on the critical behavior of the model.

8.2.1 Critical Exponents and Universal Ratios

Let us discuss the definition of several thermodynamical quantities associated to the derivates of the free energy. In the following, we adopt the notation $\langle\ldots\rangle_i$ to denote the expectation values computed with an action that has, at the end, only λ_i as coupling constant different from zero. The first quantities of interest are the expectation values of the fields ϕ_j that can be parameterized as

$$\langle\phi_j\rangle_i = -\frac{\partial f_i}{\partial\lambda_j}\bigg|_{\lambda_k=0} \equiv B_{ji}\lambda_i^{\frac{d-y_j}{y_i}}, \tag{8.11}$$

with

$$B_{ji} \sim K_j K_i^{\frac{d-y_j}{y_i}}. \tag{8.12}$$

Equivalently

$$\lambda_i = D_{ij}\left(\langle\phi_j\rangle_i\right)^{\frac{y_i}{d-y_j}}, \tag{8.13}$$

with

$$D_{ij} \sim \frac{1}{K_i K_j^{\frac{y_j}{d-y_j}}}. \tag{8.14}$$

The generalized susceptibilities are defined by

$$\hat{\Gamma}^i_{jk} = \frac{\partial}{\partial \lambda_k} \langle \phi_j \rangle_i = -\frac{\partial^2 f_i}{\partial \lambda_k \partial \lambda_j}. \tag{8.15}$$

These quantities are obviously symmetric with respect to the lower indices. For the fluctuation-dissipation theorem, they are related to the off-critical correlation functions as

$$\hat{\Gamma}^i_{jk} = \int dx \langle \phi_k(x) \phi_j(0) \rangle_i. \tag{8.16}$$

Taking out the dependence on the coupling constant λ_i, we have

$$\hat{\Gamma}^i_{jk} = \Gamma^i_{jk} \lambda_i^{\frac{d-y_j-y_k}{y_i}}, \tag{8.17}$$

with

$$\Gamma^i_{jk} \sim K_j K_k K_i^{\frac{d-y_j-y_k}{y_i}}. \tag{8.18}$$

As shown by the formulas above, the various quantities contain the metric factors K_i and their expressions are therefore not universal. However, we can consider some special combinations of these quantities in which the metric factors cancel out. Here we give some examples of the so-called *universal ratios*

$$(R_c)^i_{jk} = \frac{\Gamma^i_{ii} \Gamma^i_{jk}}{B_{ji} B_{ki}}; \tag{8.19}$$

$$(R_\chi)^i_j = \Gamma^i_{jj} D_{jj} B_{ji}^{\frac{D-4\Delta_j}{2\Delta_j}}; \tag{8.20}$$

$$R^i_\xi = \left(\Gamma^i_{ii} \right)^{1/D} \xi^0_i; \tag{8.21}$$

$$(R_A)^i_j = \Gamma^i_{jj} D_{ii}^{\frac{4\Delta_j+2\Delta_i-2D}{D-2\Delta_i}} B_{ij}^{\frac{2\Delta_j-D}{\Delta_i}}; \tag{8.22}$$

$$(Q_2)^i_{jk} = \frac{\Gamma^i_{jj}}{\Gamma^k_{jj}} \left(\frac{\xi^0_k}{\xi^0_j} \right)^{D-4\Delta_j}. \tag{8.23}$$

As the critical exponents, these pure numbers characterize the universality class of a given model. It is worth emphasizing that, from an experimental point of view, it should be simpler to measure universal amplitude ratios rather than critical exponents: in fact to determine the former quantities one needs to perform several measurements at a single, fixed value of the coupling which drives the system away from criticality whereas to determine the latter, one needs to make measurements over several decades along the axes of the off–critical couplings. Moreover, although not all of them are independent, the universal ratios are a set of numbers larger than the critical exponents and therefore permit a more precise determination of the class of universality. Finally, being universal quantities, they can be theoretically computed by analyzing the simplest representative of the class of universality under scrutiny. This may be given by a field theory and therefore all the universal quantities above can be determined by computing the one and two-point functions of such a theory. This task can be performed employing the Form Factor approach.

8.3 General Properties of the Form Factors

In order to compute the universal ratios we need to compute the correlation functions. This task can be performed using the spectral series of these quantities, based on the matrix elements of the various operators on the asymptotic states, the so-called Form Factors [5, 6]. An essential quantity for the computation of the matrix elements is the S-matrix of the problem. The S-matrix of two-dimensional integrable systems is particularly simple and can be explicitly found in many interesting cases: in fact, for the infinite number of conservation laws, the scattering processes of integrable systems are purely elastic and the n-particle S-matrix can be factorized in terms of the $n(n-1)/2$ two-body scattering amplitudes [4]. In the following, for simplicity, we mainly focus our attention on diagonal scattering theories with non-degenerate spectrum. To characterize the kinematic state of the particles we use the rapidities θ_i, that enter the dispersion relations

$$p_i^0 = m_i \cosh \theta_i, \quad p_i^1 = m_i \sinh \theta_i. \tag{8.24}$$

The two-body S matrix amplitudes depend on the difference of the rapidities $\theta_{ij} = \theta_i - \theta_j$ and satisfy the unitary and crossing symmetry equations

$$
\begin{aligned}
S_{ij}(\theta_{ij}) &= S_{ji}(\theta_{ij}) = S_{ij}^{-1}(-\theta_{ij}), \\
S_{i\bar{j}}(\theta_{ij}) &= S_{ij}(i\pi - \theta_{ij}).
\end{aligned}
\tag{8.25}
$$

Possible bound states correspond to simple poles (or higher order odd poles) of these amplitudes, placed at imaginary values of θ_{ij} in the physical strip $0 < \mathrm{Im}\theta < \pi$. Let us see how the S-matrix allows us to compute the matrix elements of the (semi)-local operators on the asymptotic states. To this aim, it is useful to introduce an algebraic formalism.

8.3.1 Faddeev-Zamolodchikov Algebra

A key assumption of the Form Factor theory is that there exist some operators, both of creation and annihilation type, $V_{\alpha_i}^{\dagger}(\theta_i)$, $V_{\alpha_i}(\theta_i)$, that implement a generalization of the usual bosonic and fermionic algebraic relations. Let us call them *vertex operators*. Denoting by α_i the quantum number that distinguishes the different types of particles of the theory, these operators satisfy the associative algebra in which enters the S-matrix

$$V_{\alpha_i}(\theta_i)V_{\alpha_j}(\theta_j) = S_{ij}(\theta_{ij})V_{\alpha_j}(\theta_j)V_{\alpha_i}(\theta_i) \tag{8.26}$$

$$V_{\alpha_i}^{\dagger}(\theta_i)V_{\alpha_j}^{\dagger}(\theta_j) = S_{ij}(\theta_{ij})V_{\alpha_j}^{\dagger}(\theta_j)V_{\alpha_i}^{\dagger}(\theta_i) \tag{8.27}$$

$$V_{\alpha_i}(\theta_i)V_{\alpha_j}^{\dagger}(\theta_j) = S_{ij}(\theta_{ji})V_{\alpha_j}^{\dagger}(\theta_j)V_{\alpha_i}(\theta_i) + 2\pi\delta_{\alpha_i\alpha_j}\delta(\theta_{ij}). \tag{8.28}$$

Any commutation of these operators can be interpreted as a scattering process. The Poincaré group, generated by the Lorentz transformations $L(\epsilon)$ and the translations T_y, acts on the operators as

$$U_L V_{\alpha}(\theta)U_L^{-1} = V_{\alpha}(\theta + \epsilon) \tag{8.29}$$

$$U_{T_y} V_{\alpha}(\theta)U_{T_y}^{-1} = e^{iP_{\mu}(\theta)y^{\mu}} V_{\alpha}(\theta). \tag{8.30}$$

Obviously the explicit form of the creation and annihilation operators depends crucially on the theory in question and their construction is an open problem for most of the models. This difficulty does not stop us however to derive the fundamental equations for the matrix elements starting from the algebraic equations given above.

The vertex operators define the space of the physical states. The vacuum $|0\rangle$ is the state annihilated by $V_{\alpha}(\theta)$,

$$V_{\alpha}(\theta)|0\rangle = 0 = \langle 0|V_{\alpha}^{\dagger}(\theta),$$

while the Hilbert space is constructed by applying the various vertex operators $V_{\alpha}^{\dagger}(\theta)$ on $|0\rangle$

$$|V_{\alpha_1}(\theta_1)\dots V_{\alpha_n}(\theta_n)\rangle \equiv V_{\alpha_1}^{\dagger}(\theta_1)\dots V_{\alpha_n}^{\dagger}(\theta_n)|0\rangle. \tag{8.31}$$

From Eq. 8.28, the one-particle states have the normalization

$$\langle V_{\alpha_i}(\theta_i)|V_{\alpha_j}(\theta_j)\rangle = 2\pi\delta_{\alpha_i\alpha_j}\delta(\theta_{ij}).$$

The algebra of the vertex operators implies that the vectors (8.31) are not all linearly independent. To select a basis of linear independent vectors we need an additional requirement: for the initial states, the rapidites must be ordered in a decreasing way

$$\theta_1 > \theta_2 > \cdots > \theta_n$$

while, for the final states in an increasing way

$$\theta_1 < \theta_2 < \cdots < \theta_n.$$

These orderings select a set of linearly independent vectors that form a basis in the Hilbert space.

8.3.2 Form Factors

In this section we expose the principles of the theory following the references [5, 6]. However, it is also useful to consult [7–10, 12–16]. Unless explicitly stated, in the following we consider the matrix elements between the *in* and *out* states of the particle with the lowest mass of local, scalar and hermitian operators $\mathcal{O}(x)$

$$_{\text{out}}\langle V(\theta_{m+1}) \ldots V(\theta_n) | \mathcal{O}(x) | V(\theta_1) \ldots V(\theta_m) \rangle_{\text{in}}. \tag{8.32}$$

We can always place the operator at the origin by using the translation operator, $U_{T_y}\mathcal{O}(x)U_{T_y}^{-1} = \mathcal{O}(x+y)$, and using Eq. 8.30, the matrix elements above are given by

$$\exp\left[i\left(\sum_{i=m+1}^{n} p_\mu(\theta_i) - \sum_{i=1}^{m} p_\mu(\theta_i)\right)x^\mu\right]$$
$$\times {}_{\text{out}}\langle V(\theta_{m+1}) \ldots V(\theta_n) | \mathcal{O}(0) | V(\theta_1) \ldots V(\theta_m) \rangle_{\text{in}}. \tag{8.33}$$

It is convenient to define the functions

$$F_n^{\mathcal{O}}(\theta_1, \theta_2, \ldots, \theta_n) = \langle 0 \mid \mathcal{O}(0) \mid \theta_1, \theta_2, \ldots, \theta_n \rangle_{\text{in}}, \tag{8.34}$$

called the *Form Factors* (FF), whose graphical representation is shown in Fig. 8.1: they are the matrix elements of an operator placed at the origin between the n-particle state and the vacuum.[1]

For local and scalar operators, the relativistic invariance of the theory implies that the FF are functions of the differences of the rapidities θ_{ij}

$$F_n^{\mathcal{O}}(\theta_1, \theta_2, \ldots, \theta_n) = F_n^{\mathcal{O}}(\theta_{12}, \theta_{13}, \ldots, \theta_{ij}, \ldots), \quad i < j. \tag{8.35}$$

The invariance under crossing symmetry (i.e., the possibility to pass from *in* to *out* states) permits to recover the most general matrix elements by an analytic continuation of the functions (8.34)

---

[1] From now on we use the simplified notation $| \ldots V(\theta_n) \ldots \rangle \equiv | \ldots \theta_n \ldots \rangle$ to denote the physical states of the particle with the lowest mass.

Fig. 8.1 Form factor of the operator $\mathcal{O}$

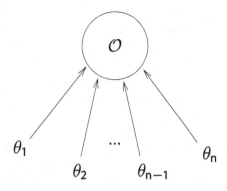

$$F_{n+m}^{\mathcal{O}}(\theta_1, \theta_2, \ldots, \theta_m, \theta_{m+1} - i\pi, \ldots, \theta_n - i\pi) = F_{n+m}^{\mathcal{O}}(\theta_{ij}, i\pi - \theta_{sr}, \theta_{kl}) \quad (8.36)$$

where $1 \le i < j \le m$, $1 \le r \le m < s \le n$, and $m < k < l \le n$.

Apart from the poles corresponding to the bound states present in all possible channels of this amplitude, the form factors $F_n^{\mathcal{O}}$ are expected to be analytic functions in the strips $0 < \mathrm{Im}\theta_{ij} < 2\pi$.

8.4 Watson's Equations

The FF of a scalar and hermitian operator $\mathcal{O}$ satisfy a set of equations, known as *Watson equations*, that assume a particularly simple form for the integrable systems

$$F_n^{\mathcal{O}}(\theta_1, \ldots, \theta_i, \theta_{i+1}, \ldots, \theta_n) = F_n^{\mathcal{O}}(\theta_1, \ldots, \theta_{i+1}, \theta_i, \ldots, \theta_n) S(\theta_i - \theta_{i+1}),$$

$$F_n^{\mathcal{O}}(\theta_1 + 2\pi i, \ldots, \theta_{n-1}, \theta_n) = e^{2\pi i\gamma} F_n^{\mathcal{O}}(\theta_2, \ldots, \theta_n, \theta_1)$$

$$= \prod_{i=2}^{n} S(\theta_i - \theta_1) F_n^{\mathcal{O}}(\theta_1, \ldots, \theta_n), \quad (8.37)$$

where γ is the semi-local index of the operator $\mathcal{O}$ with respect to the operator that creates the particles. The first equation is a simple consequence of Eq. 8.26, because a commutation of two operators is equivalent to a scattering process. Concerning the second equation, it states the nature of the discontinuity of these functions at the cuts $\theta_{1i} = 2\pi i$. The graphical representation of these equations is shown in Fig. 8.2. When $n = 2$, Eq. 8.37 reduce to

$$F_2^{\mathcal{O}}(\theta) = F_2^{\mathcal{O}}(-\theta) S(\theta),$$

$$F_2^{\mathcal{O}}(i\pi - \theta) = F_2^{\mathcal{O}}(i\pi + \theta). \quad (8.38)$$

Fig. 8.2 Graphical form of
the Watson equations

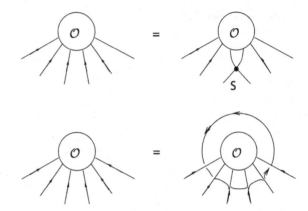

The most general solution of the Watson equations (8.37) is given by

$$F_n^{\mathcal{O}}(\theta_1, \ldots, \theta_n) = K_n^{\mathcal{O}}(\theta_1, \ldots, \theta_n) \prod_{i<j} F_{\min}(\theta_{ij}). \qquad (8.39)$$

Let us discuss the various terms entering this expression.

Minimal 2-Particle Form Factor $F_{\min}(\theta)$ is an analytic function in the region $0 \leq \mathrm{Im}\,\theta \leq \pi$, solution of the two Eq. 8.38, with neither zeros nor poles in the strip $0 < \mathrm{Im}\,\theta < \pi$, and with the mildest behavior at $|\theta| \to \infty$. These requirements determine uniquely this function, up to a normalization factor $\mathcal{N}$. Its explicit expression can be found by writing the S-matrix as

$$S(\theta) = \exp\left[\int_0^\infty \frac{dt}{t} f(t) \sinh \frac{t\theta}{i\pi} \right].$$

In fact, it is easy to see that $F_{\min}(\theta)$ is given by

$$F_{\min}(\theta) = \mathcal{N} \exp\left[\int_0^\infty \frac{dt}{t} \frac{f(t)}{\sinh t} \sin^2\left(\frac{t\pi\hat{\theta}}{2\pi} \right) \right], \quad \hat{\theta} = i\pi - \theta. \qquad (8.40)$$

Note that for interacting theories, $S(0) = -1$, and therefore the first equation in (38) forces $F_{\min}(\theta)$ to have a zero at the two-particle threshold

$$F(\theta) \simeq \theta, \quad \theta \to 0. \qquad (8.41)$$

$K_n^{\mathcal{O}}$ **Factors** The remaining factors $K_n^{\mathcal{O}}$ in (8.39) satisfy the Watson equation but with $S = 1$: this implies that they are completely symmetric functions in the variables θ_{ij}, periodic with period $2\pi i$. Therefore they can be considered as functions of the variables $\cosh \theta_{ij}$. Let us investigate other properties of the functions $K_n^{\mathcal{O}}$. They must

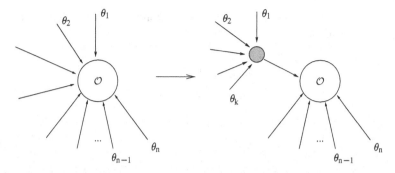

Fig. 8.3 Kinematic configuration of k-particle cluster responsible for a pole in the form factors

have all physical poles expected for the form factors. We recall that, in general, there is a simple pole in the form factors when a cluster made of k particles can reach a kinematical configuration that is equivalent to the one of a single particle, as shown in Fig. 8.3, with the pole given just by the propagator of the latter particle. If this is the general situation, for the integrable theories there is however an important simplification. In fact, for the factorization property of the S-matrix, it is sufficient to consider only the cases in which the clusters are made of $k = 2$ or $k = 3$: the poles coming from the 2-particle clusters are dictated uniquely by the bound states of the S-matrix, while those coming from the 3-particle clusters are determined by the crossing processes, although they are also related to the S-matrix (see the discussion in the next section). In conclusion, all the poles of the form factors are determined by the underlying scattering theory and they do not depend on the operator! In the light of this analysis, the functions $K_n^{\mathcal{O}}$ can be parameterized as follows

$$K_n^{\mathcal{O}}(\theta_1, \ldots, \theta_n) = \frac{Q_n^{\mathcal{O}}(\theta_1, \ldots, \theta_n)}{D_n(\theta_1, \ldots, \theta_n)}, \tag{8.42}$$

where the denominator D_n is a polynomial in $\cosh \theta_{ij}$ that is fixed only by the pole structure of the S-matrix while the information on the operator $\mathcal{O}$ is enclosed in the polynomial $Q_n^{\mathcal{O}}$ of the variables $\cosh \theta_{ij}$ placed at the *numerator*. We will come back to this important point in the later sections.

Symmetric Polynomials As shown above, the functions $K_n^{\mathcal{O}}$ are symmetric under the permutation of the rapidities of the various particles. In many case it is convenient to change variables, introducing the parameters $x_i \equiv e^{\theta_i}$, so that both numerator and denominator become symmetric polynomials in the x_i variables. A basis in the functional space of the symmetric polynomials in n variables is given by the *elementary symmetric polynomials* $\sigma_k^{(n)}(x_1, \ldots, x_n)$, whose generating function is

$$\prod_{i=1}^{n}(x + x_i) = \sum_{k=0}^{n} x^{n-k} \sigma_k^{(n)}(x_1, x_2, \ldots, x_n). \tag{8.43}$$

Conventionally all $\sigma_k^{(n)}$ with $k > n$ and with $n < 0$ are zero. The explicit expressions for the other cases are

$$\sigma_0 = 1,$$
$$\sigma_1 = x_1 + x_2 + \cdots + x_n,$$
$$\sigma_2 = x_1 x_2 + x_1 x_3 + \cdots x_{n-1} x_n,$$

$$\vdots \qquad \vdots$$

$$\sigma_n = x_1 x_2 \ldots x_n. \tag{8.44}$$

The $\sigma_k^{(n)}$ are homogeneous polynomials in x_i, of total degree k but linear in each variable.

Total and Partial Degrees of the Polynomials The polynomials $Q_n^{\mathcal{O}}(x_1, \ldots, x_n)$ in the numerator of the factor $K_n^{\mathcal{O}}$ satisfy additional conditions coming from the asymptotic behavior of the form factors. The first condition simply comes from the relativistic invariance: in fact, for a simultaneous translation of all the rapidities, the form factors of a scalar operator[2] satisfy

$$F_n^{\mathcal{O}}(\theta_1 + \Lambda, \theta_2 + \Lambda, \ldots, \theta_n + \Lambda) = F_n^{\mathcal{O}}(\theta_1, \theta_2, \ldots, \theta_n). \tag{8.45}$$

This implies the equality of the total degrees of the polynomials $Q_n^{\mathcal{O}}(x_1, \ldots, x_n)$ and $D_n(x_1, \ldots, x_n)$. Concerning the partial degree with respect to each variable, it is worth anticipating a result discussed in Sect. 8.9: in order to have a power-law behavior of the two-point correlation function of the operator $\mathcal{O}(x)$, its form factors must behave for $\theta_i \to \infty$ at most as $\exp(k\theta_i)$, where k is a constant (independent of i), related to the conformal weight of the operator $\mathcal{O}$.

8.5 Recursive Equations

The poles in the FF induce a set of recursive equations that are crucial for the explicit determination of these functions. As a function of the difference of the rapidities θ_{ij}, the FF have two kinds of simple poles.[3]

Kinematical Poles The first kind of singularity does not depend on whether the model has bound states. It is in fact associated to the kinematical poles at $\theta_{ij} = i\pi$ that come from the one-particle state realized by the 3-particle clusters. In turn, these processes correspond to the crossing channels of the S-matrix, as shown in Fig. 8.4. The residues at these poles give rise to a recursive equation that links the n-particle and the $(n-2)$-particle form factors

---

[2] For the form factors of an operator $\mathcal{O}(x)$ of spin s, the equation generalizes as $F_n^{\mathcal{O}}(\theta_1 + \Lambda, \theta_2 + \Lambda, \ldots, \theta_n + \Lambda) = e^{s\Lambda} F_n^{\mathcal{O}}(\theta_1, \theta_2, \ldots, \theta_n)$.

[3] There could be also higher order poles, in correspondence with the higher order poles of the S-matrix. Their discussion is however beyond the scope of these lectures.

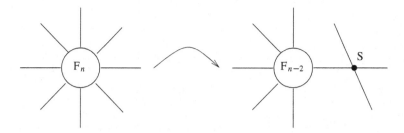

Fig. 8.4 Recursive equation of the kinematical poles

$$-i \lim_{\tilde{\theta} \to \theta} (\tilde{\theta} - \theta) F_{n+2}^{\mathcal{O}}(\tilde{\theta} + i\pi, \theta, \theta_1, \theta_2, \ldots, \theta_n)$$

$$= \left(1 - e^{2\pi i \gamma} \prod_{i=1}^{n} S(\theta - \theta_i)\right) F_n^{\mathcal{O}}(\theta_1, \ldots, \theta_n). \tag{8.46}$$

Let us denote concisely by $\mathcal{C}$ the map between $F_{n+2}^{\mathcal{O}}$ and $F_n^{\mathcal{O}}$ established by the recursive equation

$$F_{n+2}^{\mathcal{O}} = \mathcal{C} F_n^{\mathcal{O}}. \tag{8.47}$$

Bound State Poles There is another family of poles in F_n if the S-matrix has simple poles related to the bound states. These poles are at the values of θ_{ij} corresponding to the resonance angles. Let $\theta_{ij} = i u_{ij}^k$ be one of these poles, associated to the bound state A_k present in the channel $A_i \times A_j$. For the S-matrix we have

$$-i \lim_{\theta \to i u_{ij}^k} (\theta - i u_{ij}^k) S_{ij}(\theta) = \left(\Gamma_{ij}^k\right)^2, \tag{8.48}$$

where Γ_{ij}^k is the on-shell 3-particle vertex and for the residue of the form factor F_{n+1} involving the particles A_i and A_j we have

$$-i \lim_{\epsilon \to 0} \epsilon F_{n+1}^{\mathcal{O}}(\theta + i \bar{u}_{ik}^j - \epsilon, \theta - i \bar{u}_{jk}^i + \epsilon, \theta_1, \ldots, \theta_{n-1}) = \Gamma_{ij}^k F_n^{\mathcal{O}}(\theta, \theta_1, \ldots, \theta_{n-1}), \tag{8.49}$$

where $\bar{u}_{ab}^c \equiv (\pi - u_{ab}^c)$. This equation sets up a recursive structure between the $(n+1)$ and the n particle form factors, as shown in Fig. 8.5. Let us denote by $\mathcal{B}$ the map between F_{n+1}' and $F_n^{\mathcal{O}}$ set by this recursive equation

$$F_{n+1}^{\mathcal{O}} = \mathcal{B} F_n^{\mathcal{O}}. \tag{8.50}$$

When the theory presents bound states, it is possible to show that the two kinds of recursive equation are compatible, so that it is possible to reach the $(n+2)$-particle FF by the n-particle FF either using directly the recursive Eq. 8.46 or applying twice the recursive Eq. 8.49. In terms of the mappings $\mathcal{B}$ and $\mathcal{C}$ we have $\mathcal{C} = \mathcal{B}^2$.

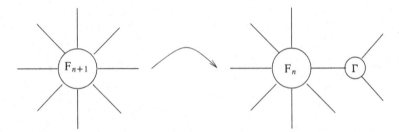

Fig. 8.5 Recursive equation of the bound state poles

8.6 The Operator Space

At the critical point, one can identify the operator space of a quantum field theory in terms of the irreducible representations of the Virasoro algebra [2, 3]. An extremely interesting point is the characterization of the operator content also away from criticality. As argued below, this can be achieved by means of the form factor theory [12, 14]: although this identification is based on different principles than the conformal theories, nevertheless it allows us to shed light on the classification problem of the operators.

Let us start our discussion with some general considerations. In the form factor approach, an operator $\mathcal{O}$ is defined once all its matrix elements $F_n^{\mathcal{O}}$ are known. Notice the particular nature of all the functional equations—the recursive and Watson's equations—satisfied by the form factors: (1) they are all linear; (2) they do not refer to any particular operator! This implies that, given a fixed number n of asymptotic particles, the solutions of the form factor equations form a linear space. The classification of the operator content is then obtained by putting the vectors of this linear space in correspondence with the operators [12, 14].

Kernel Solutions Among the functions of these linear spaces, there are those belonging to the kernel of the operators $\mathcal{B}$ and $\mathcal{C}$: these are the functions $\hat{F}_n^{(i)}$ and $\hat{F}_n^{(j)}$ that satisfy

$$\mathcal{B}\hat{F}_n^{(i)} = 0,$$
$$\mathcal{C}\hat{F}_n^{(j)} = 0. \tag{8.51}$$

Their general expression is given in Eq. 8.39 but, in this case, the function K_n does not contain poles that give rise to the recursive equations. Hence each of the functions $\hat{F}_n^{(i)}$ and $\hat{F}_n^{(j)}$ is simply a symmetric polynomials in the x_i variables. The vector space of the form factors that belong to the kernels can be further specified by assigning the total and partial degrees of these polynomials.

A non-vanishing kernel of the operators $\mathcal{B}$ and $\mathcal{C}$ has the important consequence that at each level n, if $\tilde{F}_n$ is a reference solution of the recursive equation and $\hat{F}_n$ a function of any of the two kernels, the most general form factor can be written as

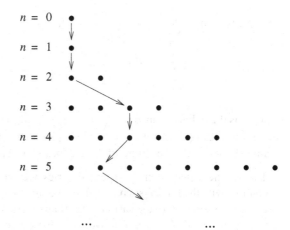

Fig. 8.6 Vector spaces of the solutions of the form factor equations (the number of dots at each level is purely indicative). An operator is associated to the sequence of its matrix elements F_n

$$F_n = \tilde{F}_n + \sum_i \alpha_i \hat{F}_n. \tag{8.52}$$

Therefore the identification of each operator is obtained by specifying at each level n the constants α_i. If we graphically represent by dots the linearly independent solutions at the level n of the form factor equations, we have the situation of Fig. 8.6. In this graphical representation, each operator is associated to a well-defined path on this lattice, with each step $(n + 1) \rightarrow n$ (or $(n + 2) \rightarrow n$) ruled by the operator $\mathcal{B}$ (or $\mathcal{C}$). We will see explicit examples of this operator structure when we will discuss the form factors of the Ising and the Sinh-Gordon models.

8.7 Correlation Functions

Once we have determined the form factors of a given operator, its correlation functions can be written in terms of the spectral series using the completeness relation of the multi-particle states

$$1 = \sum_{n=0}^{\infty} \int \frac{d\theta_1 \dots d\theta_n}{n!(2\pi)^n} |\theta_1, \dots, \theta_n\rangle \langle \theta_1, \dots, \theta_n|. \tag{8.53}$$

For instance, for the two-point correlation function of the operator $\mathcal{O}(x)$ in the euclidean space, we have

$$\langle \mathcal{O}(x)\mathcal{O}(0) \rangle = \sum_{n=0}^{\infty} \int \frac{d\theta_1 \dots d\theta_n}{n!(2\pi)^n} \langle 0|\mathcal{O}(x)|\theta_1, \dots, \theta_n \rangle_{\text{in in}} \langle \theta_1, \dots, \theta_n|\mathcal{O}(0)|0 \rangle$$

$$= \sum_{n=0}^{\infty} \int \frac{d\theta_1 \dots d\theta_n}{n!(2\pi)^n} \mid F_n(\theta_1 \dots \theta_n) \mid^2 \exp\left(-mr \sum_{i=1}^{n} \cosh \theta_i \right),$$

(8.54)

where r is the radial distance $r = \sqrt{x_0^2 + x_1^2}$ (Fig. 8.7). Similar expressions, although more complicated, hold for the n-point correlation functions. It is worth making some comments to clarify the nature of these expressions and their advantage.

- The integrals that enter the spectral series are all convergent. This is in sharp contrast with the formalism based on the Feynman diagrams, in which one encounters the divergences of the various perturbative terms. In a nutshell, the deep reason of this difference between the two formalisms can be expressed as follows. The Feynman formalism is based on the quantization of a *free* theory and on the *bare* unphysical parameters of the Lagrangian. What the renormalization procedure does is to implement the change from the bare to the physical parameters (such as, the physical value of the mass of the particle). But the form factors employ *ab initio* all the physical parameters of the theory and for this reason the divergences of the perturbative series are absent.
- If the S-matrix depends on a coupling constant, as it happens in the Sinh-Gordon model or in other Toda field theories, each matrix element provides the exact resummation of all terms of perturbation theory.
- If the correlation functions do not have particularly violent ultraviolet singularities (this is the case, for instance, of the correlation functions of the relevant fields), the corresponding spectral series has an extremely fast convergent behavior for all values of mr. In the infrared region, that is for large values of mr, this is evident from the nature of the series, because its natural parameter of expansion is e^{-mr}. The reason of the fast convergent behavior also in the ultraviolet region $mr \to 0$ is twofold: the peculiar behavior of the n-particle phase space in two-dimensional theories and a further enhancement of the convergence provided by the form factors [11]. To better understand this aspect, consider the Fourier transform of the correlator

$$G(x) = \langle \mathcal{O}(x)\mathcal{O}(0) \rangle = \int \frac{d^2 p}{(2\pi)^2} e^{ip \cdot x} \hat{G}(p).$$

(8.55)

The function $\hat{G}(p)$ can be written as

$$\hat{G}(p) = \int_0^{\infty} d\mu^2 \rho(\mu^2) \frac{1}{p^2 + \mu^2},$$

(8.56)

where $\rho(k^2)$ is a relativistic invariant function called the *spectral density*

Fig. 8.7 Spectral
representation of the
two-point correlation
functions

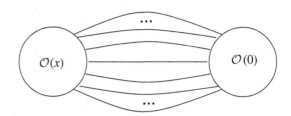

$$\rho(k^2) = 2\pi \sum_{n=0}^{\infty} \int d\Omega_1 \dots d\Omega_n \delta^2(k - P_n) |\langle 0|\mathcal{O}(0)|\theta_1, \dots, \theta_n\rangle|^2$$

$$d\Omega = \frac{dp}{2\pi E} = \frac{d\theta}{2\pi} \quad , \quad P_n^{(0)} = \sum_{k=0}^{n} \cosh\theta_k \quad , \quad P_n^{(1)} = \sum_{k=0}^{n} \sinh\theta_k.$$

Since $1/(p^2 + \mu^2)$ is the two-point correlation function of the euclidean free theory
with mass μ, i.e., the propagator, Eq. 8.56 shows that the two-point correlation
function can be regarded as a linear superposition of the free propagators weighted
with the spectral density $\rho(\mu^2)$. Notice that the contribution given by the single
particle state of mass m in the spectral density is given by

$$\rho_{1part}(k^2) = \frac{1}{2\pi}\delta(k^2 - m^2). \qquad (8.57)$$

To analyze the behavior of $\rho(k^2)$ by varying k^2, let us make the initial approximation
to take all the matrix elements equal to 1. In this way, each term of the spectral series
coincides with the n-particle phase space

$$\Phi_n(k^2) \equiv \int \prod_{k=1}^{n} d\Omega_k \delta^2(k - P_n). \qquad (8.58)$$

In two dimensions, the phase space goes to zero when $k^2 \to \infty$ as

$$\Phi_n(k^2) \simeq \frac{1}{(2\pi)^{n-2}} \frac{1}{(n-2)!} \frac{1}{k^2} \left(\log\frac{k^2}{m^2}\right)^{n-2}, \qquad (8.59)$$

whereas for $d > 2$ it diverges as

$$\Phi_n(k^2) \sim k^{\frac{n(d-2)-d}{2}}. \qquad (8.60)$$

On the other hand, $\Phi_n(k^2) = 0$ if $k^2 < (nm)^2$ and near the threshold values we have

$$\Phi(k^2) \simeq A_n \left(\sqrt{k^2} - nm\right)^{\frac{n-3}{2}}. \qquad (8.61)$$

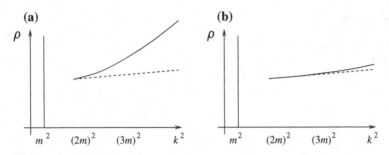

Fig. 8.8 Plot of the spectral series in a model in $d = 4$ (**a**), and in $d = 2$ (**b**). The contribution of the two-particle state is given by the dashed line. In $d = 4$ this do not provide a good approximation of $\rho(k^2)$ for large values of k^2 while in $d = 2$ it gives very often an excellent approximation of this quantity

Hence, we see that for pure reasons related to the phase space we have two different scenarios for the quantum field theories in two dimensions and in higher dimensions: while in $d > 2$ surpassing the various thresholds the spectral density receives contributions that are more divergent, in $d = 2$ they are all of the same order and all go to zero at large value of the energy. Hence, for $d > 2$ it is practically impossible to approximate the spectral density for large values of k^2 by using the first terms of the series, relative to the states with few particles, whereas in $d = 2$ this is perfectly plausible. If we now include in the discussion also the form factors, one realizes that the situation is even better in $d = 2$! In fact, from the general expression (8.39) and for the vanishing of $F_{\min}(\theta_{ij})$ at the origin Eq. 8.41, the form factors vanish at the n-particle thresholds as

$$|\langle 0|\mathcal{O}(0)|\theta_1, \ldots, \theta_n\rangle|^2 \simeq \left(\sqrt{k^2} - nm\right)^{n(n-1)}, \quad \theta_1 \simeq \ldots \simeq \theta_n \simeq 0, \quad (8.62)$$

while, for large values of their rapidities, they typically tend to a constant.[4] This scenario implies that the spectral density of the correlation functions of the two-dimensional integrable models usually flatten more at the thresholds and therefore becomes a very smooth function varying k^2. For all these reasons, the spectral density can be approximated with a great accuracy just taking the first terms of the series, even for large values of k^2, therefore leading to a fast convergence also in the ultraviolet region (Fig. 8.8).

8.8 Form Factors of the Stress–Energy Tensor

The stress–energy tensor is an operator that plays an important role in quantum field theory and its form factors have special properties [8, 15]. For its conservation law $\partial_\mu T^{\mu\nu}(x) = 0$, this operator can be written in terms of an auxiliary scalar field $A(x)$ as

---

[4] This is what usually happens for the form factors of the strongly relevant operators.

$$T_{\mu\nu}(x) = \left(\partial_\mu \partial_\nu - g_{\mu\nu}\Box\right) A(x). \tag{8.63}$$

In the light-cone coordinates, $x^\pm = x^0 \pm x^1$, its components are

$$T_{++} = \partial_+^2 A, \quad T_{--} = \partial_-^2 A,$$

$$\Theta = T_\mu^\mu = -\Box A = -4\partial_+ \partial_- A.$$

Introducing the variables $x_j = e^{\theta_j}$ and the elementary symmetric polynomials $\sigma_i^{(n)}$, it is easy to see that

$$F_n^{T_{++}}(\theta_1, \ldots, \theta_n) = -\frac{1}{4}m^2 \left(\frac{\sigma_{n-1}^{(n)}}{\sigma_n^{(n)}}\right)^2 F_n^A(\theta_1, \ldots, \theta_n),$$

$$F_n^{T_{--}}(\theta_1, \ldots, \theta_n) = -\frac{1}{4}m^2 \left(\sigma_1^{(n)}\right)^2 F_n^A(\theta_1, \ldots, \theta_n), \tag{8.64}$$

$$F_n^\Theta(\theta_1, \ldots, \theta_n) = m^2 \frac{\sigma_1^{(n)} \sigma_{n-1}^{(n)}}{\sigma_k} F_n^A(\theta_1, \ldots, \theta_n).$$

Solving for F_n^A, we have

$$F_n^{T_{++}}(\theta_1, \ldots, \theta_n) = -\frac{1}{4} \frac{\sigma_{n-1}^{(n)}}{\sigma_1^{(n)} \sigma_n^{(n)}} F_n^\Theta(\theta_1, \ldots, \theta_n),$$

$$F_n^{T_{--}}(\theta_1, \ldots, \theta_n) = -\frac{1}{4} \frac{\sigma_1^{(n)} \sigma_n^{(n)}}{\sigma_{n-1}^{(n)}} F_n^\Theta(\theta_1, \ldots, \theta_n). \tag{8.65}$$

Hence, the whole set of the form factors of $T_{\mu\nu}$ can be recovered by the form factors of the trace Θ. This is a scalar operator and therefore its form factors depend on the differences of the rapidities $\theta_{ij} = \theta_i - \theta_j$. Moreover, since the form factors of T_{--} and T_{++} must have the same singularities of those of Θ, $F_n^\Theta(\theta_1, \ldots, \theta_n)$ (for $n > 2$) has to be proportional to the combination $\sigma_1^{(n)} \sigma_{n-1}^{(n)}$ of the elementary symmetric polynomials. This combination corresponds to the relativistic invariant given by the total energy and momentum of the system.

For the normalization of these matrix elements, the recursive structure reduces the problem of finding the normalization of the form factors of $\Theta(x)$ on the 1 and 2-particle states, i.e., $F_1^\Theta(\theta)$ and $F_2^\Theta(\theta_{12})$. The normalization of $F_2^\Theta(\theta_{12})$ can be determined by using the total energy of the system

$$E = \frac{1}{2\pi} \int\limits_{-\infty}^{+\infty} dx^1 T^{00}(x). \tag{8.66}$$

Computing the matrix element of both terms of this equation on the asymptotic states $\langle\theta'|$ and $|\theta\rangle$, for the left hand side we have

$$\langle\theta'|E|\theta\rangle = 2\pi\, m\cosh\theta\delta(\theta'-\theta).$$

On the other hand, taking into account that $T^{00} = \partial_1^2 A$ and using the relation

$$\langle\theta'|\mathcal{O}(x)|\theta\rangle = e^{i(p^\mu(\theta')-p^\mu(\theta))x_\mu}F_2^{\mathcal{O}}(\theta,\theta'-i\pi)$$

that holds for any hermitian operator $\mathcal{O}$, we obtain

$$F_2^{\partial_1^2 A}(\theta_1,\theta_2) = -m^2(\sinh\theta_1 + \sinh\theta_2)^2 F_2^A(\theta_{12}).$$

From Eqs. 8.64 and 8.66 it follows that the normalization of F_2^Θ is given by

$$F_2^\Theta(i\pi) = 2\pi m^2. \tag{8.67}$$

However, there is no particular constraint on the one-particle form factor of $\Theta(x)$ coming from general considerations

$$F_1^\Theta = \langle 0\,|\,\Theta(0)\,|\,\theta\rangle. \tag{8.68}$$

This is a free parameter of the theory, related to the intrinsic ambiguity of $T^{\mu\nu}(x)$, since this tensor can always be modified by adding a total divergence (see [15]).

8.9 Ultraviolet Limit

In the ultraviolet limit, the correlation functions of the scaling operators have a power-law behavior, dictated by the conformal weight of the operator

$$G(r) = \langle\mathcal{O}(r)\mathcal{O}(0)\rangle \simeq \frac{1}{r^{4\Delta}}\,,\quad r\to 0. \tag{8.69}$$

One may wonder how the spectral series (8.54), that is based on the exponential terms e^{-kmr}, is able to reproduce a power law in the limit $r\to 0$. The answer to this question comes from an interesting analogy noticed in [12].

Feynman Gas Note that the formula (8.54) is formally similar to the expression of the grand-canonical partition function of a fictitious one-dimensional gas

$$\mathcal{Z}(mr) = \sum_{n=0}^{\infty} z^n Z_n. \tag{8.70}$$

To set up the vocabulary of this analogy, let us identify the coordinates of the gas particles with the rapidities θ_i, while the Boltzmann weight relative to the potential of the gas with the modulus square of the form factors

$$e^{-V(\theta_1,\ldots,\theta_n)} \equiv |\langle 0|\mathcal{O}(0)|\theta_1,\ldots,\theta_n\rangle|^2. \tag{8.71}$$

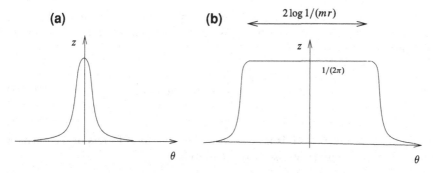

Fig. 8.9 Plot of the fugacity as a function of θ (a), for finite values of $(m\,r)$ (b), in the limit $(m\,r) \to 0$

Finally, let us identify the fugacity of the gas with

$$z(\theta) = \frac{1}{2\pi} e^{-mr\cosh\theta}. \tag{8.72}$$

The only difference with respect to the standard case of a gas is the coordinate dependence of the fugacity of this gas. Although the coordinates of the particles of this gas span the infinite real axis, the effective volume of the system is however determined by the region in which the fugacity (8.72) is significantly different from zero, as shown in Fig. 8.9. Note that $z(\theta)$ is a function that rapidly goes to zero outside a finite interval and, in the limit $mr \to 0$, presents a plateau of height $z_c = 1/(2\pi)$ and width

$$L \simeq 2\log\frac{1}{mr}.$$

The equation of state of a one-dimensional gas is given by

$$\mathcal{Z} = e^{p(z)L},$$

where $p(z)$ is the pressure as a function of the fugacity. Following this analogy, for the two-point correlation function in the limit $(mr) \to 0$ we have

$$G(r) = \mathcal{Z} = e^{p(z_c)L} \simeq e^{2p(z_c)\log 1/(mr)} = \left(\frac{1}{mr}\right)^{2p(z_c)}, \tag{8.73}$$

i.e., a power law behavior! Moreover, comparing with the short-distance behavior of the correlator given in Eq. 8.69, there is an interesting result: the conformal weight can be expressed in terms of the pressure of this fictitious one-dimensional gas, evaluated at the plateau value of the fugacity

$$2\Delta = p(1/2\pi). \tag{8.74}$$

Apart from the thermodynamics of the Feynman gas, the conformal weight of the operators can be also extracted by applying the sum rule given by the so-called Δ-theorem [17]

$$\Delta = -\frac{1}{2\langle\mathcal{O}\rangle} \int\limits_0^\infty dr\, r \langle \Theta(r)\mathcal{O}(0)\rangle. \tag{8.75}$$

To compute this quantity, it is necessary to know the form factors of the operator $\mathcal{O}(x)$ and the trace of the stress–energy tensor $\Theta(x)$.

c-Theorem Sum Rule An additional control of the ultraviolet limit of the theory is provided by the sum-rule of the c-theorem [18, 19]: it gives the central charge of the conformal field theory associated to the ultraviolet limit of the massive theory through the integral

$$c = \frac{3}{2} \int\limits_0^\infty dr\, r^3 \langle \Theta(r)\Theta(0)\rangle_c.$$

Using the spectral representation of this correlator we have

$$c = \sum_{n=1}^\infty c_n, \tag{8.76}$$

where the n-particle contribution is

$$c_n = \frac{12}{n!} \int\limits_0^\infty \frac{d\mu}{\mu^3} \int\limits_{-\infty}^\infty \frac{d\theta_1}{2\pi} \cdots \frac{d\theta_n}{2\pi}$$

$$\times \delta\left(\sum_{i=1}^n \sinh\theta_i\right) \delta\left(\sum_{i=1}^n \cosh\theta_i - \mu\right) |\langle 0|\Theta(0)|\theta_1,\ldots,\theta_n\rangle|^2. \tag{8.77}$$

Usually this series presents a very fast convergence, see for instance [13]. This permits to obtain rather accurate estimates of the central charge c, with an explicit check of the entire formalism of the S-matrix and form factors. It is easy to understand the reason of this fast convergence by studying the integrand, shown in Fig. 8.10 : the term r^3 kills the singularity of the correlator at short distance (therefore the integrand vanishes at the origin), while it weights the correlator more at large distances. But this is just the region where few terms of the spectral series are very efficient in approximating the correlation function with high accuracy.

Asymptotic Behavior Finally, following Ref. [9], let us discuss the upper bound on the asymptotic behavior of the form factors dictated by the ultraviolet behavior of the correlator (8.69). To establish this bound, let us start by noting that in a massive theory for the p-th moment of the correlation function we have

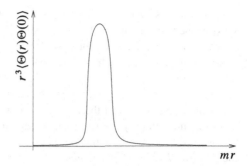

Fig. 8.10 Plot of the integrand $r^3 \langle \Theta(r)\Theta(0) \rangle$ in the c-theorem sum rule

$$M_p \equiv \int d^2x \, |x|^p \langle \mathcal{O}(x)\mathcal{O}(0) \rangle_c \quad < +\infty \quad \text{if} \quad p > 4\Delta_{\mathcal{O}} - 2. \tag{8.78}$$

Employing now the spectral representation of the correlator (8.55) and integrating on p, μ, and x, we get

$$M_p \sim \sum_{n=1}^{\infty} \int_{\theta_1 > \cdots > \theta_n} d\theta_1 \ldots d\theta_n \frac{|F_n^{\mathcal{O}}(\theta_1,\ldots,\theta_n)|^2}{\left(\sum_{k=1}^{n} m_k \cosh\theta_k\right)^{p+1}} \delta\left(\sum_{k=1}^{n} m_k \sinh\theta_k\right). \tag{8.79}$$

Equation 8.78 can be now used to find an upper limit on the real quantity y_Φ, defined by

$$\lim_{|\theta_i| \to \infty} F_n^{\mathcal{O}}(\theta_1,\ldots,\theta_n) \sim e^{y_\Phi |\theta_i|}. \tag{8.80}$$

In fact, taking the limit $\theta_i \to +\infty$ in the integrand of (8.79), the delta-function forces some other rapidities to move to $-\infty$ as $-\theta_i$. Because the matrix element $F_n^{\mathcal{O}}(\theta_1,\ldots,\theta_n)$ depends on the differences of the rapidities, it contributes to the integrand with the factor $e^{2y_\Phi |\theta_i|}$ in the limit $|\theta_i| \to \infty$. Hence, Eq. 8.78 leads to the condition

$$y_{\mathcal{O}} \leq \Delta_{\mathcal{O}}. \tag{8.81}$$

This equation provides information on the partial degree of the polynomial $Q_n^{\mathcal{O}}$. Note, however, that this conclusion may not apply for non-unitary theories because not all terms of the expansion on the intermediate states are necessarily positive in this case.

8.10 The Ising Model at $T \neq T_c$

In this section we present the form factors and the correlation functions of the relevant operators $\epsilon(x)$, $\sigma(x)$ and $\mu(x)$ of the two-dimensional Ising model when the temperature T is away from its critical value [7]. For a review of the class of universality of the two-dimensional Ising model see Ref. [20]. For the duality of the model, we can discuss equivalently the case $T > T_c$ or $T < T_c$. Suppose to be in the high-temperature phase where the scattering theory of the off-critical model involves only one particle with an S-matrix $S = -1$. There are no bound states. The particle A can be considered as created by the magnetization operator $\sigma(x)$, so that it is odd under the Z_2 symmetry of the Ising model, with its mass given by $m = |T - T_c|$.

Let us now employ the form factor equations to find the matrix elements of the various operators on the multi-particle states. The first step is the determination of the function $F_{\min}(\theta)$ that satisfies

$$F_{\min}(\theta) = - F_{\min}(-\theta)$$
$$F_{\min}(i\pi - \theta) = F_{\min}(i\pi + \theta). \tag{8.82}$$

The minimal solution is

$$F_{\min}(\theta) = \sinh \frac{\theta}{2}. \tag{8.83}$$

8.10.1 The Energy Operator

Let us discuss initially the form factors of the energy operator $\epsilon(x)$ or, equivalently, those of the trace of the stress–energy tensor, since the two operators are related by

$$\Theta(x) = 2\pi m \epsilon(x). \tag{8.84}$$

This is an even operator under the Z_2 symmetry and therefore it has matrix elements only on states with an even number of particles, F_{2n}^{Θ}. The recursive equation of the kinematical poles are particularly simple

$$-i \lim_{\tilde{\theta} \to \theta} (\tilde{\theta} - \theta) F_{2n+2}^{\Theta}(\tilde{\theta} + i\pi, \theta, \theta_1, \theta_2, \ldots, \theta_{2n})$$
$$= \left(1 - (-1)^{2n}\right) F_{2n}^{\Theta}(\theta_1, \ldots, \theta_{2n}) = 0. \tag{8.85}$$

Taking into account the normalization of the trace operator $F_2^{\Theta}(i\pi) = 2\pi m^2$, the simplest solution of all these equations is

$$F_{2n}^{\Theta}(\theta_1, \ldots, \theta_{2n}) = \begin{cases} -2\pi i m^2 \sinh \frac{\theta_1 - \theta_2}{2} , & n = 2 \\ 0 , & \text{otherwise.} \end{cases} \tag{8.86}$$

In the light of the discussion in Sect. 8.6, note that the identification of the operator Θ with this specific sequence of form factors is equivalent to put equal to zero all coefficients of the kernel solutions $F_{2n}^{(i)}$ at all the higher levels.

An explicit check that (8.86) is the correct sequence of the form factors of the trace operator comes from its two-point correlation function and from the c-theorem. For the correlator we get

$$
\begin{aligned}
G^{\Theta}(r) = \langle \Theta(r)\Theta(0) \rangle &= \frac{1}{2} \int \frac{d\theta_1}{2\pi} \frac{d\theta_2}{2\pi} |F_2^{\Theta}(\theta_{12})|^2 e^{-mr(\cosh\theta_1 + \cosh\theta_1)} \\
&= \frac{m^4}{2} \int d\theta_1 d\theta_2 \sinh^2 \frac{\theta_1 - \theta_2}{2} e^{-mr(\cosh\theta_1 + \cosh\theta_2)} \\
&= \frac{m^4}{4} \int d\theta_1 d\theta_2 \left[\cosh(\theta_1 - \theta_2) - 1\right] e^{-mr(\cosh\theta_1 - \cosh\theta_2)} \qquad (8.87) \\
&= m^4 \left(\left[\int d\theta \cosh\theta e^{-mr\cosh\theta}\right]^2 - \left[\int d\theta e^{-mr\cosh\theta}\right]^2 \right) \\
&= m^4 \left(K_1^2(mr) - K_0^2(mr) \right),
\end{aligned}
$$

where, in the last line, we used the integral representation of the modified Bessel functions

$$
K_{\nu}(z) = \int_0^{\infty} dt \cosh \nu t\, e^{-z\cosh t}.
$$

Hence, we have

$$
G^{\Theta}(r) = \langle \Theta(r)\Theta(0) \rangle = m^4 \left[K_1^2(mr) - K_0^2(mr) \right]. \qquad (8.88)
$$

whose plot is in Fig. 8.11. This function has the correct ultraviolet behavior associated to the energy operator

$$
G^{\Theta}(r) \to \frac{m^2}{|x|^2}, \quad |x| \to 0. \qquad (8.89)
$$

Substituting the expression above in the c-theorem, we get the correct value of the central charge of the Ising model

$$
c = \frac{3}{2} \int_0^{\infty} dr\, r^3 \langle \Theta(r)\Theta(0) \rangle = \frac{1}{2}. \qquad (8.90)
$$

8.10.2 Magnetization Operators

In the high-temperature phase, the order parameter $\sigma(x)$ is odd under the Z_2 symmetry while the disorder operator $\mu(x)$ is even. Hence, $\sigma(x)$ has matrix elements

Fig. 8.11 Plot of the
two-point correlation
function of the trace of the
stress–energy tensor for the
thermal Ising model

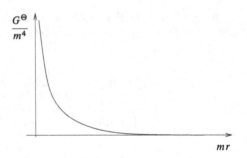

on states with an odd number of particles, F_{2n+1}^{σ}, whereas $\mu(x)$ on an even number,
F_{2n}^{μ}. In writing down the residue equations relative to the kinematical poles, we have
to take into account that the operator μ has a semi-local index equal to $1/2$ with
respect to the operator $\sigma(x)$ that creates the asymptotic states. Denoting by F_n the
form factors of these operators (for n even they refer to $\mu(x)$ while for n odd to
$\sigma(x)$), we have the recursive equation

$$-i \lim_{\tilde{\theta} \to \theta} (\tilde{\theta} - \theta) F_{n+2}(\tilde{\theta} + i\pi, \theta, \theta_1, \theta_2, \dots, \theta_{2n}) = 2 F_n(\theta_1, \dots, \theta_{2n}). \qquad (8.91)$$

As for any form-factor equation, these equations admit an infinite number of solutions, that can be obtained by adding all possible kernel solutions at each level. The
minimal solution is the one chosen to identify the form factors of the order and
disorder operators

$$F_n(\theta_1, \dots, \theta_n) = H_n \prod_{i<j}^{n} \tanh \frac{\theta_i - \theta_j}{2}. \qquad (8.92)$$

The normalization coefficients satisfy the recursive equation

$$H_{n+2} = i H_n.$$

The solutions with n even are therefore fixed by choosing $F_0 = H_0$, namely with a
non-zero value of the vacuum expectation of the disorder operator

$$F_0 = \langle 0 | \mu(0) | 0 \rangle = \langle \mu \rangle, \qquad (8.93)$$

while those with n odd are determined by the real constant F_1 relative to the one-
particle matrix element of $\sigma(x)$

$$F_1 = \langle 0 | \sigma(0) | A \rangle. \qquad (8.94)$$

Adopting the conformal normalization of both operators

$$\langle \sigma(x)\sigma(0) \rangle = \langle \mu(x)\mu(0) \rangle \simeq \frac{1}{|x|^{1/4}} \ , \quad |x| \to 0, \qquad (8.95)$$

it is possible to show that $F_0 = F_1$ and the vacuum expectation value F_0 can also computed

$$F_0 = F_1 = 2^{1/3} e^{-1/4} A^3 m^{1/4}, \tag{8.96}$$

where $A = 1.282427..$ is called the Glasher constant. Vice versa, if we choose $F_0 = F_1 = 1$ (as we do hereafter), for the ultraviolet behavior of the correlation functions we have

$$\langle \sigma(x)\sigma(0) \rangle = \langle \mu(x)\mu(0) \rangle \simeq \frac{2^{-1/3} e^{1/4} A^{-3}}{|x|^{1/4}} = \frac{0.5423804\ldots}{|x|^{1/4}}, \quad |x| \to 0. \tag{8.97}$$

There are several ways to check the correct identification of the form factors of the order/disorder operators. A direct way is to employ the Δ-theorem [17]. In fact, using the matrix elements of $\mu(x)$ and $\Theta(x)$, we can compute their correlator, following the same procedure as in Eq. 8.87

$$\langle \Theta(r)\mu(0) \rangle = \frac{1}{2} \int \frac{d\theta_1 \, d\theta_2}{2\pi \, 2\pi} F^\Theta(\theta_{12}) \bar{F}^\mu(\theta_{12}) e^{-mr(\cosh \theta_1 + \cosh \theta_2)}$$
$$= -m^2 \langle \mu \rangle \left[\frac{e^{-2mr}}{2mr} + Ei(-2mr) \right], \tag{8.98}$$

where

$$Ei(-x) = -\int_x^\infty \frac{dt}{t} e^{-t}.$$

Substituting this correlator in the formula of the Δ-theorem Eq. 8.75, one obtains the correct value of the conformal weight of the disorder operator

$$\Delta = -\frac{1}{2\langle \mu \rangle} \int_0^\infty dr r \langle \Theta(r)\mu(0) \rangle = \frac{1}{4\pi} \int_0^\infty d\theta \frac{\sinh^2 \theta}{\cosh^3 \theta} = \frac{1}{16}. \tag{8.99}$$

Another way to determine the conformal weight of the magnetization operators consists of solving the thermodynamics of the Feynman gas associated to the form factors [12]. Using the nearest-neighbor approximation, the pressure of this gas satisfies the integral equation

$$z_c^{-1} = 2\pi = \int_0^\infty dx \tanh^2 \frac{x}{2} e^{-px}, \tag{8.100}$$

whose numerical solution is

$$p \simeq 0.12529\ldots \tag{8.101}$$

Comparing with the exact value

$$p = 2\Delta = \frac{1}{8} = 0.125, \tag{8.102}$$

we see that the relative error is less than one part in a thousand! This result confirms the validity of the form factor solution for the magnetization operators and, furthemore, it explicitly shows the convergence property of the spectral series.

8.10.3 The Painlevé Equation

The two-point correlation functions of the magnetization operators are given by

$$\langle \mu(r)\mu(0) \rangle = \sum_{n=0}^{\infty} g_{2n}(r)$$

$$\langle \sigma(r)\sigma(0) \rangle = \sum_{n=0}^{\infty} g_{2n+1}(r),$$

where

$$g_n(r) = \frac{1}{n!} \int \left[\prod_{k=1}^{n} \frac{d\theta_k}{2\pi} e^{-mr\cosh\theta_k} \right] \prod_{i<j} \tanh^2 \frac{\theta_{ij}}{2}.$$

These expressions can be further elaborated: posing $u_i = e^{\theta_i}$ and using

$$\tanh^2 \frac{\theta_i - \theta_j}{2} = \left(\frac{u_i - u_j}{u_i + u_j} \right)^2,$$

we get

$$\prod_{i<j} \tanh^2 \frac{\theta_{ij}}{2} = \prod_{i<j} \left(\frac{u_i - u_j}{u_i + u_j} \right)^2 = \det W, \tag{8.103}$$

where the matrix elements of the operator W are

$$W_{ij} = \frac{2\sqrt{u_i u_j}}{u_i + u_j}.$$

Combining the two correlators

$$G^{(\pm)}(r) = \langle \mu(r)\mu(0) \rangle \pm \langle \sigma(r)\sigma(0) \rangle = \sum_{n=0}^{\infty} \lambda^n g_n(r) \tag{8.104}$$

(with $\lambda = \pm 1$) and using (8.103) we obtain

$$G^{(\pm)}(r) = \sum_{n=0}^{\infty} \frac{\lambda^n}{n!} \int \left[\prod_{k=1}^{n} \frac{d\theta_k}{2\pi} e^{-mr \cosh \theta_k} \right] \det W. \qquad (8.105)$$

The last expression is nothing else but the Fredholm determinant of an integral operator V, whose kernel is

$$V(\theta_i, \theta_j, r) = \frac{E(\theta_i, r) E(\theta_j, r)}{u_i + u_j}$$

$$E(\theta_i, r) = (2u_i e^{-mr \cosh \theta_i})^{1/2}.$$

Hence

$$G^{(\pm)}(r) = \text{Det}(1 + \lambda V). \qquad (8.106)$$

The remarkable circumstance that the correlation functions are expressed in terms of the Fredholm determinant of an integral operator is crucial for studying their properties. The detailed discussion is beyond the scope of this chapter and here we simply present the main conclusions.

First of all, the expression given in Eq. 8.106 permits to solve *exactly* the thermodynamics of the Feynman gas associated to the form factors of the correlation function $G^{(+)}(r)$. The exact expression of the pressure of the Feynman gas is given by [7]

$$p(z) = \frac{1}{4} \int \frac{dp}{2\pi} \log \left[1 + \left(\frac{2\pi z}{\sinh \pi p} \right)^2 \right]$$

$$= \frac{1}{4\pi} \arcsin(2\pi z) - \frac{1}{4\pi^2} \arcsin^2(2\pi z).$$

Substituting in this formula the plateau value of the fugacity, $z = z_c = 1/(2\pi)$, one obtains the exact value of the conformal weight of the magnetization operators, $p = 2\Delta = 1/8$.

Secondly, using the Fredholm determinant (8.106), it is possible to show that the correlators can be concisely written as [21–23]

$$\left(\begin{array}{c} \langle \mu(r)\mu(0) \rangle \\ \langle \sigma(r)\sigma(0) \rangle \end{array} \right) = \left(\begin{array}{c} \cosh \frac{\Psi(s)}{2} \\ \sinh \frac{\Psi(s)}{2} \end{array} \right) \exp \left[-\frac{1}{4} \int_{s}^{\infty} dt\, t \left[\left(\frac{d\Psi}{dt} \right)^2 - \sinh^2 \Psi \right] \right] \qquad (8.107)$$

$(s = mr)$, where $\Psi(s)$ is a solution of the differential equation

$$\frac{d^2\Psi}{ds^2} + \frac{1}{s} \frac{d\Psi}{ds} = 2 \sinh(2\Psi), \qquad (8.108)$$

with boundary conditions

$$\Psi(s) \simeq -\log s + \text{constant} \ , \quad s \to 0$$
$$\Psi(2) \simeq 2/\pi \, K_0(2s) \ , \quad s \to \infty .$$

(8.109)

With the substitution $\eta = e^{-\Psi}$, the differential equation becomes the celebrated Painlevé differential equation of the third kind

$$\frac{\eta''}{\eta} = \left(\frac{\eta'}{\eta}\right)^2 - \frac{1}{s}\left(\frac{\eta'}{\eta}\right) + \eta^2 - \frac{1}{\eta^2}.$$

(8.110)

This equation has been originally obtained by Wu, McCoy, Tracy, and Barouch [21, 22] by studying the scaling limit of the lattice Ising model. It has also ben derived by Jimbo, Miwa, and Ueno [23] by using the monodromy theory of the differential equations.

8.11 The Ising Model in a Magnetic Field

The Ising model in a magnetic field has quite a rich S-matrix [4]: it has 8 massive exitations and 36 elastic scattering amplitudes, some of them with higher-order poles. In addition to the functional and recursive equations, the form factors of this theory also satisfy other recursive equations related to the higher poles of the S-matrix. The relative formulas can be found in the papers by Delfino, Mussardo and Simonetti [9, 10]. Here we only report the main results about the form factors of the energy operator $\epsilon(x)$ and of magnetization operator $\sigma(x)$. In this theory, the latter operator is proportional to the trace

$$\Theta(x) = 2\pi h(2 - 2\Delta_\sigma)\sigma(x).$$

(8.111)

Relying on the fast convergence of the spectral series, for the correlation functions of these operators we can focus our attention on the 1 and 2-particle form factors. To begin with, let us fix some notation. For the S-matrix of the particles A_a and A_b we have

$$S_{ab}(\theta) = \prod_{\alpha \in \mathcal{A}_{ab}} (f_\alpha(\theta))^{p_\alpha} ,$$

(8.112)

where

$$f_\alpha(\theta) \equiv \frac{\tanh \frac{1}{2}(\theta + i\pi\alpha)}{\tanh \frac{1}{2}(\theta - i\pi\alpha)}.$$

(8.113)

The set of the numbers $\mathcal{A}_{ab}$ and their multiplicity p_α can be found in Tables 8.1 and 8.2, where we use the notation

Table 8.1 *S*-matrix of the Ising model in a magnetic field at $T = T_c$. The factors $\left(f_{\gamma/30}(\theta)\right)^{p_\gamma}$ in $S_{ab}(\theta)$ correspond to $(\gamma)^{p_\gamma}$ ($p_\gamma = 1$ is omitted). The upper index **c** in (γ) denotes the particle A_c that appears as bound state of $A_a A_b$ at $\theta = i\pi\gamma/30$ in the amplitudes $S_{ab}(\theta)$

a	b	S_{ab}
1	1	$\overset{1}{(20)}\,\overset{2}{(12)}\,\overset{3}{(2)}$
1	2	$\overset{1}{(24)}\,\overset{2}{(18)}\,\overset{3}{(14)}\,\overset{4}{(8)}$
1	3	$\overset{1}{(29)}\,\overset{2}{(21)}\,\overset{4}{(13)}\,\overset{5}{(3)}(11)^2$
1	4	$\overset{2}{(25)}\,\overset{3}{(21)}\,\overset{4}{(17)}\,\overset{5}{(11)}\,\overset{6}{(7)}(15)$
1	5	$\overset{3}{(28)}\,\overset{4}{(22)}\,\overset{6}{(14)}\,\overset{7}{(4)}(10)^2(12)^2$
1	6	$\overset{4}{(25)}\,\overset{5}{(19)}\,\overset{7}{(9)}(7)^2(13)^2(15)$
1	7	$\overset{5}{(27)}\,\overset{6}{(23)}\,\overset{8}{(5)}(9)^2(11)^2(13)^2(15)$
1	8	$\overset{7}{(26)}\,\overset{8}{(16)^3}(6)^2(8)^2(10)^2(12)^2$
2	2	$\overset{1}{(24)}\,\overset{2}{(20)}\,\overset{4}{(14)}\,\overset{5}{(8)}\,\overset{6}{(2)}(12)^2$
2	3	$\overset{1}{(25)}\,\overset{3}{(19)}\,\overset{6}{(9)}(7)^2(13)^2(15)$
2	4	$\overset{1}{(27)}\,\overset{2}{(23)}\,\overset{7}{(5)}(9)^2(11)^2(13)^2(15)$
2	5	$\overset{2}{(26)}\,\overset{6}{(16)^3}(6)^2(8)^2(10)^2(12)^2$
2	6	$\overset{2}{(29)}\,\overset{3}{(25)}\,\overset{5}{(19)^3}\,\overset{7}{(13)^3}\,\overset{8}{(3)}(7)^2(9)^2(15)$
2	7	$\overset{4}{(27)}\,\overset{6}{(21)^3}\,\overset{7}{(17)^3}\,\overset{8}{(11)^3}(5)^2(7)^2(15)^2$
2	8	$\overset{6}{(28)}\,\overset{7}{(22)^3}(4)^2(6)^2(10)^4(12)^4(16)^4$

$$(\gamma) \equiv f_{\frac{\gamma}{30}}(\theta).$$

Notice that several amplitudes have higher-order poles that can be explained in terms of the multi-scattering processes.

It is convenient to parameterize the two-particle form factors of this theory as

$$F_{ab}^{\mathcal{O}}(\theta) = \frac{Q_{ab}^{\Phi}(\theta)}{D_{ab}(\theta)} F_{ab}^{min}(\theta), \qquad (8.114)$$

where $D_{ab}(\theta)$ and $Q_{ab}^{\mathcal{O}}(\theta)$ are polynomials in $\cosh\theta$: the latter is fixed by the singularities of the *S*-matrix, the former depends on the operator $\mathcal{O}(x)$. The minimal form factors can be written as

$$F_{ab}^{min}(\theta) = \left(-i\sinh\frac{\theta}{2}\right)^{\delta_{ab}} \prod_{\alpha \in \mathcal{A}_{ab}} (G_\alpha(\theta))^{p_\alpha}, \qquad (8.115)$$

where

$$G_\alpha(\theta) = \exp\left\{2\int_0^\infty \frac{dt}{t}\frac{\cosh\left(\alpha - \frac{1}{2}\right)t}{\cosh\frac{t}{2}\sinh t}\sin^2\frac{(i\pi - \theta)t}{2\pi}\right\}. \qquad (8.116)$$

Table 8.2 Continuation of S-matrix of the Ising model in a magnetic field at $T = T_c$

a	b	
3	3	$\overset{2}{(22)}\,\overset{3}{(20)}^3\,\overset{5}{(14)}\,\overset{6}{(12)}^3\,\overset{7}{(4)}\,(2)^2$
3	4	$\overset{1}{(26)}\,(16)^3\,\overset{5}{(6)}^2\,(8)^2\,(10)^2\,(12)^2$
3	5	$\overset{1}{(29)}\,\overset{3}{(23)}\,(21)^3\,\overset{4}{(13)}^3\,\overset{7}{(5)}\,(3)^2\,\overset{8}{(11)}^4\,(15)$
3	6	$\overset{2}{(26)}\,\overset{3}{(24)}^3\,(18)^3\,\overset{6}{(8)}^3\,(10)^2\,\overset{8}{(16)}^4$
3	7	$\overset{3}{(28)}\,\overset{5}{(22)}^3\,(4)^2\,(6)^2\,(10)^4\,(12)^4\,(16)^4$
3	8	$\overset{5}{(27)}\,\overset{6}{(25)}^3\,\overset{8}{(17)}^5\,(7)^4\,(9)^4\,(11)^2\,(15)^3$
4	4	$\overset{1}{(26)}\,\overset{4}{(20)}^3\,\overset{6}{(16)}^3\,\overset{7}{(12)}^3\,\overset{8}{(2)}\,(6)^2\,(8)^2$
4	5	$\overset{1}{(27)}\,\overset{3}{(23)}^3\,\overset{5}{(19)}^3\,\overset{8}{(9)}^3\,(5)^2\,(13)^4\,(15)^2$
4	6	$\overset{1}{(28)}\,\overset{4}{(22)}^3\,(4)^2\,(6)^2\,(10)^4\,(12)^4\,(16)^4$
4	7	$\overset{2}{(28)}\,\overset{4}{(24)}^3\,\overset{7}{(18)}^5\,\overset{8}{(14)}^5\,(4)^2\,(8)^4\,(10)^4$
4	8	$\overset{4}{(29)}\,\overset{5}{(25)}^3\,\overset{7}{(21)}^5\,(3)^2\,(7)^4\,(11)^6\,(13)^6\,(15)^3$
5	5	$\overset{4}{(22)}^3\,\overset{5}{(20)}^5\,\overset{8}{(12)}^5\,(2)^2\,(4)^2\,(6)^2\,(16)^4$
5	6	$\overset{1}{(27)}\,\overset{2}{(25)}^3\,\overset{7}{(17)}^5\,(7)^4\,(9)^4\,(11)^4\,(15)^3$
5	7	$\overset{1}{(29)}\,\overset{3}{(25)}^3\,\overset{6}{(21)}^5\,(3)^2\,(7)^4\,(11)^6\,(13)^6\,(15)^3$
5	8	$\overset{3}{(28)}\,\overset{4}{(26)}^3\,\overset{5}{(24)}^5\,\overset{8}{(18)}^7\,(8)^6\,(10)^6\,(16)^8$
6	6	$\overset{3}{(24)}^3)\,\overset{6}{(20)}^5\,\overset{8}{(14)}^5\,(2)^2\,(4)^2\,(8)^4\,(12)^6$
6	7	$\overset{1}{(28)}\,\overset{2}{(26)}^3\,\overset{5}{(22)}^5\,\overset{8}{(16)}^7\,(6)^4\,(10)^6\,(12)^6$
6	8	$\overset{2}{(29)}\,\overset{3}{(27)}^3\,\overset{6}{(23)}^5\,\overset{7}{(21)}^7\,(5)^4\,(11)^8\,(13)^8\,(15)^4$
7	7	$\overset{2}{(26)}^3\,\overset{4}{(24)}^5\,\overset{7}{(20)}^7\,(2)^2\,(8)^6\,(12)^8\,(16)^8$
7	8	$\overset{1}{(29)}\,\overset{2}{(27)}^3\,\overset{4}{(25)}^5\,\overset{6}{(23)}^7\,\overset{8}{(19)}^9\,(9)^8\,(13)^{10}\,(15)^5$
8	8	$\overset{1}{(28)}^3\,\overset{3}{(26)}^5\,\overset{5}{(24)}^7\,\overset{7}{(22)}^9\,\overset{8}{(20)}^{11}\,(12)^{12}\,(16)^{12}$

For large values of the rapidity, we have

$$G_\alpha(\theta) \sim \exp(|\theta|/2), \quad |\theta| \to \infty, \tag{8.117}$$

independently of the index α.

From the analysis of the singularities of the form factors, one can arrive to the following expression of the denominator

$$D_{ab}(\theta) = \prod_{\alpha \in \mathcal{A}_{ab}} (\mathcal{P}_\alpha(\theta))^{i_\alpha}\,(\mathcal{P}_{1-\alpha}(\theta))^{j_\alpha}, \tag{8.118}$$

Table 8.3 Central charge given by the partial sum of the form factors entering the c-theorem. $c_{ab..}$ denotes the contribution of the state $A_a A_b...$ The exact result is $c = 1/2$

c_1	0.472038282
c_2	0.019231268
c_3	0.002557246
c_{11}	0.003919717
c_4	0.000700348
c_{12}	0.000974265
c_5	0.000054754
c_{13}	0.000154186
c_{partial}	0.499630066

Table 8.4 Conformal weights $\Delta_\mathcal{O}$ given by the partial sum of the form factors of the correlation functions entering the Δ-theorem. $\Delta_{ab..}$ denotes the contribution of the state $A_a A_b...$ The exact values are $\Delta_\sigma = 1/16 = 0.0625$ and $\Delta_\varepsilon = 1/2$

	σ	ϵ
Δ_1	0.0507107	0.2932796
Δ_2	0.0054088	0.0546562
Δ_3	0.0010868	0.0138858
Δ_{11}	0.0025274	0.0425125
Δ_4	0.0004351	0.0069134
Δ_{12}	0.0010446	0.0245129
Δ_5	0.0000514	0.0010340
Δ_{13}	0.0002283	0.0065067
Δ_{partial}	0.0614934	0.4433015

where

$$i_\alpha = n + 1, \quad j_\alpha = n, \quad \text{if } p_\alpha = 2n + 1;$$
$$i_\alpha = n, \quad \quad j_\alpha = n, \quad \text{if } p_\alpha = 2n, \tag{8.119}$$

having introduced the notation

$$\mathcal{P}_\alpha(\theta) \equiv \frac{\cos \pi\alpha - \cosh \theta}{2 \cos^2 \frac{\pi\alpha}{2}}. \tag{8.120}$$

Both quantities $F_{ab}^{min}(\theta)$ and $D_{ab}(\theta)$ are normalized to be equal to 1 when $\theta = i\pi$. The polynomials of the numerator can be expressed as

$$Q_{ab}^\mathcal{O}(\theta) = \sum_{k=0}^{N_{ab}^\mathcal{O}} c_{ab,\mathcal{O}}^{(k)} \cosh^k \theta. \tag{8.121}$$

The condition $\left[F_{ab}^\mathcal{O}(\theta)\right]^* = F_{ab}^\mathcal{O}(-\theta)$ follows from the monodromy condition satisfied by the form factors and from the property $S_{ab}^*(\theta) = S_{ab}(-\theta)$. This means that the coefficients $c_{ab,\mathcal{O}}^{(k)}$ are real numbers and their values identify the different operators.

The degrees of the polynomials are fixed by the conformal weight of the operators and, both for $\sigma(x)$ and $\epsilon(x)$, we have in particular $N_{11}^\Phi \leq 1$. Therefore the initial conditions of the recursive equation for the form factors of the two relevant operators

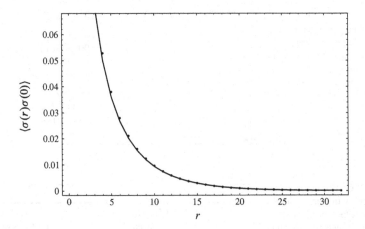

Fig. 8.12 Plot of the correlation function $\langle \sigma(r)\sigma(0)\rangle$ for the Ising model in a magnetic field. The continuos line is the determination obtained with the first 8 form factors, while the dots are the numerical determination of the correlators obtained by a Monte Carlo simulation, for details see [9]

consists of *two* free parameters, i.e., the coefficients $c^{(0)}_{11,\mathcal{O}}$ and $c^{(1)}_{11,\mathcal{O}}$. Furthemore, it can be checked that the number of free parameters does not increase implementing the form factor equations. Consider, for instance the condition $N^{\mathcal{O}}_{12} \leq 2$, that seems to imply three new coefficients $c^{(k)}_{12,\mathcal{O}}$ ($k = 1, 2, 3$) for $F^{\mathcal{O}}_{12}(\theta)$. However, the amplitudes $S_{11}(\theta)$ and $S_{12}(\theta)$ have three common bound states. This circumstance gives rise to three equations

$$\frac{1}{\Gamma^c_{11}}\mathrm{Res}_{\theta=iu^c_{11}} F^{\Phi}_{11}(\theta) = \frac{1}{\Gamma^c_{12}}\mathrm{Res}_{\theta=iu^c_{12}} F^{\Phi}_{12}(\theta), \qquad c = 1, 2, 3$$

that permit to fix the three coefficients $c^{(k)}_{12,\mathcal{O}}$ in terms of the two coefficients $c^{(k)}_{11,\mathcal{O}}$.

There is an additional piece of information on the numerator Q_{ab} of the operator $\Theta(x)$. In fact, the conservation law $\partial_\mu T^{\mu\nu} = 0$ implies that the polynomials Q^{Θ}_{ab} contain the factor

$$\left(\cosh\theta + \frac{m^2_a + m^2_b}{2m_a m_b}\right)^{1-\delta_{ab}}. \tag{8.122}$$

The determination of the coefficients $c^{(k)}_{ab}$ and the one-particle form factors of the two operators $\sigma \sim \Theta$ and ϵ has been done in the papers cited at the end of these lectures and their values can be found there.

Employing these lowest form factors one can compute the correlation functions and perform some non-trivial checks by applying the sum rules of the c-theorem and Δ-theorem. The relative results are given in the Tables 8.3 and 8.4. A successful check of the correlation function $\langle \sigma(r)\sigma(0)\rangle$ has also been done versus the numerical determination of this function, as shown in Fig. 8.12. An experimental confirmation of these results have been presented in Ref. [24].

8.12 Conclusions

In these lectures we present the main results about correlation functions in (1+1) integrable quantum field theories. Key quantities are the Form Factors of local operators. They can be used to classify the operator content of the quantum field theories away from the critical points and to efficiently compute the correlation functions through the spectral series. Although it was not discussed here, it is worth stressing that they can be also used to investigate the breaking of integrability and to understand interesting phenomena as the confinement of topological excitations, as shown in [25–27], or the decay of higher mass excitations [28].

References

1. Privman, V., Hohenberg, P.C., Aharony, A.: Universal critical-point amplitude relations. In: Domb, C., Green, M. S. (eds.) Phase Transitions, vol. 14. Academic Press, NY
2. Belavin, A.A., Polyakov, A.M., Zamolodchikov, A.B.: Infinite conformal symmetry in two-dimensional quantum field theory. Nucl. Phys. B **241**, 333 (1984)
3. Friedan, D., Qiu, Z., Shenker, S.: Conformal invariance, unitarity, and critical exponents in two dimensions. Phys. Rev. Lett. **52**, 1575 (1984)
4. Zamolodchikov, A.B.: Integrable field theory from conformal field. Adv. Stud. Pure Math. **19**, 641 (1989)
5. Karowski, M., Weisz, P.: Exact form-factors in $(1 + 1)$-dimensional field theoretic models with soliton behavior. Nucl. Phys. B **139**, 455 (1978)
6. Smirnov, F.: Form factors in completly integrable models of quantum field theory. World Scientific, Singapore (1992)
7. Yurov, V.P., Zamolodchikov, Al.B.: Correlation functions of integrable 2-D models of relativistic field theory. Ising model. Int. J. Mod. Phys. A **6**, 3419 (1991)
8. Zamolodchikov, Al.B.: Two point correlation function in scaling Lee-Yang model. Nucl. Phys. B **348**, 619 (1991)
9. Delfino, G., Mussardo, G.: The Spin spin correlation function in the two-dimensional Ising model in a magnetic field at $T = T_c$. Nucl. Phys. B **455**, 724 (1995)
10. Delfino, G., Simonetti, P.: Correlation functions in the two-dimensional Ising model in a magnetic field at $T = T_c$. Phys. Lett. B **383**, 450 (1996)
11. Cardy, J., Mussardo, G.: Universal properties of self-avoiding walks from two-dimensional field theory. Nucl. Phys. B **410**, 451 (1993)
12. Cardy, J.L., Mussardo, G.: Form-factors of descendent operators in perturbed conformal field theories. Nucl. Phys. B **340**, 387 (1990)
13. Fring, A., Mussardo, G., Simonetti, P.: Form-factors for integrable Lagrangian field theories, the sinh-Gordon theory. Nucl. Phys. B **393**, 413 (1993)
14. Koubek, A., Mussardo, G.: On the operator content of the sinh-Gordon model. Phys. Lett. B **311**, 193 (1993)
15. Mussardo, G., Simonetti, P.: Stress–energy tensor and ultraviolet behavior in massive integrable quantum field theories. Int. J. Mod. Phys. A **9**, 3307 (1994)

16. Essler, F., Konik, R.: Applications of massive integrable quantum field theories to problems in condensed matter physics, in from fields to strings: circumnavigating theoretical physics. World Scientific, Singapore (2005)

17. Delfino, G., Simonetti, P., Cardy, J.L.: Asymptotic factorization of form-factors in two-dimensional quantum field theory. Phys. Lett. B **387**, 327 (1996)

18. Zamolodchikov, A.B.: Irreversibility of the flux of the renormalization group in a 2D field theory. JETP Lett. **43**, 730 (1986)

19. Cardy, J.L.: The central charge and universal combinations of amplitudes in two-dimensional theories away from criticality. Phys. Rev. Lett. **60**, 2709 (1988)

20. Delfino, G.: Integrable field theory and critical phenomena. The Ising model in a magnetic field. Jour. Phys. A **37**, R45 (2004)

21. Wu, T.T., McCoy, B., Tracy, C., Barouch, E.: Spin-spin correlation functions for the two-dimensional Ising model: exact theory in the scaling region. Phys. Rev. B **13**, 316 (1976)

22. McCoy, B., Tracy, C., Wu, T.T.: Painlevé functions of the third kind. Jour. Math. Phys. **18**, 1058 (1977)

23. Jimbo, M., Miwa, T., Ueno, K.: Monodromy preserving deformation of linear ordinary differential equations with rational coefficients : i general theory and τ-function. Phys D **2**, 306 (1981)

24. Coldea, R., Tennant, D.A., Wheeler, E.M., Wawrzynska, E., Prabhakaran, D., Telling, M., Habicht, K., Smeibidl, P., Kiefer, K.: Quantum criticality in an Ising chain: experimental evidence for emergent E_8 symmetry. Science **327**, 177 (2010)

25. Delfino, G., Mussardo, G., Simonetti, P.: Nonintegrable quantum field theories as perturbations of certain integrable models. Nucl. Phys. B **473**, 469 (1996)

26. Delfino, G., Mussardo, G.: Nonintegrable aspects of the multifrequency sine-gordon model. Nucl. Phys. B **516**, 675 (1998)

27. Controzzi, D., Mussardo, G.: On the mass spectrum of the two-dimensional O(3) sigma model with theta term. Phys. Rev. Lett. **92**, 021601 (2004)

28. Delfino, G., Grinza, P., Mussardo, G.: Decay of particles above threshold in the Ising field theory with magnetic field. Nucl. Phys. B **737**, 291 (2006)